AF566968

Extracellular Matrix and the Liver

APPROACH TO GENE THERAPY

Extracellular Matrix and the Liver

APPROACH TO GENE THERAPY

Edited by

Isao Okazaki
Tokai University School of Medicine
Kanagawa, Japan

Yoshifumi Ninomiya
Okayama University Medical School
Okayama, Japan

Scott L. Friedman
Mount Sinai School of Medicine
New York, New York

Kyuichi Tanikawa
Kurume University School of Medicine
Fukuoka, Japan

Amsterdam Boston London New York Oxford Paris
San Diego San Francisco Singapore Sydney Tokyo

Cover image photos: Front cover: Expressions of intermediate filaments and myofilaments in HSC and rMF. (For more details see Chapter 7, Figure 4; see also Chapter 20, Figure 4, *In situ* hybridization of MMP-13 mRNA.) Back cover: *In situ* hybridization using antisense and sense probes. (For more details see Chapter 20, Figure 5.)

This book is printed on acid-free paper. 

Copyright © 2003, Elsevier Science (USA).

All Rights Reserved.
No part of this publication may be reproduced or transmitted in any form or by any means, electronic or mechanical, including photocopy, recording, or any information storage and retrieval system, without permission in writing from the publisher.

Requests for permission to make copies of any part of the work should be mailed to: Permissions Department, Academic Press, 6277 Sea Harbor Drive, Orlando, Florida 32887-6777

Academic Press
An imprint of Elsevier Science.
525 B Street, Suite 1900, San Diego, California 92101-4495, USA
http://www.academicpress.com

Academic Press
84 Theobald's Road, London WC1X 8RR, UK
http://www.academicpress.com

Library of Congress Catalog Card Number: 2002113023

International Standard Book Number: 0-12-525251-X

PRINTED IN CHINA
02 03 04 05 06 7 6 5 4 3 2 1

CONTENTS

PART II
Cells Responsible for Extracellular Matrix Production

PART III
Activation Mechanism of Hepatic Stellate Cells and Signal Transduction

PART IV

Basic Science of Matrix Metalloproteinases and Tissue Inhibitor Metalloproteinases

CONTRIBUTORS

Numbers in parentheses indicate the pages on which the authors' contributions begin.

EIJIRO ADACHI[1] (39), Department of Live Sciences, Graduate School of Arts and Sciences, The University of Tokyo, Tokyo 153-8902, Japan

KUNIHARU AKITA (391), First Department of Internal Medicine, Gifu University School of Medicine, Gifu 500-8705, Japan

MASAO ARAI (361), Department of Community Health, Tokai University School of Medicine, Kanagawa 259-1193, Japan

MICHAEL J. P. ARTHUR (347), Liver Research Group, Division of Infection, Inflammation, and Repair, University of Southampton, Hampshire SO16 6YD, United Kingdom

ULRIKE BENBOW[2] (277), Department of Medicine, Dartmouth Medical School, Hanover, New Hampshire 03755

AMY D. BRADSHAW (75), Department of Vascular Biology, The Hope Heart Institute, Seattle, Washington 98104

ROLF BREKKEN (75), Department of Vascular Biology, The Hope Heart Institute, Seattle, Washington 98104

CONSTANCE E. BRINCKERHOFF (277), Departments of Medicine and Biochemistry, Dartmouth Medical School, Hanover, New Hampshire 03755

ANTHONY J. DONAGHY (405), A.W. Morrow Gastroenterology and Liver Centre, Royal Prince Alfred Hospital, Camperdown 2050, New South Wales, Australia

[1]Present address: Department of Anatomy, Kitasato University, School of Medicine, Kanagawa 228-8555, Japan.

[2]Present address: Bristol Heart Institute, University of Bristol, Bristol BS9 1HD.

SCOTT L. FRIEDMAN (155), Division of Liver Diseases, Mount Sinai School of Medicine, New York, New York 10029

JIRO FUJIMOTO (443), First Department of Surgery, Hyogo College of Medicine, Nishinomiya 663-8501, Japan

ALBERT GEERTS (189), Laboratory for Molecular Liver Cell Biology, Free University of Brussels (VUB), 1090 Brussels, Belgium, and Department of Medical Cell Biology and Centre for Liver Research, University of Newcastle-upon-Tyne, United Kingdom

TARO HAYAKAWA (309), Department of Biochemistry, Aichi-Gakuin University, School of Dentistry, Nagoya 464-8650, Japan

TOSHIHIKO HAYASHI (39), Department of Live Sciences, Graduate School of Arts and Sciences, The University of Tokyo, Tokyo 153-8902, Japan

SASWATI HAZRA (179), USC-UCLA Research Center for Alcoholic Liver and Pancreatic Diseases, USC Research Center for Liver Diseases, Keck School of Medicine of the University of Southern California, Los Angeles, California 90033, and VA Greater Los Angeles Healthcare System, Los Angeles, California 90073

MOTOHIRO HIROSE (39), Department of Live Sciences, Graduate School of Arts and Sciences, The University of Tokyo, Tokyo 153-8902, Japan

SHIGENARI HOZAWA (361), Department of Community Health, Tokai University School of Medicine, Kanagawa 259-1193, Japan

YASUTADA IMAMURA (39), Department of Live Sciences, Graduate School of Arts and Sciences, The University of Tokyo, Tokyo 153-8902, Japan

YUTAKA INAGAKI (233), Department of Internal Medicine and Division of Clinical Research, National Kanazawa Hospital, Kanazawa 920-8650, Japan

TERUHIKO INOUE (251), Department of Pathology, Miyazaki Medical College, Miyazaki 889-1692, Japan

KEN-ICHI IYAMA (23), Department of Surgical Pathology, Kumamoto University School of Medicine, Kumamoto 860-8556, Japan

DAISUKE KAJIMURA (39), Department of Live Sciences, Graduate School of Arts and Sciences, The University of Tokyo, Tokyo 153-8902, Japan

TAKANORI KIHARA (39), Department of Live Sciences, Graduate School of Arts and Sciences, The University of Tokyo, Tokyo 153-8902, Japan

KOJI KIMATA (55), Institute for Molecular Science of Medicine, Aichi Medical University, Aichi 480-1195, Japan

THOMAS KNITTEL[3] (105), Department of Internal Medicine, University of Göttingen, 37075 Göttingen, Germany.

NORIO KOIDE (55), Department of Laboratory Medicine, Okayama University Medical School, Okayama 700-9559, Japan

SOICHI KOJIMA (391), Laboratory of Molecular Cell Sciences, RIKEN, Wako, Japan

HIROAKI KOSUGI (39), Department of Live Sciences, Graduate School of Arts and Sciences, The University of Tokyo, Tokyo 153-8902, Japan

DAN LI (155), Division of Liver Diseases, Mount Sinai School of Medicine, New York, New York 10029

FABIO MARRA (207), Dipartimento di Medicina Interna, Università degli Study di Firenze, I-50134 Firenze, Italy

KATSUYA MARUYAMA (361), Department of Internal Medicine, Kurihama National Hospital, Kanagawa 239-0841, Japan

RICHARD MAYNE (3), Department of Cell Biology, University of Alabama at Birmingham, Birmingham, Alabama 35294

TAKEO MIYAHARA (179), USC-UCLA Research Center for Alcoholic Liver and Pancreatic Diseases, USC Research Center for Liver Diseases, Keck School of Medicine of the University of Southern California, Los Angeles, California 90033, and VA Greater Los Angeles Healthcare System, Los Angeles, California 90073

KAZUNORI MIZUNO (39), Department of Live Sciences, Graduate School of Arts and Sciences, The University of Tokyo, Tokyo 153-8902, Japan

HISATAKA MORIWAKI (391), First Department of Internal Medicine, Gifu University School of Medicine, Gifu 500-8705, Japan

KENTA MOTOMURA (179), USC-UCLA Research Center for Alcoholic Liver and Pancreatic Diseases, USC Research Center for Liver Diseases, Keck School of Medicine of the University of Southern California, Los Angeles, California 90033, and VA Greater Los Angeles Healthcare System, Los Angeles, California 90073

KAZUKI NABESHIMA (251), Department of Pathology, Miyazaki Medical College, Miyazaki 889-1692, Japan

ATSUHITO NAKAO (233), Allergy Research Center, Juntendo University School of Medicine, Tokyo 113-8421, Japan

[3]Present address: DeveloGen AG, 37079 Göttingen, Germany.

KOICHI NAKAZATO (39), Department of Live Sciences, Graduate School of Arts and Sciences, The University of Tokyo, Tokyo 153-8902, Japan

TOMOYUKI NEMOTO (233), Department of Functional Biology, Kyoto University Graduate School of Biostudies, Kyoto 606-8502, Japan

KATRIN NEUBAUER (135), Department of Internal Medicine, Section of Gastroenterology and Endocrinology, University of Göttingen, Göttengen D-37075, Germany

MAKI NIIOKA (361), Department of Community Health, Tokai University School of Medicine, Kanagawa 259-1193, Japan

TOSHIRO NIKI (189), Department of Pathology, University of Tokyo, Tokyo 113-8655, Japan

YOSHIFUMI NINOMIYA (3, 23), Department of Molecular Biology and Biochemistry, Okayama University Graduate School of Medicine and Dentistry, Okayama 700-8558, Japan

AKIKO OKADA (299), Sekiguchi Bio-matrix Signaling Project, Exploratory Research for Advanced Technology, Japan Science and Technology Corporation, c/o. Aichi Medical University, Aichi 480-1195, Japan

HIDEYUKI OKANO (361), Department of Physiology, Keio University School of Medicine, Tokyo 160-8582, Japan

ISAO OKAZAKI (3, 361), Department of Community Health, Tokai University School of Medicine, Kanagawa 259-1193, Japan

MASATAKA OKUNO (391), First Department of Internal Medicine, Gifu University School of Medicine, Gifu 500-8705, Japan

MASAO OMATA (431), Department of Gastroenterology, University of Tokyo, Tokyo 113-8655, Japan

MASSIMO PINZANI (207), Dipartimento di Medicina Interna, Università degli Studi di Firenze, I-50134 Firenze, Italy

GIULIANO RAMADORI (135), Department of Internal Medicine, Section of Gastroenterology and Endocrinology, University of Göttingen, Göttengen D-37075, Germany

KRISTA ROMBOUTS (189), Laboratory for Molecular Liver Cell Biology, Free University of Brussels (VUB), 1090 Brussels, Belgium

E. HELENE SAGE (75), Department of Vascular Biology, The Hope Heart Institute, Seattle, Washington 98104

TETSURO SAMESHIMA (251), Department of Neurosurgery, Miyazaki Medical College, Miyazaki 889-1692, Japan

MICHIO SATA (89), Research Center for Innovative Cancer Therapy and Second Department of Medicine, Kurume University School of Medicine, Kurume 830-0011, Japan

MOTOHARU SEIKI (299), Department of Cancer Cell Research, Institute of Medical Science, University of Tokyo, Tokyo 113-8655, Japan

HONGYUN SHE (179), USC-UCLA Research Center for Alcoholic Liver and Pancreatic Diseases, USC Research Center for Liver Diseases, Keck School of Medicine of the University of Southern California, Los Angeles, California 90033, and VA Greater Los Angeles Healthcare System, Los Angeles, California 90073

YOSHIYA SHIMAO (251), Department of Pathology, Miyazaki Medical College, Miyazaki 889-1692, Japan

YASUSHI SHIRATORI (431), Department of Gastroenterology, University of Tokyo, Tokyo 113-8655, Japan

YOSHIHIKO SUGIOKA (361), Department of Environmental Health, Tokai University School of Medicine, Kanagawa 259-1193, Japan

TERUMI TAKAHARA (333), Third Department of Internal Medicine, Toyama Medical and Pharmaceutical University, Toyama 930-0194, Japan

SEIICHIRO TAKAHASHI (39), Department of Live Sciences, Graduate School of Arts and Sciences, The University of Tokyo, Tokyo 153-8902, Japan

YASUSHI TAKEDA (39), Department of Live Sciences, Graduate School of Arts and Sciences, The University of Tokyo, Tokyo 153-8902, Japan

KYUICHI TANIKAWA (89), International Institute for Liver Research, Kurume Research Center, Kurume 839-0861, Japan

HIDEKAZU TSUKAMOTO (179), USC-UCLA Research Center for Alcoholic Liver and Pancreatic Diseases, USC Research Center for Liver Diseases, Keck School of Medicine of the University of Southern California, Los Angeles, California 90033, and VA Greater Los Angeles Healthcare System, Los Angeles, California 90073

TAKAHIRO UEKI (443), First Department of Surgery, Hyogo College of Medicine, Nishinomiya 663-8501, Japan

TAKATO UENO (89), Research Center for Innovative Cancer Therapy and Second Department of Medicine, Kurume University School of Medicine, Kurume 830-0011, Japan

AKIHARU WATANABE (333), Third Department of Internal Medicine, Toyama Medical and Pharmaceutical University, Toyama 930-0194, Japan

TETSU WATANABE (3, 361), Department of Environmental Health, Tokai University School of Medicine, Kanagawa 259-1193, Japan

SHIGANG XIONG (179), USC-UCLA Research Center for Alcoholic Liver and Pancreatic Diseases, USC Research Center for Liver Diseases, Keck School of Medicine of the University of Southern California, Los Angeles, California 90033, and VA Greater Los Angeles Healthcare System, Los Angeles, California 90073

TOSHIKAZUI YADA (55), Institute for Molecular Science of Medicine, Aichi Medical University, Aichi 480-1195, Japan

HIROKO YAMANO (39), Department of Live Sciences, Graduate School of Arts and Sciences, The University of Tokyo, Tokyo 153-8902, Japan

KYOTO YAMASHITA (309), Department of Biochemistry, Aichi-Gakuin University, School of Dentistry, Nagoya 464-8650, Japan

YUTAKA YATA (333), Third Department of Internal Medicine, Toyama Medical and Pharmaceutical University, Toyama 930-0194, Japan

TOMOKO YONEZAWA (3), Department of Molecular Biology and Biochemistry, Okayama University Graduate School of Medicine and Dentistry, Okayama 700-8558, Japan

HARUHIKO YOSHIDA (431), Department of Gastroenterology, University of Tokyo, Tokyo 113-8655, Japan

MINURA YOSHIDA (189), Department of Biotechnology, University of Tokyo, Tokyo 113-8655, Japan

KIWAMU YOSHIKAWA (39), Department of Live Sciences, Graduate School of Arts and Sciences, The University of Tokyo, Tokyo 153-8902, Japan

LI PING ZHANG (333), Third Department of Internal Medicine, Toyama Medical and Pharmaceutical University, Toyama 930-0194, Japan

PREFACE

The most commonly accepted view of liver fibrosis is one of a gradually progressive disease that sometimes leads to liver cirrhosis. Liver transplantation is not the ultimate treatment for liver cirrhosis because the number of donors is limited. Since the introduction of interferon, lamivudine, ribavirin, and other excellent drugs, patients with HBV- or HCV-positive chronic hepatitis and/or liver cirrhosis have shown dynamic improvement in liver fibrosis. As a large percentage of such patients show a poor response to medication, however, the development of new treatments is eagerly awaited throughout the world. This situation and recent progress in molecular biology have encouraged researchers to develop gene therapies for the treatment of liver cirrhosis.

The Japanese Society for Connective Tissue Research and the Japan Society of Hepatology organized the "International Conference on New Strategies for the Treatment of Liver Cirrhosis" at the 33rd Annual Meeting of the Japanese Society for Connective Tissue Research, held June 6–9, 2001, in Tokyo. Professors and other researchers active in the front lines of basic and clinical fields of science gathered and discussed this issue. This volume is designed to convey the achievements of this international conference. The editors invited a number of distinguished scientists to contribute manuscripts.

The 24 chapters have been divided into six sections addressing the general mechanisms of formation and degradation of the extracellular matrix compared with those in the liver and the possibility of gene therapy. It is hoped the reader will gain some appreciation of what has been accomplished to date, as well as an understanding of the many problems that lie ahead in the development of new treatment. The editors also hope this volume will contribute to the further development of new strategies for the treatment of liver cirrhosis.

Isao Okazaki
Yosifumi Ninomiya
Scott L. Friedman
Kyuichi Tanikawa

PART I

Basic Science of Extracellular Matrix

CHAPTER 1

New Insights into the Extracellular Matrix

Isao Okazaki,* Tomoko Yonezawa,[†] Tetsu Watanabe,[‡]
Richard Mayne,[§] and Yoshifumi Ninomiya[†]

*Departments of [‡]Environmental Health and *Community Health, Tokai University School of Medicine, Kanagawa 259-1193, Japan; [†]Department of Molecular Biology and Biochemistry, Okayama University Graduate School of Medicine and Dentistry, Okayama 700-8558, Japan; and [§]Department of Cell Biology, University of Alabama at Birmingham, Birmingham, Alabama 35294*

I. INTRODUCTION

The extracellular matrix (ECM) forms both the classical "ground substance" produced by mesenchymal tissues and is also the major component of basement membranes. However, it also possesses several physiological and pathological functions. Most components of the extracellular matrix are large macromolecules with domains for various functions affecting differentiation, cell motility, adhesion, and other activities of cells. The extracellular matrix refers to the insoluble protein complex between cells that is composed of families of macromolecules consisting of collagens, elastin, glycoproteins, and proteoglycans (Timpl, 1996). In addition, many other macromolecules are functionally related to the extracellular matrix, such as (i) cell adhesion molecules together with their receptors, (ii) members of the transforming growth factor-β (TGF-β) superfamily, and (iii) degradative enzymes such as metalloproteinases, as well as other proteases, including processing enzymes. Collagens are characterized by at least one triple helical domain, whereas laminins, fibronectins, and tenascins are all glycoproteins. Proteoglycans such as biglycan, decorin, or aggrecan, in addition to their

Copyright 2003, Elsevier Science (USA). All rights reserved.

glycosaminoglycan chains, are often ligands for cytokines as well as binding strongly to many other extracellular matrix macromolecules. The extracellular matrix receptors of the integrin family are the major receptors for matrix macromolecules and often act as mechanoreceptors, transmitting mechanical signals to the cytoskeleton, microtubules, intermediate filaments, and microfilaments (Wang *et al.*, 1993).

This chapter introduces the biochemical components of the extracellular matrix in order to help understand the other chapters in this volume. Moreover, several current topics of extracellular matrix research are mentioned, specifically including those that involve the extracellular matrix present in the liver.

II. BIOCHEMISTRY OF EXTRACELLULAR MATRIX

A. Collagens

The characteristic structure of all collagens is the presence of one or more domains of triple helical conformation. The three subunit polypeptide chains each contain a $(\text{Gly-X-Y})_n$ repetitive motif required for successful helix formation. Reports demonstrate that more than 20 distinct types of collagens exist in human (Fitzgerald and Bateman, 2001; Koch *et al.*, 2001; Myllyharju and Kivirikko, 2001).

Collagens are classified into four subfamilies; (i) fibrillar collagens (types I, II, III, V, and XI), (ii) FACIT collagens (types IX, XII, XIV, XVI, XIX, XX, and XXI), short chain collagens (types VIII and X), (iii) basement membrane collagen (type IV), as well as the multiplexins (types XV and XVIII), and (iv) collagens with transmembrane domains called MACITs (membrane-associated collagens with interrupted triple helices) (types XIII and XVII). Other collagens, which cannot be classified easily, include types VI and VII (Olsen and Ninomiya, 1999). The collagen types in each classification, the composition of chains in each collagen type, the nomenclature of genes, and their chromosomal locations are shown in Table I.

1. Fibrillar Collagens

The major helical region of types I, II, III, V, and XI are all the same length (295 nm), which is necessary for the packing of individual collagen molecules to form a fibril. These five types of collagen in various combinations form all the collagen fibrils that demonstrate periodic banding as visualized by electron microscopy. It is thought that the organization of different collagen types is critical in determining fibril structure and diameter, this perhaps finally determining the overall tensile strength of the matrix. For example, bone collagen

TABLE I Collagens

Subfamily	Type	α chain	Molecule	Gene	Chromosome
Fibrillar	I	α1(I)	$(\alpha1)_2\alpha2$	COL1A1	17q21-q22
		α2(I)	$(\alpha1)_3$	COL1A2	7q21-q22
	II	α1(II)	$(\alpha1)_3$	COL2A1	12q13-q14
	III	α1(III)	$(\alpha1)_3$	COL3A1	2q24.3-q3117
	V	α1(V)	α1α2α3	COL5A1	9q34.2-q39
		α2(V)	$(\alpha1)_2\alpha2$	COL5A2	2q24.3-q31
		α3(V)		COL5A3	19p13.2
	XI	α1(XI)	α1α2α3	COL11A1	1p21
		α2(XI)		COL11A2	6p212
		α1(II)		COL2A1	12q13-q14
FACIT	IX	α1(IX)	α1α2α3	COL9A1	6q12-q14
		α2(IX)		COL9A2	1p32.3-p33
		α3(IX)		COL9A3	20q13.3
	XII	α1(XII)	$(\alpha1)_3$	COL12A1	6q12-q14
	XIV	α1(XIV)	$(\alpha1)_3$	COL14A1	8q23
	XVI	α1(XVI)	$(\alpha1)_3$	COL16A1	1p314-p35
	XIX	α1(XIX)	$(\alpha1)_3$	COL19A1	6q12-q14
	XX	α1(XX)	$(\alpha1)_3$	COL20A1	?
	XXI	α1(XXI)		COL21A1	6p11-p12
Short	VIII	α1(VIII)	$(\alpha1)_2\alpha2$	COL8A1	3q11.1-q13.2
		α2(VIII)		COL8A2	1p32.3-p34.3
	X	α1(X)	$(\alpha1)_3$	COL10A1	6q21q22
Basement membrane	IV	α1(IV)	$(\alpha1)_2\alpha2$	COL4A1	13q34
		α2(IV)		COL4A2	13q34
		α3(IV)	α3α4α5	COL4A3	21q22.3
		α4(IV)		COL4A4	21q22.3
		α5(IV)	$(\alpha5)_2\alpha6$	COL4A5	Xq22
		α6(IV)		COL4A6	Xq22
Multiplexin	XV	α1(XV)	$(\alpha1)_3$	COL15A1	9q21-q22
		XVIII	$\alpha1(XVIII)(\alpha1)_3$	COL18A1	21q22.3
MACIT		XIII	$\alpha1(XIII)(\alpha1)_3$	COL13A1	10q11-ter
		XVII	$\alpha1(XVII)(\alpha1)_3$	COL17A1	10q24.3
Others		VI	$\alpha1(VI)(\alpha1)_2\alpha2$	COL6A1	21q22.3
			α2(VI)	COL6A2	21q22.3
			α3(VI)	COL6A3	2q3710
		VII	$\alpha1(VII)(\alpha1)_3$	COL7A1	3p21.1

consists mainly of type I collagen, which is required for tensile strength, whereas collagen fibrils of the intestinal mucosa are rich in type III collagen, which is required to form thin, flexible fibers. Fibril diameter, length, and susceptibility to degradation are all carefully regulated by tissues. In liver cirrhosis, large bundles of collagen fibers are synthesized, which may cause reduced distention of vascular channels at the level of sinusoids, portal veins, or hepatic veins and result in increased resistance to blood flow (Bissell *et al.*, 1990).

2. FACIT Collagens

FACIT (fibril-associated collagens with interrupted triple helices) collagens include types IX, XII, XIV, XVI, XIX (Olsen and Ninomiya, 1999), XX (Koch *et al.*, 2001), and XXI (Fitzgerald and Bateman, 2001). This group of collagens potentially serve as molecular bridges between the fibrillar collagens and other extracellular matrix components. Type IX and XII collagens have short triple helical segments, and type IX is also a proteoglycan with a unique glycosaminoglycan (GAG) side chain. These collagens do not form banded fibrils but are located on the surface of banded fibrils. Type IX collagen molecules are found in cartilage and are associated with collagen fibrils assembled from type II and type XI collagens. Type XII and XIV collagens are associated with the surface of type I collagen fibrils. Newly reported type XX collagen also belongs to the FACIT group of collagens. It is likely to bind to collagen fibrils, with the amino-terminal (NC3) domains projecting away from the fibrillar surface. NC3 domains are predicted to be about half the length of those of collagen XIV (Koch *et al.*, 2001)

3. Short Chain Collagens

Type VIII and X collagen molecules have short triple helical domains of only about 130 nm in length. Type VIII collagen is distributed in various tissues, including Descemet's membrane, vascular subendothelial matrices, heart, liver, kidney, perichondrium, and lung, as well as several malignant tumors, including astrocytoma, Ewing's sarcoma, and hepatocellular carcinoma (Olsen and Ninomiya, 1999). Type X collagen is a specific product of hypertrophic chondrocytes and is a useful marker for chondrocyte maturation to hypertrophy.

4. Basement Membrane Collagens

Type IV collagen molecules associate with other type IV molecules, glycoproteins such as laminin and nidogen, and proteoglycans such as perlecan or type XVIII collagen to form the mesh-like structure of basement membranes. Six genetically distinct collagen type IV chains have been identified and are distributed in a tissue-specific manner. In particular, basement membrane collagen containing α3(IV), α4(IV), and α5(IV) chains seems to function as a molecular sieve for the glomerulus of the kidney, for the basement membrane of the alveoli of lung, and for the basement membrane of the cerebrospinal plexus where it functions in the transport of cerebrospinal fluid (Sado *et al.*, 1998).

Petitclerc *et al.* (2000) proposed that the soluble noncollagenous, carboxy-terminal, NC1 domains of the α2(IV), α3(IV), and α6(IV) chains of human collagen type IV are involved in angiogenesis and tumor cell growth. Purified NC1 domains regulate endothelial cell adhesion and migration by distinct av and b1 integrin-dependent mechanisms. In further studies, systemic administration of recombinant α2(IV), α3(IV), and α6(IV) NC1 domains was found to inhibit angiogenesis. Another group (Kamphaus *et al.*, 2000) prepared a recombinant NC1 domain of the α2 chain of type IV collagen and called it "canstatin." This fragment strongly inhibited endothelial cell proliferation and induced apoptosis *in vitro*, as well as suppressing tumor growth *in vivo*. The mechanism of the inhibitory effect of canstatin on ERK phosphorylation at later time points was shown to involve several mitogens instead of growth factors such as vascular endothelial growth factor (VEGF) and bFGF. The apoptotic event induced by canstatin was associated with downregulation of an antiapoptotic protein called FLIP (Kamphaus *et al.*, 2000).

Shahan *et al.* (1999, 2000) demonstrated an inhibitory effect on tumor growth for the NC1 domain of the α3(IV) chain and showed that tumor cell chemotaxis is regulated by the CD47/integrin-associated protein and the integrin $\alpha_v\beta_3$, acting as receptors for the α3(IV) chain through a Ca^{2+}-dependent mechanism. This group also reported that a specific sequence comprising *residues* 185–203 of the NC1 domain of the α3(IV) chain inhibits tumor cell migration *in vitro*, and also decreased the fraction of MMP-2 bound to the plasma membrane of the tumor cells. However, the amount of the inactive forms of MMP-2 secreted into the medium was not altered by the α3(IV)185–203 peptide. Activation of this MMP-2 fraction was strongly inhibited by the peptide, and this inhibition coincided with the inhibition of expression of both MT1-MMP and β3 integrin subunit (Pasco *et al.*, 2000). In a separate study, a fragment of the α3(IV) chain was found to possess antiangiogenic activity and was called tumstatin (Maeshima *et al.*, 2001a,b). The expressed 54–132 amino acid Tum-5 domain has activity for endothelial cells mediated by $\alpha_v\beta_3$ integrin interaction in an RGD-independent manner. It caused apoptosis of endothelial cells with no significant effect on nonendothelial cells. The results suggested that the short tumstatin sequence may be a powerful inhibitor of tumor-associated angiogenesis and may be useful in cancer treatment.

5. Multiplexins

The nonfibrillar collagens type XV and type XVIII are broadly expressed in many tissues and are particularly highly expressed in internal organs. The term multiplexin defines collagens that contain multiple triple-helical domains with interruptions.

Collagen type XVIII, localized mainly in a perivascular position around blood vessels, contains multiple triple-helical domains and serves as the core protein of a heparan sulfate proteoglycan in basement membranes (Muragaki *et al.*, 1995; Oh *et al.*, 1994). Endostatin is a 20-kDa C-terminal fragment of collagen XVIII, which specifically inhibits endothelial proliferation and potently inhibits angiogenesis and tumor growth (O'Reilly *et al.*, 1997). The release of endostatin *in vivo* is likely to be mediated by an elastase and by cathepsin L activities (Felbor *et al.*, 2000; Wen *et al.*, 1999). Endostatin, produced as a recombinant protein by human 293-EBNA cells, blocks various steps in the (VEGF)-induced migration of human umbilical vein endothelial cells in a dose-dependent manner and prevents the subcutaneous growth of human renal cell carcinoma in nude mice (Yamaguchi *et al.*, 1999). The regulation of angiogenesis may be controlled by a balance of stimulators and inhibitors (Hanahan and Folkman, 1996). However, the control mechanisms for the release of angiogenesis inhibitors from their precursors needs further analysis to obtain a better understanding of the process of angiogenesis, as well as for the development of potential therapeutic applications.

6. Collagens with Transmembrane Domains-MACITS

Types XIII and XVII collagen are cell surface molecules with multiple, extracellular, triple-helical domains, connected to a cytoplasmic region by a transmembrane segment (Olsen and Ninomiya, 1999). They form a new class of cellular adhesion molecules by which cells are connected to the extracellular matrix.

Type XIII collagen is a nonfibrillar collagen that has so far largely been characterized from human cDNA and genomic clones. Hägg *et al.* (1998) cloned and characterized the primary structure of mouse type XIII collagen. It was found that cDNA clones for mouse type XIII collagen extend further in the 5′ direction than human clones. The additional sequence suggested that type XIII collagen is a transmembrane protein (Hägg *et al.*, 1998). The α1(XIII) collagen polypeptide consists of short N- and C-terminal noncollagenous domains, termed NC1 and NC4, respectively, together with three collagenous domains called COL1–3, separated by noncollagenous domains called NC2 and NC3. The cytoplasmic domain is short and is unlikely to have any enzymatic activity. However, there is a threonine residue at position 6 of this domain that could be phosphorylated. The predicted large ectodomain may interact with soluble ligands or components of the extracellular matrix or interact laterally with other components of the cell surface (Hägg *et al.*, 1998). Immunolabeling of the epidermis revealed that type XIII collagen was detected both at cell–cell contact sites and at the dermal–epidermal junction. In cultured keratinocytes, type XIII collagen epitopes were detected in focal contacts and in inter-cellular contacts (Peltonen *et al.*, 1999). In addition to its high degree of colocalization with E-cadherin, type XIII collagen is closely associated with adherens type junctions and may be a component of these junctions (Peltonen *et al.*, 1999).

Type XVII collagen is a novel transmembrane ligand of the a6b4 integrin heterodimer (Hopkinson *et al.*, 1998). Further, type XVII collagen seems to be a key component of hemidesmosomes.

7. Other Collagens

Type VI collagen molecules assemble into disulfide-bonded polymers that form beaded microfibrils. Type VI collagen binds to hyaluronan and also binds to the membrane-associated chondroitin sulfate proteoglycan NG2. It also interacts with the microfibril-associated glycoprotein-1 (Olsen and Ninomiya, 1999).

Type VII collagen constitutes the major or only component of anchoring fibrils that link epidermal basement membranes to the underlying connective tissue stroma (Burgeson, 1988).

B. Glycoproteins

Matrix glycoproteins include laminins, fibronectins, tenascins, and nidogen/entactin. Among them, laminin, fibronectin, and nidogen/entactin are all major proteins of basement membranes.

1. Laminins

Laminins are a family of related proteins (400–1000 kDa) characterized by a heterotrimeric chain assembly (α, β, and γ chains), a preferred localization in basement membranes, and a multitude of biological activities (Sasaki and Timpl, 1999). Twelve different chains ($\alpha1$, $\alpha2$, $\alpha3A$, $\alpha3B$, $\alpha4$, $\alpha5$, $\beta1$, $\beta2$, $\beta3$, $\gamma1$, $\gamma2$, and $\gamma3$) have so far been reported (Koch *et al.*, 1999). Combinations of these chains give rise to different molecules called laminins 1–12 (Koch *et al.*, 1999). Major activities include the formation of networks in basement membranes together with heterotypic binding to other basement membrane molecules to form supramolecular assemblies. Other activities include binding to cells through integrins and other receptors.

2. Fibronectin

Fibronectin is a high molecular weight glycoprotein, which has many functions involving cell adhesion, cell morphology, migration, differentiation, and cytoskeletal organization. Fibronectin is spliced differentially and is secreted as a dimer of two subunits held together by a pair of disulfide bonds located near the carboxyl termini (Hynes, 1999). Due to differential splicing, the subunits of fibronectin can vary in size between approximately 235 and 270 kD together with variable amounts of carbohydrate.

3. Nidogen/Entactin

Nidogen is a sulfated protein with a molecular mass of 158 kDa (Carlin *et al.*, 1981; Timpl *et al.*, 1983) that is susceptible to degradation by endogenous proteases, with cleavages occurring at both ends of the polypeptide chain (Dziadek *et al.*, 1985; Paulsson *et al.*, 1985). Binding with strong affinity to laminin occurs *in vivo* (Carlin *et al.*, 1983; Hogan *et al.*, 1982), with the specific binding site being localized to the carboxyl-terminal globular domain (Dziadek *et al.*, 1985; Mann *et al.*, 1989). The structure of nidogen is dumbbell shaped with three globules joined by a 15-nm-long rod and flexible link. The nidogen G3 domain binds to laminin γ1, perlecan, and fibulin 2, whereas the G2 domain binds to collagen IV, perlecan, fibulin 1, and fibronectin (Timpl, 1999).

C. Proteoglycans

Proteoglycans consist of a core protein (backbone) to which are attached numerous glycosaminoglycan side chains and oligosaccharides by N or O linkages. The classification of proteoglycans is based on the composition of glycosaminoglycan chains, such as chondroitin sufate, dermatan sulfate, keratan sulfate, and heparan sulfate. The proteoglycans called perlecan, syndecan, biglycan, and decorin are all present in the liver (Meyer *et al.*, 1992; Rescan *et al.*, 1993; Roskams *et al.*, 1995). In this book, Yada and Kimata present a detailed review of proteoglycans (Chapter 4).

D. Elastin

Elastin is secreted from the cell as a soluble protein with a molecular mass of approximately 70 kDa (tropoelastin). The tropoelastin molecule consists, for the most part, of alternating hydrophobic and cross-linking domains (Mecham, 1999). The hydrophobic domains of elastin contribute to the conformational changes of the protein as it undergoes elastic recoil. However, an intact elastic fiber is a complex structure including elastin, microfibrillar proteins, lysyl oxidase, and proteoglycans.

III. CURRENT TOPICS IN EXTRACELLULAR MATRIX RESEARCH

A. BMP-1 and Procollagen C-Proteinase Are Identical

Collagen types I, II, and III (the major fibrous components of extracellular matrix) are synthesized as procollagens with NH2- and COOH-terminal propeptides that must be cleaved to yield mature monomers capable of forming fibrils (Olsen and Ninomiya, 1999). Procollagen C-proteinase (PCP) cleaves the COOH propeptides of procollagens I, II, and III, which must occur before fibril assembly. However, it is the COOH-terminal propeptides of collagens type I, II, and III that are responsible for determining chain stoichiometry and alignment leading to specific triple helix formation. Bone morphogenetic proteins (BMP) are bone-derived factors capable of inducing ectopic bone formation. However, it was subsequently shown that BMP-1 and PCP are identical (Kessler *et al.*, 1996) and BMP-1 by itself cannot induce ectopic bone formation. After the initial isolation of BMP-1 from an osteogenetic fraction of bone (Wozney *et al.*, 1988), other proteins involved in morphogenetic patterning, such as *Drosophila* tolloid (TLD) and tolloid-related-1 (TLR-1)/Tolkin, sea urchin BP10, and SpAN, were isolated and found to be structural homologues of BMP-1 (Finelli *et al.*, 1995; Lepage *et al.*, 1992; Nguyen *et al.*, 1994; Reynolds *et al.*, 1992; Shimell *et al.*, 1991). BMP-1/TLD-like proteases may affect cell fate decisions through the activation of TGF-β-like proteins. The BMP-1 gene produces two alternatively spliced transcripts encoding BMP-1 and a longer protein, called mammalian tolloid (mTLD) with a domain structure identical to that of *Drosophila* TLD (Takahara *et al.*, 1994). Moreover, a product(s) of the BMP-1 gene will activate TGF-β-like molecules, which then induces procollagen deposition into insoluble matrix, the latter further influencing cell fate decisions by altering cell–matrix interactions (Kessler *et al.*, 1996).

Gene targeting of mTLL-1 revealed that mTLL-1 plays multiple roles in the formation of the mammalian heart and is essential for formation of the interventricular septum (Clark *et al.*, 1999). Scott *et al.* (1999) identified mammalian Tolloid-like 2 (mTLL-2) as a novel family member and compared enzymatic activities and expression domains among all four known mammalian BMP-1/TLD-like proteases, i.e., BMP-1, mammalian Tolloid (mTLD), mammalian Tolloid-like-1 (mTLL-1), and mTLL-2. Among these proteins, there are differences in their ability to process fibrillar collagen precursors and to cleave chordin, the vertebrate orthologue of short gastrulation (SOG). Differences in enzymatic activities and expression domains of the four proteases suggest that BMP-1 is the major chordin antagonist in early mammalian embryogenesis and in pre- and postnatal skeletogenesis. This topic needs to be investigated further, as these proteins participate not only in morphogenesis, but also in wound healing and in pathological conditions involving fibrosis.

B. Mechanical Stress Induces ECM Gene Expression

Gene expression of the glycoprotein tenascin C and collagen type XII is altered by extracellular mechanical stress as might be expected of two ECM proteins typical of tendons and ligaments (Chiquet-Ehrismann *et al.*, 1999; Trachslin *et al.*, 1999).

These authors showed that the responses to changes in the stretching capacity of a collagen gel, which acts as a scaffold of fibroblasts, are rapid and reversible and are reflected by changes at the mRNA level. Both the promoters for tenascin-C and the collagen XII gene contain "stretch-responsive" enhancer regions with similar "shear stress response elements" as are found in the gene for PDGF-B. Regulation seems to involve the increased production of tenascin-C and collagen XII in fibroblasts attached to a stretched collagen matrix, but suppression occurs in cells in a relaxed matrix (Chiquet *et al.*, 1999).

C. New Receptors of Collagens

Cell-to-matrix interactions play important roles in proliferation, differentiation, motility, adhesion, and other functions of cells. Integrins transmit extracellular signals to intracellular locations. Integrins are composed of one α chain and one β chain, both with membrane-spanning domains. Twenty-four dimers of 18 α chains and 8 β chains are known, and the ligands of each integrin have been reported. β_1-integrins include all known cellular receptors for native collagen, namely $\alpha_1\beta_1$, $\alpha_2\beta_1$, $\alpha_3\beta_1$, $\alpha_{10}\beta_1$, and $\alpha_{11}\beta_1$ heterodimers (Hemler, 1999).

Two integrins, $\alpha_1\beta_1$ and $\alpha_2\beta_1$, are very similar in structure and properties. Nykvist *et al.* (2000) have investigated whether both of these integrins act as receptors for the same collagens or for different collagens. They used Chinese hamster ovary (CHO) cells that lack endogenous collagen receptors and transfected CHO cells with either α_1 or α_2 integrin cDNA. Cells expressing the $\alpha_1\beta_1$ integrin were able to spread on collagen types I, III, IV, and V, but not on type II, whereas $\alpha_2\beta_1$ integrin-expressing cells were able to spread on all five collagens (Nykvist *et al.*, 2000). The integrins $\alpha_1\beta_1$, $\alpha_2\beta_1$, $\alpha_{11}\beta_1$, and $\alpha_{10}\beta_1$ that recognize collagens all possess an I domain, which is the site of binding to the collagen fibril. Moreover, it has been shown that α_1I domains, unlike α_2I domains, could attach to type XIII collagen (Nykvist *et al.*, 2000). Thus, $\alpha_1\beta_1$ and $\alpha_2\beta_1$ have different ligand-binding specificities. As cells can concomitantly express a number of different receptors, it is very difficult to identify the specific function of each integrin. However, it is clear that $\alpha_1\beta_1$ and $\alpha_2\beta_1$ integrins affect different signal transduction pathways after binding to different ligands (Nykvist *et al.*, 2000).

Nonintegrin collagen receptors have been reported called discoidin domain receptors DDR1 and DDR2, (Shrivastava *et al.*, 1997; Vogel *et al.*, 1997). DDR1 is activated by collagen types I–V, whereas DDR2 is only activated by fibrillar collagens. DDR1 and DDR2 are also only activated by native collagen molecules, whereas collagen glycosylation is only important for DDR2 stimulation. The activation process is surprisingly slow, requiring collagen treatment for 18 h to reach maximal tyrosine kinase activity. It is sustained for up to 4 days. Signal transduction is different between both receptors, and prolonged activation of DDR2 is associated with an upregulation of matrix metalloproteinase-1 (MMP-1) (Vogel *et al.*, 1997). Overexpression of DDR1 was found to occur in primary breast cancer, ovarian cancer, esophageal cancer, and pediatric brain cancer, particularly in highly invasive tumors, whereas transcripts for DDR2 were observed in the surrounding stromal cells (Vogel, 1999). Both DDR1 and DDR2 may be involved not only in tumor progression, but also in other diseases with deregulated matrix production, including lung fibrosis, liver cirrhosis, osteoporosis, and rheumatoid arthritis (Vogel, 1999).

D. Collagen-like Molecules with Triple Helical Structure

Several proteins have a collagen-like structure, but are not included in the superfamily of collagens. These include C1q and the related collectin family, the ficolin family, macrophage scavenger receptor, macrophage receptor with collagenous structure (MARCO), acetylcholinesterase (collagen Q), p200, and ectodysplasin (Myllyharju and Kivirikko, 2001; Tenner, 1999). The following four protein families are described further.

1. Collectins

The definition of a collectin is a C-type lectin family member with collagen-like sequences and carbohydrate recognition domains. These include mannose-binding lectin (MBL) (previously called mannose-binding protein or MBP), conglutinin, collectin 43, surfactant protein A, surfactant protein D, and collectin liver 1 (CL-L1) (Holmskov *et al.*, 1994). MBL plays a role in the first line of host defense, acting as an opsonin, and may directly facilitate the recognition of pathogens by phagocytes. MBL consists of 18 identical subunits of 32 kDa arranged as a hexamer of trimers. Each subunit consists of an amino-terminal region rich in cysteine, followed by 19 collagen repeats and a carbohydrate recognition domain that requires calcium to bind ligands (Sheriff *et al.*, 1994). Biological roles of the collagen motif of MBL are based on the minimal triple-helical structure, which still causes phagocytic activity as an opsonin. The collagen-like motif is the binding motif of a MBL-associated serine protease (MASP), which then activates the lectin complement pathway (Matsushita *et al.*, 1995; Nepomuceno *et al.*, 1997).

2. Ficolins

Ficolins are defined as a family of proteins with both fibrinogen-like and collagen-like domains. Originally, these molecules were purified as TGF-β1-binding proteins from porcine uterus (Ichijo *et al.*, 1991), and the overall structures of ficolin-α and -β are very similar to those of C1q and the C-type lectins. All of these molecules have short noncollagenous N-terminal sequences followed by collagen-like regions that contain short gaps, except for the C1q B chain, and C-terminal regions (Ichijo *et al.*, 1993). To date, ficolins have been identified in several kinds of vertebrates. In humans, three kinds of ficolins have been shown (Matsushita *et al.*, 2001). The first type of human ficolin was identified by different groups as elastin-binding protein EBP-37 (Harumiya *et al.*, 1995), a corticosteroid-binding protein (Edgar *et al.*, 1995), and mannose-binding lectin p35 (Matsushita *et al.*, 1996). This implies that ficolin is a multifunctional protein in serum. Serum ficolin is considered to bind to a receptor and to mediate the clearance and scavenging of the sugar residues *N*-acetyl-D-glucosamine (GlcNAc) and *N*-acetyl D-galactosamine (GalNAc) via a fibrinogen-like domain (Le *et al.*, 1998; Matsushita *et al.*, 2001). Serum human ficolin also makes a complex with MASP via its collagenous motif and has the capacity to activate the lectin complement pathway (Matsushita *et al.*, 2000).

3. Acetylcholinesterase-Associated Collagen

Acetylcholine released from motor nerve terminals is inactivated rapidly by the synaptic acetylcholinesterase (AChE), thereby limiting the action of the neurotransmitter. Much of the AChE in the neuromuscular junction is present in "asymmetric" forms. The asymmetric forms of AChE are heterooligomers in which $AChE_T$ catalytic subunits are associated with collagenous Q subunits (Krejci *et al.*, 1997). This group reported that collagen tails are encoded by a single gene, COLQ, and that the same COLQ subunits are incorporated, *in vivo*, in asymmetric forms of both AChE and butyrylcholinesterase (Krejci *et al.*, 1997). The COLQ gene is expressed in cholinergic tissues, brain, muscle, and heart and also in noncholinergic tissues such as lung and testis (Krejci *et al.*, 1997). $ColQ^{-/-}$ mice completely lacked asymmetric AChE in skeletal and cardiac muscles and brain (Feng *et al.*, 1999). Globular AChE tetramers are also absent in neonatal $ColQ^{-/-}$ muscles, suggesting a role for the ColQ gene in assembly or stabilization of AChE globular forms that do not themselves contain a collagenous subunit (Feng *et al.*, 1999).

4. p200

The protein p200 was purified from conditioned medium of cultured rat Schwann cells. It has an apparent molecular mass of approximately 200 kDa and has the ability to bind cell surface heparan sulfate proteoglycan *N*-syndecan (Chernousov *et al.*, 1996). Immunofluorescent staining of rat sciatic nerve revealed that p200 was seen in the extracellular matrix surrounding individual Schwann cell-axon units (Chernousov *et al.*, 1996). Thus, Schwann cells secrete a collagen-like adhesive protein that interacts with cells through cell surface heparan sulfate proteoglycans. A report dealing with cDNA cloning for p200 concluded that it is a potentially novel member of the collagen type V gene family, closely related to the human and mouse α3(V) chain (Chernousov *et al.*, 2000; Imamura *et al.*, 2000).

IV. EXTRACELLULAR MATRIX IN LIVER CIRRHOSIS COMPARED WITH THAT IN NORMAL LIVER

A. Extracellular Matrix in Normal Liver

Extracellular matrix proteins in normal liver are distributed mainly in the liver capsule, in the portal tracts with large and medium-sized blood vessels, and in the ducts and ductules of the biliary tree, whereas a basement membrane-like matrix is located in the sinusoidal space of Disse. Fibrillar collagens (types I, III, and V) are found by immunohistochemistry predominantly in the perivascular spaces and portal tracts and in subcapsular areas (Abrahamson and Caulfield, 1985; Arenson *et al.*, 1988; Hahn *et al.*, 1980; Martinez-Hernandez, 1984; Maher *et al.*, 1988). The interstitial extracellular matrix around large vessels and portal tracts also contains fibronectin, undulin (collagen type XIV) (Schuppan *et al.*, 1990), and other glycoconjugates. Musso *et al.* (1998) demonstrated collagen type XVIII in the liver where it is a potential precursor of endostatin (Clement *et al.*, 1999). Endostatin plays an important role in angiogenesis as mentioned earlier.

In normal liver the structure of the sinusoidal space of Disse is unique. The typical electron-dense basement membrane cannot be observed in the space of Disse (Okazaki *et al.*, 1973). However, collagen types IV, VI, and XIV, together with laminin, fibronectin, nidogen/entactin, and proteoglycans, must form a basement membrane-like structure (Rescan *et al.*, 1993; Roskams *et al.*, 1995; Schwögler *et al.*, 1994). Moreover, Geerts *et al.* (1990) reported that type I collagen and type III collagen can be present together in hybrid fibrils in the space of Disse with type V fibers forming the fibril core. Heparan sulfate is localized at the portal terminals (Geerts *et al.*, 1986), and heparin proteoglycan is observed at the venous terminals (Reid *et al.*, 1992). The low content of dermatan and chondroitin sulfate may contribute to the fenestration of the sinusoidal cells (McGuire *et al.*, 1992).

B. Extracellular Matrix in Liver Cirrhosis

Fibrosis itself plays an important role in the repair process in cases involving wound healing. In a case of extensive fibrosis, however, replacing the parenchyma of the functioning organs with fibrous tissue results in a marked functional impairment. Liver fibrosis is usually encountered following liver cell damage and inflammation of the portal tracts and is one of the main factors underlying circulatory disturbances in the liver. Perihepatocellular fibrosis with ensuing capillarization of the sinusoid and the formation of fibrous septa, which carry vessels causing shunting, will lead to a disturbance of intrahepatic microcirculation. This will cause reduced parenchymal perfusion, followed by further liver cell damage and acceleration of hepatic fibrosis, leading to a "vicious cycle" (Alcolado *et al.*, 1997; Friedman, 1993, 1997; Li and Friedman, 1999; Okazaki *et al.*, 1973).

Biochemically, the cirrhotic liver contains approximately six times as much matrix as normal liver (Schuppan, 1990). Comparison of the contents of each extracellular matrix component in normal liver with those in liver cirrhosis is shown in Table II. There is a fivefold increase in total collagen content and a three- to fivefold increase of other noncollagenous components (Rojkind *et al.*, 1979;

TABLE II Comparison of Content of Collagens and Noncollagenous Matrix Components between Normal and Cirrhotic Liver[a]

	Normal liver	Cirrhotic liver
Collagens		
Total collagen (mg/g protein)	80	470
Type I collagen (%)	40–50	60–70
Type III collagen (%)	40–50	20–30
Type IV collagen (mg/g protein)	0.64	5.20
Type IV collagen (%)	1	1–2
Type V collagen (%)	2–5(?)	5–10(?)
Type VI collagen (mg/g protein)	0.06	0.68
Type VI collagen (%)	0.1	0.2
Noncollagenous components		
Fibronectin (μg/g wet weight)	300	625
Laminin	—	—
Total glycosaminoglycans (μg uronic acid/g defatted dry weight)	98	520
Heparan sulfate (%)	63	47
Chondroitin (%)	3	3
Chondroitin 4-sulfate (%)	6	10
Chondroitin 6-sulfate (%)	8	7
Dermatan sulfate (%)	18	27
Hyaluronic acid (%)	1.5	6
Elastin (mmol desmosine + isodesmosine/g wet weight)	1.44	7.0

[a]Modified from papers of Schuppan (1990) and Gressner and Bachem (1990). Data on collagens, fibronectin, total glycosaminoglycans, and their fractions were derived from human liver, but elastin from rat liver. The proportions of different collagens were condensed from Rojkind *et al.* (1979), Seyer *et al.* (1977), and Schuppan *et al.* (1984) by Schuppan (1990). Data on fibronectin are originally based on the report of Isemura *et al.* (1984), glycosaminoglycans from Murata *et al.* (1984, 1985), and elastin from Velebny *et al.* (1983). Quantitative data on laminin are not available.

Schuppan *et al.*, 1990). Thick bands of fibrous septa in cirrhosis reflect the increased amount of type I, type III, and type IV collagen, as well as the disproportionate increase in the ratio of type I collagen to types III and IV (Rojkind *et al.*, 1979; Seyer *et al.*, 1977).

The total increase in extracellular matrix is contributed largely by the accumulation of matrix in the space of Disse, which is consistent with sinusoidal capillarization. Type IV collagen, laminin, fibronectin, and other glycoproteins all increase in the cirrhotic liver (Schuppan, 1990). A relative decrease in heparin and heparan sulfate proteoglycans and an increase in dermatan and chondroitin sulfate proteoglycans also occur (Alcolado *et al.*, 1997). The increase in decorin and biglycan reveals a potentially important role in cytokine binding, including TGF-β as Alcolado *et al.* (1997) have stressed. Hyaluronan increases more than eight-fold in cirrhosis compared to normal liver and is secreted by hepatic stellate cells (Gressner and Haarmann, 1988). The high serum level of hyaluronan is thought to reflect impaired phagocytosis by sinusoidal endothelial cells (see Chapter 4).

REFERENCES

Abrahamson, D. R., and Caulfield, J. P. (1985). Distribution of laminin with rat and mouse renal, splenic, intestinal, and hepatic basement membranes identified after the intravenous injection of heterologous antilaminin IgG. *Lab. Invest.* **52**, 169–181.

Alcolado, R., Arthur, M. J. P., and Iredale, J. P. (1997). Pathogenesis of liver fibrosis. *Clin. Sci.* **92**, 103–112.

Arenson, D. M., Friedman, S. L., and Bissell, D. M. (1988). Formation of extracellular matrix in normal rat liver: Lipocytes as a major source of proteoglycan. *Gastroenterology* **95**, 441–447.

Bissell, D. M., Friedman, S. L., Maher, J. J., and Roll, F. J. (1990). Connective tissue biology and hepatic fibrosis: Report of a conference. *Hepatology* **11**, 488–498.

Burgeson, R. E. (1988). New collagens, new concepts. *Annu. Rev. Cell. Biol.* **4**, 551–577.

Carlin, B., Durkin, M. E., Bender, B., Jaffe, R., and Chung, A. E. (1983). Synthesis of laminin and entactin by F9 cells induced with retinoic acid and dibutyryl cyclic AMP. *J. Biol. Chem.* **258**, 7729–7737.

Carlin, B., Jaffe, R., Bender, B., and Chung, A. E. (1981). Entactin: A novel basal lamina-associated glycoprotein. *J. Biol. Chem.* **256**, 5209–5214.

Chernousov, M. A., Rothblum, K., Tyler, W. A., Stahl, R. C., and Carey, D. J. (2000). Schwann cells synthesize type V collagen that contains a novel alpha 4 chain: Molecular cloning, biochemical characterization, and high affinity heparin binding of alpha 4(V) collagen. *J. Biol. Chem.* **275**, 28208–28215.

Chernousov, M. A., Stahl, R. C., and Carey, D. J. (1996). Schwann cells secrete a novel collagen-like adhesive protein that binds N-syndecan. *J. Biol. Chem.* **271**, 13844–13853.

Chiquet, M. (1999). Regulation of extracellular matrix gene expression by mechanical stress. *Matrix Biol.* **18**, 417–426.

Chiquet-Ehrismann, R., Tannheimer, M., Koch, M., Brunner, A., Spring, J., Martin, D., Baumgartner, S., and Chiquet, M. (1994). Tenascin-C expression by fibloblasts is elevated in stressed collagen gels. *J. Cell Biol.* **127**, 2093–2101.

Clark, T. G., Conway, S. J., Scott, I. A., Labosky, P. A., Winnier, G., Bundy, J., Hogan B. L. M., and Greenspan, D. S. (1999). The mammalian tolloid-like 1 gene, Tll1, is necessary for normal septation and positioning of the heart. *Development* 126, 2631–2642.

Clement, B., Musso, O., Lietard, J., and Theret, N. (1999). Homeostatic control of angiogenesis: A newly identified function of the liver? *Hepatology* 29, 621–623.

Dziadek, M., Paulsson, M., and Timpl, R. (1985). Identification and interaction repertoire of large forms of the basement membrane protein nidogen. *EMBO J.* 4, 2513–2518.

Edgar PF. (1995). Hucolin, a new corticosteroid-binding protein from human plasma with structural similarities to ficolins, transforming growth factor-beta 1-binding proteins. *FEBS Lett.* 375, 159–161.

Felbor, U., Dreier, L., Bryant, R. A., Ploegh, H. L., Olsen, B. R., and Mothes, W. (2000). Secreted cathepsin L generates endostatin from collagen XVIII. *EMBO J.* 19, 1187–1194.

Feng, G., Krejci, E., Molgo, J., Cunningham, J. M., Massoulié, J., and Sanes, J. R. (1999). Genetic analysis of collagen Q: Roles in acetylcholinesterase and butyrylcholin-esterase assembly and in synaptic structure and function. *J. Cell Biol.* 144, 1349–1360.

Finelli, A. L., Xie, T., Bossie, C. A., Blackman, R. K., and Padgett, R. W. (1995). The tolkin gene is a tolloid/BMP-1 homologue that is essential for Drosophila development. *Genetics* 141, 271–281.

Fitzgerald, J., and Bateman, J.F. (2001). A new FACIT of the collagen family: COL21A1. *FEBS Lett.* 505, 275–280.

Friedman, S. L. (1993). Seminars in medicine of the Beth Israel Hospital, Boston. The cellular basis of hepatic fibrosis. Mechanisms and treatment strategies. *N. Engl. J.* 328, 1828–1835.

Friedman, S. L. (1997). Molecular mechanisms of hepatic fibrosis and principles of therapy. *J. Gastroenterol.* 32, 424–430.

Geerts, A., Schuppan, D., Lazeroms, S., De-Zanger, R., and Wisse, E. (1990). Collagen type I and III occur together in hybrid fibrils in the space of Disse of normal rat liver. *Hepatology* 12, 233–241.

Geerts, A., Geuze, H. J., Slot, J. W., Voss, B., Schuppan, D., Schellinck, P., and Wisse, E. (1986). Immunogold localization of procollagen III, fibronectin and heparan sulfate proteoglycan on ultrathin frozen sections of the normal rat liver. *Histochemistry* 84, 355–362.

Gressner, A. M., and Haarmann, R. (1988). Hyaluronic acid synthesis and secretion by rat liver fat storing cells (perisinusoidal lipocytes) in culture. *Biochem. Biophys. Res.* 151, 222–229.

Hägg, P., Rehn, M., Huhtala, P., Väisänen, T., Tamminen, M., and Pihlajaniemi, T. (1998). Type XIII collagen is identified as a plasma membrane protein. *J. Biol. Chem.* 273, 15590–15597.

Hahn, E. G., Wick, G., Pencev, D., and Timpl, R. (1980). Distribution of basement of membrane proteins in normal and fibrotic human liver: Collagen type IV, laminin fibronectin. *Gut* 21, 63–71.

Hanahan, D., and Folkman, J. (1996). Patterns and emerging mechanisms of the angiogenic switch during tumorigenesis. *Cell* 86, 353–364.

Harumiya, S., Omori, A., Sugiura, T., Fukumoto, Y., Tachikawa, H., and Fujimoto, D. (1995). EBP-37, a new elastin-binding protein in human plasma: Structural similarity to ficolins, transforming growth factor-beta 1-binding proteins. *J. Biochem. (Tokyo)* 117, 1029–1035.

Hemler, M. E. (1999). Integrins. *In* "Guide Book to the Extracellular Matrix and Adhesion Proteins" (T. Kreis *et al.*, eds.), pp. 196–212. CRC Press, Boca Raton, FL.

Hogan, B. L. M., Taylor, A., Kurkinen, M., and Couchman, J. R. (1982). Synthesis and localization of two sulphate glycoproteins associated with basement membranes and the extracellular matrix. *J. Cell Biol.* 95, 197–204.

Holmskov, U., Malhotra, R., Sim, R. B., and Jensenius, J. C. (1994). Collectins: Collagenous C-type lectins of the innate immune defense system. *Immunol. Today* 15, 67–74.

Hopkinson, S. B., Findlay, K., deHart, G. W., and Jones, J. C. (1998). Interaction of BP180 (type XVII collagen) and alpha6 integrin is necessary for stabilization of hemidesmosome structure. *J. Invest. Dermatol.* 111, 1015–1022.

Hynes, R. (1999). Fibronectins. *In* "Guide book to the Extracellular Matrix and Adhesion Proteins" (T. Kreis *et al.*, eds.), pp. 422–424. CRC Press, Boca Raton, FL.

Ichijo, H., Hellman, U., Wernstedt, C., Gonez, L. J., Claesson-Welsh, L., Heldin, C. H., and Miyazono, K. (1993). Molecular cloning and characterization of ficolin, a multimeric protein with fibrinogen- and collagen-like domains. *J. Biol. Chem.* **268**, 14505–14513.

Ichijo, H., Ronnstrand, L., Miyagawa, K., Ohashi, H., Heldin, C. H., and Miyazono, K. (1991). Purification of transforming growth factor-beta 1 binding proteins from porcine uterus membranes. *J. Biol. Chem.* **266**, 22459–22464.

Imamura, Y., Scott, I. C., and Greenspan, D. S. (2000). The pro-a3(V) collagen chain complete primary structure, expression domains in adult and developing tissues, and comparison to the structures and expression domains of the other types V and XI procollagen chains. *J. Biol. Chem.* **275**, 8749–8759.

Kamphaus, G. D., Colorado, P. C., Panka, D. J., Hopfer, H., Ramchandran, R., Torre, A., Maeshima, Y., Mier, J. W., Sukhatme, V. P., and Kalluri, R. (2000). Canstatin, a novel matrix-derived inhibitor of angiogenesis and tumor growth. *J. Biol. Chem.* **275**, 1209–1215.

Kessler, E., Takahara, K., Biniaminov, L., Brusel, M., and Greenspan, D. S. (1996). Bone morphogenetic protein-1: The type I procollagen C-proteinase. *Science* **271**, 360–362.

Koch, M., Foley, J. E., Hahn, R., Zhou, P., Burgeson, R. E., Gerecke, D. R., and Gordon, M. K. (2001). α1(XX) collagen, a new member of the collagen subfamily, fibril-associated collagens with interrupted triple helices. *J. Biol. Chem.* **276**, 23120–23126.

Koch, M., Olson, P. F., Albus, A., Jin, W., Hunter, D. D., Brunken, W. J., Burgeson, R. E., and Champliaud, M. -F. (1999). Characterization and expression of the laminin γ3: A novel, nonbasement membrane-associated, laminin chain. *J. Cell Biol.* **145**, 605–617.

Krejci, E., Thomine, S., Boschetti, N., Legay, C., Sketelj, J., and Massoulié, J. (1997). The mammalian gene of acetylcholinesterase-associated collagen. *J. Biol. Chem.* **272**, 22840–22847.

Le, Y., Lee, S. H., Kon, O. L., and Lu, J. (1998). Human L-ficolin: plasma levels, sugar specificity, and assignment of its lectin activity to the fibrinogen-like (FBG) domain. *FEBS Lett.* **425**, 367–370.

Li, D., and Friedman, S. L. (1999). Liver fibrogenesis and the role of hepatic stellate cells: New insights and prospects for therapy. *J. Gastroenterol. Hepatol.* **14**, 618–633.

Lepage, T., Ghiglione, C., and Gache, C. (1992). Spatial and temporal expression pattern during sea urchin embryogenesis of a gene coding for a protease homologous to the human protein BMP-1 and to the product of the Drosophila-ventral patterning gene tolloid. *Development* **114**, 147–163.

Maeshima, Y., Manfredi, M., Reimer, C., Holthaus, K. A., Hopfer, H., Chandamuri, B. R., Kharbanda, S., and Kalluri, R. (2001a). Identification of the anti-angiogenic site within vascular basement membrane-derived tumstatin. *J. Biol. Chem.* **276**, 15240–15248.

Maeshima, Y., Yerramalla, U. L., Dhanabal, M., Holthaus, K. A., Barbashov, S., Kharbanda, S., Reimer, C., Manfredi, M., Dickerson, W. M., and Kalluri, R. (2001b). Extracellular matrix-derived peptide binds to $\alpha_v\beta_3$ integrin and inhibits angiogenesis. *J. Biol. Chem.* **276**, 31959–31968.

Maher, J. J., Friedman, S. L., Roll, F. J., and Bissell, D. M. (1988). Immunolocalization of laminin in normal rat liver and biosynthesis of laminin by hepatic lipocytes in primary culture. *Gastroenterology* **94**, 1053–1062.

Mann, K., Deutzmann, R., and Timpl, R. (1989). Characterization of proteolytic fragments of the laminin-nidogen complex and their activity in ligand-binding assays. *Eur. J. Biochem.* **178**, 71–80.

Mann, K., Deutzmann, R., Aumailley, M., Timpl, R., Raimondi, L., Yamada, Y., Pan, T. C., Conway, D., and Chu, M. L. (1989). Amino acid sequence of mouse nidogen, a multidomain basement membrane protein with binding activity for laminin, collagen type IV, and cells. *EMBO J.* **8**, 65–72.

Martinez-Hernandez, A. (1984). The hepatic extracellular matrix. I. Electron immunohistochemical studies in normal rat liver. *Lab. Invest.* **51**, 57–69.

Matsushita, M., and Fujita, T. (2001). Ficolins and the lectin complement pathway. *Immunol. Rev.* **180**, 78–85.

Matsushita, M., Endo, Y., and Fujita, T. (2000). Cutting edge: Complement-activating complex of ficolin and mannose-binding lectin-associated serine protease. *J. Immunol.* **164**, 2281–2284.

Matsushita, M., Endo, Y., Taira, S., Sato, Y., Fujita, T., Ichikawa, N., Nakata, M., and Mizuochi, T. (1996) A novel human serum lectin with collagen- and fibrinogen-like domains that functions as an opsonin. *J. Biol. Chem.* **271**, 2448–2454.

Matsushita, M., Ezekowitz, R. A. B., and Fujita, T. (1995). The Gly-54→Asp allelic form of human mannose-binding protein (MBP) fails to bind MBP-associated serine protease. *Biochem. J.* **311**, 1021–1023.

McGuire, R. F., Bissell, D. M., Boyles, J., and Roll, F. J. (1992) Role of extracellular matrix in regulating fenestrations of endothelial cells isolated from normal rat liver. *Hepatology* **15**, 989–997.

Mecham, R. P. (1999) Elastin. *In* "Guide book to the Extracellular Matrix and Adhesion Proteins" (T. Kreis *et al.*, eds.), pp. 414–417. CRC Press, Boca Raton, FL.

Meyer, D. H., Krull, N., Dreher, K. L., and Gressner, A. M. (1992). Biglycan and decorin gene expression in normal and fibrotic rat liver: Cellular localization and regulatory factors. *Hepatology* **16**, 204–216.

Muragaki, Y., Timmons, S., Griffith, C. M., Oh, S. P., Fadel, B., Quertermous, T., and Olsen, B. R. (1995). Mouse Coll8a1 is expressed in a tissue-specific manner as three alternative variants and is localized in basement membrane zones. *Proc. Natl. Acad. USA* **92**, 8763–8767.

Musso, O., Rehn, M., Saarela, J., Théret, N., Liétard, J., Hintikka, E., Lotrian, D., Campion, J. -P., Pihlajaniemi, T., and Clément, B. (1998). Collagen XVIII is localized in sinusoids and basement membrane zones and expressed by hepatocytes and activated stellate cells in fibrotic human liver. *Hepatology* **28**, 98–107.

Myllyharju, J., and Kivirikko, K. I. (2001) Collagens and collagen-related diseases. *Ann. Med.* **33**, 7–21.

Nepomuceno, R. R., Henschen-Edman, A. H., Burgess, W. H., and Tenner, A. J. (1997). cDNA cloning and primary structure analysis of C1qR(P), the human C1q/MBL/SPA receptor that mediates enhanced phagocytosis in vitro. *Immunity* **6**, 119–129.

Nguyen, T., Jamal, J., Shimell, M. J., Arora, K., and O'Connor, M. B. (1994). Characterization of tolloid-related-1: A BMP-1-like product that is required during larval and pupal stages of Drosophila development. *Dev. Biol.* **166**, 569–586.

Nykvist, P., Tu, H., Ivaska, J., Käpylä, J., Pihlajaniemi, T., and Heino, J. (2000). Distinct recognition of collagen subtypes by $\alpha_1\beta_1$ and $\alpha_2\beta_1$ integrins: $\alpha_1\beta_1$ mediates cell adhesion to type XIII collagen. *J. Biol. Chem.* **275**, 8255–8261.

Oh, S. P., Kamagata, Y., Muragaki, Y., Timmons, S., Ooshima, A., and Olsen, B. R. (1994). Isolation and sequencing of cDNAs for proteins with multiple domains of Gly-X-Y repeats identify a novel family of collagenous proteins. *Proc. Natl. Acad. Sci. USA.* **91**, 4229–4233.

Okazaki, I., Tsuchiya, M., Kamegaya, K., Oda, M., Maruyama, K., and Oshio, C. (1973) Capillarization of hepatic sinusoids in carbon tetrachloride-induced hepatic fibrosis. *Bibl. Anat.* **12**, 476–483.

Olsen, B. R., and Ninomiya, Y. (1999) Collagens. *In* "Guide Book to the Extracellular Matrix and Adhesion Proteins" (T. Kreis *et al.*, eds.), pp. 380–408. CRC Press, Boca Raton, FL.

O'Reilly, M. S., Boehm, T., Shing, Y., Fukai, N., Vasios, G., Lane, W. S., Flynn, E., Birkhead, J. R., Olsen, B. R., and Folkman, J. (1997) Endostatin: An endogenous inhibitor of angiogenesis and tumor growth. *Cell* **88**, 277–285.

Pasco, S., Han, J., Gillery, P., Bellon, G., Maquart, F. -X., Borel, J. P., Kefalides, N. A., and Monboisse, J. C. (2000) A specific sequence of the noncollagenous domain of the α3(IV) chain of type IV collagen inhibits expression and activation of matrix metalloproteinases by tumor cells. *Cancer Res.* **60**, 467–473.

Paulsson, M., Dziadek, M., Suchaner, C., Huttner, W. B., and Timpl, R. (1985). Nature of sulphated macromolecules in mouse Reichert's membrane: Evidence for tyrosine O-sulphate in basement membrane proteins. *Biochem. J.* **231**, 571–579.

Peltonen, S., Hentula, M., Hägg, P., Ylä-Outinen, H., Tuukkanen, J., Lakkakorpi, J., Rehn, M., Pihlajaniemi, T., and Peltonen, J. (1999). A novel component of epidermal cell-matrix and cell-cell contacts: Transmembrane protein type XIII collagen. *J. Invest. Dermatol.* **113**, 635–642.

Petitclerc, E., Boutaud, A., Prestayko, A., Xu, J., Sado Y., Ninomiya, Y., Sarras, M.P., Jr., Hudson, B.G., and Brooks, P.C. (2000) New functions for noncollagenous domains of human collagen. *J. Biol. Chem.* **275**, 8051–8061.

Reid, L. M., Fiorino, A. S., Sigal, S. H., Brill, S., and Holst, P. A. (1992). Extracellular matrix gradients in the space of Disse: relevance to liver biology. *Hepatology* **15**, 1198–1203.

Reynolds, S. D., Angerer, L. M., Palis, J., Nasir, A., and Angerer, R. C. (1992). Early mRNAs, spatially restricted along the animal-vegetal axis of sea urchin embryos, include one encoding a protein related to tolloid and BMP-1. *Development* **114**, 769–786.

Rescan, P. Y., Loréal, O., Hassell, J. R., Yamada, Y., Guillouzo, A., and Clément, B. (1993). Distribution and origin of the basement membrane component perlecan in rat liver and primary hepatocyte culture. *Am. J. Pathol.* **142**, 199–208.

Rojkind, M., Giambrone, M. -A., and Biempica, L. (1979). Collagen types in normal and cirrhotic liver. *Gastroenterology* **76**, 710–719.

Roskams, T., Moshage, H., De Vos, R., Guido, D., Yap, P., and Desmet, V. (1995). Heparan sulphate proteoglycan expression in normal human liver. *Hepatology* **21**, 950–958.

Sado, Y., Kagawa, M., Naito, I., Ueki, Y., Seki, T., Momota, R., Oohashi, T., and Ninomiya, Y. (1998) Organization and expression of basement membrane collagen IV genes and their roles in human disorders. *J. Biochem.* **123**, 767–776.

Sasaki, T., and Timpl, R. (1999) Laminins. *In* "Guide book to the Extracellular Matrix and Adhesion Proteins" (T. Kreis *et al.*, eds.), pp. 434–443. CRC Press, Boca Raton, FL.

Schuppan, D. (1990). Structure of extracellular matrix in normal and fibrotic liver: Collagens and glycoproteins. *Semin. Liver Dis.* **10**, 1–10.

Schuppan, D., Cantaluppi, M. C., Becker, J., Veit, A., Buntel, T., Tryer, D., Schuppan, F., Schmid, M., Ackermann, R., and Hahn, E. G. (1990). Undulin, an extracellular matrix glycoprotein associated with collagen fibrils. *J. Biol. Chem.* **265**, 8823–8832.

Schwögler, S., Neubauer, K., Knittel, T., Cheung, A. E., and Ramadori, G. (1994). Entactin gene expression in normal and fibrotic rat liver and in rat liver cells. *Lab. Invest.* **70**, 525–536.

Scott, I. C., Blitz, I. L., Pappano, W. N., Imamura, Y., Clark, T. G., Steiglitz, B. M., Thomas, C. L., Maas, S. A., Takahara, K., Cho, K. W. Y., and Greenspan, D. S. (1999). Mammalian BMP-1/Tolloid-related metalloproteinases, including novel family member mammalian tolloid-like 2, have differential enzymatic activities and distributions of expression relevant to patterning and skeletogenesis. *Dev. Biol.* **213**, 283–300.

Seyer, J. M., Huherson, E. T., and Kang, A. H. (1977). Collagen polymorphism in normal and cirrhotic human liver. *J. Clin. Invest.* **59**, 241–248.

Shahan, T. A., Fawzi, A., Bellon, G., Monboisse, J. C., and Kefalides, N. A. (2000). Regulation of tumor cell chemotaxis by type IV collagen is mediated by a Ca^{2+}-dependent mechanism requiring CD47 and the integrin $\alpha_v\beta_3$. *J. Biol. Chem.* **275**, 4796–4802.

Shahan, T. A., Ziaie, Z., Pasco, S., Fawzi, A., Bellon, G., Monboisse, J. C., and Kefalides, N. A. (1999). Identification of CD47/integrin-associated protein and alpha(V)beta3 as two receptors for the alpha3(IV) chain of type IV collagen on tumor cells. *Cancer Res.* **59**, 4584–4590.

Sheriff, S., Chang, C. Y., and Ezekowitz, R. A. B. (1994). Human mannose-binding protein carbohydrate recognition domain trimerizes through a triple α-helical coiled coil. *Nat. Struc. Biol.* **1**, 789–794.

Shimell, M. J., Ferguson, E. L., Childs, S. R., and O'Connor, M. B. (1991). The Drosophila dorsal-ventral patterning gene tolloid is related to human bone morphogenetic protein 1. *Cell* **67**, 469–481.

Shrivastava, A., Radziejewski, C., Campbell, E., Kovac, L., McGlynn, M., Ryan, T. E., Davis, S., Goldfarb, M. P., Glass, D. J., Lemke, G., and Yancopoulos, G. D. (1997). An orphan receptor tyrosine kinase family whose members serve as nonintegrin collagen receptors. *Mol. Cell* **1**, 25–34.

Takahara, K., Lyons, G. E., and Greenspan, D. S. (1994). Bone morphogenetic protein-1 and a mammalian tolloid homologue (mTld) are encoded by alternatively spliced transcripts which are differentially expressed in some tissues. *J. Biol. Chem.* **269**, 32572–32578.

Tenner, A. J. (1999). Membrane receptors for soluble defense collagens. *Curr. Opin. Immunol.* **11**, 34–41.

Timpl, R., Dziadek, M., Fujiwara, S., Nowak, H., and Wick, G. (1983). Nidogen, a new self-aggregating basement membrane protein. *Eur. J. Biochem.* **137**, 455–465.

Timpl, R. (1996). Macromolecular organization of basement membranes. *Curr. Opin. Cell Biol.* **8**, 618–624.

Timpl, R. (1999). Nidogen. *In* "Guide Book to the Extracellular Matrix and Adhesion Proteins" (T. Kreis *et al.*, eds.), pp. 455–457. CRC Press, Boca Raton, FL.

Trachslin, J., Koch, M., and Chiquet, M. (1999). Rapid and reversible regulation of collagen XII expression by changes in tensile stress. *Exp. Cell Res.* **247**, 320–328.

Vogel, W. (1999). Discoidin domain receptors: Structural relations and functional implications. *FASEB J.* **13**(Suppl.), S77–S82.

Vogel, W., Gish, G., Alves, F., and Pawson, T. (1997). The discoidin domain receptor tyrosine kinases are activated by collagen. *Mol. Cell* **1**, 13–23.

Wang, N., Butler, J. P., and Ingber, D. E. (1993). Mechanotransduction across the cell surface and through the cytoskeleton. *Science* **260**, 1124–1127.

Wen, W., Moses, M. A., Wiederschain, D., Arbiser, J. L., and Folkman, J. (1999). The generation of endostatin is mediated by elastase. *Cancer Res.* **59**, 6052–6056.

Wozney, J. M., Rosen, V., Celeste, A. J., Mitsock, L. M., Whitters, M. J., Kriz, R. W., Hewick, R. M., and Wang, E. A. (1988). Novel regulators of bone formation: Molecular clones and activities. *Science* **242**, 1528–1534.

Yamaguchi, N., Anand, B., Lee, M., Sasaki, T., Fukai, N., Shapiro, R., Que, I., Lowik, C., Timpl, R., and Olsen, B. R. (1999). Endostatin inhibits VEGF-induced endothelial cell migration and tumor growth independently of zinc binding. *EMBO J.* **18**, 4414–4423.

CHAPTER 2

Dynamic Regulation of Basement Membrane Collagen IV Gene Expression in Malignant Tumors

KEN-ICHI IYAMA

Department of Surgical Pathology, Kumamoto University School of Medicine, Kumamoto 860-8556, Japan

YOSHIFUMI NINOMIYA

Department of Molecular Biology and Biochemistry, Okayama University Graduate School of Medicine and Dentistry, Okayama 700-8558, Japan

Six distinct genes belonging to collagen IV are arranged in three different chromosomes, 13, 2, and X, and are located in a head-to-head arrangement. Their expression appears to be regulated by bidirectional promoters between the genes. The translation products of the six genes make three major molecules of $[\alpha1(IV)]_2\alpha2(IV)$, $\alpha3(IV)\alpha4(IV)\alpha5(IV)$, and $[\alpha5(IV)]_2\alpha6(IV)$. There is tissue-specific regulation in how supramolecular aggregates of basement membranes (BMs) are assembled. For instance, the glomerular basement membrane was reported to contain two different and independent networks: one with the $[\alpha1(IV)]_2\alpha2(IV)$ molecule and the other with the $\alpha3(IV)\alpha4(IV)\alpha5(IV)$ molecule.

In the field of surgical pathology, it is important to make a clear diagnosis to differentiate noninvasive cancers from invasive cancers. When normal breast or noninvasive ductal carcinoma was examined, $\alpha5(IV)$ and $\alpha6(IV)$ chains were stained in the BMs of mammary ducts and small lobules, capillaries, and myofibroblastic cells. In contrast, in cases of invasive ductal carcinoma, BMs containing $\alpha1(IV)$ and $\alpha2(IV)$ were stained discontinuously but $\alpha5(IV)$ and $\alpha6(IV)$ chains were absent in most of the invasive cancer cell nests. Thus, the loss of

Extracellular Matrix and the Liver
Copyright 2003, Elsevier Science (USA). All rights reserved.

staining pattern of the α5(IV) and α6(IV) chains was consistent with invasiveness of the tumors. Therefore, estimation of α5(IV)/α6(IV) expression may be a useful marker for invasive cancers.

I. INTRODUCTION

Basement membranes underlie the basal surface of epithelial cells and endothelial cells and surround the entire surface of adipocytes and muscle cells as a sheet-like structure. They also provide a boundary between organ-specific parenchymal cells and stroma underneath that also has characteristic members of extracellular matrix in specialized organs or tissues. As one of the biological functions of BMs, they maintain the architecture of most of the cells. They also provide unique filtration barriers to select certain kinds of molecules by charge and size. Further, they also play important functions in cell differentiation, proliferation, and tumor invasion and metastasis.

Progress has been made since the early 1990s in understanding the structure and function of basement membranes. More information dealing with the function of basement membranes has started accumulating by the development of gene cloning, gene targeting, transfection experiments, and protein-to-protein interactions. The basement membrane contains different sets of gene products from other parts of tissues. They are collagen IV, laminin, nidogen/entactin, and heparan sulfate proteoglycan. Collagen IV is one of the major components. It consists of six different α chains, whose unique gene structures have been reported (Hudson *et al.*, 1993). Most recently, several reports have defined the architecture of basement membranes, regulation of collagen IV gene expression, and func-tion of basement membranes. First, how the complex supramolecular structure of collagen IV is aggregated in the basement membrane has been solved, in part, with evidence that the chain specificity of network assembly is encoded by the noncollagenous NC1 domains (Borza *et al.*, 2001; Boutaud *et al.*, 2000). Second, one of the transcription factors, LMX1B, which is known as a causative gene for human and mouse Nail Patella syndrome, was found to regulate collagen IV gene expression (Morello *et al.*, 2001). Third, new functions for noncollagenous domains of human collagen IV have been identified (Petitcleric *et al.*, 2000). This implies novel integrin ligands inhibiting angiogenesis and tumor growth *in vivo*.

We are investigating BM production in normal and tumors. Several investigators noted the absence of BM components (Gusterson *et al.*, 1982; Hewitt *et al.*, 1997; Rudland *et al.*, 1993). This chapter describes what we observed in the regulation of collagen IV gene expression in normal cells, noninvasive tumors, and invasive cancers.

II. TISSUE-SPECIFIC EXPRESSION OF COLLAGEN IV GENES

The recent cloning and characterization of genes for collagen IV have provided information on the structure, expression, and function of BMs during development and diseases states. Six distinct genes belonging to this family have been identified. Three sets of human genes, COL4A1/COL4A2, COL4A3/COL4A4, and COL4A5/COL4A6, are arranged in three different chromosomes, 13, 2, and X, respectively (Fig. 1; Hudson *et al.*, 1993; Sado *et al.*, 1998). They are located in a head-to-head arrangement and their expression appears to be regulated by bidirectional promoters between the genes. The 5′ ends of the genes overlap each other, and the transcription start sites are separated only 130bp in human (Poschl *et al.*, 1988; Soininen *et al.*, 1988) and mouse (Burbelo *et al.*, 1988;

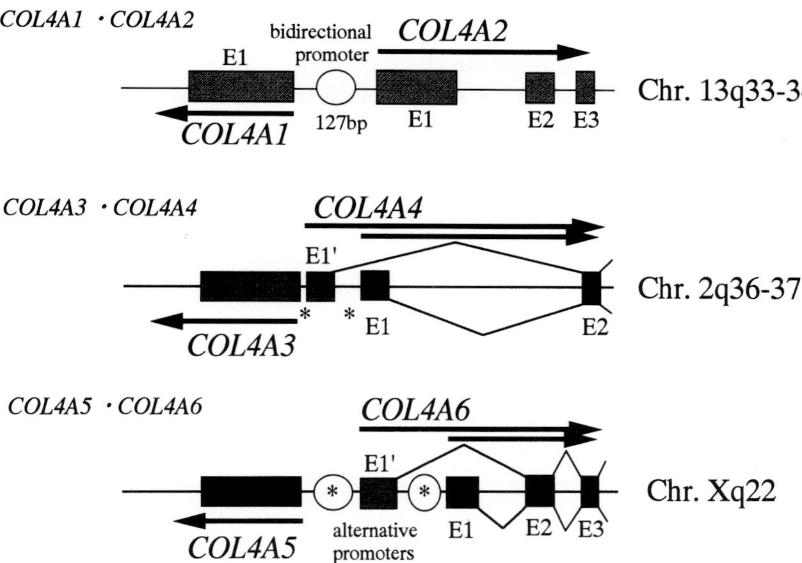

FIGURE 1 Organization of collagen IV genes. The six human collagen IV genes are located in pairs in a head-to-head manner on three different chromosomes. These six genes can be classified into two groups, which are group A (depicted by white boxes): COL4A1, COL4A3, and COL4A5 encoding the α1, α3, and α5 chains, respectively, and group B (shaded boxed): COL4A2, COL4A4, and COL4A6 encoding the α2, α4, and α6 chains, respectively. These are based on the gene structure, amino acid sequence homology, and domain structure of the translation products. Transcription of COL4A1 and COL4A2 genes is controlled by the bidirectional promoter (top). The other two sets of genes seem to be regulated by similar bidirectional promoters. Of interest was not only that transcription of COL4A4 and COL4A6 appears to be regulated by two alternative promoters, but also that regulation was in a tissue- or cell-specific manner (Sugimoto *et al.*, 1994; Momota *et al.*, 1998).

Kaytes *et al.*, 1988) α1(IV) and α2(IV) genes. Transcriptional regulation for COL4A1/COL4A2 was characterized (Fischer *et al.*, 1993; Genersch *et al.*, 1995). Interestingly, there are two alternative promoters in COL4A4 and COL4A6 genes that are regulated in a tissue-specific manner (Momota *et al.*, 1998; Sugimoto *et al.*, 1994).

In order to determine whether the translational products of the six genes are colocalized in various tissues, we raised α1–α6 chain-specific rat monoclonal antibodies against synthetic peptides reflecting sequences near the carboxy terminus of each NC1 domain (Kishiro *et al.*, 1995; Ninomiya *et al.*, 1995; Sado *et al.*, 1995). In the kidney, all basement membranes in glomeruli and tubules were stained positively by antibodies for the α1 and α2 chains, whereas the α3 and α4 chains were restricted to the glomerular BMs and some parts of the tubules. Thus, the α3 and α4 staining pattern is rather limited as compared to that of the α1 and α2 chains, but their common location suggests that they are synthesized by the same cells. In stark contrast, the expression pattern of the α5 and α6 chains is different in the glomerulus: the α6 chain is never detected in the glomerular BMs, but is positive in BMs of Bowman's capsules and some distal tubules in humans, whereas the α5 chain is densely positive in glomerular BMs and its staining pattern is similar to that obtained with α3 and α4 antibodies. These results suggested that there are two kinds of collagen IV molecules: one with α3, α4, and α5 chains and the other with α5 and α6 chains (Ninomiya *et al.*, 1995). Distribution of the six α(IV) chains in smooth muscle cell BMs and epithelial BMs was described in reports that suggested that the tissue-specific distribution of α(IV) chains is consistent with the function of BMs (Saito *et al.*, 2000; Seki *et al.*, 1998).

Four different disease states are related to dysregulated collagen IV gene expression. First, nearly 300 mutations have been identified in the COL4A5 gene in X-linked Alport patients, and several mutations in COL4A3 and COL4A4 were also reported in the autosomal type Alport syndrome (Jais *et al.*, 2000; Tryggvason and Ninomiya, 1997). Experiments indicate that any one of the three α chains has been translated abnormally into incomplete molecules, which can degraded within cells. The second is deletions at the 5′ end of COL4A6 extending to the neighboring COL4A5 gene have been reported to cause diffuse leiomyomatosis associated with Alport syndrome (Heidet *et al.*, 1995; Ueki *et al.*, 1998; Zhou *et al.*, 1993). The third is not a genetic disorder but an autoimmune disease, Goodpasture syndrome, which is mediated by anti-BM antibodies that are targeted to α3(IV)NC1 and α4(IV)NC1 domains. The uniqueness of the α3(IV)NC1 and α4(IV)NC1 domains, among the six NC1 domains, to induce severe Goodpasture disease may relate to the accessibility of epitopes in the GBM for binding of the antibody. The pathogenicity of the α4(IV)NC1 antibodies has not been clarified because the pathogenic antibodies in patients are not believed to be detected as target to the α4(IV)NC1, but are detected to the α3(IV)NC1 domain in Goodpasture nephritis (Sado *et al.*, 1998).

The last disease associated with altered type IV collagen is one in which the expression of α3(IV) and α4(IV) chains is diminished strongly within glomerular BMs in Nail Patella syndrome, which is believed to cause the renal phenotype (Morello *et al.*, 2001). The responsible gene of the human disease has been identified as a transcription factor, LMX1B gene (Chen *et al.*, 1998; Dreyer *et al.*, 1998). Morello *et al.* (2001) showed that LMX1B binds to the enhancer sequence in intron 1 of the mouse and human COL4A4 gene and upregulates reporter constructs containing the enhancer-like sequence. LMX1B directly regulates COL4A3 and COL4A4 coordinate expression required for normal renal morphogenesis, and dysfunction in glomerular BM contributes to the renal pathology.

III. SUPRAMOLECULAR AGGREGATES OF COLLAGEN IV

As described earlier, collagen IV genes are arranged as paired genes on three different chromosomal locations (Fig. 1). To make collagen molecules, three α chains are necessary (Fig. 2B). Out of six α chains, there are 56 combinations to make 6 homotrimers and 50 heterotrimers. The question is what combinations are allowed to form as native molecules in tissue and organs? We believe that there are at least three different molecules: $[\alpha1(IV)]_2\alpha2(IV)$, α3(IV)α4(IV)α5(IV), and $[\alpha5(IV)]_2\alpha6(IV)$ after double staining of different combinations using chain-specific monoclonal antibodies (Ninomiya *et al.*, 1995). We have also predicted the molecular structure according to the local amino acid sequences (Sado *et al.*, 1998). To make a molecule is not the last step for these translational products in order to function. More complex supramolecular aggregates are incorporated into the BM structure. How can they make them? Figure 2 illustrates the key steps. Three α chains (Fig. 2A) make a collagen molecule (Fig. 2B), which is called a monomer. In Fig. 2, three different kinds monomers are drawn, but two different kinds of dimers, 7S ends and NC1s, are attached, a tetramer of only one type of monomer is drawn. Combinations of these units extend the association of dimers and tetramers to make a sheet-like structure, supramolecular assembly (Figs. 2F–2H). It is intriguing that these molecules are distributed in a tissue-specific manner (see Fig. 3). For instance, glomerular BMs that exist between podcytes and endothelial cells filter through serum to create urine. This BM is composed of $[\alpha1(IV)]_2\alpha2(IV)$ and α3(IV)α4(IV)α5(IV) molecules, the latter of which is only found in specified places, including the glomerulus and alveolus.

Two reports have clarified how the supramolecular aggregates are assembled in BMs. We utilized monoclonal antibodies, Mab1, Mab3, B51, and B66, that react exclusively with the native NC1 hexamers that contain α1(IV)NC1, α3(IV)NC1, α5(IV)NC1, and α6(IV)NC1, respectively (Borza *et al.*, 2001; Boutaud *et al.*, 2000). For instance, because Mab1 can immunoprecipite native NC1 hexamers that contain α1NC1s, if the precipitates can be analyzed by the antibodies that

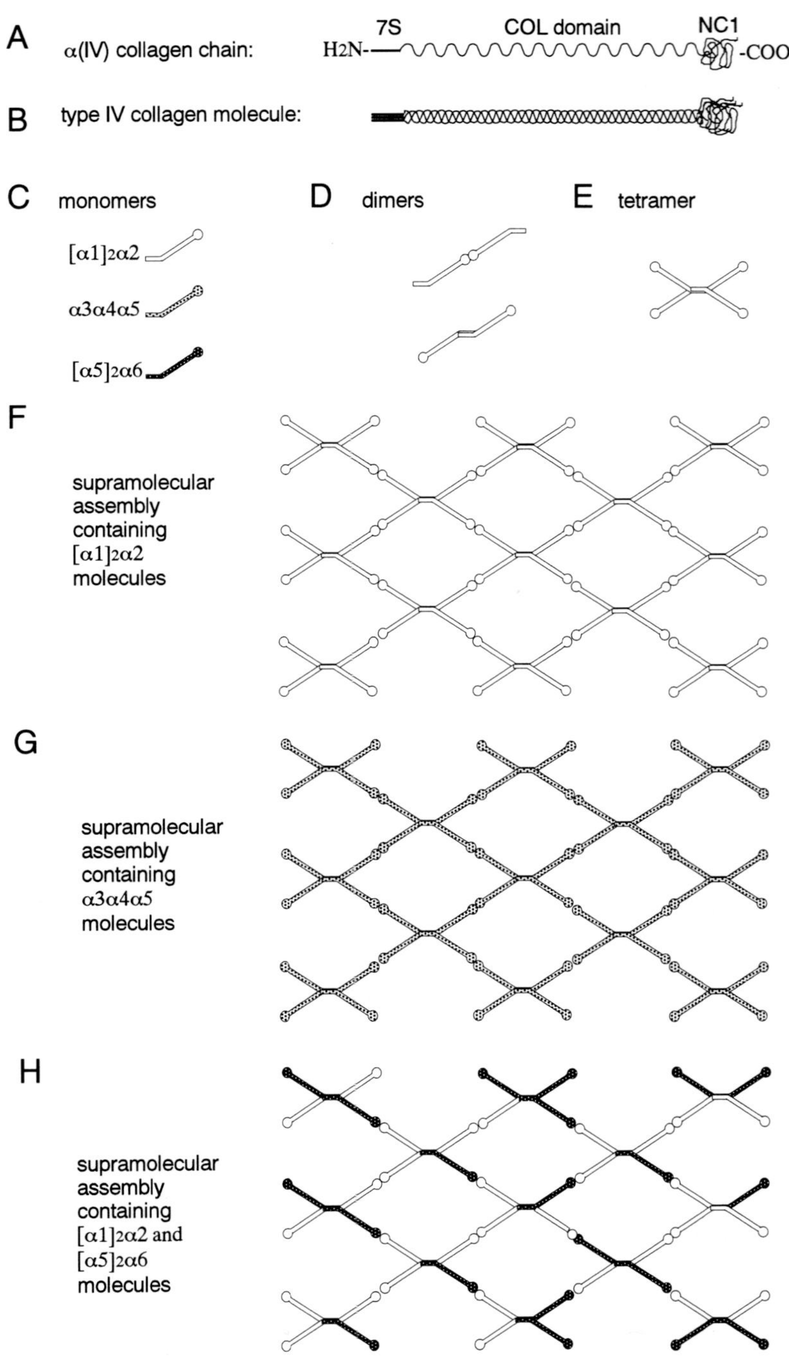
A
α(IV) collagen chain:
7S
COL domain
NC1
H2N-
-COOH
B
type IV collagen molecule:
C
monomers
[α1]2α2
α3α4α5
[α5]2α6
D
dimers
E
tetramer
F
supramolecular assembly containing [α1]2α2 molecules
G
supramolecular assembly containing α3α4α5 molecules
H
supramolecular assembly containing [α1]2α2 and [α5]2α6 molecules

[I] Classical type
α1/α2
[α1(IV)]2α2(IV)
Capillary, Hepatic sinusoid, Lymphatic, Schwann cell, Fat cell, Skeletal muscle

[II] Filtration barrier type
α1/α2/α3/α4/α5
[α1(IV)]2α2(IV) and α3(IV)α4(IV)α5(IV)
Glomerular BM, Alveolar BM

[III] Maintenance of architecture
α1/α2/α5/α6
[α1(IV)]2α2(IV) and [α5(IV)]2α6(IV)
Epithelial BM (epidermis, esophagus, mammary gland, colon, bile duct, bronchus)
Smooth muscle cell BM (artery, vein, alimentary and urinary tract)

FIGURE 3 Three basement membrane types. Three molecules consisting of α1/α2 chains, α3/α4/α5 chains, and α5/α6 chains are predicted to be present in the three different BMs, where not only molecular forms but also functional relevance can be different in various tissues and organs.

recognize the denatured NC1s using SDS–PAGE, one can predict what kinds of molecules are assembled. When GBM fractions that were digested with purified bacterial collagenase and purified were immunoprecipitated by Mab1, the materials were analyzed by Western blotting (Boutaud *et al.*, 2000). The result showed that they contained only α1 and α2NC1s. The same starting materials immunoprecipitated by Mab3 contained α3, α4, and α5NC1s. These results indicated that the NC1 hexamers were composed of the same NC1s of the same molecules. Namely, a supramolecular assembly containing $[\alpha 1(IV)]_2\alpha 2(IV)$ molecules (Fig. 2F) exists independently from that containing α3(IV), α4(IV), and α5(IV) molecules (Fig. 2G).

In contrast, NC1 fractions from aorta and bladder were immunoprecipitated by B51, B66, and Mab1 and analyzed by SDS–PAGE and resulted in all the same pattern of α1/α2/α5/α6 (Borza *et al.*, 2001). However, nonimmunoprecipitated fractions analyzed by SDS–PAGE resulted in α1/α2, α1/α2, and none, respectively.

FIGURE 2 Schematic representation of the collagen IV α chain (A) and a collagen IV molecule (B) consisting of three α chains. Three monomers, white, gray, and black (in C), consisting of α1/α2 chains, α3/α4/α5 chains, and α5/α6 chains, respectively, form dimers (D) through interactions between the amino-terminal 7S domains or the carboxy termini and tetramers (E) through the combination of the two dimers at the 7S domains. Thus, the supramolecular network (F, G, and H) is formed through the assembly of dimers and further aggregates. Some aggregates consist of only α1/α2 (F) or α3/α4/α5 molecules (G), but others of α1/α2 and α5/α6 together in the same network (H). Note that the NC1 hexamers in the assembly of $[\alpha 1(IV)]_2\alpha 2(IV)$ and $[\alpha 5(IV)]_2\alpha 6(IV)$ molecules are always α1/α2 vs α1/α2 or α1/α2 vs α5/α6, which is based on the results of Borza *et al.* (2001).

These results indicated that NC1 hexamers in aorta and bladder were a combination of $[\alpha1(IV)]_2\alpha2(IV)$ molecules alone (Fig. 2F) and a hybrid of $[\alpha1(IV)]_2\alpha2(IV)$ and $[\alpha5(IV)]_2\alpha6(IV)$ molecules (Fig. 2H), which is quite different from the GBM aggregates mentioned above.

IV. WHY ARE BASEMENT MEMBRANES IMPORTANT FOR DIAGNOSIS IN SURGICAL PATHOLOGY?

In epithelial organs such as skin, esophagus, mammary gland, uterus, and bladder, we often observe intraepithelial carcinoma or noninvasive carcinoma, where cancers grow and expand slowly but are still limited within the epithelial region. In this stage, BMs underneath epithelial layers are not disrupted at all and cancer cells do not yet invade into stroma. Because metastasis through lymphatics and veins is not established in this stage, it is possible to expect 100% remission by dissecting out the tumor regions in skin cancers, breast cancers, and uterovaginal cancers. In the field of surgical pathology, it is important to clearly differentiate noninvasive from invasive cancers.

Previous studies of invasive cancers using antibodies against the classic type, α1(IV) and α2(IV) chains or laminin, did not solve these challenges of diagnosis. Therefore, it is meaningful for surgicopathological diagnosis to investigate how the six α(IV) chains are expressed in BMs surrounding tumors that have different histopathological features, mitotic activity, and invasiveness.

V. DIFFERENTIAL EXPRESSIONS OF THE SIX α(IV) CHAINS IN MAMMARY NEOPLASIA

When normal breast was examined, α1(IV) and α2(IV) chains were stained in the BMs of mammary ducts and small lobules, capillaries, and myofibroblastic cells (Nakano *et al.*, 1999). The staining of α5(IV) and α6(IV) chains was limited to the mammary ducts and small lobules, and there was no positive staining with α3(IV) and α4(IV) chains. When noninvasive ductal carcinoma was stained, BMs were well preserved in a linear pattern in almost the same fashion as that of the normal breast, as seen in Fig. 4. In contrast, in cases of invasive ductal carcinoma shown in Fig. 5, BMs containing α1(IV) and α2(IV) were disrupted or stained, discontinuously; however, α5(IV) and α6(IV) chains were absent in most of the invasive cancer cell nests. Table I summarizes what we observed in mammary neoplasma. Thus, the loss of staining pattern of the α5(IV) and α6(IV) chains was consistent with the invasiveness determined by immunostaining.

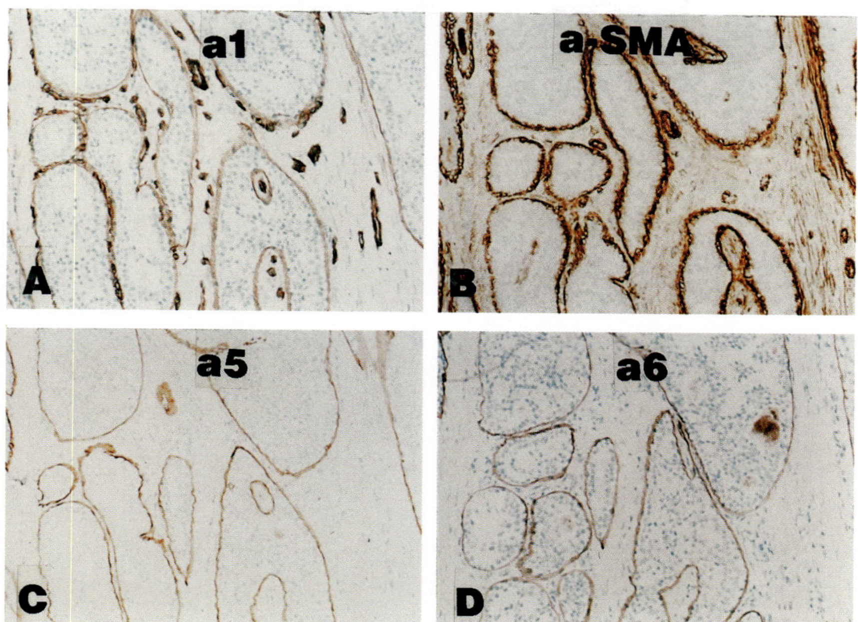

FIGURE 4 Immunostaining of α1(IV) chain (A), α-smooth muscle actin (α-SMA)(B), α5(IV) (C), and α6(IV) chains in noninvasive ductal carcinoma. (A) The α1(IV) chain is stained in the BM surrounding the ducts replaced by cancer cells and in the capillary BM in close contact with the ducts. (B) α-SMA-positive myoepithelial cells are noted in the noninvasive cancer cell nests, and capillary pericytes and fibroblastic cells with α-SMA also localized in the stroma. (C and D) Linear immunostaining of the α5(IV) and α6(IV) chains is clearly identified in the BM zone in the serial sections of A and B. Original magnification, ×100.

We also examined expression of the six COLA4 genes by *in situ* hybridization in serial tissue sections of intraductal papilloma, which is characterized as a benign tumor of two types of epithelial cells: ductal epithelial cells and myoepithelial cells (Nakano *et al.*, 1999). Intriguingly, the signal for α5(IV) and α6(IV) mRNA was observed strongly in both epithelial cells and myoepithelial cells (Table II), indicating that α5(IV) and α6(IV) producing cells are epithelial cells. We predict that there are two different networks here in the BM of mammary ducts: (1) the classic network formed by $[\alpha1(IV)]_2\alpha2(IV)$ molecules, which is of stromal origin, and (2) a hybrid network of $[\alpha1(IV)]_2\alpha2(IV)$ molecules and $[\alpha5(IV)]_2\alpha6(IV)$ molecules, which is of epithelial origin. The epithelial–myoepithelial interaction might preserve the second hybrid network of BMs (Nakano *et al.*, 1999). Table II summarizes the results of quantitative *in situ* hybridization in mammary neoplasma.

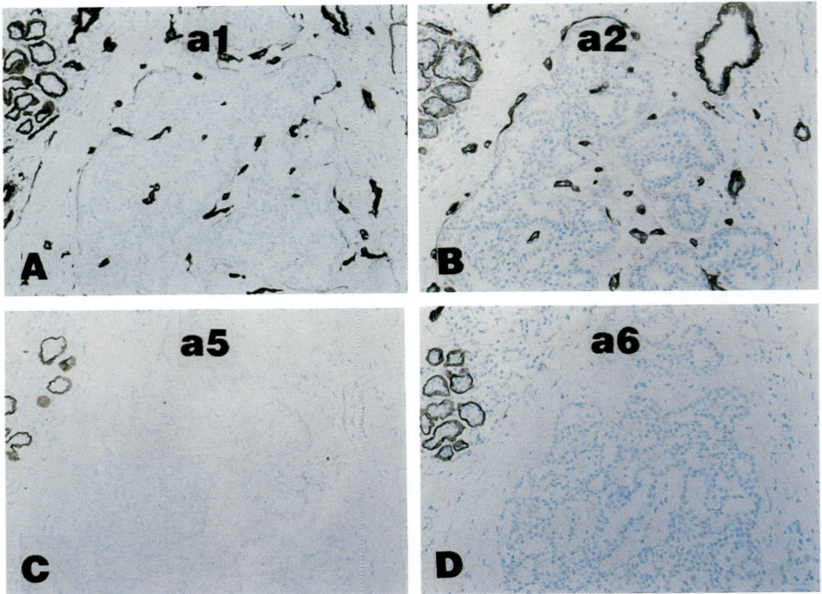

FIGURE 5 Immunostaining of α1(IV) (A), α2(IV) (B), α5(IV) (C), and α6(IV) (D) chains in invasive ductal carcinoma. (A and B) α1(IV) and α2(IV) chains are stained discontinuously in marginal areas of the cancer cell nests or are partly negatively stained. (C and D) Complete loss of α5(IV) and α6(IV) chains is noted in the cancer cell nests in the serial sections next to A and B. α1(IV), α2(IV), α5(IV), and α6(IV) chains are observed in the BMs of normal mammary ducts and small lobules (upper left in A, B, C, and D) as a positive control. Original magnification, ×140.

TABLE I Result of Immunohistochemical Studies in Mammary Neoplasia[a]

	α(IV) chain (BM)[b]		
	α1/α2	α5	α6
Normal tissue	+	+	+
Intraductal papilloma	++	++	++
Breast cancer			
Noninvasive	+	+	+
Invasive	dc	−	−

[a]Colocalization of α5(IV) and α6(IV) chains was evident in the BM of mammary ducts in normal breast and benign tumor (intraductal papilloma) and noninvasive ductal carcinoma, all of which also expressed the classic type of α1(IV) and α2(IV) chains. In contrast, in invasive ductal carcinoma without myoepithelial cells, α1(IV) and α2(IV) chains were stained discontinuously in the marginal areas of the cancer cell nests or were partly negative stained. In addition, the assembly of α5(IV) and α6(IV) chains into BM was completely inhibited in the massive cancer cell nests.

[b]−, negative; dc, discontinuous positive; +, positive; ++, strongly positive.

TABLE II Result of *in Situ* Hybridization in Mammary Neoplasia[a]

	mRNA (epithelium/stroma)[b]		
	α1/α2	α5(IV)	α6(IV)
Normal tissue	−/−	−/−	−/−
Intraductal papilloma	+/+	+/−	+/−
Breast cancer			
Noninvasive	−/++[c]	−/−	−/−
Invasive	−/++[d]	−/−	−/−

[a]In intraductal papilloma, there were highly elevated signals for α1(IV)mRNA and α2(IV)mRNA in the double-layered epithelium composed of epithelial cells and myoepithelial cells. In addition, the double-layered epithelium also clearly expressed α5(IV)mRNA and α6(IV)mRNA. In noninvasive ductal carcinoma, intensive signals for α1(IV)mRNA and α2(IV)mRNA were detected in myofibroblastic cells around the cancer cell nests as a ring-like pattern. In invasive ductal carcinoma, α5(IV)mRNA and α6(IV)mRNA were completely absent in both cancer cell nests.

[b]−, negative (0 to 10 silver grains per cell); +, mild to moderately positive (10 to 20 silver grains per cell); ++, strongly positive (>20 silver grains per cell).

[c]Ring-like pattern.

[d]Random pattern.

VI. PROTECTIVE EFFECTS OF BASEMENT MEMBRANES AGAINST CANCER INVASION DUE TO REORGANIZATION OF α(IV) CHAINS

In breast cancers, matrix-degrading enzymes such as MMPs and others can remodel extracellular matrices, including BMs. What we observed so far was, however, was that the expression of collagen IV genes already started as if they were destined to reorganize newly synthesizing BMs before tumors become malignant and invasive.

Because BMs containing collagen IV always lie between epithelia and stroma and function as selective barriers, proliferation of myofibroblastic cells might be considered as a protective effect from the stroma against cancer invasion. In doing so, they synthesize α1(IV) and α2(IV) mRNA at high levels to produce the classic network containing $[\alpha(IV)]_2\alpha2(IV)$ molecules, e.g., in noninvasive mammary cancer with expansive growth (Fig. 6). In invasive cancers with infiltrative patterns, the proliferation of myofibroblasts cannot be seen around the cancer nests. Cancer cells lose their capsules, including BM materials, become to be exposed to various kinds of growth factors, induce expression of various genes linked to tumor invasion, acquire increased cytological and nuclear atypia and architectural atypia, and successfully develop invasion to neighboring stroma (Fig. 7).

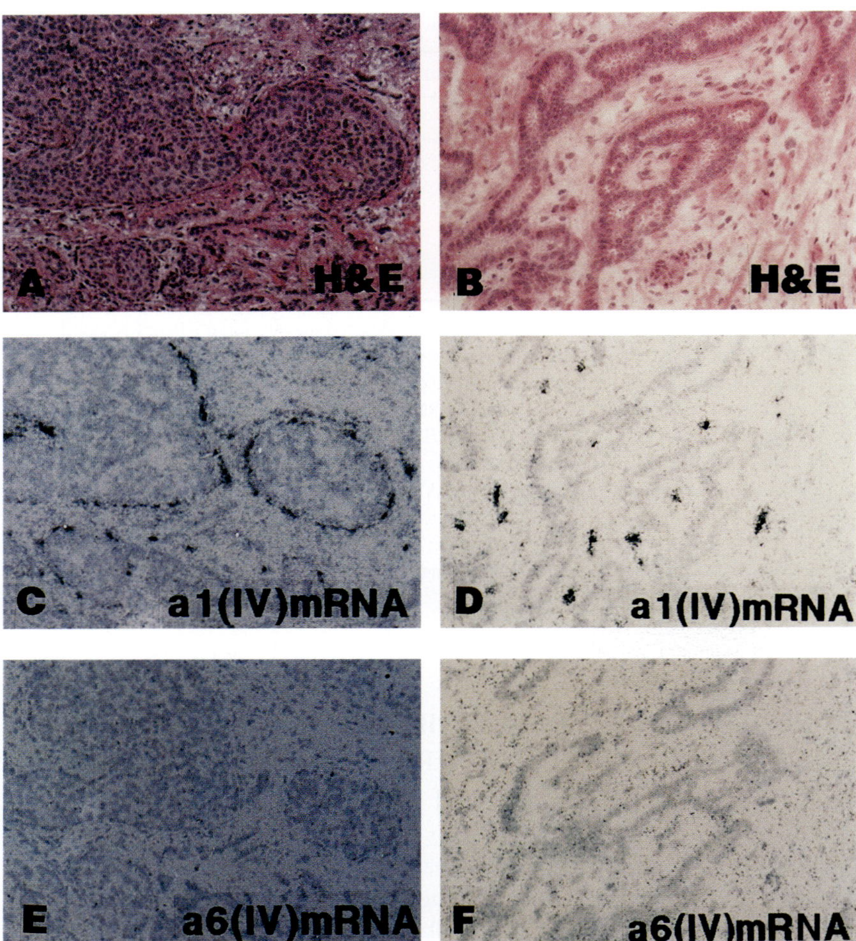

FIGURE 6 Expression of $\alpha1$(IV) and $\alpha6$(IV) mRNA in noninvasive ductal carcinoma (left: A, C, and E) and invasive ductal carcinoma (right: B, D, and F). Strong signals for $\alpha1$(IV) mRNA are detected as a ring-like pattern around the noninvasive cancer nest (C), whereas in the invasive ductal carcinoma, foci with intensive signals for $\alpha1$(IV) mRNA are distributed randomly in the stromal cells (D). No obvious signal for $\alpha6$(IV) mRNA is seen in both sections of the noninvasive (E) and invasive (F) ductal carcinoma. (A and B) Hematoxyline and eosin staining. (C and D) *In situ* hybridization. Original magnifications: ×200 in A, C, and E, and ×300 in B, D, and F.

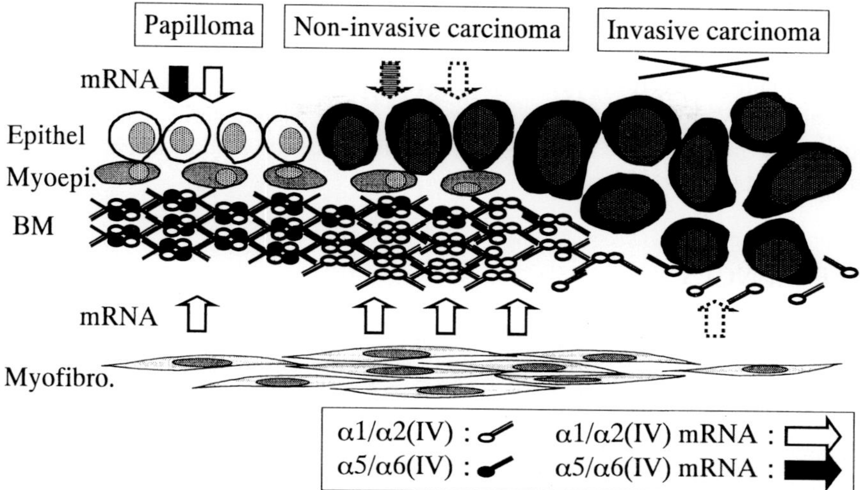

FIGURE 7 Differential tissular expressions of α(IV) chains in mammary neoplasma. In intraductal papilloma, which is a benign tumor, the signals for not only α1(IV)mRNA and α2(IV)mRNA, but also for α5(IV)mRNA and α6(IV)mRNA were identified in the mammary epithelial cells. In addition to *in situ* hybridization, immunohistochemistry using the antibody confirmed that both inner epithelial cells and outer myoepithelial cells synthesize α5(IV) and α6(IV) chains. The expression of these α(IV) chains seems to be regulated by an epithelial–myoepithelial interaction in intraductal papilloma. In noninvasive ductal carcinoma, capillary endothelial cells and periductal myofibroblastic cells around the cancer cell nests actively synthesize α1(IV) and α2(IV) chains with a ring-like pattern. These periductal stromal cells seem to take part in remodeling of the extracellular matrix, including BM. According to *in situ* hybridization findings, we could show the possibility that newly formed BM composed of α1(IV) and α2(IV) chains secreted by stromal cells plays an important role as one of the barrier systems against the activation of proteolytic enzyme and matrix degradation. The invasive potential of breast cancer may be associated with a complete loss of this barrier system.

VII. CONCLUSION

We have described specific expression of α(IV) genes, aggregates of collagen IV, and tumor BMs and their diagnostic significance. The expression of α(IV) collagen chains seems to be regulated differentially by the epithelial–myoepithelial interaction and to be associated with the invasive potential of breast cancer. To understand cancer biology, one must investigate not only cytological and architectural atypia, but also the involvement of stroma and the remodeling of extracellular matrix. For both pathologists and matrix biologists, we consider that the differential detection of BM collagen gene expression in tumor tissues can contribute to an improved surgicohistopathological diagnosis of cancer.

ACKNOWLEDGMENTS

We thank Dr. Yoshikazu Sado and Dr. Ichiro Naito for long-term collaboration on collagen IV monoclonal antibody production and related works. This work was supported in part by a grant-in-aid from the Ministry of Education, Science, Sports, and Culture of Japan.

REFERENCES

Borza, D. B., Bondar, O., Ninomiya, Y., Sado, Y., Naito, I., Todd, P., and Hudson, B. G. (2001). The NC1 domain of collagen IV encodes a novel network composed of the α1, α2, α5, and α6 chains in smooth muscle basement membranes. *J. Biol. Chem.* **276**, 28532–28540.

Boutaud, A., Borza, D. B., Bondar, O., Gunwar, S., Netzer, K. O., Singh, N., Ninomiya, Y., Sado, Y., Noelken, M. E., and Hudson, B. G. (2000). Type IV collagen of the glomerular basement membrane: Evidence that the chain specificity of network assembly is encoded by the noncollagenous NC1 domains. *J. Biol. Chem.* **275**, 30716–30724.

Burbelo, P. D., Martin, G. R., and Yamada, Y. (1988). α1(IV) and α2(IV) collagen genes are regulated by a bidirectional promoter and a shared enhancer. *Proc. Natl. Acad. Sci. USA* **85**, 9679–9682.

Chen, H., Lun, Y., Ovchinnikov, D., Kokubo, H., Oberg, K. C., Pepicelli, C. V., Gan, L., Lee, B., and Johnson, R. L. (1998). Limb and kidney defects in Lmx1b mutant mice suggest an involvement of LMX1B in human nail patella syndrome. *Nature Genet.* **19**, 51–55.

Dreyer, S. D., Zhou, G., Baldini, A., Winterpacht, A., Zabel, B., Cole, W., Johnson, R. L., and Lee, B. (1998). Mutations in LMX1B cause abnormal skeletal patterning and renal dysplasia in nail patella syndrome. *Nature Genet.* **19**, 47–50.

Fischer, G., Schmidt, C., Optitz, J., Cully, Z., Kuhn, K., and Poschl, E. (1993). Identification of a sequence element in the common promoter region of human collagen type IV genes, involved in the regulation of divergent transcription. *Biochem. J.* **292**, 687–695.

Genersch, E., Eckerskorn, C., Lottspeich, F., Herzog, C., Kuhn, K., and Poschl, E. (1995). Purification of the sequence-specific transcription factor CTCBF, involved in the control of human collagen IV genes: Subunits with homology to Ku antigen. *EMBO J.* **14**, 791–800.

Gusterson, B. A., Warburton, M. J., Mitchell, D., Ellison, M., Neville, A. M., and Rudland, P. S. (1982). Distribution of myoepithelial cells and basement membrane proteins in the normal breast and in benign and malignant breast diseases. *Cancer Res.* **42**, 4763–4770.

Heidet, L., Dahan, K., Zhou, J., Xu, Z., Cochat, P., Gould, J. D., Leppig, K. A., Proesmans, W., Guyot, C., Guillot, M., *et al.* (1995). Deletions of both α5(IV) and α6(IV) collagen genes in Alport syndrome and in Alport syndrome associated with smooth muscle tumours. *Hum. Mol. Genet.* **4**, 99–108.

Hewitt, R. E., Powe, D. G., Morrell, K., Balley, E., Leach, I. H., Ellis, I. O., and Turner, D. R. (1997). Laminin and collagen IV subunit distribution in normal and neoplastic tissues of colorectum and breast. *Br. J. Cancer* **75**, 221–229.

Hudson, B. G., Reeders, S. T., and Tryggvason, K. (1993). Type IV collagen: Structure, gene organization, and role in human diseases. Molecular basis of Goodpasture and Alport syndromes and diffuse leiomyomatosis. *J. Biol. Chem.* **268**, 26033–26036.

Jais, J. P., Knebelmann, B., Giatras, I., De Marchi, M., Rizzoni, G., Renieri, A., Weber, M., Gross, O., Netzer, K. O., Flinter, F., Pirson, Y., Verellen, C., Wieslande, J., Persson, U., Tryggvason, K., Martin, P., Hertz, J. M., Schroder, C., Sanak, M., Krejcova, S., Carvalho, M. F., Saus, J., Antignac, C.,

Smeets, H., and Gubler, M. C. (2000). X-linked Alport syndrome: Natural history in 195 families and genotype- phenotype correlations in males. *J. Am. Soc. Nephrol.* **11**, 649–657.

Kaytes, P., Wood, L., Theriault, N., Kurkinen, M., Vogeli, G., Xu, Z., Cochat, P., Gould, J. D., Leppig, K. A., Proesmans, W., Guyot, C., Guillot, M., *et al.* (1988). Head-to-head arrangement of murine type IV collagen genes. *J. Biol. Chem.* **263**, 19274–19277.

Kishiro, Y., Kagawa, M., Naito, I., and Sado, Y. (1995). A novel method of preparing rat monoclonal antibody-producing hybridomas by using rat medial iliac lymph node cells. *Cell Struct. Funct.* **20**, 151–156.

Momota, R., Sugimoto, M., Oohashi, T., Kigasawa, K., Yoshioka, H., and Ninomiya, Y. (1998). Two genes, COL4A3 and COL4A4 coding for the human α3(IV) and α4(IV) collagen chains are arranged head-to-head on chromosome 2q36. *FEBS Lett.* **424**, 11–16.

Morello, R., Zhou, G., Dreyer, S. D., Harvey, S. J., Ninomiya, Y., Thorner, P. S., Miner, J. H., Cole, W., Winterpacht, A., Zabel, B., Oberg, K. C., and Lee, B. (2001). Regulation of glomerular basement membrane collagen expression by LMX1B contributes to renal disease in nail patella syndrome. *Nature Genet.* **27**, 205–208.

Nakano, S., Iyama, K., Ogawa, M., Yoshioka, H., Sado, Y., Oohashi, T., and Ninomiya, Y. (1999). Differential tissular expression and localization of type IV collagen α1(IV), α2(IV), α5(IV), and α6(IV) chains and their mRNA in normal breast and in benign and malignant breast tumors. *Lab. Invest.* 79, 281–292.

Ninomiya, Y., Kagawa, M., Iyama, K., Naito, I., Kishiro, Y., Seyer, J. M., Sugimoto, M., Oohashi, T., and Sado, Y. (1995). Differential expression of two basement membrane collagen genes, COL4A6 and COL4A5, demonstrated by immunofluorescence staining using peptide-specific monoclonal antibodies. *J. Cell Biol.* **130**, 1219–1229.

Poschl, E., Pollner, R., and Kuhn, K. (1988). The genes for the α1(IV) and α2(IV) chains of human basement membrane collagen type IV are arranged head-to-head and separated by a bidirectional promoter of unique structure. *EMBO J.* 7, 2687–2695.

Rudland, P. S., Platt-Higgins, A. M., Wilkinson, M. C., and Fernig, D. G. (1993). Immunocytochemical identification of basic fibroblast growth factor in the developing rat mammary gland: Variations in location are dependent on glandular structure and differentiation. *J. Histochem. Cytochem.* **41**, 887–898.

Sado, Y., Kagawa, M., Kishiro, Y., Sugihara, K., Naito, I., Seyer, J. M., Sugimoto, M., Oohashi, T., and Ninomiya, Y. (1995). Establishment by the rat lymph node method of epitope defined monoclonal antibodies recognizing the six different α chains of human type IV collagen. *Histochem. Cell Biol.* **104**, 267–275.

Saito, K., Naito, I., Seki, T., Oohashi, T., Kimura, E., Momota, R., Kishiro, Y., Sado, Y., Yoshioka, H., and Ninomiya, Y. (2000). Differential expression of mouse α5(IV) and α6(IV) collagen genes in epithelial basement membranes. *J. Biochem. (Tokyo)* **128**, 427–434.

Seki, T., Naito, I., Oohashi, T., Sado, Y., and Ninomiya, Y. (1998). Differential expression of type IV collagen isoforms, α5(IV) and α6(IV) chains, in basement membranes surrounding smooth muscle cells. *Histochem. Cell Biol.* **110**, 359–366.

Soininen, R., Houtari, M., Hostikka, S. L., Prockop, D. J., and Tryggvason, K. (1988). The structural genes for α1 and α2 chains of human type IV collagen are divergently encoded on opposite DNA strands and have an overlapping promoter region. *J. Biol. Chem.* **263**, 17217–17220.

Sugimoto, M., Oohashi, T., and Ninomiya, Y. (1994). The genes COL4A5 and COL4A6, coding for basement membrane collagen chains α5(IV) and α6(IV), are located head-to-head in close proximity on human chromosome Xq22 and COL4A6 is transcribed from two alternative promoters. *Proc. Natl. Acad. Sci. USA* **91**, 11679–11683.

Tryggvason, K., and Ninomiya, Y. (1997). Alport syndrome (Hereditary nephritis). *In* "Clinical Studies in Medical Biochemistry," pp. 218–226.

Ueki, Y., Naito, I., Oohashi, T., Sugimoto, M., Seki, T., Yoshioka, H., Sado, Y., Sato, H., Sawai, T., Sasaki, F., Matsuoka, M., Fukuda, S., and Ninomiya, Y. (1998). Topoisomerase I and II consensus sequences in a 17-kb deletion junction of the COL4A5 and COL4A6 genes and immunohistochemical analysis of esophageal leiomyomatosis associated with Alport syndrome. *Am. J. Hum. Genet.* **62**, 253–261.

Zhou, J., Mochizuki, T., Smeets, H., Antignac, C., Laurila, P., Paepe, A., Tryggvason, K., and Reeders, S. T. (1993). Deletion of the paired α5(IV) and α6(IV) collagen genes in inherited smooth muscle tumors. *Science* **261**, 1167–1169.

CHAPTER 3

Regulation of Phenotypes of Human Aorta Endothelial Cells and Smooth Muscle Cells in Culture by Type IV Collagen Aggregates

TOSHIHIKO HAYASHI, MOTOHIRO HIROSE, HIROKO YAMANO, YASUSHI TAKEDA, HIROAKI KOSUGI, TAKANORI KIHARA, YASUTADA IMAMURA, KAZUNORI MIZUNO, KOICHI NAKAZATO, KIWAMU YOSHIKAWA, DAISUKE KAJIMURA, SEIICHIRO TAKAHASHI, AND EIJIRO ADACHI[1]

Department of Life Sciences, Graduate School of Arts and Sciences, The University of Tokyo, Tokyo 153-8902, Japan

I. INTRODUCTION

Type IV collagen comprises the skeletal structure of basal lamina, which functions as a physical barrier for biological macromolecules and cells. Basal lamina affects the most fundamental cell functions, including cell proliferation and differentiation. Vascular endothelial cells and smooth muscle cells produce and deposit basal lamina components in extracellular surroundings. Reciprocal interactions of the cells with the basal lamina may result in a vicious or favorable spiral. The favorable spiral can be defined as a dynamic steady state of tissue structure and function.

Incubation of the type IV collagen solution in neutral pH and physiological ionic strength causes two macroscopically different forms of type IV collagen aggregates to form, depending on the temperature of incubation. Aggregates formed at 4°C exhibit a gel property, whereas aggregates formed at 28°C are so fragile as a gel that the whole structure tends to collapse during the manipulation. Both the aggregates reconstituted from type IV collagen, the rigid and fragile gel forms, show the similar polygonal meshwork structure shown in Fig. 1.

[1]Present address: Department of Anatomy, Kitasato University, School of Medicine, Kanagawa 228-8555, Japan.

Copyright 2003, Elsevier Science (USA). All rights reserved.

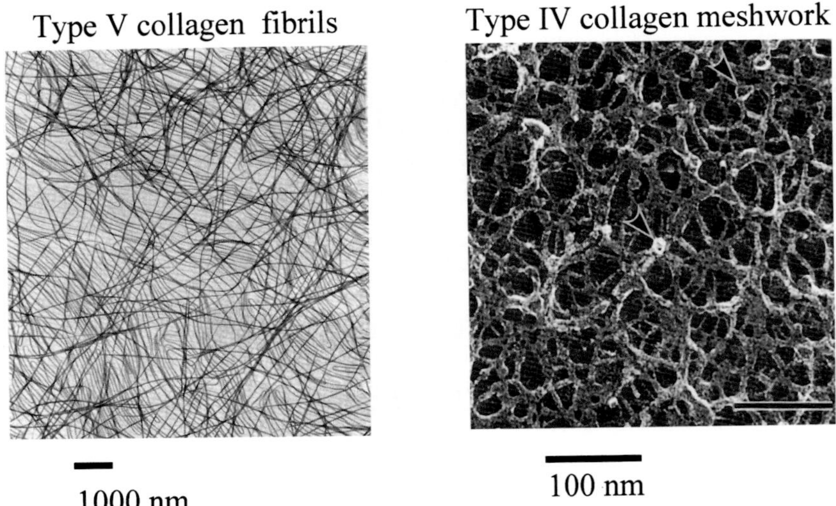

FIGURE 1 Electron micrographs of reconstituted aggregates from type IV collagen and type V collagen.

The electron micrographs are indistinguishable (Nakazato *et al.*, 1996). The polygonal meshwork structure suggests that the skeletal structure of basal lamina is composed of type IV collagen aggregates.

We previously studied the immunohistochemical tissue distribution of each type of collagen and the ultrafine structure of the tissue in comparison with the reconstituted aggregates from each type of the collagen protein isolated. From the analysis, we hypothesized a vascular tissue model for the localization of collagen types in relation to cell types as depicted in Fig. 2 (Adachi *et al.*, 1997; Hayashi

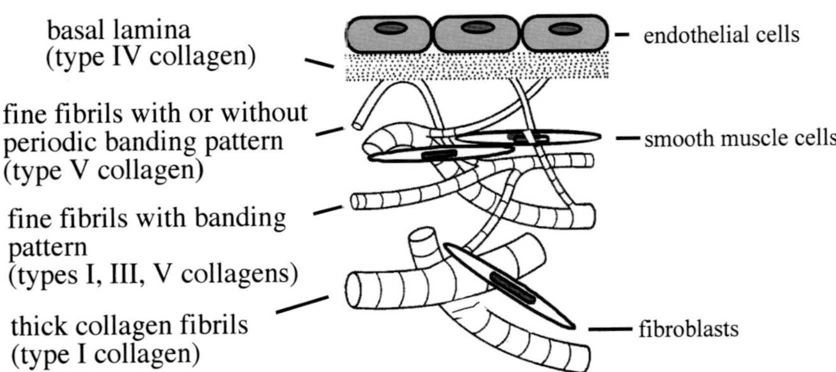

FIGURE 2 Tissue model for localizations of collagen types and cells.

et al., 1999). Electron micrographs of the aggregates reconstituted from the type IV collagen and the type V collagen are shown in Fig. 1 (Adachi *et al.*, 1997; Mizuno *et al.*, 2001; Nakazato *et al.*, 1996).

The type IV collagen aggregates with different physicochemical properties were used as culture substrata for human aorta smooth muscle cells in our previous study (Hirose *et al.*, 1999). Results indicated that when human aorta smooth muscle cells that had become fibroblast-like cells (sometimes called myofibroblast) were cultured on the type IV collagen gel, the cells acquire a contractile phenotype. The following is a summary from the previous study (Hirose *et al.*, 1999).

II. HUMAN AORTIC SMOOTH MUSCLE CELLS ON TYPE IV COLLAGEN GEL

Human aorta smooth muscle cells remain round after seeding at an early stage of culture. The cells show delayed morphological change to an elongated cell shape. Further elongation of cells touch neighboring elongated cells and eventually connect with each other. Overall connections of the cells with an elongate cell shape form multicellular meshwork after 1 or 2 days of culture on the type IV collagen gel (Fig. 3). Such meshwork formation of cells was also reported for endothelial cells by culturing on Matrigel (Vernon *et al.*, 1992).

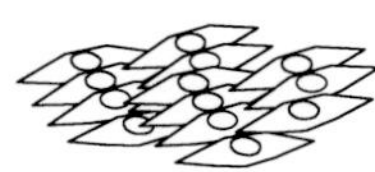

On the non-gel form

Cells have bipolar, spindle-like shape and the highest proliferative potential.

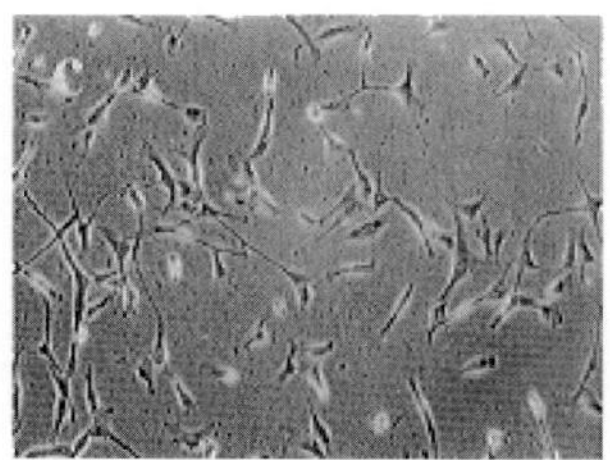

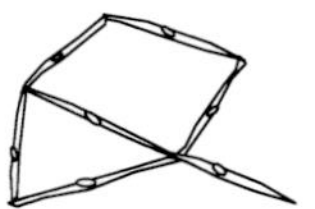

On the gel form

Cells have multicellular meshwork and the lowest proliferative potential.

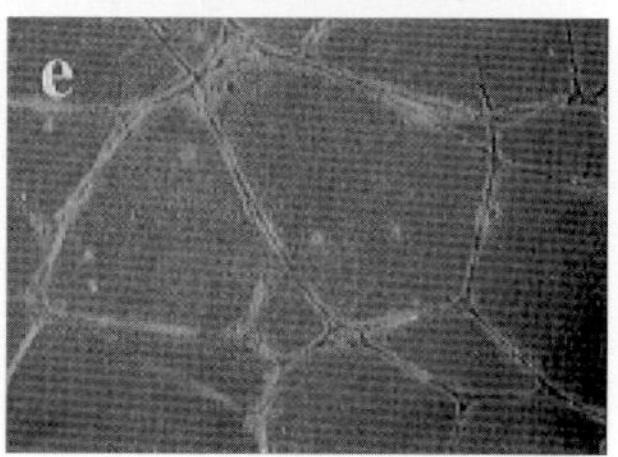

FIGURE 3 Morphology of human aorta smooth muscle cells in culture in the gel form and the nongel form of type IV collagen aggregates.

However, human aorta smooth muscle cells show a spindle-like cell shape soon after seeding on the fragile gel or nongel form of the type IV collagen aggregates. The morphology is indistinguishable from that of fibroblasts or other fibroblast-like cells cultured on other substrates, such as on the type I collagen gel. Thus, cell morphology is primarily dependent on the physical property of the type IV collagen aggregates rather than on the biochemical property of the protein. In Fig. 3, a phase-contrast micrograph of the cells cultured on the type IV collagen gel is shown in comparison with the cell cultured on the nongel form of the type IV collagen aggregates.

Cell growth is another distinct difference in the response of human aorta smooth muscle cells to the physical states of the type IV collagen aggregates whether they are in the gel form (rigid gel) or nongel form (fragile gel) (Fig. 4). On nongel type IV collagen aggregates, the cells undergo morphological change soon after seeding and proliferate to confluency most rapidly. In contrast, the gel form of type IV collagen aggregates, even in the presence of bovine fetal serum, totally represses the proliferation of the smooth muscle cells that had acquired a high proliferative potential through repeated passages in the presence of serum (Fig. 4). Smooth muscle cells with the quiescent phenotype on the type IV collagen gel express a smooth muscle myosin heavy chain, a specific marker for the contractile phenotype (Fig. 5). Furthermore, human aortic smooth muscle cells cultured on the type IV collagen gel contracted in response to the endothelin-1 added to the culture medium, showing that the type IV collagen gel has a restoring activity in

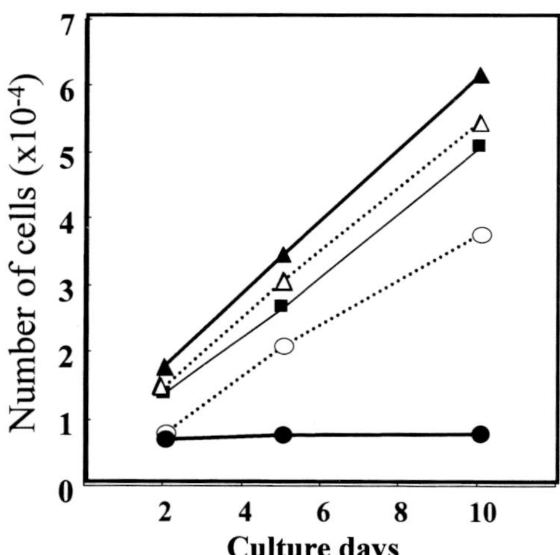

FIGURE 4 Growth curve of smooth muscle cells on type IV collagen aggregates with a nongel form: plastic dish (■), type I collagen-coated dish (△), type I collagen gel (○), type IV collagen-coated dish (▲), and type IV collagen gel (●).

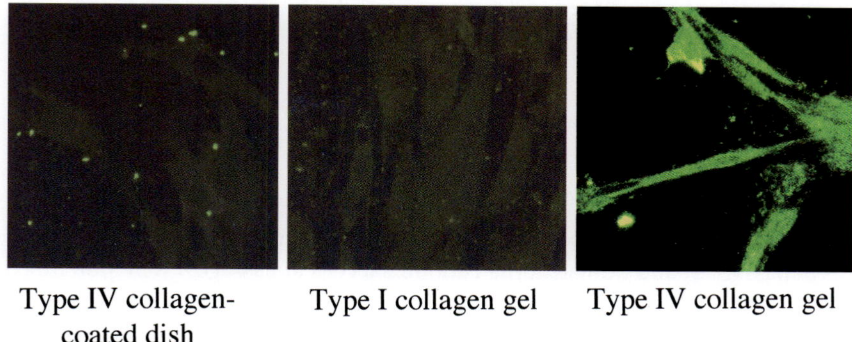

FIGURE 5 Expression of smooth muscle myosin heavy chains in myofibroblast-like aortic smooth muscle cells.

restoring the fibroblast phenotype back to the contractile phenotype. The latter phenomenon is quite marked, as it has been believed that the acquisition of the fibroblast-like, proliferative phenotype is practically irreversible.

Therefore, such previous observations on the distinct effects of the type IV collagen aggregates on human aorta smooth muscle cells have stimulated us to work further on which differences in the physicochemical properties affect the responses of the cell. Another issue is whether the type IV collagen protein is involved in regulating other cells existing in direct contact with the type IV collagen *in vivo*. The present study describes extension of the aforementioned observations on the physical property-determining effects of the type IV collagen aggregates in the following three points.

1. Human aorta smooth muscle cells cultured on the nongel form of the type IV collagen aggregates layered on top of the type I collagen gel showed essentially the same behavior cultured on the gel from type IV collagen aggregates (Hirose *et al.*, 2000). It was found that type IV collagen aggregates with continuous change in the physical property were formed from a drop of type IV collagen solution. They were used as the cell culture substrate in order to see how human aorta smooth muscle cells respond to type IV collagen aggregates with continuous change in the physical property from gel form to nongel form.
2. Other smooth muscle cell-related cells and skeletal muscle cells were examined for the effects of the type IV collagen gel on cell behaviors, morphology, and proliferation.
3. The nongel form of type IV collagen aggregates appears to be the best substrate for smooth muscle cell adherence and proliferation. The effect on the nongel form of type IV collagen aggregates was examined for human aorta endothelial cells that do not show stable adherence on any substrates used previously, especially for a prolonged period of culture.

III. PHYSICAL PROPERTY-DETERMINING EFFECTS OF TYPE IV COLLAGEN AGGREGATES

A. Experimental Procedures

1. Cultured Cells

Human aortic smooth muscle cells at passage 3, human glomerular mesangial cells at passage 3, and human aortic endothelial cells at passage 3 were purchased from Clonetics Corp. (San Diego, CA). After three to five times of passages (up to 9-13 PDL) on 100-mm dishes (Falcon, No.3003), either smooth muscle cells or mesangial cells that had become myofibroblast-like were used. Human rhabdomyosarcoma cells were purchased from Health Science Research Resources Bank, Osaka. Human dermal fibroblasts were kindly gifted by Dr. Toshio Nishiyama of Shiseido Research Center, Yokohama. Rat hepatic stellate cells were isolated from Wistar female retired rats by portal vein perfusion method as described previously in detail (Davis, 1988). In brief, phosphate-buffered saline (PBS) at 37°C was perfused at 15 ml/min for 10 min to blanch the liver. This was followed by 0.05% collagenase (Wako Pure Chemical Inc., Tokyo) perfusion at 5 ml/min for 20 min at 37°C. The liver was removed and minced in serum-free Dulbecco's Modified Eagle Medium (DMEM) (Nissui Pharmaceutical Co., Tokyo). The cell suspension was then filtered through sterile gauze and centrifuged at 50 g for 3 min to remove hepatocytes and debris. The suspension after removal of sediments containing nonparenchymal cells was layered over a 25% preformed (20,000 g for 10 min) Percoll (Pharmacia Biotech., Inc., Sweden) gradient and centrifuged at 800 g for 30 min. The gradient centrifugation produced a top layer of yellowish white oily debris with a band of cells immediately beneath. The band mainly contained stellate cells as identified with lipid droplet-laden cells. Autofluorescence at 325 nm faded rapidly under the fluorescent microscope (Model DMIRB, Leica Co., Tokyo), whereas Oil Red O (Sigma) stained the cells. From these analyses, stellate cells predominated >90%. The cells were then washed two times and resuspended in DMEM supplemented with 10% fetal bovine serum (FBS) (Cansera International Inc., Canada) and plated at a density of 1.0×10^6 cells per 100-mm tissue culture dish. Myofibroblast-like hepatic stellate cells were obtained from primary stellate cells by 14–16 passages up to 28–32 PDL on the 100-mm dishes.

2. Culture Media and Cell Culture

Rat hepatic stellate cells, human rhabdomyosarcoma cells, and human dermal fibroblasts were grown in DMEM supplemented with 10% FBS, 60 μg/ml kanamycin sulfate, and 50 ng/ml amphotericin-B. Aorta smooth muscle cells were cultured in modified MCDB131 medium (Clonetics Corp.) supplemented with 10% FBS, 10 ng/ml recombinant epidermal growth factor, 2 ng/ml recombinant basic fibroblast growth factor, 5 μg/ml insulin, 50 μg/ml gentamaicin, and 50 ng/ml amphotericin-B (designated as growth medium in this thesis). Human mesangial cells were cultured in modified MCDB131 medium containing 10% FBS, 50 μg/ml gentamaicin, and 50 ng/ml amphotericin-B. Aorta endothelial cells were grown in EBM-2 medium (Clonetics Corp.) containing 2% FBS, 10 μg/ml hydrocortisone, 2 ng/ml recombinant basic fibroblast growth factor, 5 ng/ml insulin-like growth factor-1, 5 ng/ml vascular endothelial-derived growth factor, 0.2 m*M* ascorbic acid 2-phosphate, 10 μg/ml heparin, 10 ng/ml recombinant epidermal growth factor, 50 μg/ml gentamaicin, and 50 ng/ml amphotericin-B. All the growth factors added were purchased from Clonetics Corp. All of the cell types were maintained at 37°C under humidified 5% CO_2–95% air on the 100-mm tissue culture dish. When the cells reached confluence, they were passaged at a 1:4 split ratio after they were removed from the dish with a brief exposure to 0.25% trypsin (Difco)–0.02% EDTA in PBS (-) (Nissui Pharmaceutical Co.).

3. Preparation of Substrates and Successive Cell Culture

Type I collagen was obtained from the acid extract of rat tail tendon by the method reported previously (Yamato *et al.*, 1992). The preparation of cell culture on the type I collagen gel and the successive culture were performed as described previously (Yamato *et al.*, 1992). Briefly, 6 volumes of type I collagen solution (3 mg total protein/ml) was mixed with 3 volumes of 3× concentrates of culture media and 1 volume of FBS at 4°C to give a final collagen concentration of 1.8 mg/ml. Aliquots (500 μl) of the solution were added to each well of 24-well tissue culture dishes (Falcon, No. 3047) and incubated at 37°C for gelation. One milliliter of cell suspension containing 7.2×10^3 cells was plated on each gel, and the cells were grown at 37°C with 5% CO_2–95% air. The medium was replaced every 3 days. Preparation of the type IV collagen gel was as described previously (Nakazato *et al.*, 1996). In brief, 9 volumes of type IV collagen solution (2 mg/ml) in 1m*M* HCl was mixed with 1 volume of 200 m*M* phosphate buffer containing 1.5*M* NaCl, pH 7.3, at 4°C to obtain a final collagen concentration of 1.8 mg/ml. An aliquot (500 μl) of the solution was added to each well of 24-well Falcon tissue culture dishes. The dish was incubated at 4°C for 5 days to form a gel with sufficient rigidity to accommodate cells on top of the gel. After replacing the buffer with culture media, 1 ml of cell suspension containing 7.2×10^3 cells was plated gently on each gel, and the cells were grown at 37°C with 5% CO_2–95% air. The medium was replaced every 3 days. Dishes coated with aggregated type IV collagen in the nongel form were prepared as follows. Nine volumes of type IV collagen solution (2 mg/ml) in 1 m*M* HCl was mixed with 1 volume of 200 m*M* phosphate buffer containing 1.5 *M* NaCl, pH 7.3, to give a final collagen concentration of 1.8 mg/ml. The neutralized solution was added to 24-well dishes (500 μl/well) and then dried up at 25°C. The wells were washed with PBS (–), followed by further washing with culture medium. For preparation of protein-coated dishes, type I collagen solution or type IV collagen solution in 1 m*M* HCl (100 μg/ml) was placed in 24-well dishes (250 μl/well) and allowed to adsorb on the dish surface at 37°C for 2 h. Then, the wells were washed with PBS (–), followed by further washing either with growth medium (for human aorta smooth muscle cells) or with DMEM containing 10% FBS (for human dermal fibroblasts). Cell morphology was observed with a phase-contrast microscope (Model DMIRB, Leica Co., Tokyo) at 100× magnification.

4. Cell Proliferation Assay

The cell proliferation assay system, Premix WST-1, which reflects a relative number of living cells, was used for the proliferation of human aorta endothelial cells according to the manual given by the manufacturer (TaKaRa). After the reagent was added to the well, cultured cells were incubated for 1 h and 40 m. Absorbance at 450 nm for the color developed was recorded.

5. Incorporation of Acetylated LDL

Fluorescein-labeled acetylated LDL(FITC Ac-LDL) (L-3485) was obtained from Molecular Probes, Inc. (Oregon). The endothelial cells were incubated with FITC Ac-LDL for 6 h. Cultured cells were photographed with a microscope mode changed to fluorescence with excitation at 515 nm.

B. Culture of Human Aorta Smooth Muscle Cells on Type IV Collagen Aggregates with a Continuous Change in the Physicochemical Property

A hat-like shape gel was constructed by dropping a type IV collagen solution on a cover glass. The central region has a domed structure surrounded by a broad brim-like region (Yamano *et al.*, manuscript in preparation). In the phase-contrast micrograph (Fig. 6), the brim region shows a darker background, whereas the dome region shows a light background. Human smooth muscle cells were cultured on the hat-like shape gel. Cells in the dome region remained as a round cell shape at an initial stage of culture (6 h in Fig. 6) and eventually formed a multicellular meshwork at a later stage (24 h in Fig. 6), as is seen for cells

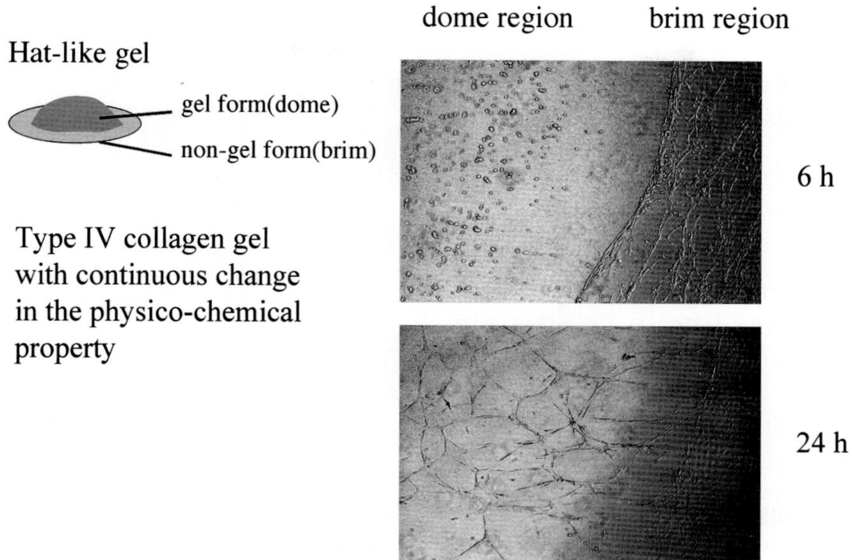

FIGURE 6 Type IV collagen gel with a domed shape.

cultured on the type IV collagen gel. However, cells at the brim region started to change cell shape soon after seeding (6 h in Fig. 6), as is so in the culture on the nongel form of the type IV collagen aggregates. Results confirmed directly that the physicochemical state of type IV collagen proteins determines the cell shape.

C. Cell Morphology of Other Smooth Muscle Cell-Related Cells and Skeletal Muscle Cells by Culturing on the Type IV Collagen Gel

Cells covering skeletal muscle cells, fetal fibroblasts, and rhabdomyosarcoma cells, as well as smooth muscle cell-related cells, such as hepatic stellate cells and kidney mesangial cells, were tested (M. Hirose *et al.*, manuscript in preparation). As shown in Fig. 7, all the cells examined showed substrate-dependent morphology on the type IV collagen gel with multicellular meshwork. Human dermal fibroblasts tended to show a weaker or shorter influence from the type IV collagen gel, yet they also took a multicellular meshwork structure for a while. The fibroblasts, thereafter, soon enter into the proliferation phase by changing the cell morphology to spindle-like, as seen for cells cultured on other substrates. In general, the length of the maintenance of the cell meshwork appeared to depend on the cell types, presumably because the cells may eventually secrete

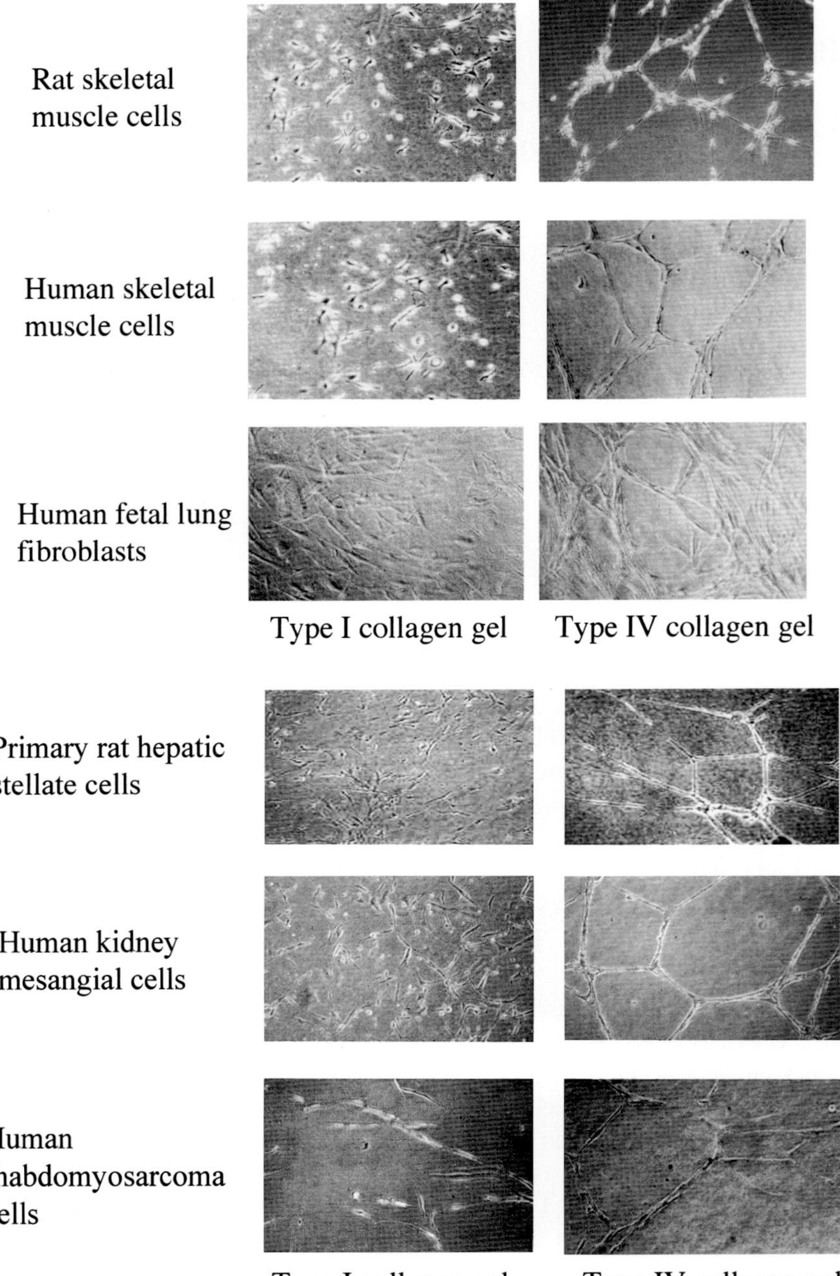

FIGURE 7 Phase-contrast micrographs of cells cultured on type I or type IV collagen gels.

some materials that affect the physicochemical property of the type IV collagen gel, such as gelatinases that may cut the type IV collagen molecules. Together with the micrographs shown in Fig. 7, it is concluded that cell morphology is determined by the physical property of the type IV collagen aggregates in the gel form. However, the nongel form of the type IV collagen aggregates did not show any particular effects on cell morphology and all the cells show a spindle-like shape typical for fibroblast-like cells as seen for other substrates.

D. Effect of Collagen Aggregates on Proliferation of Human Aorta Endothelial Cells

The effects of collagen in reconstituted aggregate or in molecular form were examined for proliferation of the human aorta endothelial cells. As shown in Fig. 8, the molecular forms of the type V collagen, type IV collagen, and type I collagen showed a distinct but small difference at day 4 of culture in the order of preferential substrate from type IV collagen, type I collagen, and type V collagen. The result is consistent with the report that the type V collagen was reported to be an antiadhesive or nonadhesive substrate in comparison with the type I collagen that is currently used as a substrate for endothelial cell culture. The effects of aggregates reconstituted from different types of collagen on the cell density were in contrast, especially at day 7 of the cell culture. The type IV collagen aggregates were

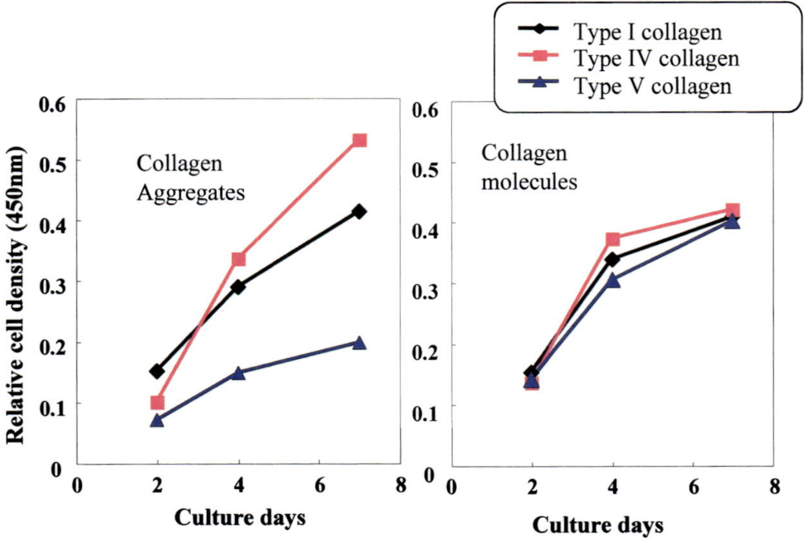

FIGURE 8 Proliferation of human aorta endothelial cells cultured on collagen aggregates.

prominent in maintaining the highest density of the cells. In contrast, the effect of the type V collagen aggregates was more prominent in repressing the cell density at a latest stage of culture (Fig. 8).

The effects of type IV collagen aggregates on cell proliferation are entirely opposite, depending on whether the aggregates are in a gel form or not, for smooth muscle cells (Fig. 4). The effects of type IV collagen aggregates on human aorta endothelial cells in culture appear to be common with smooth muscle cells in terms of cell proliferation. Human aorta endothelial cells remain round in shape and quiescent when cultured on the gel form of type IV collagen aggregates (data not shown). The nongel form of type IV collagen aggregates was the best substrate for facilitating the proliferation to the highest cell density (Fig. 8).

E. Effect of Nongel Form of Type IV Collagen Aggregates on Adherence of Human Aorta Endothelial Cells

Human aorta endothelial cells show characteristic morphology (Fig. 9), as well as facilitated proliferation, when cultured on the nongel form of type IV collagen aggregates. The difference in the overall morphology of endothelial cells among the different substrates was particularly marked after a prolonged period of culture (Fig. 9). In this sense, type IV collagen aggregates serve as a substrate with the highest density of human aorta endothelial cells presumably due to a strong

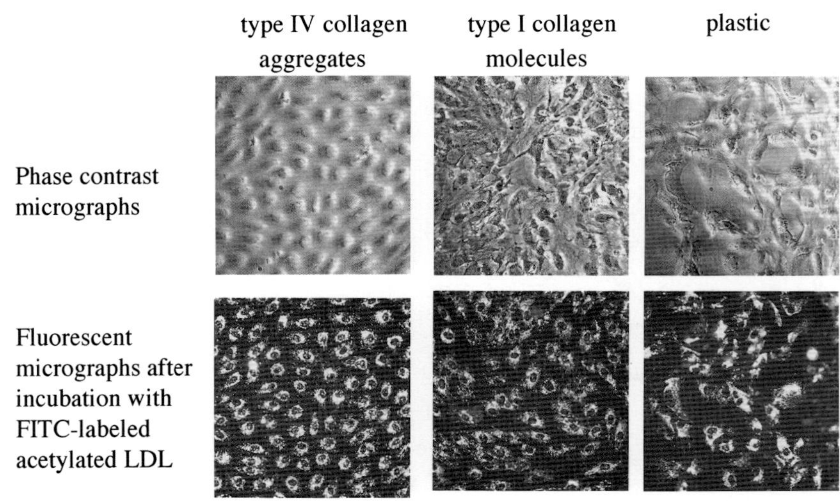

FIGURE 9 Human aortic endothelial cells cultured on the nongel form of type IV collagen aggregates.

and stable adherence of the cells to the type IV collagen aggregates. On other substrates, the endothelial cells detach from the substrate and grow on top of other cells, resulting in irregularly packed cells after a prolonged culturing period. However, human aorta endothelial cells maintained cobblestone-like cell morphology for a prolonged period of more than 3 months (Fig. 9). Figure 9 shows fluorescent micrographs of the cultured endothelial cells after incubation with FITC-labeled acetylated LDL, the material that is specifically incorporated by the living endothelial cells.

IV. PERSPECTIVES

We reported previously that fibrotic or cirrhotic liver shows an altered distribution of basement membrane-related collagen (Yoshida *et al.*, 1997). The deposition of basement membrane-associated collagen, as detected by the JK132 monoclonal antibody that recognizes only the nonhelical form of type IV collagen $\alpha 1$ chain, increased in the pathological regions of liver tissues. The distribution of type IV collagen was delocalized from the region close to epithelial or endothelial cells and colocalized with type I collagen. Furuyama *et al.* (1999) reported that the formation of basal lamina structure in cultured cells requires a balanced production of basal lamina components. Thus the excessive presence of transforming growth factor (TGF)-β inhibits the formation of basal lamina because of the overproduction of interstitial matrix components, including type I collagen. However, an appropriately low level of TGF-β stimulates the formation of basal lamina, presumably because an appropriate production of basal lamina component includes type IV collagen. More recently, we found that human cells in culture produce and deposit type IV collagen (Takeda *et al.*, 1993; Yoshikawa *et al.*, 2001), but the production is dependent on culture conditions. The level of ascorbate is especially crucial in the polypeptide conformation of the type IV collagen (Yoshikawa *et al.*, 2001). Thus, depletion of ascorbate affects the cells to secrete the type IV collagen α chains in nonhelical form. The nonhelical form of type IV collagen polypeptide might well be the basement membrane-associated collagen denoted formerly (Kino *et al.*, 1991).

Few researchers reported that type IV collagen has biological functions. In the present report, the effects of type IV collagen aggregates on cell behaviors were examined using the type IV collagen as cell culture substrates, implicating strongly the biological functions specific for type IV collagen aggregates. Type IV collagen aggregates determine the phenotypes of human aorta endothelial cells and human smooth muscle cells, depending on the physicochemical properties. Not just the biochemical nature but physicochemical properties of the type collagen aggregates are involved in the stabilization of multicellular architecture and thus homeostasis of tissue structure and function in the biological phenomena, such as repeated injury and repair of blood vessels. Some speculations from the present results may be made. First, because the hat-like structure was made from a drop of the type IV collagen solution, the concentration and aggregate structure of the type IV collagen protein are the same. However, the human aortic smooth muscle cells responded in a different way entirely, depending on the region of the gel. This implies that three-dimensional recognition by the cells, including the interaction with solvents or some unknown recognition mechanism, is involved in the resultant cell response for cell morphology and proliferation. Another point may be more relevant to biological importance. The complex matrix components can be found in basal lamina. The copresence of one or more components with the type IV collagen aggregates may further change the physicochemical property, including gel rigidity. Perlecan, a basal lamina proteoglycan, may well affect the physicochemical property of the type IV collagen aggregates through hydration or other interactions.

For a long-term culture of dermal fibroblasts, ascorbate 2-phosphate, a long sustainable ascobate derivative, has been used successfully, indicating that the production of type I collagen and collagenous proteins that require ascorbate for the syntheses contribute to the formation of a dermis-like structure (Hazeki *et al.*, 1998). One difficulty in establishment of a long culture of endothelial cells or epithelial cells with the differentiated state is to maintain the adherence of the cells densely populated on the culture dish. That is, the cells tend to detach from the dish after or close to reaching confluence. A long-term culture requires some conditions in the culture system, including the adequate establishment of cell-to-cell junctions, appropriate extracellular environments, and/or physical or chemical stimulation from outside. From the three-dimensional cell culture with the type I collagen gel, we noted previously that cultured cells change cell fundamental potentials, including proliferation, migration, and gene expressions in response to the extracellular environments (Hayashi *et al.*, 1993). Furthermore, the fibroblasts modify the distribution of collagen fibrils or contract the type I collagen gel (Yamato *et al.*, 1995). The extracellular environments thus changed by the cells in turn affect cellular activity. The repetitive interactions of dermal fibroblasts with type I collagen fibrils converge into stable dermis-like structures.

Collagen aggregates, the protein forms in tissues, showed marked effects on the density of endothelial cells. Especially the type IV collagen aggregates, underlying the endothelial tissues *in vivo* under physiological conditions, showed apparently the strongest cell adherence in terms of the maintenance of the cobblestone-like structure for a long period. In contrast, the type V collagen aggregates that exist underneath the basal lamina *in vivo* reduced the cell density. We propose an entirely new concept for the effect of type V collagen fibrils on cell adhesion. Type V collagen fibrils may interfere with cell spreading to cover the whole surface of the culture dish by adhering strongly to the cell surfaces (T. Kihara *et al.*, manuscript in preparation).

Human aortic endothelial cells cultured on type IV collagen aggregates reached the most highly packed density with a cobblestone-like structure. The tissue-like structure was maintained up to 3 months or more. Type IV collagen deposition was found homogeneously underneath cells cultured on type IV collagen aggregates (T. Kihara, manuscript in preparation). However, when the endothelial cells were cultured on other substrates, many cells were flat in shape and some cells migrated on top of neighboring cells. The cells showed little deposition or irregular distribution of type IV collagen. It appeared that type IV collagen aggregates provide the best substrate for human aorta endothelial cells to enter into a favorable spiral of the interactions between cells and the matrix to form and maintain the endothelial tissue-like structure. Finally, it should be noted that skeletal muscle stem cells fused into myotubes by culturing on the type IV collagen gel. The myotubes eventually contracted spontaneously (So *et al.*, personal communications).

Results that the incorporation activity of acetylated LDL is indistinguishable among the cells suggest that cells adhering to the substrate, regardless of the cell density, incorporated the acetylated LDL. Thus, acetylated LDL incorporation activity, a good marker for endothelial cell function, is not necessarily a good marker for endothelial tissue function. Because distributions of the cells and fluorescence are colocalized, evaluation for the tissue-like function may need other activities, such as barrier functions, against the penetration or crossing of cells or materials.

V. CONCLUSION

Proliferation and differentiation of human aorta endothelial cells and smooth muscle cells in culture are dependent on the aggregate forms or nongel and gel form as well as collagen types. The particularly prominent effects were seen for type IV collagen aggregates. Extracellular environments are one of the most fundamental elements for the maintenance or the restoration of human aorta endothelial tissues and smooth muscle tissues. Type IV collagen aggregates could comprise such tissue constituents, especially for endothelial, epithelial, muscular adipose, or neural tissues that contain the cells in direct contact with the basal lamina.

REFERENCES

Adachi, E., Hopkinson, I., and Hayashi, T. (1997). Basement-membrane stromal relationships: Interactions between collagen fibrils and the lamina densa. *Int. Rev. Cytol.* **173**, 73–156.

Davis, B.H. (1988). Transforming growth factor β responsiveness is modulated by the extracellular collagen matrix during hepatic I to cell culture. *J. Cell. Physiol.* **136**, 547–553.

Furuyama, A., Iwata, M., Hayashi, T., and Mochitate, K. (1999). Transforming growth factor-1 regulates basement membrane formation by alveolar epithelial cells *in vitro*. *Eur. J. Cell Biol.* **78**, 867–875.

Hayashi, T., Yamato, M., Adachi, E., and Yamamoto, K. (1993). Unique features of type I collagen as a regulator for fibroblast functions: Iterative interactions between reconstituted type I collagen fibrils and fibroblasts. *In* "New Functionality Materials" (T. Tsuruta, M. Doyama, M. Seno, and Y. Imanishi, eds.), Vol. B, pp. 239–246. Elsevier Science, New York.

Hayashi, T., Mizuno, K., Hirose, M., Nakazato, K., Adachi, E., Imamura, Y., Kosugi, H., and Yoshikawa, K. (1999). Cell type specific recognition of the reconstituted aggregates from isolated type I collagen, type IV collagen, and type V collagen. *In* "Proceedings Chemical Sciences" (V. Krishnan, ed.), Vol. 111, pp. 171–177 Indian Academy of Sciences, Bangalore.

Hazeki, N., Yamato, M., Imamura, Y., Sasaki, T., Nakazato, K., Yamamoto, K., Konomi, H., and Hayashi, T. (1998). Analysis of matrix components of the dermis-like structure formed in a long-term culture of human fibroblasts: Type VI collagen is a major component. *J. Biochem.* **123**, 587–595.

Hirose, M., Kosugi, H., Nakazato, K., and Hayashi, T. (1999). Restoration to a quiescent and contractile phenotype from a proliferative phenotype of myofibroblast-like human aortic smooth muscle cells by culture on type IV collagen gels. *J. Biochem.* **125**, 991–1000.

Hirose, M., Mizuno, K., Kosugi, H., Yoshikawa, K., Yamano, H., Kihara, T., and Hayashi, T. (2000). Implications for biological functions of type IV collagen gels. *Trans. Mater. Res. Soc. Japan* **25**, 907–910.

Kino, J., Adachi, E., Yoshida, T., Asamatsu, C., Nakajima, K., Yamamoto, K., and Hayashi, T. (1991). A novel chain of basement membrane-associated collagen as revealed by biochemical and immunohistochemical characterizations of the epitope recognized by a monoclonal antibody against human placenta basement membrane collagen. *Am. J. Pathol.* **138**, 911–920.

Mizuno, K., Adachi, E., Imamura, Y., Katsumata, O., and Hayashi, T. (2001). The fibril structure of type V collagen triple-helical domain. *Micron* **32**, 317–323.

Nakazato, K., Muraoka, M., Adachi, E., and Hayashi, T. (1996). Gelation of the lens capsule type IV collagen solution at a neutral pH. *J. Biochem.* **120**, 889–894.

Takeda, Y., Sasaki, T., and Hayashi, T. (1993). Quantitative analysis of type IV collagen by an enzyme immunoassay in the cell layer deposited by cultured fibroblasts. *Connect. Tissue* **25**, 183–188.

Vernon, R. B., Angello, J. C., Iruela-Arispe, M. L., Lane, T. F., and Sage, E. H. (1992). Reorganization of basement membrane matrices by cellular traction promotes the formation of cellular networks. *In vitro Lab. Invest.* **66**, 536–545.

Yamato, M., Adachi, E., Yamamoto, K., and Hayashi, T. (1995). Condensation of collagen fibrils to the direct vicinity of fibroblasts as a cause of gel contraction. *J. Biochem.* **117**, 940–946.

Yamato, M., Yamamoto, K., and Hayashi, T. (1992). Decrease in cellular potential of collagen gel contraction due to in vitro cellular aging: A new aging index of fibroblast with high sensitivity. *Connect. Tissue* **24**, 157–162.

Yoshida, T., Matsubara, O., Adachi, E., Kino, J., Asamatsu, C., Takeda, Y., and Hayashi, T. (1997). Increased deposition and altered distribution of basement membrane-related collagens in human fibrotic or cirrhotic liver. *Connect. Tissue* **29**, 189–198.

Yoshikawa, K., Takahashi, S., Imamura, Y., Sado, Y., and Hayashi, T. (2001). Secretion of non-helical collagenous polypeptides of α1(IV) and α2(IV) chains upon depletion of ascorbate by cultured human cells. *J. Biochem.* **129**, 929–936.

CHAPTER 4

Functions of Proteoglycan/ Glycosaminoglycan in Liver

TOSHIKAZU YADA,* NORIO KOIDE,† AND KOJI KIMATA*

*Institute for Molecular Science of Medicine, Aichi Medical University, Aichi 480-1195, Japan and
†Department of Laboratory Medicine, Okayama University Medical School, Okayama 700-9559, Japan

Proteoglycans/glycosaminoglycans comprise a collection of macromolecules that surround the plasma membrane of cell and constitute part of the substrate upon which cells are attached and perform their major functions. The solid frameworks needed for the structural support and functional organizations of these assemblies are made primarily of cross-linked fibers of collagen embedded in a fine meshwork of proteoglycans/glycosaminoglycans, and sometimes reinforced by deposited minerals. Proteoglycans/glycosaminoglycans are involved in maintaining the transparency of the cornea, the tensile strength of the skin and tendon, the viscoelasticity of blood vessels, the compressive properties of cartilage, and the mineralized matrix of bones. In addition, proteoglycans play key roles as storage depots for growth factors and cytokines, and by virtue of their multifunctional properties, they alter the biology of these factors.

This chapter summarizes general aspects of proteoglycans/glycosaminoglycan slightly and focuses on chondroitin sulfate/dermatan sulfate proteoglycans and hyaluronan in normal and cirrhotic liver.

I. INTRODUCTION

Although the extracellular matrix is only a small component of the liver, it has a crucial role, not only in providing a structural framework, but also in maintaining

Copyright 2003, Elsevier Science (USA). All rights reserved.

the differentiated state of the hepatocyte (Martinez-Hernandez, 1984). This role of the hepatic matrix has been demonstrated dramatically in cell culture, where the hepatocyte phenotype is dependent on the nature of the matrix substratum upon which it is cultured (Rojikind *et al.*, 1980). Furthermore, *in vivo* as a consequence of chronic injury, there are radical changes in hepatic matrix composition, distribution, excessive accumulation, transformation of sinusoids into capillaries, and appearance of basement membranes along the space of Disse (Martinez-Hernandez, 1985). The consequence of these changes is hepatic failure. They consist of mainly collagen types I, III, and IV, fibronectin, laminin, and glycosaminoglycans/proteoglycans.

A. Proteoglycan

Proteoglycans are a set of ubiquitous proteins found on cell surfaces and incorporated into extracellular matrices. Proteoglycans are a diverse set of macromolecules. Their protein cores can consist of polypeptide chains as small as 10 kDa or as large as 400 kDa; they can be soluble or insoluble; they can be membrane spanning, lipid tailed, or secreted; and they can carry only a single glycosaminoglycan chain, or well over a hundred. Traditionally, proteoglycans were found to be on the surface of cells and in the extracellular matrix, and the sequences of the core proteins deduced from cDNA clones indicate that they contain structural features that place them in one of these two locations (Table I). The core proteins of cell surface proteoglycans have a transmembrane segment, a small cytoplasmic domain, and a larger, although somewhat featureless, extracellular domain. The core proteins of extracellular matrix proteoglycans lack a transmembrane segment but have multidomain core proteins that often have homology to the structural glycoproteins of the cell surface and extracellular matrix (Table I). Some of the proteoglycans in each of these subdivisions have homologous core proteins, and therefore can be grouped into families (Table I). Unlike most protein, which are grouped into families on the basis of amino acid similarities alone, proteoglycans are also able to be defined by a common type of posttranslational modification: the glycosaminoglycan (GAG).

B. Glycosaminoglycan

GAGs are polysaccharides that are quite different from the N- and O-linked oligosaccharides found on most cell surface and secreted proteins, and GAGs strongly influence the structure and molecular interactions of the proteins to which they are attached. Given that GAGs are among the largest of protein-bound polysaccharide moieties, one might expect GAGs to exhibit an almost

TABLE I Proteoglycan Gene Families

Cell surface	Syndecan family
	Syndecan-1/syndecan/B-B4/CD138
	Syndecan-2/fibroglycan
	Syndecan-3/*N*-syndecan
	Syndecan-4/ryudocan/amphiglycan
	Glypican family
	Glypican-1/GPC1/glypican
	Glypican-2/GPC2/cerebroglycan
	Glypican-3/GPC3/OCI-5,/MXR7
	Glypican-4/GPC4/K-glypican
	Glypican-5/GPC5
	Orphans
	NG2
	Betaglycan
	RPTPβ/phosphacan
	Neuroglycan C
	CD44
Extracellular matrix	Small leucine-rich proteoglycans (SLRPs) family
	Decorin/DSPGII
	Biglycan/DSPGI
	Fibromodulin
	Lumican
	Epiphycan/PG-Lb
	Keratocan
	Mimecan
	Hyaluronan-binding proteoglycan family
	Aggrecan
	Versican/PG-M
	Neurocan
	Brevican
	Collagenous family
	Collagen IX/PG-Lt
	Collagen XII
	Collagen XVIII
	Orphans
	Perlecan
	Bamacan
	Agrin
	Serglycin

indecipherable degree of complexity. Yet their structures are surprisingly easy to understand: GAGs are linear polymers; there are no branches such as found in N-linked oligosaccharides. Except for the short linkage region by which GAGs are attached to the serine residues of protein cores, each GAG is synthesized from only two monosaccharides, strung together in strictly alternating fashion. There are only three such disaccharides repeat units that can be polymerized onto proteins in this fashion, giving rise to three basic "parent polymers" from which all protein-bound GAGs are fashioned. Following polymerization of these simple chains (the lengths of which are variable), certain types of enzymatic modifications are carried out on the sugars themselves.

C. Structures of Glycosaminoglycans

The GAGs heparin and heparan sulfate are both derived from the parent polymer [D-glucronic acid $\beta(1\rightarrow4)$D-*N*-acetyl glucosamine $\alpha(1\rightarrow4)]_n$, the disaccharides of which are then subjected to any or all of five different chemical modifications (epimerization of glucronate to iduronate, *N*-acetylation/*N*-sulfation, and *O*-sulfation at position 2 of iduronate and position 3 and/or 6 of glucosamine). Chondroitin sulfate and dermatan sulfate are also GAGs that derived from the same starting polymer [D-glucronic acid $\beta(1\rightarrow3)$D-*N*-acetyl galactosamine $\beta(1\rightarrow4)]_n$. Both may be modified by sulfation at position 2 of the uronic acid and position 4 or 6 of the amino sugar. *N*-sulfation does not occur. The distinction between chondroitin sulfate and dermatan sulfate involves epimerization of glucuronic acid to iduronic acid; if this modification is found at high frequency, the GAG is called dermatan sulfate; if not, it is called chondroitin sulfate. Keratan sulfate differs from its parent polymer, [D-galactose $\beta(1\rightarrow4)$D-*N*-acetyl glucosamine $\beta(1\rightarrow3)]_n$, only by *O*-sulfation at position 6 of glucosamine and/or position 6 of galactose. Keratan sulfate is unique in that it can be synthesized not only as an O-linked sugar attached to serine (as are other GAGs), but also as an N-linked sugar. Hyaluronan differs from other GAGs in that it is not synthesized attached to protein (so it is not a part of proteoglycans) and is not modified further from the structure of its parent polymer, [D-glucronic acid $\beta(1\rightarrow3)$D-*N*-acetyl glucosamine $\beta(1\rightarrow4)]_n$. Hyaluronan polymers can be extremely long, with molecular masses in the millions.

D. Functions of Proteoglycans

The functions of proteoglycans are only just beginning to become clear, but much progress has been made in a decade. For the sake of discussion, it is convenient to divide known functions into those that can be mediated by core proteins without the participation of GAG chains and those that can be mediated by GAG chains without the participation of core proteins. In reality, this distinction may not always be clear-cut, as some functions may involve the participation of both core protein and GAG, at least under some conditions.

A particularly good example of core protein-mediated function is the ability of small leucine-rich extracellular matrix proteoglycans to regulate collagen fibrillogenesis *in vitro*, which also appears to be an essential function of at least one of the proteoglycans (decorin) *in vivo* (Danielson *et al.*, 1997). It has also been demonstrated that decorin, biglycan, and fibromodulin among these small proteoglycans are able to interact *in vitro* with transforming growth factor (TGF)-β, which is a profibrotic key mediator in tissue fibrosis (Yamaguchi *et al.*, 1990; Hildebrand *et al.*, 1994). The affinity to TGF-β is similar for all, and it has been suggested that these proteoglycans may be able to sequester an overwhelming amount of TGF-β into the matrix and thus control its biological effects. The betaglycan core protein also binds TGF-β with rather high affinity and alters the ability of different TGF-β isoforms to interact with signaling receptors (Lopez-Casillas *et al.*, 1993). The binding of proteoglycans of the aggrecan family, aggrecan, versican, neurocan, and brevican, as well as CD44, to hyaluronan also represents a core protein-mediated interaction.

GAG-dependent functions can be loosely subdivided into two classes: the biophysical and the biochemical. The former refers to functions that depend on the unique biophysical properties of GAGs—the ability to fill space, bind and organize water molecules, and repel negatively charged molecules. The large quantities of chondroitin sulfate and keratan sulfate found on aggrecan, for example, are thought to play an important role in the hydration of cartilage. In contrast, the heparan sulfate on the kidney glomerular basement membrane proteoglycan is thought to play a role in filtration, impeding the passage of anionic serum proteins into the urine (Kanwar *et al.*, 1980).

The other functions of GAGs are those that are mediated by specific binding to proteins. Many GAG–protein interactions have been reported. For some of these proteins, all that is known is that they bind to a GAG affinity column (e.g., heparin–agarose) under physiological conditions of salt and pH. For others, affinity constants have been measured (K_d tend to be in the range of 10^{-6} to 10^{-9} M), whereas for still others, direct evidence for interaction with proteoglycans has been obtained (e.g., copurification, proteoglycan-dependent binding to cells). Only in a few cases have clear physiological functions been associated with GAG–protein interactions, but those case that have been well studied include (i) regulation of the activity of protease and antiproteases, (ii) regulation of cellular response to growth factors, (iii) regulation of cell–cell and cell–matrix interactions, (iv) regulation of extracellular matrix assembly and structure, and (v) immobilization of diffusible molecules. In these interactions, proteins that bind GAGs tend to recognize contiguous stretchs of 6–14 monosaccarides (Lander *et al.*, 1994). Given the number of covalent modifications that are possible on each disaccharide unit, the number of GAG sequences with which proteins can potentially interact is very large, especially for GAGs of the heparin/heparan sulfate family (which can exhibit any of five different modifications per disaccharide, alone or in various combinations). Studies on antithrombin III have demonstrated that proteins can be extremely selective for some GAG sequences (Lam *et al.*, 1976;

Lindahl, 1984). This protease inhibitor binds with high affinity only to heparin or heparan sulfate species that contain at least one pentasaccharide in which a strictly specified pattern of modifications (including the relatively uncommon 3-*O*-sulfation of glucosamine) is present. Absence of this pentasaccharide reduces GAG affinity approximately 1000-fold and accounts for the very different ability of different tissue heparan sulfates to stimulate antithrombin activity. To date, no other molecules have been shown to exhibit such strong sequence-dependent selectivity for naturally occurring GAG species (although 10- to 20-fold differences have been reported, but studies on the binding of growth factors to GAG fragments suggest that additional examples of strong sequence-selective binding *in vivo* will eventually be established) (Habuchi *et al.*, 1992; Ashikari *et al.*, 1995; San Antonio *et al.*, 1993; Guimond *et al.*, 1993; Lortat-Jacob *et al.*, 1995; Turnbull *et al.*, 1992).

II. CHONDROITIN SULFATE/DERMATAN SULFATE PROTEOGLYCANS IN LIVER RETICULIN FIBER

The hepatic extracellular matrix plays an important role in liver function in the normal state. Among them, the basement membrane-like extracellular matrix in the space of Disse has a crucial role for the hepatocyte to express various functions. They consist of collagen types I, III, and IV, fibronectin, laminin, and GAGs/proteoglycans (Bianchi *et al.*, 1984; Geets *et al.*, 1986, 1990; Martinez-Hernandez, 1984, 1985; Voss *et al.*, 1986). It is believed that the excessive and disorganized deposition of ECM components may be directly responsible for the deranged hepatic function characteristic of the cirrhotic state. Not only in the cirrhosis state but also in regeneration, changes of the ECM components in the space of Disse were observed (Yada *et al.*, 1996). These matrix components were partially prepared as a fiber complex by Koide *et al.* (1989) as reticulin fiber.

Developments in cell culture technology have made possible long-term survival and partial maintenance of differentiated functions of hepatocytes in primary cultures. In particular, hepatotropic growth factors in combination with other soluble factors, i.e., nutrient supplements, hormones, and trace elements, allow the maintenance of normal adult hepatocytes in serum-free medium. Adult rat hepatocytes in primary culture are known to form a monolayer, but rapidly lose differentiated functions on adherent substrates such as collagen and anchorage glycoproteins. Furthermore, the extracellular matrix and its components are used as culture substrata to improve the attachment and survival of the cultured hepatocytes; materials used as substrata include a biomatrix composed of fibers isolated from liver tissue, collagens, laminin, and fibronectin (Rojikind *et al.*, 1980). However, the full maintenance of the differentiated functions of hepatocytes *in vitro* is still difficult, and in all cases reported, albumin secretion decreased gradually when the cells were cultured for 4 or 5 days. Hepatocytes *in vivo* are surrounded by framework connective tissues in the space of Disse, namely reticulin fibers, to which the hepatocytes are probably anchored and which are thought to play a crucial role for hepatocyte to express many differentiated functions.

A. Spheroid Formation of Adult Rat Hepatocytes on Proteoglycan Substrata

We have isolated reticulin fibers from liver tissues by the procedure described by Rojikind *et al.* (1980), with a minor modification. We have also characterized matrix components from the reticulin fibers and examined their abilities to support the expression of liver-specific functions of hepatocytes in primary culture as substrata. Interestingly, when the proteoglycan fraction of reticulin fibers was used as the substrate, hepatocytes formed floating spherical aggregates (spheroid) (Shinji *et al.*, 1988). A spheroid 120–150 μm in diameter was composed of five to six cell layers with a surface layer of relatively thinner cells and an inner layer of cuboidal cells. Electron microscopy revealed that each hepatocyte, even at the second week of culture, had a large round nucleus and abundant cytoplasmic organelles as observed in hepatocytes *in vivo*, and in each, cells were in close contact with junctional complexes such as desmosomes. The junctional complexes and bile canaliculus-like structures, both of which were scarcely found in monolayer culture, were also identified between hepatocytes near the surface layer (Koide *et al.*, 1990). Cells in the spheroid showed only low growth activity. Albumin production by the spheroids increased up to 1.5 μg/μg DNA/day (180 μg/mg protein/day) during the first 6 days and remained constant thereafter (Koide *et al.*, 1989). In contrast, albumin production by the monolayer decreased markedly after 4 days (Koide *et al.*, 1989). The spheroid culture appears to be more suitable than the monolayer in studying differentiated functions of adult rat hepatocyte.

Components of the isolated proteoglycan fraction migrated as a double band at 200 and 108 kDa on SDS–polyacrylamide gel electrophoresis stained by Alcian blue and Coomassie blue, respectively (Fig. 1, lanes a and c). After digestion with chondroitinase ABC to remove most of the chondroitin sulfate/dermatan sulfate (CS/DS) chains, there was no staining with Alcian blue (Fig. 1, lane b) and there was single core protein band in Coomasie blue, which was of 45 kDa (Fig. 1, lane d). Antibodies raised against decorin were used for immunoblotting of the proteoglycans. The antibodies recognized the 108-kDa diffused band before chondroitinase digestion and the 45- to 47-kDa band after chondroitinase digestion, respectively. These results suggested that one of the proteoglycans was decorin (108 kDa). Another proteoglycan, the 200-kDa component, was thought to be biglycan based on the mobilities of both the intact form and the core molecular form on SDS–polyacrylamide gel electrophoresis. Their glycosaminoglycan

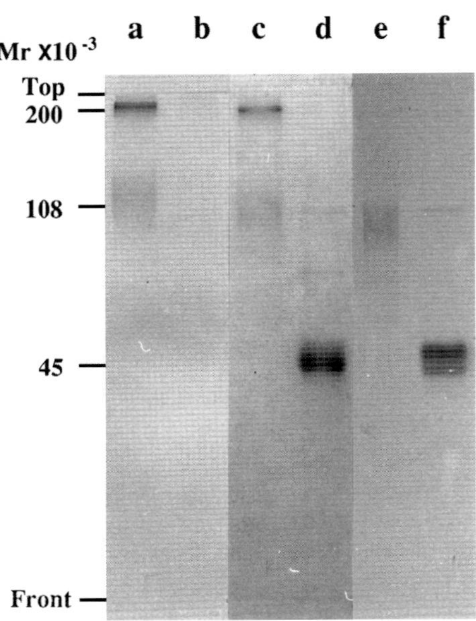

FIGURE 1 SDS–PAGE and Western blotting of human liver reticulin fiber proteoglycans before and after chondroitinase ABC digestion. Proteoglycans purified by ion-exchange chromatography from the extract of human reticulin fiber were subjected with (b, d, and f) or without (a, c, and e) chondroitinase ABC digestions and then separated by SDS–PAGE under reducing conditions. Gels were stained with Alcian blue, pH 1.0 (a and b), or C.B.B (c and d). Proteoglycans were transferred to a nitrocellulose membrane, and membranes were stained with anti-PG40 (anti-human decorin; e and f).

compositions of the proteoglycans were shown as a relatively high content of iduronic acid, and this result suggested that the proteoglycans bore a dermatan sulfate/chondroitin sulfate hybrid chain(s).

In order to determine which parts of the dermatan sulfate proteoglycan (DSPG) molecules are responsible for the effect, an experiment was conducted where the dermatan sulfate chains were selectively removed with chondroitinase ABC, and the resultant core proteins were used as substrata. This enzymatic removal of GAG moiety from proteoglycan fractions completely abolished the effects. Other purified chondroitin sulfate and/or dermatan sulfate proteoglycans, aggrecan, versican/PG-M, decorin (from rat skin), but not perlecan from EHS tumor, were also responsible for the spheroid formation, and enzymatic removal of these GAG moieties also abolished the effect as same as the liver DSPGs. As a result, we concluded that chondroitin sulfate/dermatan sulfate moieties of these proteoglycans were important for spheroid formation. However, GAG alone, chondroitin sulfate, dermatan sulfate, heparan sulfate, hyaluronan, chondroitin, and heparin were shown to have partial or little effect for spheroid formation as substrata, described previously (Shinji *et al.*, 1988). Together, the results indicate that neither dermatan sulfate-free core proteins nor core protein-free dermatan sulfate chains are capable of supporting spheroid formation as substrata and that the molecular form consisting of dermatan sulfate chains attached to a core protein is essential for its activity. It was thought to be important that dermatan sulfate chains were immobilized on the surface of culture dishes. *In vivo*, proteoglycans are fixed topologically onto other ECM molecules by the specific association of core proteins with them; the CS/DS chain positioned extracellularly might also provide a microenvironment suitable for maintaining hepatic functions.

B. Identification of the Principal Cellular Site of Liver CS/DS Proteoglycan Synthesis

In view of the possible functional importance of small proteoglycans, it seemed desirable to identify the type of liver cell responsible for their production.

The biosynthesis of GAGs by parenchymal and nonparenchymal liver cell has been studied extensively by metabolic labeling of GAG chains with [^{35}S]sulfate or [^{3}H]glucosamine (Gressner and Schafer, 1989). GAGs produced by freshly isolated hepatocytes are almost entirely intracellularly and pericellularly localized heparan sulfate (Meyer *et al.*, 1990), according for about 90% of the total. Only 10–20% of GAGs are secreted into the medium. There is no evidence for the synthesis of hyaluronan by parenchymal cells. Hepatocytes probably do not contribute to the synthesis of ECM proteoglycans/GAGs. Among the nonparenchymal, sinusoidal liver cell types, hepatic stellate cells (HSC), synonymously termed Ito cells, fat-storing cells, perisinusoidal lipocytes, and vitamin A-storing cells have been recognized as the main important cellular sites of proteoglycan production in liver (Bissell *et al.*, 1990; Ramadori, 1991). These cells, devoted in healthy liver primarily to the storage and metabolism of retinoids, are of stellate shape. They are localized in the space of Disse in deep recesses between adjacent hepatocytes; the cell processes surround the endothelial lining of the sinusoid. Numerous immunohistochemical studies have shown that these cells proliferate (Geerts *et al.*, 1991), transform via transitional cells into myofibroblast-like cells, and produce significant quantities of interstitial collagens (Friedman, 1990), structural glycoproteins, and proteoglycans (Gressner and Bachem, 1990) in the area of necroinflammation. In comparison with cultured HSC, hepatocytes and Kupffer cells produce below one-tenth of GAGs (Gressner and Schafer, 1989). The profiles of ^{35}S-labeled GAGs in HSC and Kupffer cells are significantly different from that of parenchymal liver cells. In the latter cell type, HS contributed about 90% and CS and DS comprised less than 10% of total labeled GAGs (Gressner and Schafer, 1989). In the medium of HSC cultures, DS and CS were found to be the major fractions, whereas only trace amounts of HS were measured. HSC secreted nearly 11 times more labeled CS and even 900 times more DS into the medium than the corresponding hepatocytes (Gressner and Schafer, 1989). Hyaluronan synthesis was measured almost exclusively in HSC (Gresner and Haarmann, 1988). These data suggest that HSC are the most active cell type in the synthesis of GAGs in ECM.

Cultured HSCs transform into myofibroblast-like cells spontaneously by long-term culture (Gressner, 1991). The synthesis of GAGs in these cells increased several-fold during transformation of HSCs into myofibroblast-like cells (Gressner, 1991).

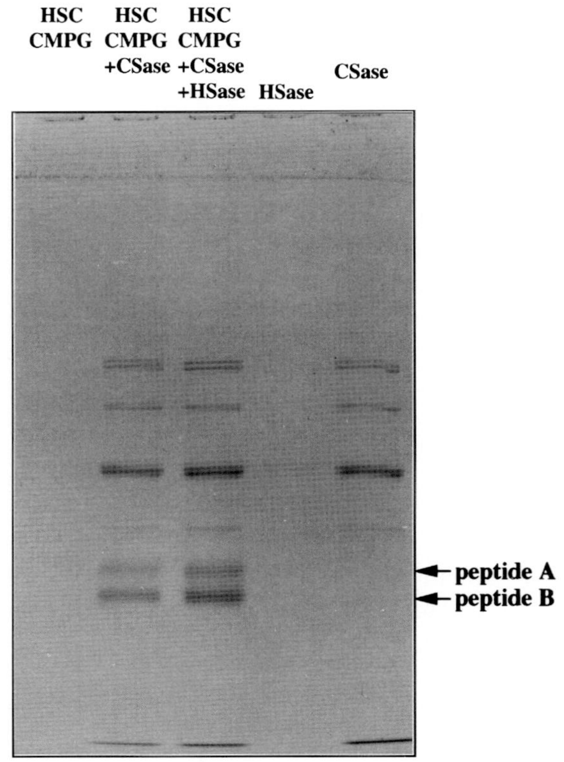

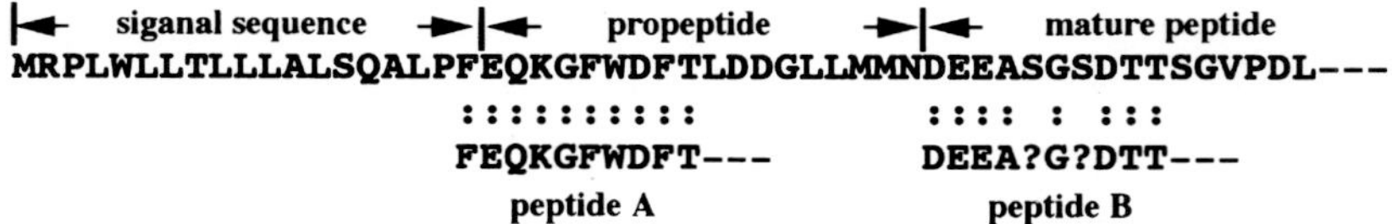

FIGURE 2 Identification of proteoglycan secreted from transformed HSC culture. Transformed HSC secreted predominantly mature and immature biglycans in the medium. HSC, hepatic stellate cell; CMPG, proteoglycan fraction from culture medium; CSase, chondroitinase ABC; HSase, heparitinase.

Thus, myofibroblast-like cells have been identified as the principal cellular source of GAGs in fibrotic liver, as deduced from their ability *in vitro* to synthesize and secrete large amounts of and a broad spectrum of sulfated GAGs (Arenson *et al.*, 1988; Gressner, 1991). Actually, the pattern of secreted GAGs by transformed HSC, mainly of the chondroitin sulfate and dermatan sulfate, resembles that established in the fibrotic liver extracellular matrix (Gressner and Bachem, 1990).

Their core proteins, however, have not yet been characterized. We isolated proteoglycan from the culture medium of transformed HSC by the combination of CsCl isopycnic ultracentrifugation and anion-exchange chromatography. Then the proteoglycan fraction digested with chondroitinase and/or heparitinase digestion was subjected to SDS–PAGE. The bands of proteoglycans core were subjected to amino acid sequencing (Fig. 2). Data show that the sequence from peptide A and peptide B matched the one of the propeptide (but with an additional phenylalanyl residue at N terminus) and the one of the mature peptide sequence from rat biglycan. The amino acid sequence from rat decorin was not detected by sequencing at all. These results indicate that predominant proteoglycans in the culture medium of transformed HSC were mature and immature forms of biglycans (Fig. 2). HSC clearly expressed both biglycan and decorin transcripts (Meyer *et al.*, 1992). The steady-state levels of their mRNAs increased threefold (biglycan) and fourfold (decorin) during primary culture (Meyer *et al.*, 1992). Myofibroblast-like cells (transformed HSC after the second passage) contained dramatically reduced levels of decorin mRNA and lower levels of biglycan mRNA compared with primary cultures (Meyer *et al.*, 1992). These results indicate that biglycan and decorin are expressed mainly by HSC *in vitro* and that transformed HSC secreted predominantly biglycan as proteoglycan.

In order to identify the subsinusoidal decorin and biglycan producing cell *in vivo*, we cloned rat cDNAs with these DSPGs and then used cDNAs for *in situ* hybridization. *In situ* hybridization revealed constitutive biglycan mRNA expression in normal liver tissue with the signals distributed over the nonparenchymal cells encompassing the periportal tracts (Fig. 3). Decorin gene expression was also

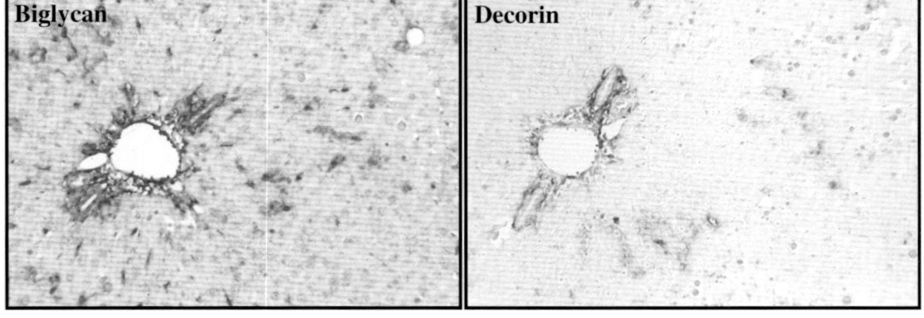

FIGURE 3 *In situ* hybridization of biglycan and decorin mRNA in normal rat liver.

detectable in normal liver tissue in a few cells of the nonparenchymal type scattered in the periportal tracts (Fig. 3). Immunocytochemical detection of desmin performed in parallel showed an identical scattered distribution pattern of the HSC as the main source for both of the proteoglycans. These results explained the *in vitro* results.

C. Changes of GAG Composition in Fibrotic Liver Tissue

Fibrotic diseases are characterized by a disproportionate increase and disordered deposition of the extracellular matrix, resulting in distortion and loss of organ function after tissue injury.

The amount of GAGs in normal liver is low (about 100 μg uronic acid/g defatted dry weight) (Kojima *et al.*, 1982; Murata *et al.*, 1985; Stuhlsatz *et al.*, 1982). In cirrhotic liver tissue the concentration is elevated frequently more than 5-fold (Murata *et al.*, 1985). Four main types of GAGs are known in normal liver, which are, in order of decreasing concentrations (fractions): heparan sulfate (HS), dermatan sulfate (DS), chondroitin sulfate (CS), and a small fraction of unsulfated chondroitin (Murata *et al.*, 1998; Stuhlsatz *et al.*, 1982). KS has not been identified in liver so far (Gressner *et al.*, 1977; Stuhlsatz *et al.*, 1982). In liver fibrosis, there is a preponderant deposition of DS and CS accompanied by a much smaller increase of HS (Murata *et al.*, 1985; Stuhlsatz *et al.*, 1982). Hyaluronan is present only in trace amounts in normal liver but increases 6- to 10-fold in the cirrhotic ECM (Stuhlsatz *et al.*, 1982). It is important to note that the major fraction of GAGs in normal liver, i.e., HS, decreases relatively in the cirrhotic organ due to only a 2-fold absolute increase. The most prominent relative increase concerns DS and CS (about 10-fold each), whereas unsulfated chondroitin increases only 2-fold in cirrhotic human liver. The expression of biglycan and decorin transcripts was detected in normal liver tissue. Not only GAG but also the relative expression of DSPG cores transcripts increased strongly during liver fibrogenesis by common bile duct ligation or thioacetoamide administration (Meyer *et al.*, 1992). These results suggest that pathophysiologically important proteoglycans in fibrotic liver are decorin and biglycan.

Cytokines have a crucial role in the pathophysiology of fibrosis independent of the system involved. Fibrotic diseases are likely the product of an overwhelming repair process after tissue injury. The "normal scenario" after injury shows an initial expression of proinflammatory cytokines, such as interleukin-1, interleukin-6, and tumor necrosis factor-α, which help the immune system eliminate the injurious agent. Shortly afterward, other cytokines and growth factors, among them TGF-β, connective tissue growth factor, and platelet-derived growth factor, are expressed to limit the inflammation in a subtle balance. During the development of fibrosis, however, the profibrotic cytokines are overexpressed and induce overwhelming repair and accumulation of ECM. One of the key profibrotic cytokines is TGF-β, which is chemotactic for fibroblasts, transforms HSCs to myofibroblast-like cells, induces the synthesis of matrix proteins and glycoproteins, and inhibits collagen degradation by the induction of protease inhibitors and the reduction of metalloproteases.

Decorin carries a single glycosaminoglycan chain and is widely distributed in mesenchymal tissues, associated and bound to collagen, which gains stability through this interaction (Danielson *et al.*, 1997). One of the key features of decorin knockout mice is fragile skin, probably due to irregularly shaped collagen (Danielson *et al.*, 1997). Biglycan has two glycosaminoglycan chains and is localized closely around cells. The precise physiological role of biglycan is still under discussion; various interactions with collagen and glycoproteins have been suggested.

Decorin, biglycan, and fibromodulin are able to interact with TGF-β, which is a profibrotic key mediator in tissue fibrosis and is a natural inhibitor for TGF-β signaling *in vitro* (Yamaguchi *et al.*, 1990; Hildebrand *et al.*, 1994). The affinity to TGF-β is similar for all. However, TGF-β has been shown to upregulate the expression of decorin and biglycan *in vitro*. This would suggest the presence of an autocrine–paracrine regulatory loop: the bound decorin might serve as a "reservoir" for a sustained release of this cytokine and could thus contribute to a local persistence of the fibrogenic process (Bachem *et al.*, 1992). For decorin, *in vitro* data have been confirmed in different disease models in animals, all of them TGF-β mediated. Decorin was successful in reducing experimental pulmonary fibrosis induced with bleomycin given either repeatedly as proteoglycan or once as gene using adenoviral gene transfer (Kolb *et al.*, 2001). In contrast, biglycan was effective *in vitro* to inhibit TGF-β but failed to reduce the fibrotic tissue response *in vivo* (Kolb *et al.*, 2001). Data suggest that the differences in tissue distribution are responsible for the different effects on TGF-β bioactivity *in vivo* and that decorin and biglycan might participate in these process differentially.

Taken together, these data point to the potentially significant role that proteoglycans might fulfill in the regulation of hepatocyte functions and in the maintenance of the supramolecular organization of the extracellular matrix in normal and diseased liver during the process of fibrogenesis.

III. HYALURONAN AS A DIAGNOSTIC MARKER FOR LIVER CIRRHOSIS

A. Hyaluronan

Hyaluronan (HA) is a widely distributed negatively charged polysaccharide, first isolated from the vitreous humor (Mayer and Palmer, 1934). It has been attributed with many biological functions, such as space filling and lubrication, as well as other more specific effects on cell functions. Its simple chemical structure of repeating disaccharide units of *N*-acetyl-D-glucosamine and D-glucronic acid linked by β(1-4) and β(1-3) glycosidic bonds, respectively, belies the range of unique viscoelastic and physiological properties of this polysaccharide. It is found in all vertebrates, but is yet to be detected in invertebrates and other lower animals. It is synthesized by most cells of the human body and is found throughout the tissue in a wide range of concentrations; there is an estimated 15 g HA per 70 kg human (Laurent and Fraser, 1991). HA is of vital importance, as there are no known viable individuals lacking the ability to synthesis this polysaccharide, and is demonstrated by the early embryonic lethality of hyaluronan synthase-2 knockout mice (Spicer and Nguyen, 1999).

HA of vertebrate organisms can exist in many stets, in a variety of sizes, in extracellular forms, free in the circulation, loosely associated with cells and tissues, tightly intercalated within proteoglycan-rich matrices, such as that of cartilage, bound by receptors to cell surfaces, or even in several intracellular locations.

HA is a major component of the extracellular matrix during embryogenesis. Its appearance and removal at specific locations are precisely controlled during this process (Toole *et al.*, 1984). HA also surrounds migrating and dividing cells during tissue regeneration and remodeling. The concentration of HA often decreases subsequent to these processes, when differentiation of the cells comes to the fore (Toole *et al.*, 1984). HA is an excellent hydrated matrix through which cells can migrate (Laurent and Fraser, 1992), and such hydrated pathways can facilitate cell passage through barriers of fibers and other cells (Toole, 1981, 1991). This and other actions are mediated by its interaction with other ECM components, receptors, and intracellular proteins (Cheung *et al.*, 1999; Knudson *et al.*, 1999). While there are numerous papers describing these interactions of HA with proteins in the extracellular space, very little is known about those proteins that deal with this ECM component when it is time for its disposal.

B. Catabolism of Hyaluronan

The clearance of HA from mammals is brought about almost entirely by its catabolism rather than excretion at three main levels (Fraser and Laurent, 1998): (1) degradation locally in the tissue where it is synthesized (10–30%); (2) degradation in the lymph nodes after displacement from the tissues (50–90%); and (3) clearance from the blood of that which is not metabolized by the lymph node, liver, kidneys, and spleen, in that order of importance. In humans, one-third of HA (5 g) is turned over daily. Its catabolism, under normal conditions, is a multistep intracellular process. The first step in the process occurs in lysosomes with the action of hyaluronidase cleaving the HA to oligosaccharides, which are then cleaved sequentially by β-D-glucronidase and β-*N*-acetyl-D-hexosaminidase, resulting in the monosaccharides D-glucronic acid and *N*-acetyl-D-glucosamine (Roden *et al.*, 1989). These monosaccharides are then released into the cytoplasm for further processing to CO_2, ammonia, acetate, and lactate (Roden *et al.*, 1989).

C. Pathophysiological Mechanisms for Increased Serum Hyaluronan Levels

Circulating connective tissue macromolecules are eliminated efficiently by receptor-mediated endocytosis in liver endothelial cells (LEC). In patients with a seriously diseased liver, e.g., in liver cirrhosis or when a transplanted liver is about to be rejected, dysfunctional LEC or altered blood perfusion through the liver results in significantly increased serum levels of connective tissue macromolecules and other substances that are normally taken up by LEC. With the aid of high-affinity antibodies or other kinds of binding proteins, it has been possible to measure a number of connective tissue macromolecules in the blood, and some of these substances have been used to diagnose serious liver disease. However, the metabolic patterns of these molecules are such that it is difficult to extrapolate directly on the basis of increased serum levels to conclude that the main underlying cause of increased serum levels is dysfunctional LEC. Elevated serum levels of many connective tissue proteins may reflect increased synthesis, increased release from cell or tissue depots, and/or decreased removal by LEC. Therefore, to use the measurement of serum connective tissue molecule levels as indicators of LEC function, it is imperative to know the catabolic routes of these substances in health and disease. Due to the fact that we presently know more about the total metabolic pattern of HA than of other connective tissue macromolecules that are cleared solely by LEC, serum HA should be used preferentially to monitor the function of LEC, particularly in liver transplantation and cirrhosis.

Simple assays for serum hyaluronan determination make it possible to use it as a diagnostic marker for diseases. Because hyaluronan is a component of liver fibrosis and is cleared from the blood by LECs, most investigations have been made in liver diseases and increased serum level were found in these patients. Similar increases in serum hyaluronan levels found in chronic liver diseases of different etiologies and pathophysiologies, such as primary biliary cirrhosis and chronic viral hepatitis, suggest a common mechanism for these increases. Although the pathophysiological mechanisms involved are not fully understood, both an increase in hepatic production and a decrease in hepatic removal might be involved. Although the role of intrahepatic hyaluronan production in the elevated serum levels is unclear, there are some arguments in favor of this mechanism. Hyaluronan is present in basement membranes and collagen fibers, which are known to develop in chronic liver diseases, and its liver concentrations increase markedly in the fibrotic liver (Gressner and Bachem, 1990). Actually,

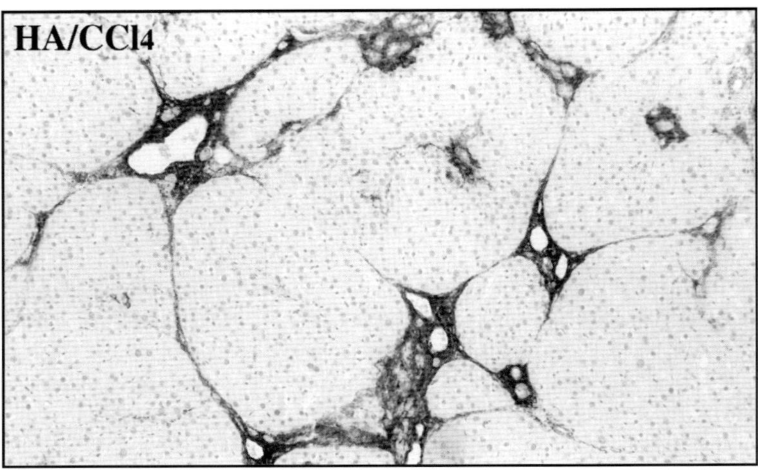

FIGURE 4 Hyaluronan accumulation in CCl_4-induced cirrhotic rat liver. Strong staining by biotinylated hyaluronan-binding protein was observed in increased fibrotic connective tissues by cirrhosis.

large amount of hyaluronan accumulation was observed in connective tissues in CCl_4-induced cirrhotic liver (Fig. 4). The ability of HSC to synthesize and secrete hyaluronan has been documented, and Kupffer cells may be involved in stimulating hyaluronan production through the secretion of various mediators (Bachem *et al.*, 1989; Gressner and Haarmann, 1988). Increases in hyaluronan production may be caused by the induction of HSC proliferation and the synthesis of extracellular matrix components by inflammation via cytokines, TGF-β, and so on. Among the three hyaluronan synthases, HAS1 and HAS2 were predominant in transformed rat HSC. These two synthases in transformed HSC may contribute to the increases of hyaluronan production in fibrotic liver.

Nevertheless, as blood hyaluronan is cleared and degraded by endothelial cells, serum levels might be related to pathological mechanisms, which affect sinusoidal cell function and impair cellular uptake of circulating hyaluronan. This mechanism is supported by the observation that *in vitro* endothelial cells lose the ability to metabolize hyaluronan (Erikksson *et al.*, 1983). The significant correlation between serum hyaluronan and histological stage could correspond to morphological changes in sinusoidal endothelial cells in patients with late-stage disease. Such changes in endothelial cell structure were found in primary biliary cirrhosis (Babbs *et al.*, 1990). Hyaluronan receptors may be lost when the sinusoidal endothelial cell fenestrations disappear during the development of a true basement membrane in chronic liver diseases. Results suggest that the measurement of serum hyaluronan levels is useful for evaluating the morphological changes in sinusoidal endothelial cells that accompany hepatic sinusoidal capillarization (Ueno *et al.*, 1993). Zhou *et al.* (2000) reported that the two main hyaluronan clearance systems in lymphatic tissues and liver were mediated by the same HA receptor, hyaluronan receptor for endocytosis (HARE), which was discovered in LECs. Decreases in serum hyaluronan uptake may be caused by a dysfunction of HARE or loss of HARE expression in injured LECs.

REFERENCES

Arenson, D. M., Friedman, S. L., and Bissell, D. M. (1988). Formation of extracellular matrix in normal rat liver: lipocytes as a major source of proteoglycan. *Gastroenterology* **95**, 441–447.

Ashikari, S., Habuchi, H., and Kimata, K. (1995). Characterization of heparan sulfate oligosaccharides that bind to hepatocyte growth factor. *J. Biol. Chem.* **270**, 29586–29593.

Babbs, C., Haboubi, N. Y., Mellor, J. M., Smith, A., Rowan, B. P., and Warnes, Y. W. (1990). Endothelial cell transformation in primary biliary cirrhosis: A morphological and biochemical study. *Hepatology* **11**, 723–729.

Bachem, M. G., Meyer, D. H., Melchior, R., Sell, K. M., and Gressner, A. M. (1992). Activation of rat liver fat-storing cells by myofibroblast-like cell-derived transforming growth factors: A potential mechanism of self perpertuation in liver fibrogenesis. *J. Clin. Invest.* **89**, 12–27.

Bachem, M. G., Riess, U., Melchior, U., Sell, K. M., and Gressner, A. M. (1989). Transforming growth factors (TGFα and TGFβ1) stimulate chondroitin sulfate and hyaluronate synthesis in cultured rat liver fat storing cells. *FEBS Lett.* **257**, 134–137.

Bianchi, F. B., Biangini, G., Ballardini, G., Cenacchi, G., Faccani, A., Pisi, E., Laschi, R., Liotta, L., and Garbisa, S. (1984). Basement membrane production by hepatocytes in chronic liver disease. *Hepatology* **4**, 1167–1172.

Bissell, D. M., Friedman, S. L., Maher, J. J., and Roll, F. J. (1990). Connective tissue biology and hepatic fibrosis: Report of a conference. *Hepatology* **11**, 488–498.

Cheung, W. F., Cruz, T. F., and Turley, E. A. (1999). Receptor for hyaluronan-mediated motility (RHAMM), a hyaladherin that regulates cell responses to growth factors. *Biochem. Soc. Trans.* **27**, 135–142.

Danielson, K., Baribault, H., Holmes, D., Graham, H., Kadler, K., and Iozzo, R. (1997). Targeted disruption of decorin leads to abnormal collagen fibril morphology and skin fragility. *J. Cell Biol.* **136**, 729–743.

Eriksson, S., Fraser, J. R. E., Laurent, T. C., Pertorf, H., and Smedsrod, B. (1983). Endothelial cells are the site of uptale and degradation of hyaluronic acid in the liver. *Exp. Cell Res.* **144**, 223–228.

Fraser, J. R. E., and Laurent, U. B. G. (1998). The turnover of hyaluronan. *In* "Connective Tissue Biology: Integration and Reductionism" (R. K. Reed and K. Rubin, eds.), pp. 49–69. Portland Press, London.

Friedman, A. S. L. (1990). Cellular sources of collagen and regulation of collagen production in liver. *Semin. Liver Dis.* **10**, 20–29.

Geerts, A., Geuze, H.J., Slot, J.-W., Voss, B., Schuppan, D., Schellinck, P., and Wisse, E. (1986). Immunogold localization of procollagen III, fibronectin and heparan sulfate proteoglycan on ultrathin frozen sections of the normal rat liver. *Histochemistry* **84**, 355–362.

Geerts, A., Lazou, J. M., De Bleser, P., and Wisse, E. (1991). Tissue distribution, quantitation and proliferation kinetics of fat-storing cells in carbon tetrachloride-injured rat liver. *Hepatology* **13**, 1193–1202.

Geerts, A., Schuppan, D., Lazeroms, S., Zanger, R. D., and Wisse, E. (1990). Collagen type I and III occur together in hybrid fibrils in the space of Disse of normal rat liver. *Hepatology* **12**, 233–241.

Gressner, A. M. (1991). Time-related distribution profiles of sulfated glycosaminoglycans in cells, cell surfaces and media of cultured rat liver fat-storing cells. *Proc. Soc. Exp. Biol. Med.* **196**, 307–315.

Gressner, A. M., and Bachem, M. G. (1990). Cellular sources of noncollagenous matrix proteins: role of fat-storing cells in fibrogenesis. *Semin. Liv. Dis.* **11**, 30–46.

Gressner, A. M., and Haarmann, R. (1998). Hyaluronic acid synthesis and secretion by rat liver fat storing cells in culture. *Biochem. Biophys. Res. Commun.* **151**, 222–229.

Gressner, A. M., and Schafer, S. (1989). Comparison of sulfated glycosaminoglycan and hyaluronate synthesis and secretion in cultured hepatocyte, fat storing cells, and Kupffer cells. *J. Chem. Clin. Biochem.* **27**, 141–149.

Gressner, A. M., Pazen, H., and Greiling, H. (1977). The biosynthesis of glycosaminoglycans in normal rat liver and in response to experimental hepatic injury. *Hoppe Seylers Z. Physiol. Chem.* **358**, 825–833.

Guimond, S., Maccarana, M., Olwin, G. G., Lindahl, U., and Rapraeger, A. D. (1993). Activating and inhibitory heparin sequences for FGF-2 (basic FGF): Distinct requirements for FGF-1, FGF-2, and FGF-4. *J. Biol. Chem.* **268**, 23906–23914.

Habuchi, H., Suzuki, S., Saito, T., Tamura, T., Harada, T., Yoshida, K., and Kimata, K. (1992). Structure of a heparan sulfate oligosaccharide that binds to basic fibroblast growth factor. *Biochem. J.* **285**, 805–813.

Hildebrand, A., Romaris, M., Rasmussen, L. M., Heinegard, D., Twardzik, D. R., Border, W. A., and Ruoslahti, E. (1994). Interaction of the small interstitial proteoglycans biglycan, decorin and fibromodulin with transforming growth factor beta. *Biochem. J.* **302**, 527–534.

Kanwar, Y., Linker, A., and Farquhar, M. (1980). Increased permeability of glomerular basement membrane to ferritin after removal of glycosaminoglycans (heparan sulfate) by enzyme digestion. *J. Cell Biol.* **86**, 688–693.

Kjellen, L., and Lindahl, U. (1991). Proteoglycans-structures and interactions. *Annu. Rev. Biochem.* **60**, 443–476.

Kolb, M., Margetts, P. J., Sime, P. J., and Gauldie, J. (2001). Proteoglycans decorin and biglycan differentially modulate TGF-beta-mediated fibrotic responses in the lung. *Am. J. Physiol. Lung Cell Mol. Physiol.* **280**, L1327–1334.

Kunudson, C. B., Nofal, G. A., Pamintuan, L., and Aguiar, D. J. (1999). The chondrocyte pericellular matrix: A model for hyaluronan-mediated cell-matrix interactions. *Biochem. Soc. Trans.* **27**, 142–147.

Koide, N., Sakaguchi, K., Koide, Y., Asano, K., Kawaguchi, M., Matsushima, H., Takenami, T., Shinji, T., Mori, M., and Tsuji, T. (1990). Formation of multicellular spheroids composed of adult rat hepatocytes in dishes with positively charged surfaces and under other nonadherent environments. *Exp. Cell Res.* **186**, 227–235.

Koide, N., Shinji, T., Tanabe, T., Asano, K., Kawaguchi, M., Sakaguchi, K., Koide, Y., Mori, M., and Tsuji, T. (1989). Continued high albumin production by multicellular spheroids of adult rat hepatocyte formed in the presence of liver-derived proteoglycans. *Biochem. Biophys. Res. Commun.* **161**, 385–391.

Kojima, J., Nakamura, N., Kanatani, M., and Akiyama, M. (1982). Glycosaminoglycans in 3′-methyl-4-dimethylaminoazobenzen-induced rat hepatic cancer. *Cancer Res.* **42**, 2857–2860.

Lam, L. H., Silvert, J. E., and Rosenberg, R. D. (1976). The separation of active and inactive forms of heparin. *Biochem. Biophys. Res. Commun.* **69**, 570–577.

Lander, A. D. (1994). Targeting the glycosaminoglycan-binding sites on proteins. *Chem. Biol.* **1**, 73-78.

Laurent, T. C., and Fraser, J. R. E. (1992). Hyaluronan. *FASEB J.* **6**, 2397–2404.

Lindahl, U., Thunberg, L., Backsrom, G., Risenfeld, J., Nordling, K., and Bjork, I. (1984). Extension and structural variability of the antithrombin-binding sequence in heparin. *J. Cell Biol.* **259**, 12368–12376.

Lopea-Casillas, F., Wrana, J. L., and Massague, J. (1993). Betaglycan presents ligand to the TGF beta signaling receptor. *Cell* **73**, 1435–1444.

Lortat-Jacob, H., Turnbull, J., and Grimaud, J. (1995). Molecular organization of the interferon gamma-binding domain in heparan sulphate. *Biochem. J.* **310**, 497–505.

Martinez-Hernandez, A. (1984). The hepatic extracellular matrix. I. Electron immunohistochemical studies in normal rat liver. *Lab. Invest.* **51**, 57–74.

Martinez-Hernandez, A. (1985). The hepatic extracellular matrix. II. Electron immunohistochemical studies in rats with CCl_4-induced cirrhosis. *Lab. Invest.* **53**, 166–186.

Meyer, D. H., Krull, N., Dreher, K. L., and Gressner, A. M. (1992). Biglycan and decorin gene expression in normal and fibrotic rat liver: Cellular localization and regulatory factors. *Hepatology* **16**, 204–216.

Meyer, D. H., Zimmermann, T., Muller, D., Franke, H., Dargel, R., and Gressner, A. M. (1990). The synthesis of glycosaminoglycans in isolated hepatocytes during experimental liver fibrogenesis. *Liver* **10**, 94–105.

Meyer, K., and Palmer, J. W. (1934). The polysaccharide of the vitreous humor. *J. Biol. Chem.* **107**, 629–634.

Murata, K., Ochiai, Y., and Akashio, K. (1985). Polydispersity of acidic glycosaminoglycan components in human liver and the changes at different stages in liver cirrhosis. *Gastroenterology* **89**, 1248–1257.

Ramadori, G. (1991). The stellate cell (Ito-cell, fat-storing cell, lipocyte, perisinusoidal cell) of the liver. *Virch. Arch. B. Cell Pathol.* **61**, 147–158.

Roden, L., Campbell, P., Fraser, J. R., Laurent, T. C., Pertoft, H., and Thompson, J. N. (1989). Enzymeic pathways of hyaluronan catabolism. *Ciba Found. Symp.* **143**, 60–76.

Rojikind, M., Gatmaintan, Z., Mackensen, S., Ponce, P., and Reid, L. M. (1980). Connective tissue biomatrix: Its isolation and utilization for long-term cultures of normal rat hepatocytes. *J. Cell Biol.* **87**, 255–263.

San Antonio, J. D., Slover, J., Lawler, J., Karnovsky, M. J., and Lander, A. D. (1993). Specificity in the interactions of extracellular matrix proteins with subpopulations of the glycosaminoglycan heparin. *Biochemistry* **32**, 4746–4755.

Shinji, T., Koide, N., and Tsuji, T. (1988). Glycosaminoglycans partially substitute for proteoglycans in spheroid formation of adult rat hepatocytes in primary culture. *Cell Struct. Funct.* **13**, 179–188.

Spicer, A. P., and Nguyen, T. K. (1999). Mammalian hyaluronan synthases: Investigation of functional relationships in vivo. *Biochem. Soc. Trans.* **27**, 109–115.

Stuhlsatz, H. W., Viethaus, S., Gressner, A. M., and Greiling, H. (1982). In structural carbohydrates in the liver. *Falk Symp.* **34**, 135–136.

Toole, B. P. (1981). Glycosaminoglycans in morphogenesis. *In* "Cell Biology of Extracellular Matrix" (E. D. Hay, ed.), pp. 259–294. Plenum Press, New York.

Toole, B. P. (1991). Proteoglycans and hyaluronan in morphogenesis and differentiation. morphogenesis. *In* "Cell Biology of Extracellular Matrix" (E. D. Hay, ed.), pp. 305–341. Plenum Press, New York.

Toole, B. P., Goldberg, R. L., Chi-Rosso, G., Underhill, C. B., and Orkin, R. W. (1984). Hyaluronate-cell interactions. *In* "The Role of Extracellular Matrix in Development" (R. C. Trelstad, ed.), pp. 43–66. A. R. Liss, Inc., New York.

Turnbull, J. E., Fernig, D. G., Ke, Y., Wilkinson, M. C., and Gallagher, J. T. (1992). Identification of the basic fibroblast growth factor binding sequence in fibroblast heparan sulfate. *J. Biol. Chem.* **267**, 10337–10341.

Ueno, T., Inuzuka, T., Torimura, T., *et al.*, (1993). Serum hyaluronate reflects hepatic sinusoidal cappilarization. *Gastroenterology* **105**, 475–481.

Voss, B., Glossel, J., Cully, Z., and Kresse, H. (1986). Immunocytochemical investigation on the distribution of small chondroitin sulfate-dermatan sulfate proteoglycan in the human. *J. Histochem. Cytochem.* **34**, 1013–1019.

Yada, T., Koide, N., and Kimata, K. (1996). Transient accumulation of perisinusoidal chondroitin sulfate proteoglycans during liver regeneration and development. *J. Histochem. Cytochem.* **44**, 969–980.

Yamaguchi, Y., Mann, D. M., and Ruoslahti, E. (1990). Negative regulation of transforming growth factor-beta by the proteoglycan decorin. *Nature* **346**, 281–284.

Zhou, B., Weigel, J. A., Fauss, L., and Weigel, P. H. (2000). Identification of the hyaluronan receptor for endocytosis (HARE). *J. Biol. Chem.* **275**, 37733–37741.

CHAPTER 5

SPARC, a Matricellular Protein That Regulates Cell–Matrix Interaction: Implications for Vascular and Connective Tissue Biology

E. HELENE SAGE, AMY D. BRADSHAW, AND ROLF BREKKEN
Department of Vascular Biology, The Hope Heart Institute, Seattle, Washington 98104

SPARC (secreted protein acidic and rich in cysteine) belongs to the matricellular class of secreted glycoproteins. Although structurally dissimilar, these proteins regulate interactions between cells and their extracellular matrix and feature prominently in morphogenesis, development, injury, and repair. SPARC has been shown to (i) inhibit the cell cycle; (ii) disrupt cell adhesion; (iii) inactivate cellular responses to certain growth factors; (iv) regulate extracellular matrix and matrix metalloproteinase production; (v) bind to specific collagens, and (vi) promote a rounded cell shape through dissolution of focal adhesions and reorganization of the actin cytoskeleton. SPARC-null mice are viable but exhibit phenotypic abnormalities associated with the eye, connective and adipose tissues, bone, and wound healing. Moreover, cells cultured from SPARC-null *vs* wild-type tissues showed significantly accelerated cell cycles, diminished production of collagen and transforming growth factor β-1, enhanced adhesion, and/or altered levels of cadherins and matricellular proteins. The expression of SPARC in remodeling tissues, as a consequence of normal development or response to injury, coupled with its multiple effects on cells of the vessel wall, has been consistent with our proposal that SPARC subserves a fundamental role in vascular morphogenesis and cellular differentiation.

Extracellular Matrix and the Liver
Copyright 2003, Elsevier Science (USA). All rights reserved.

I. INTRODUCTION

One of the unsolved problems in biology concerns the creation of three-dimensional form from cells and their associated extracellular matrix. An important element of tissue or organ morphogenesis is the selective effects of secreted proteins and other macromolecules on cellular behavior and differentiation. In addition to the changes that occur during embryogenesis, the restructuring and remodeling of tissues as a result of injury or disease rely on several classes of regulatory proteins/proteoglycans that constitute the extracellular environment or that function as cell surface-associated molecules. This chapter focuses on a prototypic member of one of these classes, the matricellular protein SPARC (secreted protein acidic and rich in cysteine), and its role in cell and tissue biology.

II. MATRICELLULAR PROTEINS

SPARC, together with tenascin C, thrombospondins 1 and 2, osteopontin, and several other proteins, as well as certain proteoglycans (Sage, 2001), belongs to a group of secreted macromolecules that interact with cell surface receptors, growth factors, proteinases, and/or extracellular matrix components, but in most cases do not function as extracellular scaffolding (Bornstein, 1995). These proteins, which are structurally unrelated, are characteristic of remodeling tissues and appear collectively to modulate interactions between cells and the extracellular matrix or milieu (Bornstein, 2001; Murphy-Ullrich, 2001; Sage, 2001; Sage and Bornstein, 1991). Disruption of the matricellular genes mentioned earlier has resulted in phenotypes that initially appeared subtle. Closer examination, however, revealed characteristics that were not only surprising, but in many cases were dependent on injury or challenge (Brekken *et al.*, 2000; Crawford *et al.*, 1998; Denhardt *et al.*, 2001; Kyriakides *et al.*, 1998; Yan and Sage, 1999).

As a prototypic matricellular protein, SPARC (also known as osteonectin and BM-40) is prominent during embryonic morphogenesis, as well as in tissues undergoing normal turnover or repair (Bradshaw and Sage, 2001; Ringuette *et al.*, 1992). It has been implicated in the regulation of tumor progression and might be diagnostic of invasive meningiomas (Ledda *et al.*, 1997; Rempel and Gutierrez, 1999). Various functions described for SPARC, as a result of experiments conducted on different types of cells *in vitro*, are listed in Table I. The perceived complexity of activities attributed to SPARC can be summarized as two rather broad functions: (i) inhibition of cell adhesion and (ii) inhibition of cell proliferation. The former is achieved in part *via* a tyrosine phosphorylation–dependent pathway involving focal adhesion proteins and *via* the regulation of extracellular matrix by SPARC (Brekken and Sage, 2000; Murphy-Ullrich *et al.*, 1995). In contrast, inhibition of proliferation is achieved through specific mechanisms in different types of cells, e.g., by diminution of the levels of cyclin A protein, phosphorylation of

TABLE I Functions Attributed to SPARC *in Vitro*

Inhibits cell spreading
Diminishes focal contacts
Inhibits cell cycle in late G1
Regulates activity of PDGF, VEGF, and FGF-2
Releases bioactive peptides through proteolysis
Undergoes nuclear translocation
Regulates production of extracellular matrix and TGF-β

retinoblastoma protein, and/or the activity of cyclin E-associated cyclin-dependent kinase (cdk)-2; by inhibition of phosphorylation of the mitogen-activated protein kinases Erk 1 and 2; by the inhibition of phosphorylation of fibroblast growth factor (FGF)-receptor 1 and vascular endothelial growth factor (VEGF)-receptor 1; and by a direct binding interaction between SPARC and platelet-derived growth factor (PDGF)-B chain or VEGF (Kupprion *et al.*, 1998; Motamed *et al.*, 2002; Raines *et al.*, 1992). Thus, there appear to be several potential pathways by which SPARC interacts with cells. To date, a specific cell surface/transmembrane receptor for SPARC has not been found. It is known, however, that SPARC binds to cell surface-associated proteins, including vitronectin, fibronectin, types I and IV collagen, cytokeratin 18, and serum albumin (Brekken and Sage, 2000). It is also a possibility that SPARC acts as an antagonist of integrin–extracellular matrix interactions or of interactions between certain of the intercellular adhesion proteins termed cadherins.

A diagram of SPARC with its three identified modules is shown in Fig. 1. The structures of the follistatin-like and EC modules have been solved by X-ray crystallography (Hohenester *et al.*, 1997), and the functions of certain subdomains within each module have been demonstrated by the use of synthetic peptides. It is important to note that SPARC presents two “faces”—one of which interacts with the cell surface and the other with the extracellular matrix. Thus, the proposed function of a matricellular protein as a mediator of cell–matrix interactions could be achieved.

SPARC belongs to a gene family that includes SC-1/hevin, QR1, testican, tsc36, and SPARC-related gene (Brekken and Sage, 2000; Yan and Sage, 1999). The translation products each have the three signature modules, with SC-1 exhibiting the highest similarity to SPARC (61% in the EC domain). Using degenerate primers representing the most highly conserved regions in SPARC, SC1, and QR1, we identified a 300-bp SC1 cDNA in a primary polymerase chain reaction screen of a mouse brain library. This cDNA was used to obtain a full-length clone, which hybridized to a 3.2-kb mRNA expressed at moderate levels (relative to brain) in mouse heart, adrenal gland, and lung and at lower levels in other organs (Soderling *et al.*, 1997). In contrast to SPARC, which is abundant in cultured cells,

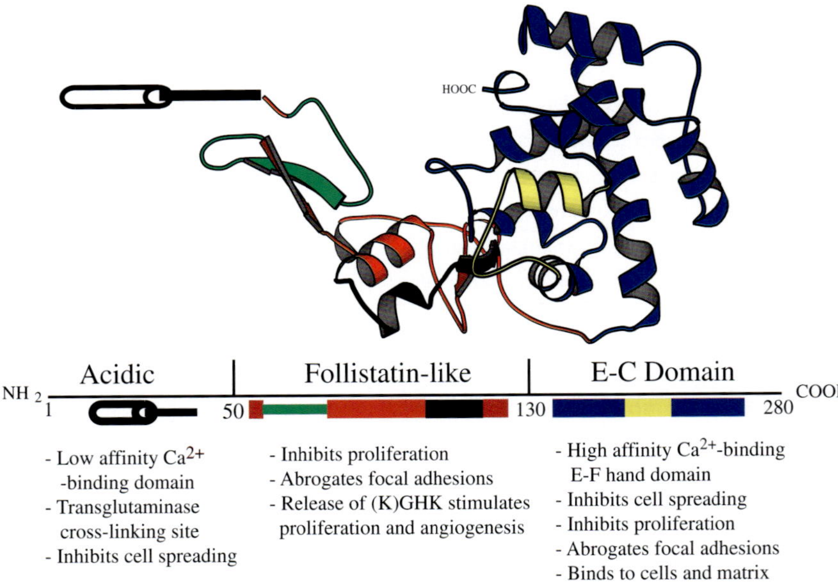

FIGURE 1 Structure of SPARC protein: A ribbon diagram derived from crystallographic data shows the three modules of SPARC. Activities that have been identified for different domains are shown beneath the designated amino acids. The follistatin-like domain contains a growth-inhibitory peptide (green) and an angiogenic peptide (KGHK, amino acids 114–130) (black). The E-C domain contains peptide 4.2 (amino acids 255–274) (yellow). From Brekken and Sage (2000), with permission.

SC1 mRNA was not evident in endothelial cells, smooth muscle cells, or fibroblasts cultured from rat, bovine, mouse, and human tissues. We noted a consistent distribution of SC1 mRNA in vessels of all organs that we examined, although some capillaries and large vessels were negative. Although the expression patterns of SC1 and SPARC are not coincident, there is appreciable overlap (Soderling *et al.*, 1997; R. Brekken *et al.*, unpublished experiments). Secreted SC1 might therefore be available to cells that are unable to produce SPARC, e.g., SPARC-null mice. Interestingly, monoclonal antibodies against mouse recombinant SC1 that react with angiogenic vessels have been produced in our laboratory (Fig. 2; R. Brekken *et al.*, unpublished experiments). Because SC1 shares both structural and functional properties with SPARC, and there is some overlap with respect to tissue-specific expression, it is not unreasonable to propose that SC1 could compensate functionally for SPARC, at least in certain tissues. This possibility is enhanced by the observation that bioactive peptides can be produced by the proteolytic cleavage of SPARC (Sage, 1997) and, presumably, of homologous regions in SC1.

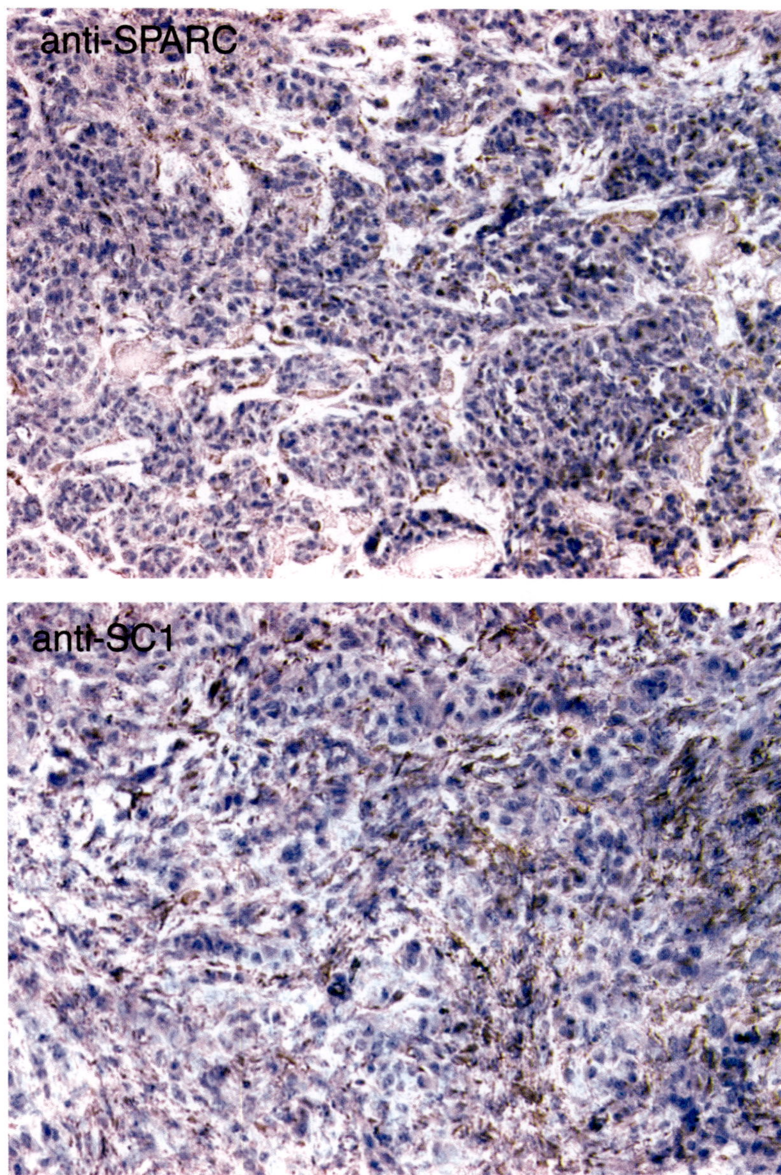

FIGURE 2 Immunohistochemical reactivity of murine monoclonal anti-SPARC and anti-SC1 antibodies on a NCI-H358 human nonsmall cell lung carcinoma xenograft. Eight micrometer frozen sections of NCI-H358 human tumor cells grown in SCID mice were stained according to the indirect immunoperoxidase technique and were counterstained with hematoxylin. Both blood vessels and tumor stroma, but not the tumor cells themselves, display immunoreactivity with anti-SPARC (top) and anti-SC1 antibodies (bottom).

III. SPARC-NULL MICE

Dr. C. Howe performed targeted disruption of the SPARC gene by homologous recombination in 129/SVJ ES cells, and Dr. Evans's laboratory performed a similar experiment with a different targeting vector (reviewed in Yan and Sage, 1999). Cataractogenesis with 100% penetration was seen in two different strains of mice maintained in separate facilities (also in a 129/SVJ×C57B1/6J strain). Crosses between heterozygotes produced a Mendelian distribution of wild-type, heterozygous, and homozygous mutants; neither SPARC mRNA nor protein was detected in the SPARC-null animals. A striking finding, consistent with our previous observations that SPARC changes cell shape, was that the epithelial cells and secondary fibers of lenses in SPARC-null mice showed grossly altered shapes, relative to those of wild-type littermates. Other characteristics of SPARC null mice include early-onset osteopenia, accelerated dermal wound healing, enhanced invasion and angiogenesis in subcutaneous sponge implants, and alterations in adipose and connective tissues (Table II). Ablation of SPARC in both *Caenorhabditis elegans* and *Xenopus laevis* has also produced morphological abnormalities that are

TABLE II Phenotypes Associated with the Inactivation of SPARC

Organism	Technique	Phenotype
Caenorhabditis elegans[a]	cRNA interference	Embryonic lethality
		Lack of gut granules
		Abnormal gonads
Xenopus laevis[b]	Antibody block	Bent, shortened embryonic axes
		Abnormal eye development
Mus musulus	Gene-targeted deletion	Cataracts[c]
		Accelerated dermal wound healing[d]
		Abnormal dermal collagen fibrils[e]
		Kinked tails[e]
		Osteopenia[f]
		Increased deposition of fat[g]

[a]From Fitzgerald and Schwarzbauer (1998).
[b]From Purcell et al. (1993).
[c]From Yan and Sage, (1999).
[d]From Bradshaw *et al.* (2002a).
[e]From Bradshaw *et al.* (2002b).
[f]From Delaney *et al.* (2000).
[g]From Bradshaw and Sage (2001).

consistent with roles for SPARC in the regulation of cell proliferation, migration, and differentiation (Table II).

Cells cultured from SPARC-null mice exhibit altered morphologies. For example, mesangial cells (a smooth muscle-like cell in the renal glomerulus) show differences in the configuration of the actin cytoskeleton, are more difficult to detach by trypsin treatment, and have more focal contacts relative to wild-type cells (Bradshaw and Sage, 2001). This observation is consistent with the *counteradhesive* properties of SPARC on cultured cells and argues for consistency between experiments performed on cells with exogenous SPARC *vs* those with a genetic deletion of SPARC. Earlier studies on cultured endothelial cells showed that exogenous SPARC decreased their production of fibronectin and thrombospondin 1 and enhanced that of plasminogen activator inhibitor-1 (reviewed in Brekken and Sage, 2000). Francki *et al.* (1999) showed that SPARC-null mesangial cells produced significantly diminished levels of type I collagen and TGF-β1 mRNA and protein relative to wild-type cells. A positive autocrine loop, in which TGF-β1 modulates the levels of type I collagen mRNA, was proposed in this study. This mechanism could account for the diminished production of the extracellular matrix seen in SPARC-null mice with bleomycin-induced pulmonary fibrosis (Strandjord *et al.*, 1999) and would be consistent with certain of the alterations in connective tissue characteristic of mice that develop in the absence of SPARC (see later).

As described in Table II, mice lacking SPARC exhibit abnormal connective tissue. A striking phenotype is evident in skin. Although SPARC-null animals do not appear to have overt blistering disorders, the skin of the animals is more fragile and easily torn. These properties of SPARC-null skin could reflect a deficiency in dermal–epidermal anchoring or collagen fibril stability. Sections of skin show an increased disorganization of the collagen fibrils in SPARC-null relative to wild-type tissue. Interestingly, SPARC-null animals have curly tails, a characteristic that might additionally reflect defective collagen fibers. Thrombospondin 2-null mice also display abnormal collagen fibrils and hyperflexible tails that can be knotted (Kyriakides *et al.*, 1998).

Excisional wounds made in the dorsa of SPARC-null and wild-type mice were measured over time to determine the rates of healing. A significant decrease in the size of the SPARC-null wounds was observed at day 4 and was maximal at day 7 (Bradshaw *et al.*, 2002a). No substantial differences in the amount of granulation tissue or in the deposition of specific extracellular matrix components were detected in SPARC-null *vs* wild-type wounds, nor were there significant differences in the percentage of proliferating cells. However, SPARC-null dermal cells migrated more rapidly than wild-type cells in assays that involved partial removal of cell monolayers. As SPARC is known to diminish the levels and/or activity of growth factors that enhance dermal healing, as well as certain matrix metalloproteinases implicated in skin repair, the absence of SPARC might provide a more

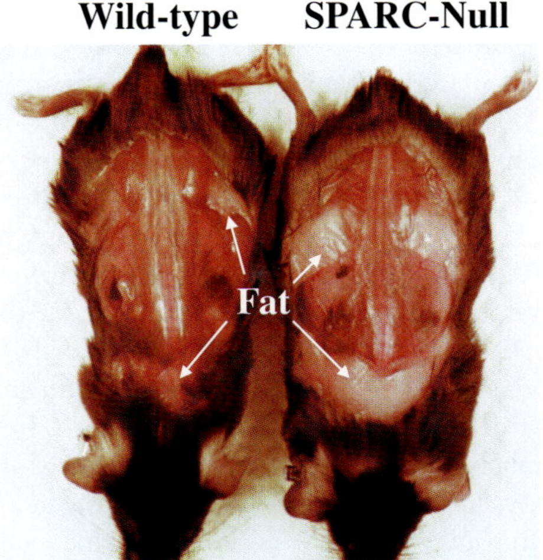

FIGURE 3 Adipose tissue is increased in SPARC-null mice. Although wild-type and SPARC-null mice of the same age do not display substantial differences in overall weight, increased fat deposits are observed in animals lacking SPARC. The wild-type and SPARC-null mice shown here were 7 months of age and weighed 29 g (wild type) and 30 g (SPARC-null), respectively.

favorable environment for cellular interaction with extracellular matrix that results in accelerated healing.

Increased amounts of dermal fat were observed in SPARC-null mice, a factor that might also contribute to their enhanced wound healing. Figure 3 shows overviews of wild-type and SPARC-null animals. The substantial increase in adipose mass is reflected in the dermal, subdermal, and interorgan fat deposits of mice lacking SPARC. Preliminary data show a two-fold increase in circulating levels of leptin in SPARC-null animals, but the mice are not overtly obese (Fig. 3) (Bradshaw *et al.*, 2002). If SPARC has a role in fat deposition, SPARC-null mice could be a model for the regulation of lean body mass that has potentially significant implications for angiogenesis (Folkman, 1995).

At first glance, a relationship among dermal collagen fibrillogenesis, adiposity, and cataract formation appears difficult to conceive. The lens is a highly specialized tissue, consisting of one cell type, the lens epithelial cell, which differentiates into fiber cells that elongate, lose their nuclei, and become transparent. Interestingly, the lens epithelium is a single layer of nucleated cells, which lie immediately adjacent to the lens capsule, a well-developed basement membrane that controls lens permeability as well as several aspects of lens cell differentiation. Because the lens lacks blood vessels and lymphatics, the issue of permeability is an important one, as nutrients and growth/differentiation factors present in the aqueous and vitreous humors must gain access to the cells of the lens. SPARC is present in both humoral compartments, as well as the lens epithelium and lens capsule (Yan and Sage, 1999). Although data have indicated that the proliferation of lens epithelial cells during development and in young mice proceeds at similar rates in both wild-type and SPARC-null mice, there are substantial changes in the shape of the cells and fibers and at the interface between the anterior epithelium and the lens capsule (Yan *et al.*, 2002; Norose *et al.*, 2000). Thus, it is possible, in analogy to the changes seen in SPARC-null skin, that the lack of SPARC in the lens affects (i) the composition of the basement membrane *vis a vis* type IV collagen; (ii) the activity and/or the distribution of factors known to affect the differentiation of the lens, such as FGF-2; or (iii) the adhesion between lens cells and the capsule. These possibilities are currently under investigation in our laboratory.

IV. PERSPECTIVES

SPARC is a matricellular protein that affects the manner in which cells interact with the extracellular matrix, including the growth factors with which the matrix is associated. Hence SPARC exerts both counteradhesive and antiproliferative effects on cells *in vitro*, and presumably *in vivo*. Characteristics of SPARC-null mice reflect differences in cell–cell interactions, cell–extracellular matrix interactions, cell cycle, and/or cell differentiation, some of which appear to be tissue specific. Some of the more exaggerated phenotypes *in vitro*, e.g., a markedly enhanced proliferation relative to wild-type cells, might reflect situations *in vivo* related to injury or disease. A potential role for SPARC in the secretion or assembly of collagen fibrils also seems worthy of consideration. In this context, abnormal collagen fibers have been reported in tissues of mice lacking thrombospondin 2 (Kyriakides *et al.*, 1998), decorin (Danielson *et al.*, 1997), and fibromodulin/lumican (Ezura *et al.*, 2000), all of which resulted in fragile skin with compromised mechanical properties. Moreover, thrombospondin 2-null mice exhibited an increased vasculature in several organs, including skin, a result consistent with the known properties of this protein as an angiogenesis inhibitor (Kyriakides *et al.*, 1998). Another, but perhaps related, role for SPARC appears to lie in the regulation of adipogenesis. Tartare-Deckert *et al.* (2001) demonstrated that SPARC mRNA is increased in several models of obesity in mice, although further experiments are needed to reconcile these findings with those described in SPARC-null animals. Disregulated angiogenesis and/or extracellular matrix production, hallmarks of SPARC function, are likely to emerge as influential players in the control of adipocyte differentiation, as well as other physiological processes.

ACKNOWLEDGMENTS

The author acknowledges former and current members of the laboratory for their ideas, insight, and dedication to SPARC. Work cited from the authors' laboratory is currently supported by National

Institutes of Health grants GM40711, DK07467, HL59475, and HL10352 and by NSF grant EEC 04-150.

REFERENCES

Bornstein, P. (1995). Diversity of function is inherent in matricellular proteins: An appraisal of thrombospondin 1. *J. Cell Biol.* **130**, 503–506.

Bornstein, P. (2001). Thrombospondins as matricellular modulators of cell function. *J. Clin. Invest.* **107**, 929–934.

Bradshaw, A. D., and Sage, E. H. (2001). SPARC, a matricellular protein that functions in cellular differentiation and tissue response to injury. *J. Clin. Invest.* **107**, 1049–1054.

Bradshaw, A. D., Reed, M. J., and Sage, E. H. (2002a). SPARC-null mice exhibit accelerated cutaneous wound closure. *J. Histochem. Cytochem.* **50**, 1–10.

Bradshaw, A. D., Puolakkainen, P., Dasgupta, J., Davidson, J. M., Wight, T. N., and Sage E. H. (2002b). SPARC-null mice display abnormalities in the dermis characterized by decreased collagen fibril diameter and reduced tensile strength. Submitted for publication.

Bradshaw, A. D., Graves, D. C., and Sage, E. H. (2002c). SPARC-null mice exhibit increased adiposity without significant differences in overall body weight. Submitted for publication.

Brekken, R. A., and Sage, E. H. (2001). SPARC, a matricellular protein: At the crossroads of cell-matrix communication. *Matrix Biol.* **19**, 815–827.

Crawford, S. E., Stellmach, V., Murphy-Ullrich, J. E., Ribeiro, S. M., Lawler, J., Hynes, R. O., Boivin, G. P., and Bouck, N. (1998). Thrombospondin-1 is a major activator of TGF-beta1 in vivo. *Cell* **93**, 1159–1170.

Danielson, K. G., Baribault, H., Holmes, D. F., Graham, H., Kadler, K. E., and Iozzo, R. V. (1997). Targeted disruption of decorin leads to abnormal collagen fibril morphology and skin fragility. *J. Cell Biol.* **136**, 729–743.

Delaney, A. M., Amling, M., Priemel, M., Howe, C., Baron, R., and Canalis, E. (2000). Osteopenia and decreased bone formation in osteonectin-deficient mice. *J. Clin. Invest.* **105**, 915–923.

Denhardt, D. T., Noda, M., O'Regan, A. W., Pavlin, D., and Berman, J. S. (2001). Osteopontin—a means to cope with environmental insults: Regulation of inflammation, tissue remodeling, and cell survival. *J. Clin. Invest.* **107**, 1055–1061.

Ezura, Y., Chakravarti, S., Oldberg, A., Chervoneva, I., and Birk, D. E. (2000). Differential expression of lumican and fibromodulin regulate collagen fibrillogenesis in developing mouse tendons. *J. Cell Biol.* **151**, 779–788.

Fitzgerald, M. C., and Schwarzbauer, J. E. (1998). Importance of the basement membrane protein SPARC for viability and fertility in *Caenorhabditis elegans*. *Curr. Biol.* **8**, 1285–1288.

Folkman, J. (1995). Angiogenesis in cancer, vascular, rheumatoid and other disease. *Nature Med.* **1**, 27–31.

Francki, A., Bradshaw, A. D., Bassuk, J., Howe, C., Couser, W., and Sage, E. H. (1999). SPARC regulates the expression of collagen type I and TGF-β1 in mesangial cells. *J. Biol. Chem.* **274**, 32145–32152.

Hohenester, E., Maurer, P., and Timpl, R. (1997). Crystal structure of a pair of follistatin-like and EF-hand calcium-binding domains in BM-40. *EMBO J.* **16**, 3778–3786.

Kupprion, C., Motamed, K., and Sage, E. H. (1998). SPARC inhibits the mitogenic effect of vascular endothelial growth factor (VEGF) on microvascular endothelial cells. *J. Biol. Chem.* **273**, 29635–29640.

Kyriakides, T.R., Zhu, Y.-H., Smith, L.T., Bain, S.D., Yang, Z., Lin, M.T., Danielson, K.G., Iozzo, R.V., LaMarca, M., McKinney, C.E., Ginns, E.I., and Bornstein, P. (1998). Mice that lack thrombospondin 2 display connective tissue abnormalities that are associated with disordered collagen fibrillogenesis, an increased vascular density, and a bleeding diathesis. *J. Cell Biol.* **140**, 419–430.

Ledda, M. F., Adris, S., Bravo, A. I., Kairiyama, C., Bover, L., Chernajovsky, Y., Mordoh, J., and Podhajcer, O. L. (1997). Suppression of SPARC expression by antisense RNA abrogates the tumorgenicity of human melanoma cells. *Nature Med.* **3**, 171–176.

Motamed, K., Blake, D. J., Angello, J., Allen, B., Rapraeger, A., Hauschka, S., and Sage, E. M. (2002). Fibroblast growth factor receptor-1 mediates the inhibition of microvascular endothelial cell proliferation and the promotion of skeletal myoblast differentiation by SPARC. Submitted for publication.

Murphy-Ullrich, J. E. (2001). The de-adhesive activity of matricellular proteins: Is intermediate cell adhesion an adaptive state? *J. Clin. Invest.* **107**, 785–790.

Murphy-Ullrich, J. E., Lane, T. F., Pallero, M. A., and Sage, E. H. (1995). SPARC mediates focal adhesion disassembly in endothelial cells through a follistatin-like region and the calcium-binding EF-hand. *J. Cell Biochem.* **57**, 341–350.

Norose, K., Lo, W.-K., Clark, J. I., Sage, E. H., and Howe, C. C. (2000). Lenses of SPARC-null mice exhibit an abnormal cell surface-basement membrane interface. *Exp. Eye Res.* **71**, 295–307.

Purcell, L., Gruia-Gray, J., Scanga, S., and Ringuette, M. (1993). Developmental anomalies of *Xenopus* embryos following microinjection of SPARC antibodies. *J. Exp. Zool.* **265**, 153–164.

Raines, E., Lane, T. F., Iruela-Arispe, L., Ross, R., and Sage, E. H. (1992). The extracellular glycoprotein SPARC interacts with platelet-derived growth factor (PDGF) AB and PDGF BB and inhibits the binding of PDGF to its receptors. *Proc. Natl. Acad. Sci. USA* **89**, 1281–1285.

Rempel, S. A., Ge, S., and Gutierrez, J. A. (1999). SPARC: A potential diagnostic marker of invasive meningiomas. *Clin. Cancer Res.* **5**, 237–241.

Ringuette, M., Drysdale, T., and Liu, F. (1992). Expression and distribution of SPARC in early *Xenopus laevis* embryos. *Roux's Arch. Dev. Biol.* **202**, 4–9.

Sage, E. H. (1997). Pieces of eight: Bioactive fragments of extracellular proteins as regulators of angiogenesis. *Trends Cell Biol.* **7**, 182–186.

Sage, E. H. (2001). Regulation of interactions between cells and extracellular matrix: A command performance on several stages. *J. Clin. Invest.* **107**, 781–783.

Sage, E. H., and Bornstein, P. (1991). Minireview: Extracellular proteins that modulate cell-matrix interactions: SPARC, tenascin, and thrombospondin. *J. Biol. Chem.* **266**, 14831–14834.

Soderling, J. A., Reed, M. J., Corsa, A., and Sage, E. H. (1997). Cloning and expression of murine SC1, a gene product homologous to SPARC. *J. Histochem. Cytochem.* **45**, 823–835.

Strandjord, T. P., Madtes, D. K., Weiss, D. J., and Sage, E. H. (1999). Collagen accumulation is decreased in SPARC-null mice with bleomycin-induced pulmonary fibrosis. *Am. J. Physiol.* **277**, L628–L635.

Tartare-Deckert, S., Chavey, C., Monthousel, M.-N., Gautier, N., and Van Obberghen, E. (2001). The matricellular protein SPARC/osteonectin as a newly identified factor upregulated in obesity. *J. Biol. Chem.*

Yan, Q., and Sage, E. H. (1999). SPARC, a matricellular protein with important biological functions. *J. Histochem. Cytochem.* **47**, 1495–1505.

Yan, Q., Wight, T., Clark, J. I., and Sage, E. H. (2002). Alterations in the lens capsule contribute to cataractogenesis in SPARC-null mice. *J. Cell Science* **115**, 2747–2756.

PART II

Cells Responsible for Extracellular Matrix Production

CHAPTER 6

Cells Responsible for Extracellular Matrix Production in the Liver

TAKATO UENO,* MICHIO SATA,* AND KYUICHI TANIKAWA†

*Research Center for Innovative Cancer Therapy and Second Department of Medicine, Kurume University School of Medicine, Kurume 830-0011, Japan and †International Institute for Liver Research, Kurume Research Center, Kurume 839-0861, Japan

I. INTRODUCTION

Hepatic fibrosis is a dynamic process from chronic liver disease to cirrhosis. Both a marked increase in production and a decrease in degradation cause excessive accumulation of extracellular matrix (ECM). In addition, hepatocytes and nonparenchymal cells such as hepatic stellate cells (HSC) (fat-storing cells, lipocytes, Ito cells), Kupffer cells, and sinusoidal endothelial cells are deeply involved in the process. This chapter emphasizes the microenvironment of hepatic fibrosis and hepatic cells responsible for ECM production, focusing on the recent progress in this area.

II. MICROENVIRONMENT IN THE NORMAL LIVER

ECM components such as collagens, glycoproteins, and proteoglycans are located in the liver (Gressner, 1998; Martinez-Hernandez and Amenta, 1993). A number of different collagens, including types I, III, IV, V, and VI, are present in the liver. Glycoproteins, including fibronectin, laminin, and tenascin, are also present, as

Copyright 2003, Elsevier Science (USA). All rights reserved.

well as proteoglycans such as heparan sulfate, perlecan, and syndecan. Some of the proteoglycans bind to hyaluronan, forming molecules with varying molecular weights. As shown in Fig. 1, collagens, glycoproteins, and proteoglycans are localized in connective tissues in the portal tracts and central veins, as well as in the Disse spaces. In the subendothelial sites of the Disse spaces in the normal liver, a typical basement membrane is not found (Fig. 2). Type IV collagen and glycoproteins such as fibronectin and proteoglycans are located in the Disse spaces (Figs. 1 and 3). These components, which are produced by intrahepatic cells such as HSC, sinusoidal endothelial cells, and hepatocytes (Fig. 4), are involved in the maintenance and function of the intrahepatic cells (Li and Friedman, 1999). In addition, sinusoidal endothelial cells produce ECM components such as type IV collagen, proteoglycans, and urokinase type plasminogen activator, which activates latent type transforming growth factor-β1 (TGF-β1). Fibroblasts, vascular endothelial cells, and biliary epithelial cells in portal tracts and myofibroblasts around the central veins are known to produce ECM components, including basement membrane components (Gressner, 1998) (Fig. 4).

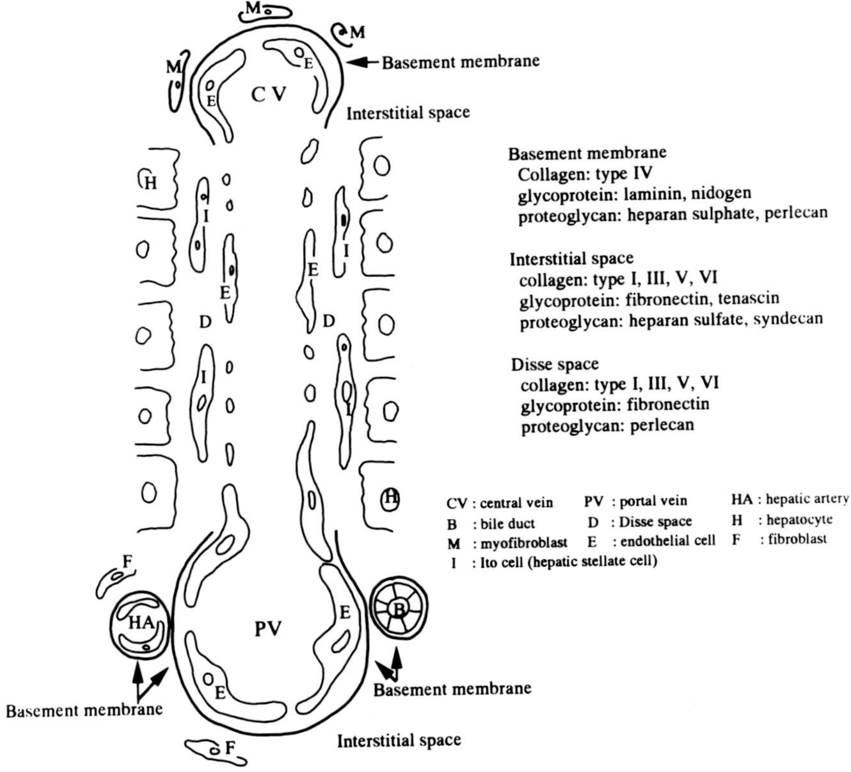

FIGURE 1 Extracellular matrix components in normal liver.

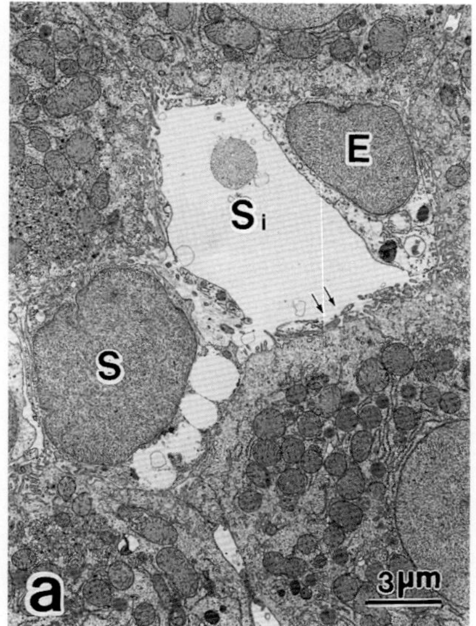

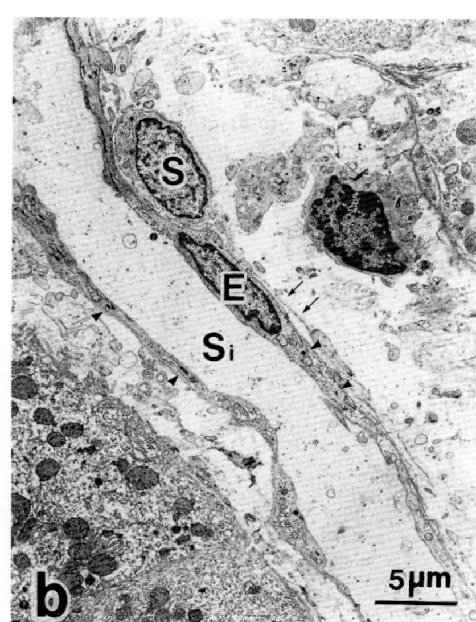

FIGURE 2 Transmission electron micrographs showing hepatic sinusoidal area in human liver. (a) Normal liver. Sinusoidal endothelial cell (E) shows many fenestrae (arrows) without Weibel–Palade bodies in the cytoplasm. However, there are no basement membranes on the basal side of the cell. In the Disse space, the hepatic stellate cell (S) containing some fat droplets in the cytoplasm is shown. (b) Cirrhotic liver. Sinusoidal endothelial cells (E) show no fenestrae but contain many Weibel–Palade bodies in the cytoplasm (arrowheads) and continuous basal membranes (arrows) on their basal side. Disse spaces are filled with collagen fibers. A typical hepatic sinusoidal capillarization is shown. Si, sinusoid; H, hepatocyte; S, hepatic stellate cell.

III. LIVER INJURY AND CHANGES IN MICROENVIRONMENT

A. Accumulation of Inflammatory Cells in Injured Areas

Liver injury is induced by various agents such as viruses, bacteria, drugs, alcohol, and bile, and inflammatory cells containing neutrophils, platelets, lymphocytes, and monocytes, including Kupffer cells, are mobilized in injured areas (Fig. 5). Kupffer cells are characteristic macrophages (monocytes) in the liver. Activated monocytes could be capable of generating monokines such as interleukin-1 (IL-1), platelet-derived growth factor (PDGF), fibroblast growth factor (FGF), and tumor necrosis factor α (TNF-α) (Kovacs, 1991; Nathan, 1984). TNF α and IL-1, which are secreted initially, by the inflammatory cells, upregulate adhesion molecules containing intercellular adhesion molecule-1 (ICAM-1) on sinusoidal endothelial cells. Thus cell–cell contact between inflammatory cells and sinusoidal

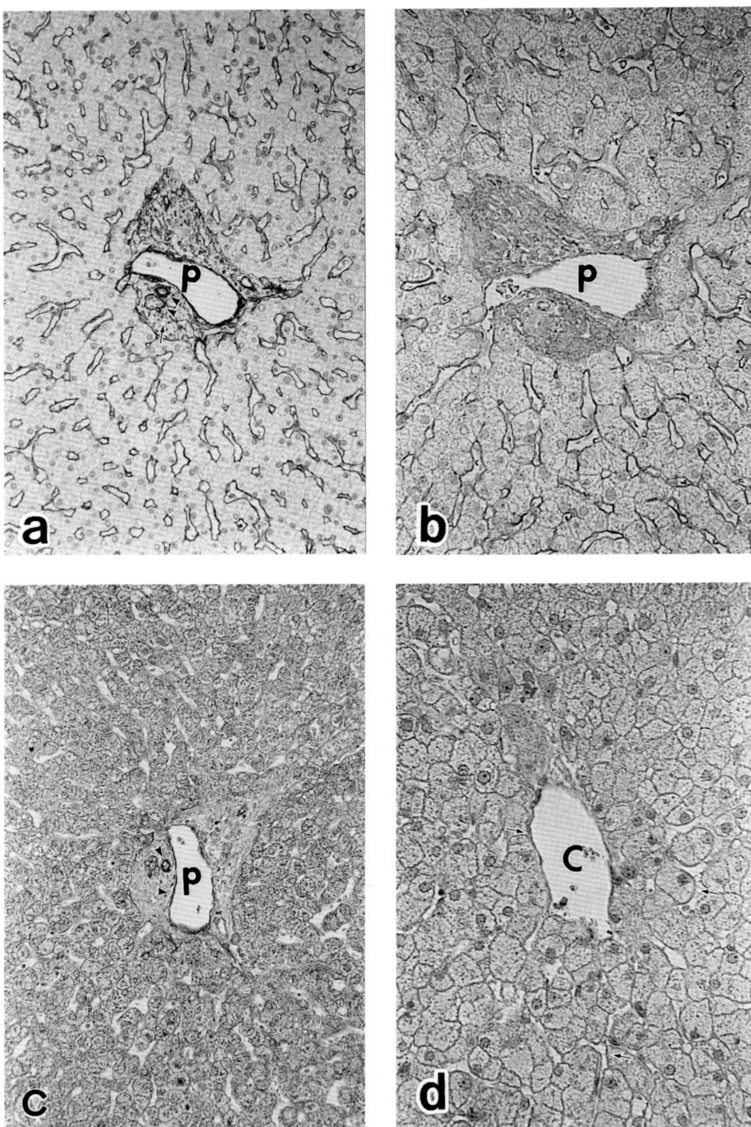

FIGURE 3 Immunolocalization of extracellular matrix components in normal human liver. (a) Type IV collagen is located along the sinusoidal walls and basement membranes around vessels and the bile duct. Furthermore, it is also visible in connective tissues around the portal vein (×64). (b) Fibronectin. Localization of fibronectin, which is one of the glycoproteins, is similar to that of type IV collagen (×64). (c) Laminin is located in accordance with basement membranes around vessels, but is not visible along the sinusoidal walls (×64). (d) Proteoglycan is located around a central vein and along the sinusoidal walls (×64). P, portal vein; C, central vein.

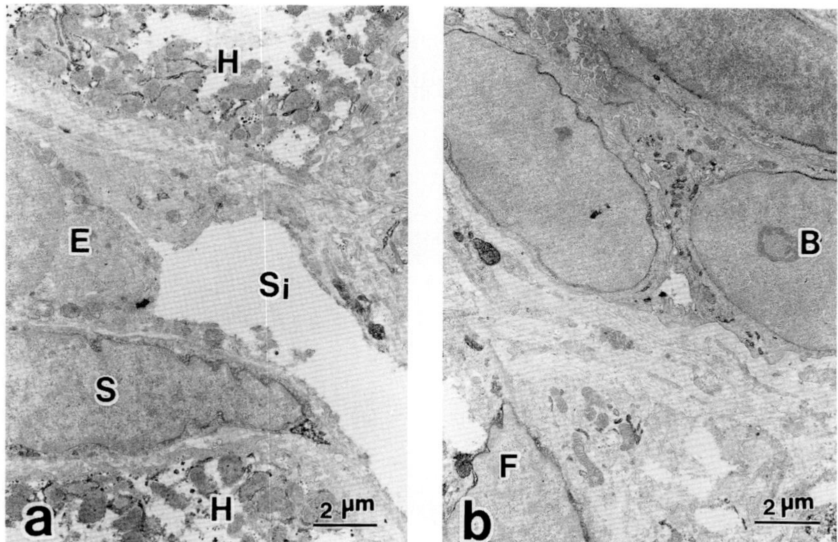

FIGURE 4 Immunolocalization of prolyl hydroxylase in human cirrhotic liver. Prolyl hydroxylase is an important key enzyme, which catalyzes the 4-hydroxylation of proline residues in procollagens. Proline residues convert to hydroxyproline residues by prolyl hydroxylase; i.e., cells containing prolyl hydroxylase have the potential to produce procollagens. (a) An electron micrograph showing the hepatic sinusoidal area. Prolyl hydroxylase is located in the rough endoplasmic reticula of HSC (S), endothelial cell (E), and hepatocytes (H). Si, sinusoid. (b) An electron micrograph showing the portal area. Prolyl hydroxylase is located in the rough endoplasmic reticula of biliary epithelial cells (B) and the fibroblast (F).

endothelial cells becomes possible (Ohira *et al.*, 1995). In addition, TNF-α seems to modulate fibroblast proliferation by increasing the number of cell receptors for epidermal growth factor (EGF) (Palomombella *et al.*, 1987) and possibly by including the production of IL-1 (Le *et al.*, 1987). In turn, IL-1 decreases the cell receptor density for TNF-α rapidly (Holtmann and Wallach, 1987). PDGF and IL-1 have concentration-dependent stimulatory effects on HSC proliferation. Furthermore, TNF-α also exhibits a similar mitogenic effect (Matuoka *et al.*, 1989).

B. ECM Destruction by Inflammatory Cells and Proliferation of ECM Producing Cells

Matrix metalloproteinases (MMPs) produced by Kupffer cells among others degrade the ECM located in the Disse spaces, allowing inflammatory cells to come into direct contact with hepatocytes, which are thus injured by these cells, and the outcome is necrosis and apoptosis (Fig. 6). In liver diseases, especially in the

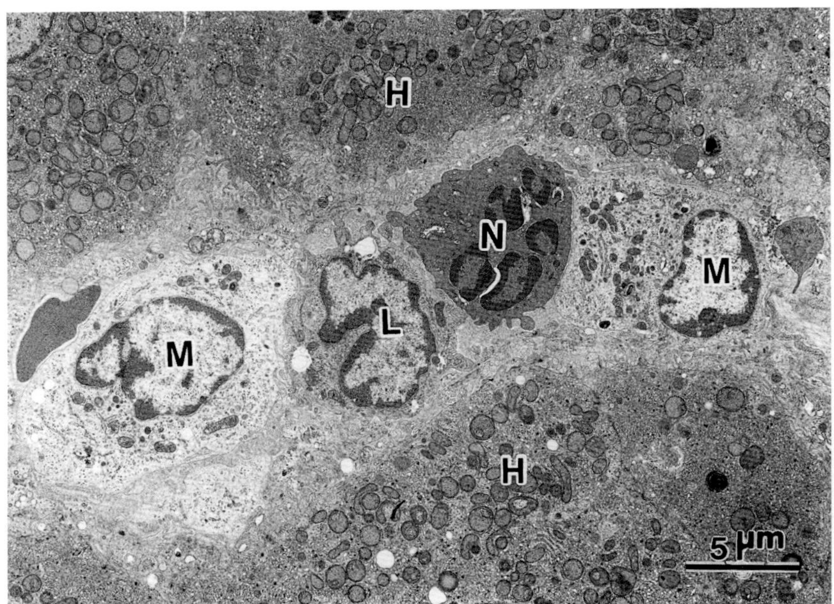

FIGURE 5 An electron micrograph showing a hepatic sinusoidal area 1 day after treatment in a dimetylnitrosamine-treated fibrosis model rat. Inflammatory cells such as neutrophil (N), monocytes (M), and lymphocyte (L) are aggregated in the hepatic sinusoid. H, hepatocyte.

early stage of liver injury, MMPs 2 and 9 are upregulated (Milani *et al.*, 1994). These MMPs belong to a type IV collagenase and digest basement membrane components, including type IV collagen. HSC are a key source of MMP-2. They also express a tissue inhibitor of metalloproteinases (TIMP)-1 and -2 (Fig. 7) (Herbst *et al.*, 1997). In fibrogenesis, TIMPs inhibit MMP activities by the formation of stoichiometric 1:1 complexes. MMP-9 is secreted locally by Kupffer cells (Fig. 8) (Sujaku *et al.*, 1998). The sources of MMP-1, which plays a crucial role in degrading the excess interstitial matrix in advanced liver disease, are Kupffer cells and HSC. An increased activity of MMP-1 has been observed in the early stage of fibrosis and its activity reduces once the fibrosis has progressed (Maruyama *et al.*, 1982). Changes in the microenvironment of the Disse spaces result in phenotypic changes in all resident liver cells.

Integrins are the most important adhesion molecules, which mediate cell–matrix interactions such as adhesion and migration (Clark and Brugge, 1995; Torimura *et al.*, 1999). There are at least 16 different α subunits and 8 different β subunits, and more than 20 different $\alpha\beta$ combinations have been described (Table I). All integrins are heterodimers consisting of a noncovalently associated α subunit of 140–180 kDa with a β subunit of 90–110 kDa (Hynes, 1992). Each integrin subunit has a large extracellular domain, a single

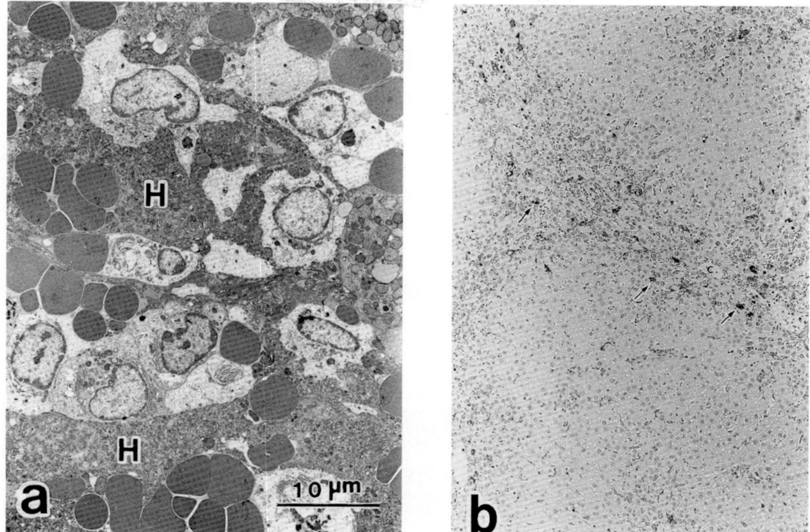

FIGURE 6 An electron micrograph and a micrograph showing an injured liver tissue 4 days after treatment in a dimetylnitrosamine-treated fibrosis model rat. (a) Inflammatory cells contact hepatocytes (H) directly. (b) Apoptotic cells (arrows) increase in injured liver tissues. C, central vein (×52).

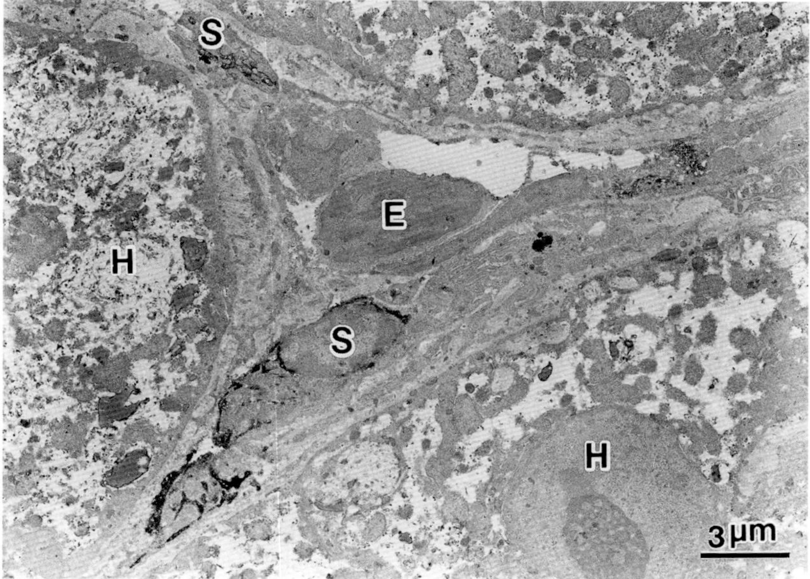

FIGURE 7 An electron micrograph showing TIMP-1 immunolocalization in human cirrhotic liver. TIMP-1 is located in the cytoplasm of HSC (S) and sinusoidal endothelial cells (E). H, hepatocyte.

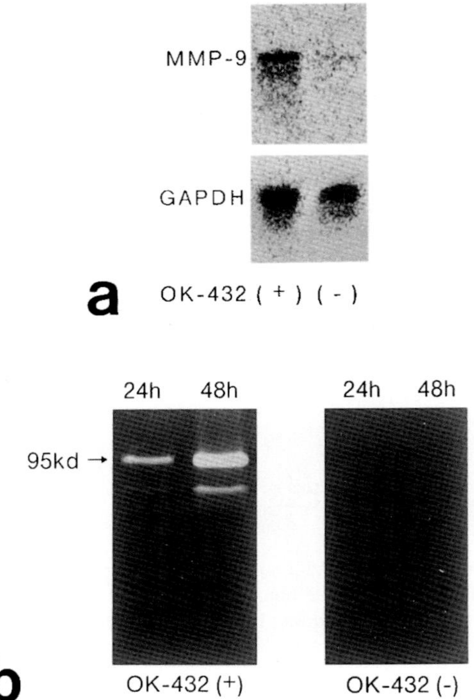

FIGURE 8 MMP-9 expression in rat Kupffer cells treated by OK-432, which is a bacterial cell preparation manufactured by treating *Streptococcus pyrogenes* Su strains in the presence of penicillin for 24 to 48 h. (a) Northern blot analysis of MMP-9 mRNA and GAPDH mRNA. In OK-432-stimulated rat Kupffer cells, MMP-9 mRNA is detected. However, in OK-432-nontreated Kupffer cells, no signals are detected. Lane 1, OK-432 (+); lane 2, OK-432 (−). (b) MMP-9 activity by zymography. In OK-432-treated Kupffer cells, 95-kDa and less than 95-kDa bands are detected 24 and 48 h after OK-432 treatment. These bands are clearly increased 48 h after OK-432 treatment. In OK-432-non-treated Kupffer cells, these bands are hardly detected.

membrane-spanning region, and a short cytoplasmic domain (Yamada and Geiger, 1997). The α-cytoplasmic domains contain highly divergent amino acid sequences, whereas the β subunits show partial sequence conservation. The β-cytoplasmic domains are necessary and sufficient to target integrins to focal adhesions in a ligand-independent manner, whereas the α-cytoplasmic domains regulate the specificity of ligand-dependent interactions (LaFlamme *et al.*, 1994).

Integrins are membrane receptors that transduce extracellular signals in liver (Torimura *et al.*, 1999). In particular, integrin ligands contain an arginine–glycine–aspartate (RGD) tripeptide sequence. The common RGD

TABLE I The Integrin Family

	Subunits	Ligands[a]
α1β2	CD49a/CD29, VLA-1	Collagen, laminin
α2β1	CD49b/CD29, VLA-2	Collagen, laminin
α3β1	CD49c/CD29, VLA-3	Epiligrin
α4β1	CD49d/CD29, VLA-4	VCAM-1, FN(CS-1), MAdCAM-1
α5β1	CD49e/CD29, VLA-5	FN(RGD)
α6β1	CD49f/CD29, VLA-6	Laminin
α6β4	CD49f/CD104	Laminin-5
α7β1		Laminin
α8β1		FN
α9β1		Tenascin
αVβ1	CD51	VN, FN
αLβ2	CD11a/CD18, LFA-1	ICAM-1, -2, -3
αMβ2	CD11b/CD18, Mac-1, CR3	iC3b, Fbg, FX, ICAM-1
αXβ2	CD11c/CD18, p150, 95	Fbg, iC3b
αIIbβ3	CD41/CD61, GPIIb/IIIa	Fbg, FN, vWF, VN
αVβ3	CD51/CD61, vitronectin receptor	VN, Fbg, vWF, TSP, FN
αVβ5		VN
αVβ6		FN
αVβ8		VN
α4β7		MAdCAM-1, VACM-1
αEβ7		E-Cadherin

[a]FN, fibronectin; VN, vitroectin; VCAM-1, vascular cell adhesion molecule-1; ICAM-1, intercellular adhesion molecule-1; Fbg, fibrinogen; FX, factor X; vWF, von Willebrand; TSP thrombospondin; MAdCAM-1, mucosal adhesion cell adhesion molecule-1.

tripeptide sequence at many integrin ligands has raised the possibility of using competitive RGD antagonists to block integrin-mediated pathways in fibrogenesis (Iwamoto *et al.*, 1998; Li and Friedman, 1999).

Intracellular regulators modify external integrin ligand-binding properties in a process termed "inside-out signaling" (Yamada, 1997). Inside-out signal transduction is speculated to be as follows: (1) activation of protein kinase C (PKC) following receptor–G-protein interactions; (2) activation of GTP-Rho, which is very important for inside-out signaling by activated PKC; (3) activation of a serine/threonine kinase, ROCK-1, or related kinases by activated Rho (Yamada, 1997); and (4) activation of integrins by ROCK-1 or related kinases. Thus, activated integrins are able to bind to extracellular matrices (Matui *et al.*, 1996).

An even more intriguing item is the identification of discoidin domain receptor (DDR) 2 mRNA in activated HSC, raising the possibility that this receptor may mediate interactions between HSC and the surrounding interstitial matrix (O'Toole *et al.*, 1994). Furthermore, local proliferation of Kupffer cells, an influx of blood macrophages (monocytes), and aggregation of platelets in necroinflammatory areas follow the initial events in hepatic injury and necrosis. Kupffer cells are also important in stimulating the release of retinoids by HSC (Friedman and Arthur, 1989). Activated Kupffer cells and platelets regulate HSC proliferation through TGF-α, EGF, and PDGF. In addition, HSC transformation into myofibroblast-like HSC is stimulated by TGF-β, TGF-α, and TNF-α (Gressner and Bachem, 1994; Gressner and Chunfang, 1995). Platelets mobilized in injured liver are a rich source of potentially important mediators such as PDGF, EGF, and TGF-β (Gressner, 1998; Li and Friedman, 1999). The ECM producing cells such as HSC proliferate in the presence of PDGF and EGF and are induced to produce ECM components by TGF-β1. TGF-β1 initiates signaling through heteromeric complexes of transmembrane type I and type II serine/threonine kinase receptors. Activated TGF-β type I receptor phosphorylates receptor-regulated Smads2 and 3. Then, they oligomerize with the common mediator Smad4 and translocate to the nucleus, where they direct transcription to affect the response of the cell to TGF-β (Heldin *et al.*, 1997). In the nucleus, the CAGA box of the TGF-β response element is the binding site of the Smad3/Smad4 complex, and the binding of the complex appears to be required for TGF-β-induced collagen upregulation

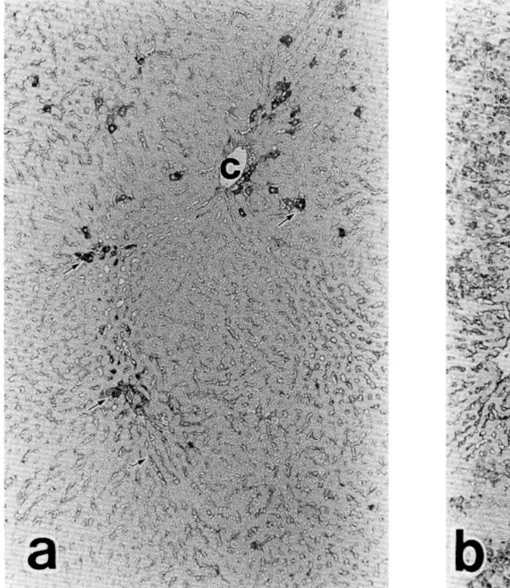

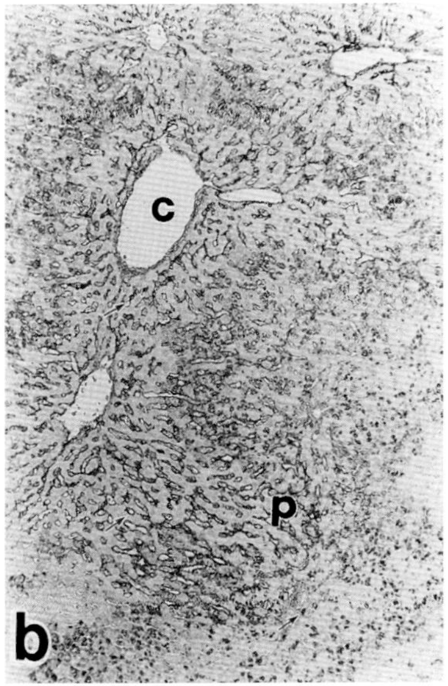

FIGURE 9 Immunolocalization of TGF-β1 and fibronectin in injured liver tissues 1 day after treatment in a dimetylnitrosamine-treated fibrosis model rat. (a) TGF-β1 (arrows) expresses in the injured area (day 1). (b) Fibronectin (arrows) also expresses in the injured area and along the sinusoidal walls near the injured area (day 2). P, portal vein; C, central vein (×32).

(Zhang *et al.*, 2000). In the liver, TGF-β1 enhances collagen accumulation significantly and also induces HSC proliferation (Reimann *et al.*, 1997).

C. Cell Sources of ECM in Hepatic Fibrosis

In the early stage of liver injury, damaged sinusoidal endothelial cells and HSC induce the production of cellular fibronectin (Fig. 9) and enhance the migration and proliferation of inflammatory cells and cell-to-cell interactions further. In experimental hepatic fibrosis induced by ligation of the biliary duct in rats, mRNA expression of the fibronectin isoform EIIIA was detected in sinusoidal endothelial cells within 12 h of injury, but was undetectable in normal sinusoidal endothelial cells. In contrast, the EIIIB form was restricted to HSC and was expressed only after 12–24 h. Both forms were expressed minimally in hepatocytes. EIIIA fibronectin was increased markedly within 2 days of injury and showed a sinusoidal distribution, and this component accelerated the conversion of quiescent HSC to myofibroblast-like HSC (Fig. 10) (Jarnagin *et al.*, 1994).

The identification of HSC as the key cellular source of ECM in the liver has been a major advance (Housset and Guechot, 1999; Inuzuka *et al.*, 1990; Lang and Brenner, 1999; Loreal *et al.*, 1997). The calculated amount of collagen

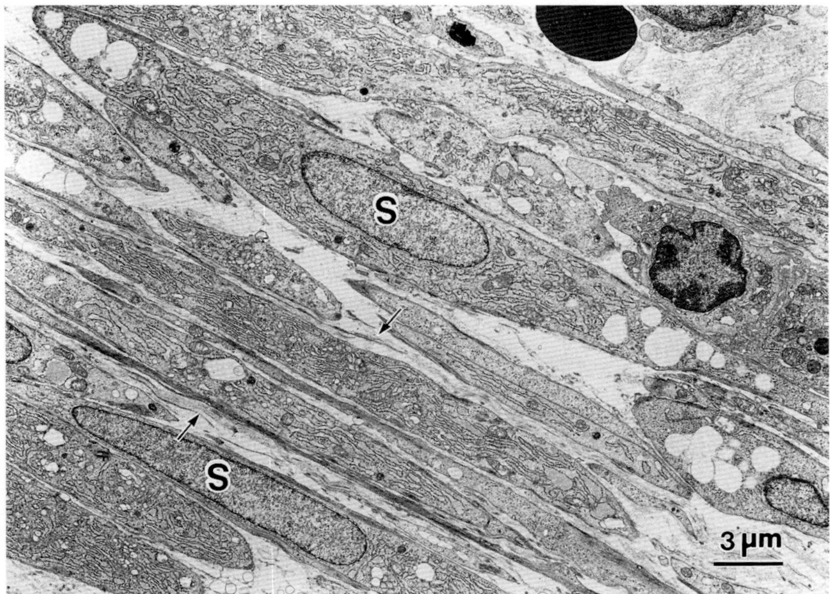

FIGURE 10 An electron micrograph showing rat injured liver 4 days after acute injury by a heat needle. Numerous myofibroblast-like HSC (S) are visible. Arrows: basement membrane-like structure.

synthesized by HSC was 10-fold greater than that produced by hepatocytes and over 20-fold greater than that produced by sinusoidal endothelial cells in normal rats (Loreal *et al.*, 1997). Highly active and synthetic myofibroblast-like HSC produce the majority of the ECM components, such as collagens (types I, III, IV, V, and XIV) (Milani *et al.*, 1990a,b), glycoproteins (fibronectin, laminin, merosin, tenascin, nidogen, and so on) (Cenacchi *et al.*, 1983; Van Eyken *et al.*, 1992; Weiner *et al.*, 1992), proteoglycans (heparan, dermatan, and chondroitin sulfates, perlecan, syndecan, biglycan, and decorin) (Arenson *et al.*, 1988; Gallai *et al.*, 1996), and hyaluronan (Gressner and G., 1989). Hepatocytes were once thought to be the main source of ECM until the 1980s (Diegelmann *et al.*, 1983). However, studies using immunohistochemistry, *in situ* hybridisation, and cell culture have clearly established the role of HSC in ECM production. Liver cirrhosis is characterized by a 2- to 20-fold overall increase of ECM components (Table II) (Tanikawa, 1994). They increase markedly in the hepatic lobules and portal tracts.

D. Role of Apoptosis in Hepatic Fibrosis

Gressner *et al.* (1996) showed that hepatic myofibroblast-like HSC could participate in the process of hepatocyte apoptosis by paracrine feedback loops involving TGF-β (Gressner *et al.*, 1996). This is additional evidence for the perpetuation of fibrogenesis. They have also shown that activated HSC express Fas/Apo-1 (CD95) (Gong *et al.*, 1998). Activated HSC are susceptible to soluble Fas ligand (sFasL), and Gressner *et al.* (1996) related the effects to the expression levels of Fas/Apo-1. Activated HSC show a dose-dependent apoptotic reaction upon exposure to sFasL. Conversely, in liver tissues allowed to recover from fibrosis spontaneously, the number of apoptotic bodies increases (Issa *et al.*, 2001). These results support

TABLE II Comparison in Concentration of Extracellular Matrix Components between Normal Liver and Cirrhotic Liver

Extracellular matrix	Normal (μg/g)	Cirrhosis (μg/g)
Type I collagen	2000	15,000
Type III collagen	2000	8,000
Type IV collagen	500	7,000
Type V collagen	900	2,000
Type VI collagen	6	60
Fibronectin	300	630
Heparan sulfate	62	245
Chondroitin sulfate	14	90
Dermatan sulfate	18	138
Hyaluronate	1.5	32

the view that myofibroblast-like HSC are a target of Fas ligand-mediated apoptosis in the liver.

In addition, liver-specific natural killer cells, i. e., pit cells, could play an important role in this process (Vanderkerken *et al.*, 1993). These cells act as the carrier of Fas ligand (Oshima *et al.*, 1996). Further studies have to be performed in order to determine the molecular details of the switch of apoptotic sensitivity when HSC are transformed to myofibroblast-like HSC. It is conceivable that modulation of the elimination pathway of myofibroblast-like HSC in injured liver determines the fibrogenic response of this organ significantly.

IV. CONCLUSION

Recent research has greatly increased our knowledge concerning ECM components, sources, cytokines, and cell–matrix interactions in hepatic fibrosis. However, there are still some unanswered questions surrounding the origin of HSC or myofibroblast-like HSC, the relationship between retinoids and activation of HSC, and the involvement of apoptosis in liver injury and repair, that should be resolved. Answers to these questions reached through new techniques such as gene therapy and new drug delivery systems will undoubtedly lead to advances in treating patients with chronic liver disease and fibrosis.

REFERENCES

Arenson, D. M., Friedman, S. L., and Bissell, D. M. (1988). Formation of extracellular matrix in normal rat liver: Lipocytes as a major source of proteoglycan. *Gastroenterology* **95**, 441–447.

Cenacchi, G., Ballardini, G., Badiali De, G. L., Busachi, C. A., Del Rosso, M., Bianchi, F. B., Bianchi, G., and Laschi, R. (1983). Relationship between connective tissue cells and fibronectin in a sequential model of experimental hepatic fibrosis. *Virch. Arch. B Cell Pathol. Incl. Mol. Pathol.* **43**, 75–84.

Clark, E. A., and Brugge, J. S. (1995). Integrins and signal transduction pathways: The road taken. *Science* **268**, 233–239.

Diegelmann, R. F., Guzelian, P. S., Gay, R., and Gay, S. (1983). Collagen formation by the hepatocyte in primary monolayer culture and in vivo. *Science* **219**, 1343–1345.

Friedman, S. L., and Arthur, M. J. (1989). Activation of cultured rat hepatic lipocytes by Kupffer cell conditioned medium: Direct enhancement of matrix synthesis and stimulation of cell proliferation via induction of platelet-derived growth factor receptors. *J. Clin. Invest.* **84**, 1780–1785.

Gallai, M., Kovalszky, I., Knittel, T., Neubauer, K., Armbrust, T., and Ramadori, G. (1996). Expression of extracellular matrix proteoglycans perlecan and decorin in carbon-tetrachloride-injured rat liver and in isolated liver cells. *Am. J. Pathol.* **148**, 1463–1471.

Gong, W., Pecci, A., Roth, S., Lahme, B., Beato, M., and Gressner, A. M. (1998). Transformation-dependent susceptibility of rat hepatic stellate cells to apoptosis induced by soluble Fas ligand. *Hepatology* **28**, 492–502.

Gressner, A. M. (1998). The cell biology of liver fibrogenesis—an imbalance of proliferation, growth arrest and apoptosis of myofibroblasts. *Cell Tissue Res.* **292**, 447–452.

Gressner, A. M., and Bachem, M. G. (1994). Cellular communications and cell-matrix interactions in the pathogenesis of fibroproliferative diseases: Liver fibrosis as a paradigm. *Ann. Biol. Clin.* **52**, 205–226.

Gressner, A. M., and Chunfang, G. (1995). A cascade-mechanism of fat storing cell activation forms the basis of the fibrogenic reaction of the liver. *Verh. Dtsch. Ges. Pathol.* **79**, 1–14.

Gressner, A. M., and Schäfer, S. (1989). Comparison of sulphated glycosaminoglycan and hyaluronate synthesis and secretion in cultured hepatocytes, fat storing cells, and Kupffer cells. *J. Clin. Chem. Clin. Biochem.* **27**, 141–149.

Gressner, A. M., Polzar, B., Lahme, B., and Mannherz, H. G. (1996). Induction of rat liver parenchymal cell apoptosis by hepatic myofibroblasts via tranforming growth factor beta. *Hepatology* **23**, 571–581.

Heldin, C. H., Miyazomo, K., and Dijke, P. (1997). TGF beta signalling from cell membrane to nucleus through SMAD proteins. *Nature* **390**, 465–471.

Herbst, H., Wege, T., Milani, S., Pellegrini, G., Orzechowski, H., Bechstein, W., Neuhaus, P., Gressner, A., and Schuppan, D. (1997). Tissue inhibitor of metalloproteinase-1 and -2 RNA expression in rat and human liver fibrosis. *Am. J. Pathol.* **150**, 1647–1659.

Holtmann, H., and Wallach, D. (1987). Down regulation of the receptors for tumor necrosis factor by interleukin 1 and 4 beta-phorbol-12-myristate-13-acetate. *J. Immunol.* **139**, 1161–1167.

Housset, C., and Guechot, J. (1999). Hepatic fibrosis: Physiopathology and biological diagnosis. *Pathol. Biol.* **47**, 886–894.

Hynes, R. O. (1992). Integrins: Versatility, modulation, and signaling in cell adhesion. *Cell* **69**, 11–25.

Inuzuka, S., Ueno, T., Torimura, T., Sata, M., Abe, H., and Tanikawa, K. (1990). Immunohistochemistry of the hepatic extracellular matrix in acute viral hepatitis. *Hepatology* **12**, 249–256.

Issa, R., Williams, E., Trim, N., Kendall, T., Arthur, M. J. P., Reichen, J., Benyon, R. C., and Iredale, J. P. (2001). Apoptosis of hepatic stellate cells: Involvement in resolution of biliary fibrosis and regulation by soluble growth factors. *Gut* **48**, 548–557.

Iwamoto, H., Sasaki, H., and Nawata, H. (1998). Inhibition of integrin signaling with Arg-Gly-Asp motifs in rat hepatic stellate cells. *J. Hepatol.* **29**, 752–759.

Jarnagin, W. R., Rockey, D. C., Koteliansky, V. E., Wang, S. S., and Bissell, D. M. (1994). Expression of variant fibronectins in wound healing: Cellular source and biological activity of the EIIIA segment in rat hepatic fibrogenesis. *J. Cell Biol.* **127**, 2037–2048.

Kovacs, E. (1991). Fibrogenic cytokines: The role of immune mediators in the development of scar tissue. *Immunol. Today* **12**, 17–23.

LaFlamme, S., Thomas, L., Yamada, S., and Yamada, K. (1994). Single subunit chimeric integrins as mimics and inhibitors of endogenous integrin functions in receptor localization, cell spreading and migration, and matrix assembly. *J. Cell Biol.* **126**, 1287–1298.

Lang, A., and Brenner, D. A. (1999). Gene regulation in hepatic stellate cell. *Ital. J. Gastroenterol. Hepatol.* **31**, 173–179.

Le, J. M., Weinstein, D., Gubler, U., and Vilcek, J. (1987). Induction of membrane-associated interleukin 1 by tumor necrosis factor in human fibroblasts. *J. Immunol.* **138**, 2137–2142.

Li, D., and Friedman, S. L. (1999). Liver fibrogenesis and the role of hepatic stellate cells: New insights and prospectes for therapy. *J. Gastroenterol. Hepatol.* **14**, 618–633.

Loreal, O., Clement, B., and Deugnier, Y. (1997). Hepatic fibrogenesis. *Rev. Prat.* **47**, 482–486.

Martinez-Hernandez, A., and Amenta, P. S. (1993). The hepatic extracellular matrix. I. Components and distribution in normal liver. *Virch. Arch. A Pathol. Anat.* **423**, 1–11.

Maruyama, K., Feinman, L., Okazaki, I., and Lieber, C. (1982). Mammalian collagenase increases in early alcoholic liver disease and decreases with cirrhosis. *Life Sci.* **30**, 1379–1384.

Matui, T., Amano, M., Yamamoto, T., Chihara, K., Nakatuka, M., Ito, N., Nakano, T., Okawa, K., Iwamatu, A., and Kaibuchi, K. (1996). Rho-associated kinase, a novel serine/threonine kinase, as a putative target for the small GTP binding protein Rho. *EMBO J.* **15**, 2208–2216.

Matuoka, M., Pham, N. T., and Tukamoto, H. (1989). Differential effects of interleukin-1α, tumor necrosis factor α, and transforming growth factor β1 on cell proliferation and collagen formation by cultured fat-storing cells. *Liver* **9**, 71–78.

Milani, S., Herbst, H., Schuppan, D., Grappone, C., Pellegrini, G., Pinzani, M., Casini, A., Calabro, A., Ciancio, G., and Stefanini, F. (1994). Differential expression of matrix-metalloproteinase-1 and -2 genes in normal and fibrotic human liver. Am. *J. Pathol.* **144**, 528–537.

Milani, S., Herbst, H., Schuppan, D., Kim, K. Y., Riecken, E. O., and Stein, H. (1990a). Procollagen expression by nonparenchymal rat liver cells in experimental biliary fibrosis. *Gastroenterology* **98**, 175–184.

Milani, S., Herbst, H., Schuppan, D., Surrenti, C., Riecken, E. O., and Stein, H. (1990b). Cellular localization of type I, III and IV procollagen gene transcripts in normal and fibrotic human liver. *Am. J. Pathol.* **137**, 59–70.

Takemura, R., and Werb, Z. (1984). Secretory products of macrophases and their physiological functions. *Am. J. Physiol.* **246**, C1–C9.

Ohira, H., Ueno, T., Torimura, T., Tanikawa, K., and Kasukawa, R. (1995). Leukocyte adhesion molecules in the liver and plasma cytokine levels in endotoxin-induced rat liver injury. *Scand. J. Gastroenterol.* **30**, 1027–1103.

Oshimi, Y., Oda, S., Honda, Y., Nagata, S., and Miyazaki, S. (1996). Involvement of Fas ligand and Fas-mediated pathway in the cytotoxity of human natural killer cells. *J. Immunol.* **157**, 2909–2915.

O'Toole, T. E., Katagiri, Y., Faull, R. J., Peter, K., Tamura, R., Quaranta, V., Loftus, J. C., Shattil, S. J., and Ginsberg, M. H. (1994). Integrin cytoplasmic domains mediate inside-out signal transduction. *J. Cell Biol.* **124**, 1047–1059.

Palomombella, V. J., Yamashiro, D. J., Maxfield, F. R., Decker, S. J., and Vilcek, J. (1987). Tumor necrosis factor increases the number of epidermal growth factor receptors on human fibroblasts. *J. Biol. Chem.* **262**, 1950–1954.

Reimann, T., Hempel, U., Krautwald, S., Axmann, A., Scheibe, R., Seidel, D., and Wenzel, K. W. (1997). Transforming growth factor-beta1 induces activation of Ras, Raf-1, MEK and MAPK in rat hepatic stellate cells. *FEBS Lett.* **403**, 57–60.

Sujaku, K., Ueno, T., Torimura, T., Sata, M., and Tanikawa, K. (1998). Effects of OK-432 on matrix metalloproteinase-9 expression and activity in rat cultured Kupffer cells. *Hepatol. Res.* **10**, 91–100.

Tanikawa, K. (1994). Serum marker for hepatic fibrosis and related liver pathology. *Path. Res. Pract.* **190**, 960–968.

Torimura, T., Ueno, T., Kin, M., Sakisaka, S., Tanikawa, K., and Sata, M. (1999). β1-Integrin mediates cell adhesion, haptotaxis, and chemokinesis of hepatoma cells. *Med. Electron Microsc.* **32**, 2–10.

Vanderkerken, K., Bouwens, L., De Neve, W., Van den Berg, K., Baekeland, M., Delens, N., and Wisse, E. (1993). Origin and differentiation of hepatic natural killer cells (pit cells). *Hepatology* **18**, 919–925.

Van Eyken, P., Geerts, A., De Bleser, P., Lazou, J. M., Vrijsen, R., Sciot, R., Wisse, E., and Desmet, V. J. (1992). Localization and cellular source of the extracellular matrix protein tenascin in normal and fibrotic rat liver. *Hepatoplogy* **15**, 909–916.

Weiner, F. R., Shah, A., Biempica, L., Zarn, M. A., and Czaja, M. J. (1992). The effects of hepatic fibrosis on Ito cell gene expression. *Matrix* **12**, 36–43.

Yamada, K. M. (1997). Integrin signaling. *Matrix* **16**, 137–141.

Yamada, K. M., and Geiger, B. (1997). Molecular interactions in cell adhesion complexes. *Curr. Opin. Cell Biol.* **9**, 76–85.

Zhang, W., Ou, J., Inagaki, Y., Greenwel, P., and Ramirez, F. (2000). Synergic cooperation between Sp1 and Smad3/Smad4 mediates transforming growth factor beta 1 stimulation of alpha 2(1)-collagen (COL1A2) transcription. *J. Biol. Chem.* **275**, 39237–39245.

CHAPTER 7

Different Hepatic Cell Populations of the Fibroblast Lineage with Fibrogenic Potential

THOMAS KNITTEL[1]

Department of Internal Medicine, University of Göttingen, 37075 Göttingen, Germany

Hepatic stellate cells (HSC) are regarded as the principal matrix-producing cell of damaged liver. However, other cell types of the fibroblast lineage might also been involved in liver tissue repair and fibrogenesis. This chapter focuses on analysis of the phenotypical and functional properties of cells of the fibroblast lineage termed rat liver myofibroblasts (rMF) in comparison to HSC. Employing the discrimination features defined by these *in vitro* studies, the localization of HSC and rMF was analyzed in normal and acutely as well as chronically carbon tetrachloride-injured livers.

HSC and rat liver myofibroblasts were discernible by morphological criteria and growth behavior. While prolonged subcultivation of rat liver myofibroblasts was achieved, HSC were maintained in culture at maximum until passage 2. HSC were characterized by the expression of glial fibrillary acidic protein (GFAP), desmin, and vascular cell adhesion molecule-1, which were almost completely absent in rat liver myofibroblasts. With respect to synthetic properties, HSC and rat liver myofibroblasts displayed mostly overlapping properties but four striking differences.

The complement-activating protease P100 and the protease inhibitor α2-macroglobulin were expressed preferentially by HSC, whereas interleukin

[1]Present address: DeveloGen AG, 37079 Göttingen, Germany.

Copyright 2003, Elsevier Science (USA). All rights reserved.

(IL)-6 coding mRNAs and the extracellular matrix protein fibulin-2 were almost exclusively detectable in rat liver myofibroblasts.

In normal livers, HSC (desmin/GFAP + cells) were distributed in the hepatic parenchyma, whereas rMF [desmin/smooth muscle α-actin (SMA) +, GFAP negative cells colocalized with fibulin-2] were located in the portal field, the walls of central veins, and only occasionally in the parenchyma. Acute liver injury was characterized almost exclusively by an increase in the number of HSC, whereas the amount of rMF was nearly unchanged. In early stages of fibrosis, HSC and rMF were detected within the developing scars. In advanced stages of fibrosis, HSC were mainly present at the scar–parenchymal interface, whereas rMF accounted for the majority of the cells located within the scar. At every stage of fibrogenesis, rMF, in contrast to HSC, were detected only occasionally in the hepatic parenchyma.

Data demonstrate that morphologically and functionally different fibroblastic populations, HSC and rat liver myofibroblasts, can be derived from liver tissue, suggesting that HSC do not represent the single matrix-producing cell type of the fibroblast lineage in the liver. HSC and rMF are present in normal and diseased livers at distinct compartments and respond differentially to tissue injury. Acute liver injury is followed by an almost exclusive increase in the number of HSC, whereas in chronically injured livers not only HSC but also rMF are involved in scar formation. As HSC and rat liver myofibroblasts display overlapping but also different characteristics with respect to growth behavior and gene expression, both cell types should be included into therapeutic strategies targeting matrix-producing cells during fibrogenesis.

I. INTRODUCTION

A. Cell Types of the Fibroblast Lineage Involved in Tissue Repair

Tissue repair reactions include inflammation, formation of granulation tissue, and scar constitution. Granulation tissue develops from the connective tissue surrounding the damaged area and its cellular components are mainly endothelial cells and inflammatory cells, as well as fibroblasts and myofibroblasts (Desmouliere and Gabbiani, 1994; Gabbiani, 1996; Schmitt-Graff *et al.*, 1994). In normal and diseased tissues, fibroblasts and myofibroblasts are heterogeneous with respect to ultrastructural features and expression patterns of cytoskeletal proteins. Using the intermediate filaments vimentin and desmin, as well as the myofilament smooth muscle α-actin (SMA), Gabbiani and co-workers defined four cytoskeletal phenotypes among fibroblast/myofibroblasts: phenotype V, cells positive for vimentin only; phenotype VA, cells positive for vimentin and SMA;

phenotype VAD, cells positive for vimentin, SMA, and desmin; and phenotype VD representing cells positive for vimentin and desmin (Desmouliere and Gabbiani, 1994; Gabbiani, 1996; Sappino *et al.*, 1990). During wound contraction and fibrocontractive diseases, fibroblastic cells express SMA in a temporal or permanent manner, thereby acquiring typical myofibroblastic features (Desmouliere, 1995; Desmouliere and Gabbiani, 1995). Although the precise origin of myofibroblast has not been fully elucidated, they might derive from local fibroblasts, preexisting myofibroblasts, or other cell types, which have the potential to acquire a myofibroblastic phenotype like the mesangial cell of the kidney and hepatic stellate cells (Desmouliere and Gabbiani, 1995; Gabbiani, 1996).

B. Hepatic Stellate Cells and Their Role during Liver Tissue Repair

HSC, also designated as Ito cells, fat-storing cells, or lipocytes, are situated in the space of Disse and are identified by the presence of intracytoplasmatic lipid vacuoles containing vitamin A (Friedman, 1996; Mathew *et al.*, 1996; Pinzani, 1995; Ramadori, 1991). HSC play a major role in vitamin A metabolism and are regarded as the principal matrix-producing cell of the liver, thereby playing a major role in liver tissue repair and fibrosis. Liver fibrogenesis represents the hepatic tissue repair reaction in response to a chronic injury and is characterized by an increased and altered deposition of extracellular matrix (ECM) components, which results from a shifted balance between connective tissue production and degradation (Gressner and Bachem, 1995). Fibrogenesis is initiated by hepatocyte damage, leading to a recruitment of inflammatory blood cells and platelets, as well as activation of Kupffer cells with the subsequent release of different cytokines. HSC seem to be the primary target cells for inflammatory stimuli. During liver fibrogenesis, and also during primary culture, HSC proliferate, transform into myofibroblast-like cells (activated HSC), and synthesize large amounts of connective tissue components, certain enzymes involved in matrix degradation, and growth factors such as transforming growth factor-β, platelet-derived growth factor (PDGF), or insulin like growth factor-1 (IGF-1) (Gressner and Bachem, 1995) (Figs. 1 and 2).

C. Characteristics of HSC with Respect to Their Cellular Origin

HSC exhibit characteristics common to smooth muscle cells and myofibroblasts. HSC in primary culture show positive staining reactions to the intermediate filaments vimentin (De Leeuw *et al.*, 1984) and desmin (Tsutsumi *et al.*, 1987)

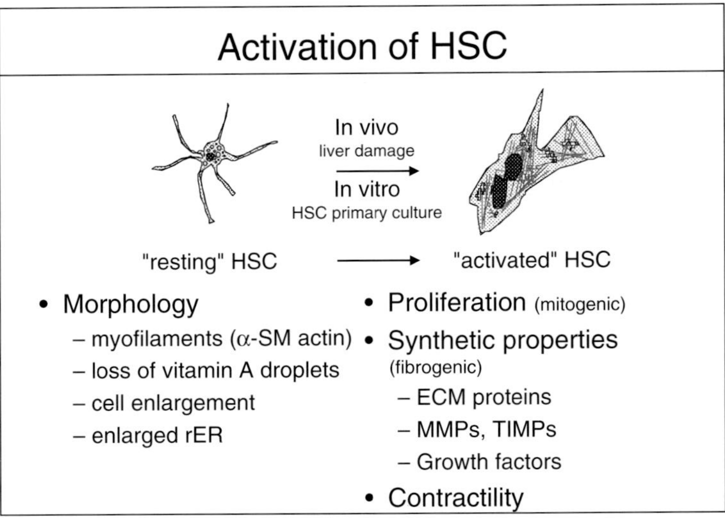

FIGURE 1 Activation of hepatic stellate cells (HSC): A key event during fibrogenesis. Basic features of the activation process of HSC.

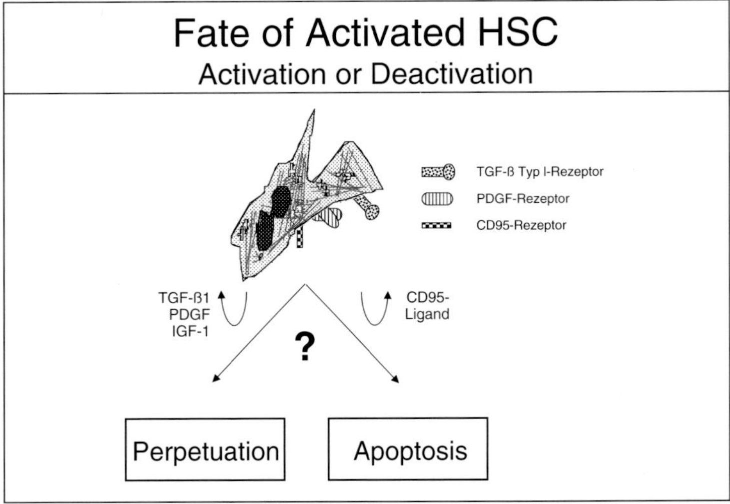

FIGURE 2 The fate of activated hepatic stellate cells. Activated HSC might undergo a self-perpetuation of the activation process triggered by an autocrine action of certain growth factors. Alternatively, a further activation might be inhibited through an induction of apoptosis mediated by the CD95 ligand.

and, after transition into myofibroblast-like cells after several days of culture termed activation, also to the myofilament SMA (Ramadori *et al.*, 1990). This *in vitro* activation process strongly resembles the morphological and functional changes observed in HSC *in vivo* during liver fibrogenesis and therefore HSC primary cultures are commonly used as an *in vitro* model to study the role of those cells during hepatic tissue repair (Fig. 1).

In addition to desmin and smooth muscle α-actin expression, which suggested a myogenic origin of these cells, HSC have been shown to express certain intermediate filaments and adhesion molecules typically found in neural cells such as glial fibrillary acidic protein, nestin, and neural cell adhesion molecule (Knittel *et al.*, 1996a; Neubauer *et al.*, 1996; Niki *et al.*, 1999). Interestingly, GFAP expression is present in resting HSC and is downregulated during HSC activation, whereas NCAM and nestin expression are undetectable in resting HSC and are induced during the activation process. Altogether, HSC display features typically found in fibroblasts and smooth muscle cells, but in addition, these cells share several characteristics with neural cells in particular with glia cells.

D. Other Liver Cell Types Potentially Involved in Liver Tissue Repair

Although myofibroblast-like (activated) HSC are believed to represent the principal fibroblastic cell type involved in liver fibrogenesis, activated HSC might undergo programmed cell death (Fig. 2). Therefore, the hypothesis is valid that other cell types of the fibroblast lineage, such as portal fibroblasts or vascular myofibroblasts, might also have fibrogenic potential and could replace activated HSC in the diseased liver. Indeed, as assessed by the expression of SMA and desmin and by their ultrastructural morphology, a heterogeneity of phenotypic features among myofibroblasts was noted in a few studies analyzing the diseased liver (Bhunchet and Wake, 1992; Enzan *et al.*, 1994; Schmitt-Graff *et al.*, 1993; Tang *et al.*, 1994; Tuchweber *et al.*, 1996) (Fig. 3). In areas of tissue injury and active fibrogenesis, SMA/desmin positive cells and SMA positive, desmin negative cells with a myofibroblast-like appearance, as well as typical fibroblasts, were detected. However, the identification of these cells was critical because marker proteins specific for HSC and other fibroblast-like cells present in the liver were not available at that time and therefore SMA positive cells detected in injured liver were often classified as activated HSC. Because the synthetic properties of HSC have been analyzed thoroughly during the last decade (Friedman, 1996; Pinzani, 1995; Ramadori, 1991) and because additional cell type-specific marker proteins have been identified (Knittel *et al.*, 1996a; Neubauer *et al.*, 1996), a reevaluation

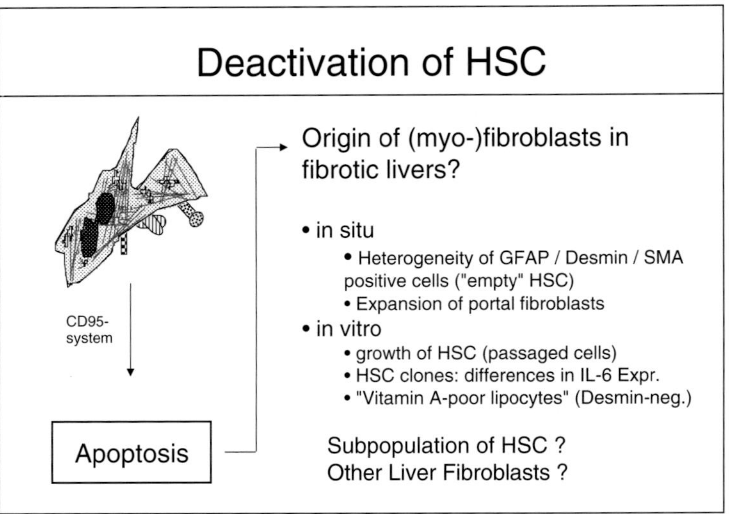

FIGURE 3 Origin of myofibroblasts in the fibrotic liver. As activated HSC might undergo apoptosis, fibroblasts different from HSC might exist in the liver and could be involved in scar formation.

of the role of HSC in comparison to other hepatic fibroblast populations during liver tissue repair became possible.

II. APPROACHES FOR THE IDENTIFICATION AND FURTHER CHARACTERIZATION OF THE DIFFERENT HEPATIC CELL POPULATIONS WITH FIBROGENIC POTENTIAL

Because a differentiation of the cell populations of the fibroblast lineage *in situ* often remains critical, an *in vitro* approach was used in recent studies as a first step to discriminate these cell populations under phenotypical and functional aspects (Knittel *et al.*, 1999c). Due to the hitherto unknown different growth kinetics of HSC and rMF, cultures were established, which were highly enriched with the distinct cell populations, in detail, HSC at different stages of differentiation and different subsets of rMF. As assessed by the expression pattern of different cytoskeletal proteins, cell surface molecules, ECM proteins, proteases, protease inhibitors, and cytokines, HSC and rMF were clearly discernible. Employing the discrimination criteria defined by these *in vitro* studies (Knittel *et al.*, 1996a), localization of the different fibroblast populations was analyzed in fibrotic liver illustrating clear differences in the tissue distribution pattern (Knittel *et al.*, 1999b).

A. Isolation and Cultivation of HSC and rMF

HSC were isolated from rat liver and kept in primary culture as described (Knittel *et al.*, 1992a,b,c, 1996a,b,c, 1997a,b,c). As assessed by morphology and by the expression of SMA of GFAP, HSC were fully "activated" at 7 days of primary culture and later, whereas cells cultured for 2 days were classified as "resting" HSC/HSC at an early stage of activation (Knittel *et al.*, 1996a,c; Neubauer *et al.*, 1996). Furthermore, freshly isolated HSC were plated for 7 days, released from culture plates by trypsinization, and replated at a 1/4 split ratio. Passaged HSC were analyzed at confluency, which usually took place within 7 days (first passage). In addition, HSC of passage 1 were subcultured using the same experimental conditions until passage 2. The purity of freshly isolated cells and cultured cells was assessed as stated earlier (Knittel *et al.*, 1992a,b,c, 1996a,b,c, 1997a,b,c).

Rat liver myofibroblasts were obtained by an outgrowth of primary nonparenchymal liver cells cultures. Briefly, the liver was digested enzymatically with pronase and collagenase. During stirring in the final enzyme-containing solution, the pH was not corrected to physiological levels in order to eliminate HSC. Nonparenchymal liver cells were separated by a Nycodenz density gradient, and the fraction comprising Kupffer cells and sinusoidal endothelial cells was purified further by centrifugal elutriation according to Knook *et al.* (1977) and De Leeuw et al. (1982, 1983) as described in detail (Knittel *et al.*, 1999c). Cells were cultured in DME supplemented with 15% FCS, 100 U/ml penicillin, 100 μg/ml streptomycin and 1% L-glutamine. At confluency, which was usually reached within 7–10 days, cells were released from the culture plates by trypsinization and were replated at a 1/4 split ratio. rMF were again passaged using the same experimental conditions at confluency. rMF were subcultured for several passages.

B. Morphological Features and Growth Kinetics of HSC and rMF

In agreement with previous studies (Knittel *et al.*, 1992b,c, 1996a,b,c, 1997a,c; Neubauer *et al.*, 1996; Ramadori *et al.*, 1990, 1991, 1992), HSC at 2 days after plating displayed numerous vitamin A-containing vacuoles located around the nucleus and had a star-like appearance. At 7 days of primary culture, HSC showed a myofibroblast-like morphology characterized by cell enlargement and reduction of the size of intracellular vacuoles (Knittel *et al.*, 1999c). HSC in first passage displayed the phenotypical characteristics of activated cells, whereas HSC of second passage were characterised by further cell enlargement, mostly a complete loss of vitamin A-containing vacuoles and intracellular condensation, probably of the cytoskeleton. HSC were maintained in culture at maximum until passage 2, which was terminated by spontaneous cell death, as evidenced by trypan blue staining.

Although the split ratio from the first to the second passage was reduced and cells were cultured in second passage for several weeks, HSC did not form a confluent monolayer but were always present as single giant cells without any proliferation capacity (Knittel *et al.*, 1999c).

In contrast to HSC, prolonged cultivation of rMF was achievable. In previous studies we observed that rMF were present in primary cultures of sinusoidal endothelial cells and Kupffer cells prepared from rat livers (unpublished observation). These rMF appeared after several days of cultivation and showed high proliferating activity, thereby representing a significant cell population in late-stage cultures. To study these cells, primary cultures of nonparenchymal liver cells were established from an elutriation fraction, which appeared to be enriched with rMF, but which also contained endothelial cells and Kupffer cells. Within a few days, rMF were detectable in the cultures and formed a confluent cell layer after prolonged cultivation. Cells in primary culture were not studied in detail because these cultures contained significant amounts of Kupffer cells and endothelial cells were not present, probably due to the cultivation conditions. rMF were passaged several times and were studied in detail between passages 1 and 6. In some cultures, about 3% Kupffer cells were still present in passage 1; however, starting from passage 2, rMF cultures were Kupffer cell free. rMF showed either a fibroblast-like phenotype characterized as spindle-shaped morphology or a myofibroblast-like appearance (Knittel *et al.*, 1999c). Upon further passaging, cells exhibiting the myofibroblast-like phenotype seemed to increase; however, until passage 6, both cell types were present within a single culture. In total, three independent rMF cultures were established, which showed similar growth kinetics but were slightly different in the amount of cells showing a fibroblast-like or myofibroblast-like phenotype.

C. Expression of Cytoskeletal Proteins by HSC and rMF

To characterize the differences of rMF and HSC in more detail, the expression of cytoskeletal proteins, namely the intermediate filaments vimentin, desmin, and GFAP and the myofilament SMA, was studied by immunocytochemistry (Fig. 4) and by Western and Northern blot analyses (Knittel *et al.*, 1999c). As reported (De Leeuw *et al.*, 1984; Neubauer *et al.*, 1996; Ramadori *et al.*, 1990; Tsutsumi *et al.*, 1987), activated HSC were vimentin, desmin, and SMA positive and showed a weak GFAP-specific immunoreaction (Fig. 4).

In accordance to the different morphological features, rMF were divergent with respect to vimentin, desmin, and SMA expression. In general, three different phenotypes were present, classified as rMF-I, rMF-II, and rMF-III. rMF-I were vimentin positive but SMA and desmin negative, cells of phenotype II were

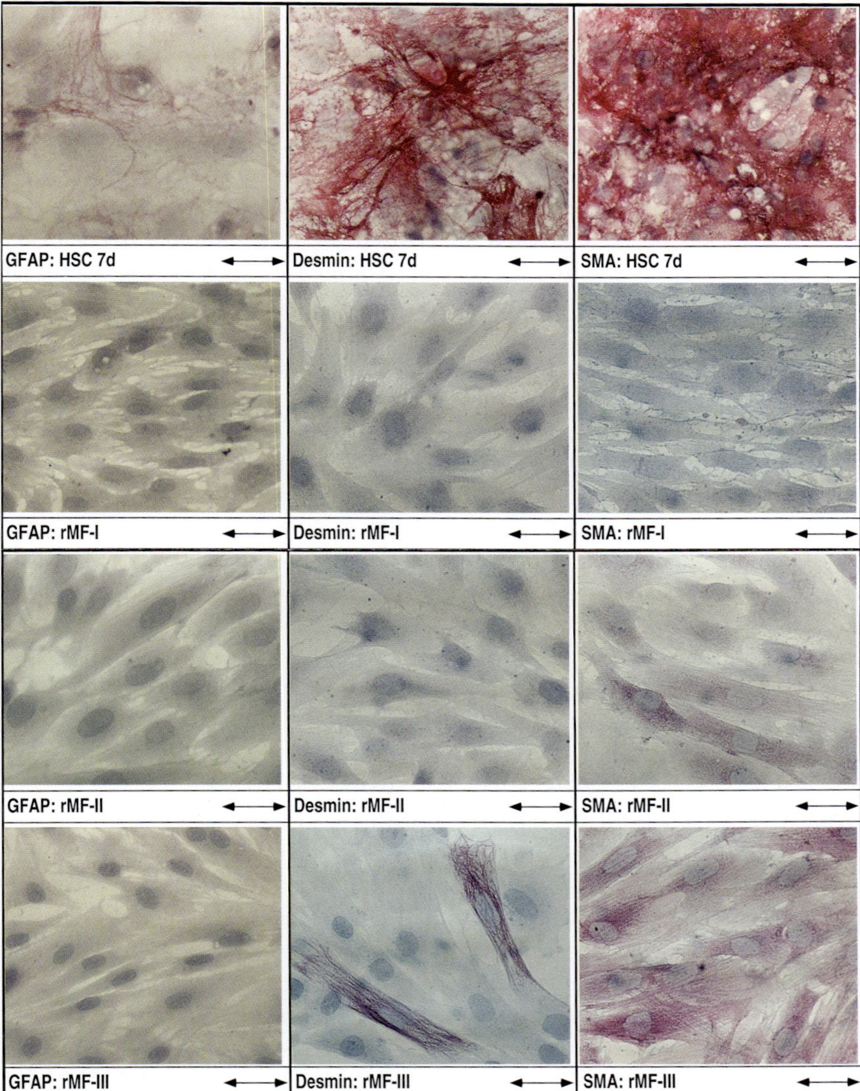

FIGURE 4 Expression of intermediate filaments and myofilaments in HSC and rMF. HSC 7 days after plating (HSC 7d) and rMF were incubated with antibodies directed against glial fibrillary acidic protein (GFAP), desmin and smooth muscle α-actin (SMA), or mouse Ig as negative controls (data not shown). Immunoreactive material was detected using the APAAP staining method. In the case of rMF, three different staining patterns were detected, which were classified as rMF-I (vimentin ++, SMA −, desmin −), rMF-II (vimentin ++, SMA +, desmin –), and rMF-II (vimentin ++, SMA ++, desmin +) (+, positive; ++, strongly positive; −, negative). Bar: 57 μm.

vimentin and SMA positive but desmin negative, and rMF-III were vimentin, SMA, and desmin positive. Analyzing three different cell preparations from passage 1 throughout passage 6, rMF were 100% positive for vimentin (V+), 3–13% of the cells were desmin positive (D+), and about 50–100% of the rMF were positive for SMA (A+). Although some rMF cultures showed the typical myofibroblast-like morphology, SMA expression varied within these cells from low to high expression levels, with the latter displaying typical stress fibers. Under all circumstances, GFAP expression was undetectable in rMF not only by immunocytochemistry (Fig. 4), but also by Western and Northern blotting (Knittel *et al.*, 1999c). In an attempt to dissect the rMF subpopulations, cells were plated in 96-well plates statistically at a concentration of one cell per well. Using this approach, we failed to generate rMF clones specific for the individual phenotype mostly because of low proliferative activity, but also due to changes in the expression pattern of cytoskeletal proteins upon further passaging.

In conclusion, by immunocytochemistry, V+, VA+, and VAD+ cells were present in rMF cultures according to the classification proposed by Gabbiani and co-workers (Desmouliere and Gabbiani, 1994; Gabbiani, 1996; Sappino *et al.*, 1990). Under quantitative aspects, the predominant phenotype of rMF was reflected by VA+ cells, whereas activated HSC were always VAD+. In addition, rMF did not express GFAP, which was detectable at low levels in activated HSC. Therefore, desmin and GFAP seem to represent cytoskeletal marker proteins, enabling a differentiation between rMF and HSC.

D. Expression of Cell Adhesion Molecules by HSC and Rat Liver Myofibroblasts

In addition to cytoskeletal proteins, HSC and rat liver myofibroblasts were classified according to their cell adhesion molecule expression, namely I-CAM-1, V-CAM-1, and N-CAM (Knittel *et al.*, 1999c). Activated HSC expressed all three types of cell adhesion molecules, which were detectable as immunoreactive material located on the cell surface and in form of their specific messengers (Knittel *et al.*, 1999c). In accordance with previous studies, N-CAM expression by HSC was induced during the *in vitro* activation process, whereas I-CAM-1 and V-CAM-1 were already present in resting cells (Knittel *et al.*, 1996a, 1999a).

Rat liver myofibroblasts were 100% I-CAM-1 positive, and V-CAM-1 was detected only in individual, mostly spindle-shaped cells, which account for about 2% of the rat liver myofibroblast population (Knittel *et al.*, 1999c). N-CAM expression was undetectable in some rat liver myofibroblasts cultures, whereas others expressed N-CAM at low levels as assessed by immunocytochemistry (Knittel *et al.*, 1999c). By Northern blot analysis, rat liver myofibroblasts and HSC expressed I-CAM-1-specific messengers at similar quantities, whereas N-CAM, particularly V-CAM-1 coding transcripts, was severalfold lower in rat liver myofibroblasts compared to HSC (Knittel *et al.*, 1999c). Therefore, V-CAM-1 and in part also N-CAM provide additional marker proteins to differentiate activated HSC and rat liver myofibroblasts. However, the use of V-CAM-1 is limited because the antibodies available commercially are only directed against human V-CAM-1, thereby displaying only a weak immunoreactivity with rat material even with rat HSC (Knittel *et al.*, 1999c).

III. EXPRESSION OF EXTRACELLULAR MATRIX PROTEINS, PROTEASES, AND PROTEASE INHIBITORS BY HSC AND rMF

Cells of fibroblastic origin play an essential role in connective tissue metabolism through the synthesis of ECM components and enzymes involved in matrix degradation. Therefore, the expression pattern of the latter proteins was analyzed in rMF and HSC, which are known to express several of the genes (Bachem *et al.*, 1993; Knittel *et al.*, 1992b,c, 1996b,c, 1997b,c; Leyland *et al.*, 1996; Ramadori *et al.*, 1991, 1992; Schwögler *et al.*, 1994).

Both cell populations expressed considerable amounts of transcripts coding for structural glycoproteins such as fibronectin, tenascin, thrombospondin-1, thrombospondin-2, laminin, and entactin (Knittel *et al.*, 1999c). Furthermore, messengers specific for proteoglycans, as shown in the case of biglycan and mRNAs specific for collagens, namely collagens type I, III, and IV, were detected in both cell populations. In accordance with previous results comparing HSC with skin fibroblast and arterial smooth muscle cells (Schwögler *et al.*, 1992), HSC and rat liver myofibroblasts contained tenascin-specific messengers of different size known to arise from alternative splicing. While HSC expressed tenascin-specific mRNAs of 7.2 kb in size, rat liver myofibroblasts contained predominantly messengers of 8.7 kb in size and of 7.2 kb only in minor quantities.

ECM protein-specific messengers were detected in activated HSC and rat liver myofibroblasts. However, the expression of proteases and protease inhibitors was markedly different (Knittel *et al.*, 1999c). Although both cell types contained mRNAs coding for t-PA, u-PA-specific transcripts were mainly present in HSC. In addition, messengers coding for the complement activating protease P100 and α_2-macroglobulin were mainly detectable in HSC and present in rat liver myofibroblasts at levels close to the detection limit. In order to demonstrate that the differences observed at the RNA level also hold up on the protein level, rat liver myofibroblasts and HSC were labeled endogenously, and α_2-macroglobulin, as well as ECM proteins fibronectin, tenascin, entactin, laminin, and collagen type IV, was immunoprecipitated from cell layer lysates and culture supernatants (Knittel *et al.*, 1999c). As reported (Andus *et al.*, 1987), α_2-macroglobulin synthesis and secretion were present in activated HSC and were only detectable after prolonged exposure in rat liver myofibroblasts at low levels. Furthermore, these experiments demonstrated that rat liver myofibroblasts in comparison to HSC displayed higher levels of fibronectin synthesis, whereas other ECM proteins, such as collagen type IV, laminin, and entactin, were detected in both cell populations at similar quantities (Knittel *et al.*, 1999c). In addition, collagen type I expression of HSC and rat liver myofibroblasts was studied at the protein level using Western blot analysis (Knittel *et al.*, 1999c). In agreement with previous studies (Knittel *et al.*, 1992b), the proα1(I) and proα2(I) chains of collagen type I were present in activated HSC but also in rat liver myofibroblast-derived samples (Knittel *et al.*, 1999c). As assessed by densitometric analysis of two independent experiments, collagen type I production of rat liver myofibroblasts was at least threefold higher compared to activated HSC (HSC day 7 after plating). Collectively, immunoprecipitation experiments and Western blot analysis confirmed Northern blot data and indicated that the ECM proteins fibronectin and collagen type I deposited in a fibrillar matrix are synthesized in higher amounts by rat liver myofibroblasts compared to HSC and that basement membrane proteins, such as collagen type IV, laminin, or entactin, are produced by rat liver myofibroblasts and HSC at similar levels.

IV. EXPRESSION OF CYTOKINES AND CELL MEMBRANE RECEPTORS BY HSC AND rMF

Cells of the fibroblast lineage are cellular targets for a number of cytokines; however, these cells are also known to secrete fibrogenic mediators. In particular, HSC have been shown to express various cytokines (Gressner, 1996), including members of the TGF family, such as TGF-β1 and bone morphogenetic protein-6 (BMP-6) (Knittel *et al.*, 1997a), and have been suggested to secrete IL-6 testing an HSC line developed after spontaneous immortalization of a HSC primary culture (Greenwel *et al.*, 1993).

By Northern blot analysis, TGF-β1 and BMP-6-specific messengers were detected in both cell populations (Knittel *et al.*, 1999c), whereas IL-6-specific mRNAs were nearly exclusively present in rMF and were detectable in HSC-derived samples only at low levels (Fig. 5). Because IL-6 expression in the liver is believed to arise mainly from liver macrophages, Kupffer cells and rMF were compared with respect to their IL-6 expression level (Fig. 5). On a microgram RNA basis, rMF displayed severalfold stronger IL-6 expression levels compared to unstimulated Kupffer cells, illustrating that rMF might represent an important cellular source of IL-6 synthesis. In addition, IL-6 expression by rat liver myofibroblasts was studied by *in situ* hybridization in order to clarify whether all rat liver myofibroblasts or only a subpopulation of rat liver myofibroblasts express IL-6-specific messengers. As shown previously (Knittel *et al.*, 1999c) most of the

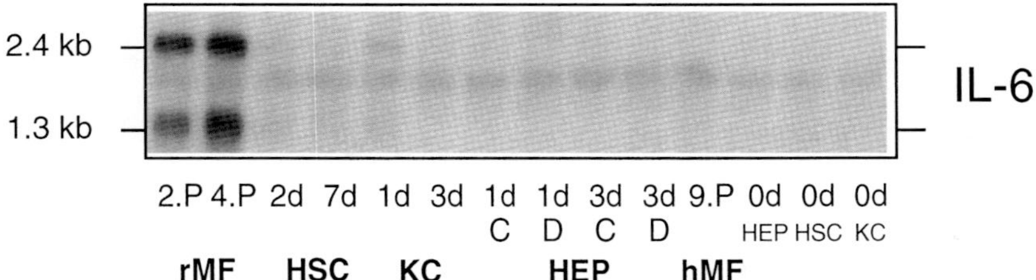

FIGURE 5 Expression of interleukin-6 by rMF in comparison to other liver cells. Total RNA was purified from rMF of passages 2–4 (2.P-4.P), rat HSC at 2 days (2), and 7 days (7d) after plating, rat Kupffer cells (KC) at 1 day (1d) and 3 days (3d) after plating, rat hepatocytes (HEP) cultured with (D) or without (C) dexamethasone at 1 day (1d) and 3 days (3d) after plating, human myofibroblast (hMF), or freshly isolated rat HSC/KC/HEP (0d). Total RNA (5 μg) was size selected by 1% agarose gel electrophoresis, and filters were hybridized using specific cDNA probes. Note that among the liver cells, rMF express large amounts of IL-6-specific messengers.

rat liver myofibroblasts were labeled heavily by an IL-6 antisense probe; however, similar to data on SMA, desmin, or V-CAM-1 expression, a subpopulation of rat liver myofibroblasts was IL-6 negative.

In addition to soluble mediators, the expression of cell membrane receptors such as the β1-integrin and the cation-dependent mannose 6-phosphate receptor (M-6-P-Rec) was studied (Knittel *et al.*, 1999c). At least for these cell membrane receptors, no major differences were noted in their expression by HSC and rMF.

V. FIBULIN-2: A NOVEL MARKER FOR THE DIFFERENTIATION OF HSC AND RAT LIVER MYOFIBROBLASTS

An additional striking difference of HSC and rat liver myofibroblasts was noted in the case of fibulin-2 expression. Fibulin-2 belongs to the larger family of ECM proteins, including fibulin-1, fibrillins, and latent TGF-β-binding proteins, which are characterized by long tandem arrays of EGF-like modules (Pan *et al.*, 1993). Fibroblasts are regarded as the typical cellular source for biosynthesis in which fibulin-2 is deposited with fibronectin into a fibrillar matrix (Sasaki *et al.*, 1996).

By employing RAP-PCR technology, fibulin-2 was identified through its presence in rat liver myofibroblast-derived samples and its absence in HSC-derived samples (Knittel *et al.*, 1999c). By Northern blot hybridization (Knittel *et al.*, 1999c), fibulin-2-specific messengers were detected as a large band migrating near the 28S RNA in accordance with the size of 4.5 kb reported for the mouse system

(Pan *et al.*, 1993). Fibulin-2 coding mRNAs were present almost exclusively in rat liver myofibroblasts, and HSC displayed only a very weak band in the case of HSC in the first passage. Using Western blot analysis (Knittel *et al.*, 1999c), fibulin-2 was detected at its expected molecular mass of 180 kDa (Sasaki *et al.*, 1996) solely in rat liver myofibroblasts, whereas in HSC, fibulin-2 was below the detection limit. By immunofluorescence, rat liver myofibroblasts were clearly fibulin-2 positive, which was present as a meshwork (Fig. 6) resembling the deposition of fibronectin. In contrast to rat liver myofibroblasts, HSC cultures were almost fibulin-2 negative (Fig. 6); however, single fibulin-2 positive cells or sometimes even clusters of fibulin-2 positive cells (Fig. 6) were detectable in HSC cultures after prolonged cultivation (day 7 onward).

The analysis of other hepatic cell populations, including hepatocytes, Kupffer cells, and sinusoidal endothelial cells using Northern blot and immunofluorescence, indicated that fibulin-2 was not present in the latter cell populations, both in their freshly isolated state and on different days after plating (Knittel *et al.*, 1999c).

Analyzing the cell populations on a single cell level indicated that more than 95% of the rat liver myofibroblasts were fibulin-2 positive. Using double immunofluorescence against desmin and SMA, the fibulin-2 positive rat liver myofibroblasts were either desmin positive or desmin negative (Knittel *et al.*, 1999c). The same was true in the case of SMA in that fibulin-2-expressing cells were SMA positive. However, the few fibulin-2 negative rat liver myofibroblasts were either desmin positive/negative or SMA positive/negative, thereby not allowing a classification of the fibulin-2 negative cells as SMA/desmin positive/negative cells (Knittel *et al.*, 1999c). Analyzing fibulin-2 positive cells in HSC cultures in detail demonstrated

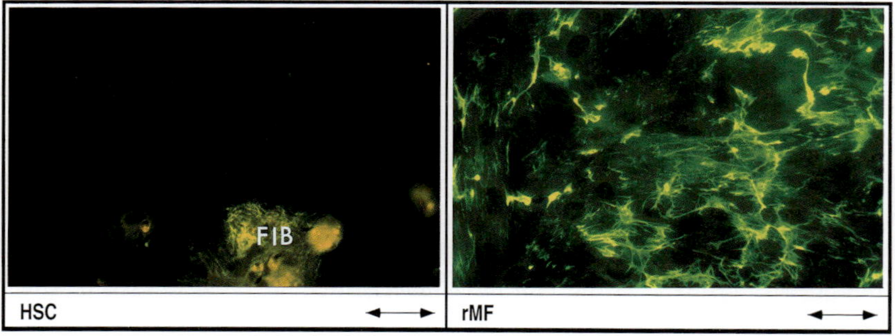

FIGURE 6 Detection of fibulin-2 in rMF by indirect immunofluorescence. HSC 7 days after plating (HSC) and rat liver myofibroblasts (rMF) of passage 2 were incubated with antibodies directed against fibulin-2. Immunoreactive material was detected using indirect immunofluorescence. "FIB" labeling indicates fibulin-2-positive cells in HSC cultures present as cell clusters. Bar: 57 μm.

that these cells, as observed in the case of rat liver myofibroblasts, were desmin positive or negative and SMA positive or negative (Knittel *et al.*, 1999c).

VI. DISTRIBUTION OF HSC AND rMF IN RAT LIVER TISSUE

As demonstrated by the aforementioned *in vitro* studies, HSC and rMF represent morphologically and functionally different cell populations of the fibroblast lineage. At present, activated HSC are regarded as the principal matrix-producing cell of diseased liver; however, the rMF population might also have fibrogenic potential. Therefore, the distribution of activated HSC and rMF within fibrotic liver tissue was analyzed, and special attention has been paid to the fibrotic septa using marker proteins identified by the cell culture experiments (Knittel *et al.*, 1999b). Because secreted proteins, such as proteases or cytokines, appear to be less suitable in identifying individual cells within a tissue by immunohistochemistry, cytoskeletal proteins and adhesion molecules were employed in the first step.

A. Distribution of HSC and rMF in Normal Rat Liver as Assessed by Cytoskeletal Proteins and Cell Adhesion Molecules

In accordance with published data, HSC, identified by the presence of GFAP, desmin, or V-CAM-1, were detected in the hepatic parenchyma within the perisinusoidal space (Knittel *et al.*, 1996a, 1999a; Neubauer *et al.*, 1996) (Table I). In addition, a GFAP-specific staining reaction was also noted in the portal field around the hepatic artery in colocalization with N-CAM (Knittel *et al.*, 1999b), which might correspond to the nerve endings as suggested previously (Knittel *et al.*, 1996a). Furthermore, desmin was also detectable in the portal field within the walls of the hepatic artery or the portal vein and the portal interstitium (Table I). V-CAM-1-specific staining reactions were additionally present in a sinusoidal pattern in the parenchyma due to its expression by other sinusoidal liver cells, such as endothelial cells or Kupffer cells (Knittel *et al.*, 1999a).

In contrast to GFAP/desmin/V-CAM-1, SMA and N-CAM-specific immunoreactivity was absent within the hepatic parenchyma (Table I). SMA was localized within the walls of the portal vein or the hepatic artery (Knittel *et al.*, 1999b). Occasionally, SMA and also desmin were detectable within the walls of major central veins. N-CAM immunoreactivity was noted around the hepatic artery and in minor amounts within the wall of the portal vein (Knittel *et al.*, 1999b). Bile ducts were negative for all the proteins mentioned earlier, but

TABLE I Overall Distribution Pattern of Immunohistochemical Markers in Normal and Diseased Rat Livers with Respect to Their Potential Cellular Counterpart or Tissue Compartment[a]

	Normal liver		Acute liver injury		Chronic liver injury	
Marker	HSC parenchyma	MF portal tract central vein	HSC necrotic area	MF portal tract	HSC parenchyma scar interface	MF scar
Vimentin	xxx	xxx	xxx	xxx	xxx	xxx
Desmin	xxx	xxx	xxx	xxx	xxx	xx
GFAP	xxx	o	xx	o	xx	x[c]
SMA	o	xxx	x	xxx	xx	xxx
V-CAM	xx	o	xxx	o	xx	x
N-CAM	o	x	x	xx	xx	xxx
Fibulin-2	x[b]	xxx	o	xxx	x[b]	xxx

[a]For further explanations, see text. Please note that positivity corresponds to the cellular compartment and/or the tissue compartment. xxx, strong positive; xx, positive; x, weak positive; o, negative.
[b]Fibulin-2 positivity most likely derived from rMF.
[c]GFAP positivity most likely derived from HSC.

were labeled using a vimentin-specific antibody, which was also reactive with other nonparenchymal liver cell populations (Knittel *et al.*, 1999b).

As outlined earlier, HSC and rMF analyzed *in vitro* were different with respect to the expression of intermediate filaments, myofilaments, and cell adhesion molecules. In detail, resting HSC, which are present in normal liver tissue, were GFAP, desmin, and V-CAM-1 positive. The latter proteins were completely absent in rMF (GFAP) or detectable only in a small subpopulation of rMF (desmin) or single positive cells (V-CAM-1). These *in vitro* results are in perfect agreement with the *in situ* data mentioned earlier, demonstrating that GFAP, desmin, and V-CAM-1 positive cells corresponding to HSC are present within the perisinusoidal space (Table I). SMA and N-CAM were found *in situ* in vessels walls and might correspond to the rMF population, as *in vitro* rMF were N-CAM and SMA positive, whereas resting HSC were N-CAM and SMA negative (Table I). Desmin positivity is found *in situ* in HSC but also in rMF located in the walls of the blood vessels and the portal interstitium (Table I).

B. Distribution of HSC and rMF in Normal Liver as Assessed by Fibulin-2 and Interleukin-6 Expression

In the former study (Knittel *et al.*, 1999c), the ECM protein fibulin-2 was found to be almost exclusively present in the rMF populations and absent in HSC of different stages of activation and other liver cell populations. Therefore, the

Differences between activated HSC and rMF in vitro

- Intermediate filaments and myofilaments
  - HSC: +: VAD; G (+) rMF: +: V, VA, VAD; G -
- Adhesion molecules
  - HSC: +: NIV rMF: I, (N), ((V))
- ECM proteins
  - HSC: Ten. 7.2 kb Fibulin-2 neg. rMF: Ten. 7.2 + 8.7 kb Fibulin-2 pos.
- Proteases and anti-proteases
  - HSC: α2-M +, P100 + rMF: α2-M -, P100 -
- Cytokines
  - HSC: IL-6 - rMF: IL-6 +

FIGURE 7 Overview of the differences between HSC and rMF in vitro. V, vimentin; A, smooth muscle α-actin; D, desmin; G, glial fibrillary acidic protein; I, ICAM-1; V, VCAM-1; N, NCAM; Ten, tenascin; α2-M, α_2-macroglobulin. +, positive; −, negative.

distribution of fibulin-2 was studied in normal liver. Fibulin-2 immunoreactivity was detectable in the walls of the portal vessels, mostly in the hepatic artery (Knittel *et al.*, 1999b), and bile ducts were fibulin-2 negative. In the hepatic parenchyma and the walls of central veins (Knittel *et al.*, 1999b), fibulin-2-specific immunoreactions were present but at considerable lower quantities compared with the portal field. The exclusive nature of fibulin-2 deposition is further underlined by a comparison with fibronectin, which is present in all hepatic compartments, including bile ducts.

Double immunofluorescence using desmin and SMA as markers for HSC/rMF and von Willebrand factor (vWF) as a marker for venous vessels due to its reactivity with endothelial cells (Knittel *et al.*, 1995) indicated that a colocalization of fibulin-2 and arterial and venous vessels was evident (Knittel *et al.*, 1999b).

Of particular interest appears the observation that the fibulin-2 deposition in the parenchyma also colocalized with vWF and, in some cases, also with desmin, suggesting that small vessels located within the parenchyma and their surrounding rMF were recognizable by the use of fibulin-2 antibodies. In summary, the differences noted for HSC and rMF studied *in vitro* also hold up when analyzing the *in situ* situation and it has been proposed that fibulin-2 immunoreactivity is a highly specific marker for rMF, which are localized within the portal field and the hepatic parenchyma, including the central veins (Knittel *et al.*, 1999b).

A major difference between HSC and rMF *in vitro* was noted in the case of IL-6 expression (Knittel *et al.*, 1999c). Consequently, the cellular localization of

IL-6-specific messengers was studied, as the use of a number of commercially available antibodies directed against IL-6 was not successful. In normal liver, IL-6-specific transcripts were present in the hepatic parenchyma, probably in Kupffer cells, but also in cells present within portal vessel walls and within the portal interstitium, most likely corresponding to rMF (Knittel *et al.*, 1999b).

C. Overall Distribution of HSC and rMF in Normal Rat Liver Tissue

It is known that, apart from HSC, other cells of the fibroblast lineage, e.g., myofibroblasts, fibroblasts, and smooth muscle cells, are detectable in normal rat within the portal field and around central veins (Bhunchet and Wake, 1992; Herbst *et al.*, 1997; Ramadori *et al.*, 1990; Tang *et al.*, 1994; Tuchweber *et al.*, 1996). In these studies, fibroblast-like cells were identified by the presence of desmin or SMA; however, desmin positive/SMA negative subpopulations or desmin negative–collagen expressing cells have been described, thereby demonstrating a heterogeneity of these fibroblast-like cells (Herbst *et al.*, 1997; Seifert *et al.*, 1994; Tang *et al.*, 1994; Tuchweber *et al.*, 1996). In the hepatic parenchyma of normal rat liver, HSC were identified through their positive staining reaction to GFAP and/or desmin; however, Niki and co-workers demonstrated that HSC positive only for GFAP or desmin were also present, again illustrating a heterogeneity of the HSC population (Niki *et al.*, 1996).

Altogether, data indicate that in normal liver, a clear differentiation of HSC from myofibroblasts, fibroblasts, and smooth muscle cells is only possible through their localization within the hepatic lobule. The ECM protein fibulin-2 could be helpful in differentiating at least the HSC population from other cells of fibroblastic origin. *In vitro* fibulin-2 synthesis was more or less absent in HSC, but present in the rMF population (Knittel *et al.*, 1999c). *In situ* fibulin-2 colocalized with desmin or SMA positive cells in the portal field and around central veins, thereby identifying the rMF population of these compartments even more precisely than desmin or SMA alone. Furthermore, fibulin-2 was codistributed in the hepatic parenchyma with small vessels identified by vWf/desmin staining. The latter result suggests that fibulin-2 represents the first marker protein, which identifies vascular rMF present in the hepatic parenchyma and differentiates them from HSC.

D. Distribution of HSC and rMF in Acutely Injured Rat Liver

During the course of a single liver injury, HSC are known to proliferate and to be present in higher amounts in the necrotic area (Friedman, 1996; Hautekeete and Geerts, 1997; Pinzani, 1995). By using the CCl_4 model characterized by hepatocellular necrosis, inflammation, and regeneration, resulting in complete restitution, the distribution of HSC (identified as desmin, GFAP, or V-CAM-1 positive cells) and rMF (identified by SMA, N-CAM, or fibulin-2 positivity) was analyzed (Knittel *et al.*, 1999b).

At 24 and 48 h after CCl_4 application, an increase in the number of either GFAP or desmin positive cells was evident in the necrotic area compared to not injured tissue regions (Knittel *et al.*, 1999b) in accordance with previous studies (Knittel *et al.*, 1996a, 1999a; Neubauer *et al.*, 1996). The same was true in the case of V-CAM-1; however, one has to take into consideration that apart from HSC, other cells, such as sinusoidal endothelial cells and Kupffer cells, might be labeled (Knittel *et al.*, 1999a). A semiquantitative analysis comparing the amount of desmin, GFAP, or V-CAM-1 positive cells in injured versus uninjured tissue areas was performed at different time points after CCl_4 application. This analysis indicated (Knittel *et al.*, 1999b) that an expansion of desmin, GFAP, or V-CAM-1 positive cells started at 12 h, reached maximal levels at 24–48 h, and declined thereafter to control levels at 96 h as reported (Saile *et al.*, 1997).

In contrast, only a few N-CAM or SMA positive cells were detectable in the necrotic area (Table I). Furthermore, fibulin-2 deposition or single fibulin-2 positive cells were not increased significantly in this compartment. Except for a stronger immunoreactivity of the vessels against SMA, N-CAM, and fibulin-2, no major differences were noted in diseased livers compared to controls throughout the complete time span (Knittel *et al.*, 1999b).

In summary, following a single liver injury comprising tissue necrosis, inflammation, and regenerative events, an expansion of the GFAP/desmin/V-CAM-1 positive, SMA/N-CAM/fibulin-2 negative cell population, thereby matching the criteria of HSC, was noted. In contrast, the SMA/N-CAM/fibulin-2 positivity, regarded as rMF, was close to the baseline level of normal liver, suggesting that in this injury model, primarily an expansion of the HSC population takes place. This observation is in agreement with the few studies analyzing the kinetics of desmin or SMA positive cells in this particular model (Ballardini *et al.*, 1988; Burt *et al.*, 1986; Geerts *et al.*, 1991; Ramadori *et al.*, 1990); however, in the case of SMA, the results contrast with others (Johnson *et al.*, 1992). Since in the latter study SMA positive HSC were already detected in normal liver, a SMA overstaining might have caused the divergent results.

E. Distribution of HSC and rMF in Fibrotic Rat Liver as Assessed by Cytoskeletal Proteins and Cell Adhesion Molecules

Presently, HSC, in particular myofibroblast-like (activated) HSC, are regarded as the principal matrix producing cell of diseased liver. Because the rMF population might also have fibrogenic potential and because specific marker proteins have been identified to differentiate even activated HSC from rMF (e.g., fibulin-2), the distribution of HSC and rMF within fibrotic liver tissue was analyzed (Knittel *et al.*, 1999b). Special attention has been paid to different stages of fibrogenesis (advanced fibrosis displaying massive scars versus early stages of fibrogenesis displaying small scars) and to fibrotic scars.

Early stages of fibrogenesis displaying small scars were characterized by the presence of SMA, desmin, and N-CAM positive cells within the scar (Knittel *et al.*, 1999b) and by the presence of desmin, GFAP positive cells located at the scar–parenchymal interface, and within the hepatic parenchyma (Knittel *et al.*, 1999b). In advanced stages of fibrosis displaying larger scars (Fig. 8), desmin and GFAP positive cells were located in the liver parenchyma, at the scar parenchyma interface but within the scar only in minor amounts (Fig. 8). As expected, SMA and N-CAM immunoreactivity was detected predominantly within the scar and in the hepatic parenchyma only in minor amounts. However, in contrast to GFAP and desmin, antibodies directed against SMA stained the full thickness of the scar with equal intensity (Fig. 8, Table I).

F. Distribution of HSC and rMF in Fibrotic Rat Liver Using Fibulin-2, P100, and Interleukin-6 as Marker Proteins

As fibulin-2 appeared to represent the most valuable marker to differentiate between HSC and rMF, its localization was analyzed in early and advanced stages of fibrogenesis (Knittel *et al.*, 1999b). In developing scars, a fine fibulin-2 positivity was noted, which colocalized with vWF, desmin, or SMA using double immunofluorescence techniques. However, in the case of desmin, desmin positive cells were also detected, which were not codistributed with fibulin-2. Fibulin-2 was additionally present in minor quantities in the hepatic parenchyma and was also in this compartment in part colocalized with vWF, desmin, or SMA positive cells.

In advanced stages of fibrosis displaying massive scars, fibulin-2-specific staining reactions were present mostly throughout the full thickness of the scar with more or less equal intensity, sometimes more pronounced in the center of the scar. Within this septa fibulin-2 deposition mapped to SMA positive cells, in part also to vWF positive material, representing newly formed vessels (Knittel *et al.*, 1995). In contrast to SMA, a difference was noted in the case of desmin. Desmin positive cells were colocalized with fibulin-2 mainly at the septal parenchymal, whereas in the center of the scar fibulin-2 positivity accompanied by desmin immunoreactivity was less evident (Knittel *et al.*, 1999b).

Because IL-6 might be a useful marker protein for rMF, the distribution of IL-6-specific transcripts was studied not only in normal but also in fibrotic livers

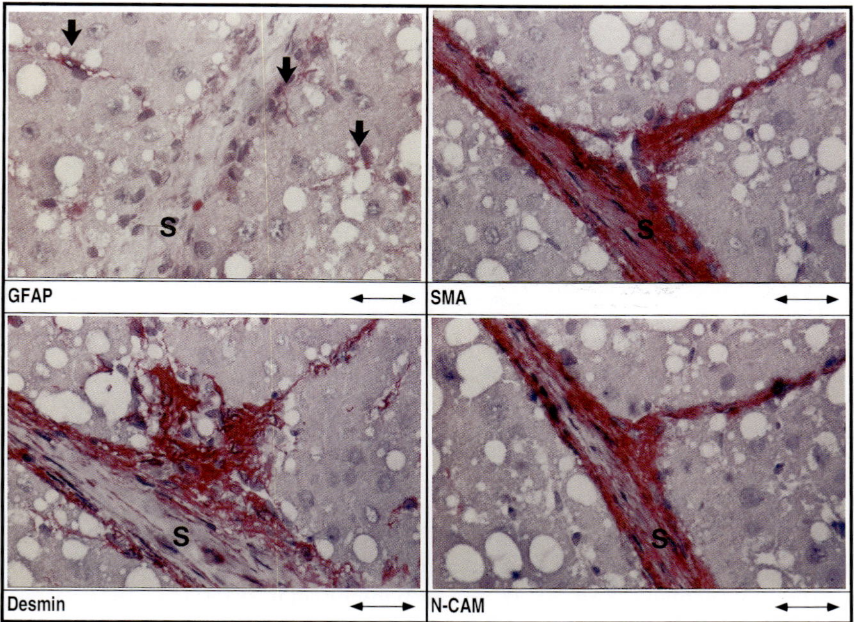

FIGURE 8 Localization of HSC and rMF in chronically damaged (fibrotic) rat livers: Distribution of GFAP, SMA, desmin, and N-CAM positive cells. Sections of chronically damaged (fibrotic) rat livers were incubated with antibodies directed against GFAP, SMA, desmin, and N-CAM. Immunoreactive material was detected using the APAAP staining method. Bar: 43 μm. Note the fibrotic scar (S) present within each section.

(Knittel *et al.*, 1999b). Nonparenchymal liver cells displaying an IL-6-specific hybridization signal were detected in the parenchyma and the fibrotic scars of early and advanced stages of fibrogenesis. In the case of pronounced septa, IL-6-specific messengers were present in cells located at the septal parenchymal interface, as well as in the center of the scar. However, because monocytes and macrophages (ED-1 and ED2 positive cells) are distributed in the latter compartments, the IL-6 hybridization pattern is not indicative for rMF.

In addition to the former experiments, the localization of P100 mRNA positive cells was studied, because the *in vitro* experiments demonstrated that expression of P100 was mainly detectable in HSC and present in rMF at levels close to the detection limit. As reported for normal liver (Knittel *et al.*, 1997b), P100-specific mRNAs were detectable in chronically damaged livers in hepatocytes identified by the size of the cells and nonparenchymal liver cells exhibiting a smaller nucleus. Interestingly, P100 positive cells were noted within the hepatic lobule, the scar–parenchyma interface, but were nearly absent within the scar (Fig. 9).

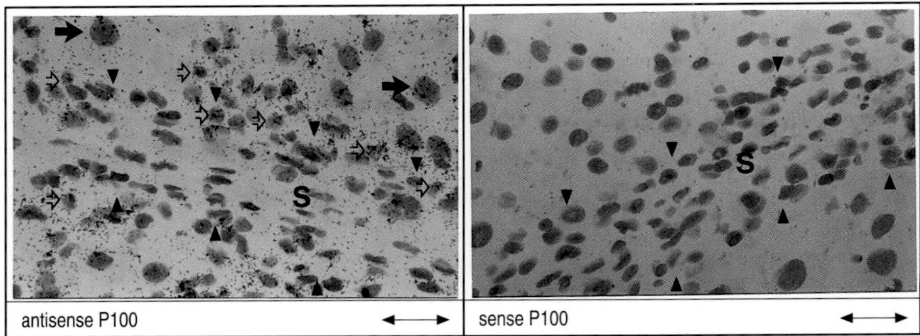

FIGURE 9 Cellular localization of P100 mRNA expressing cells. Liver tissue was hybridized with ^{35}S-labeled P100-specific antisense RNA probes and, to detect unspecific signals, with sense RNA probes. Open arrows indicate P100 positive nonparenchymal liver cells and close arrows show P100 positive hepatocytes, which are also P100 positive. Triangles illustrate the septal–parenchymal interface. S, septum. Bar: 43 μm.

In conclusion, the analysis of fibrotic liver tissue indicates that desmin/GFAP positive, P100 expressing cells, classified as HSC, were mainly present in the hepatic parenchyma and at the scar–parenchymal interface. In contrast, SMA/fibulin-2/IL-6 positive and desmin/GFAP negative, P100 nonexpressing cells, classified as rMF, accounted for the majority of the cells located within the scar, thereby displaying clear differences in the tissue distribution of the different fibroblast populations (Table I). Collectively, these observations suggest that upon chronic tissue injury, apart from HSC, fibroblastic cells present within the portal field or around the central veins start to proliferate, expand into the hepatic parenchyma, and are thereby deeply involved in scar formation (Fig. 10). Because HSC are known to undergo apoptosis following activation *in vitro* but also *in vivo* (Saile *et al.*, 1997), the rMF population could overtake the scar constitution in advanced stages of fibrosis.

VII. HSC AS A MODEL SYSTEM TO STUDY THE ROLE OF CELLS OF THE FIBROBLAST LINEAGE DURING TISSUE REPAIR

The use of HSC to study the functional characteristics and biological role of fibroblasts during tissue injury has several advantages. Commonly, cultures of fibroblastic cells are established by enzymatic digestion of the tissue or by outgrowth from small tissue parts. Under both conditions, cells have to be cultured for prolonged periods and are often passaged serially until experiments are performed. In contrast, the experimental model of HSC has the unique advantage of isolating fibroblastic cells from a tissue in considerable quantities and in

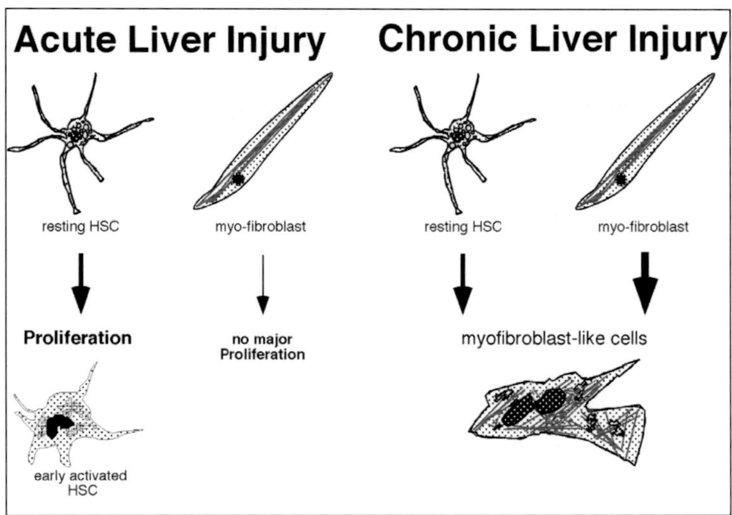

FIGURE 10 Role of cells of the fibroblast lineage during hepatic tissue repair. For further explanations, see text.

relative purities, sometimes exceeding 90%. Furthermore, in primary culture, HSC transform from a so-called resting into a myofibroblast-like (activated) cell and display basic characteristics attributed to fibroblastic cells upon tissue injury, such as proliferation, activation, differential gene expression, and apoptosis (Friedman, 1996; Knittel *et al.*, 1992b, 1996c, 1997c; Pinzani, 1995; Ramadori, 1991; Saile *et al.*, 1997). This *in vitro* situation strongly resembles the morphological and functional changes observed in HSC *in vivo* during liver fibrogenesis. Therefore, HSC primary cultures provide an attractive *in vitro* model to study the role of those cells during hepatic tissue repair and to design or to examine novel antifibrotic treatment strategies (Friedman, 1996; Pinzani, 1995; Ramadori, 1991).

However, in the future, a clear distinction of the hepatic cell types of the fibroblast lineage studied *in vitro* has to be performed. The differences in morphology, growth behavior, and gene expression between HSC and rat liver myofibroblasts are clearly evident (Knittel *et al.*, 1999c) and propose that in a number of reports, which were performed on severalfold passaged HSC, not HSC but rat liver myofibroblasts have been analyzed. It seems reasonable to suggest that upon passaging of these cultures the rat liver myofibroblast population has expanded dramatically and became the exclusive cell population present in the severalfold passaged cell cultures in the latter reports. However, using the marker proteins defined by Knittel *et al.* (1999c), a clear differentiation should be possible in upcoming studies.

A. Special Biosynthetic Properties of HSC

HSC and rMF were not only discernible by their different morphology, growth kinetics, and expression pattern of cytoskeletal proteins, but also by their synthetic capacity, suggesting different biological roles in tissue homeostasis. Although the majority of the proteins analyzed were so far expressed at similar levels, marked differences were noted with respect to the synthesis of IL-6, fibulin-2, α_2-macroglobulin, and certain proteases, such as u-PA and P100. HSC synthesized considerable higher amounts of proteases and of the protease inhibitor α_2-macroglobulin compared to rMF. Because HSC were located primarily at the scar–parenchyma interface, an area of active tissue remodeling, HSC could fulfil more extended functions in connective tissue metabolism than rMF, which were characterized by predominant matrix synthesis and were located in the center of the granulation tissue. As indicated by the differential expression of α_2-macroglobulin, a serum protein with strong cytokine-binding properties, the susceptibility to fibrogenic cytokines may also be different among the fibroblast populations. Because α_2-macroglobulin is known to reduce TGF-β1-induced collagen synthesis in fibroblastic cells of the liver (Tiggelman *et al.*, 1997), TGF-β1 might induce matrix production severalfold stronger in α_2-macroglobulin negative rMF compared to α_2-macroglobulin positive HSC, thereby resulting in a preferential matrix deposition around rMF located within the granulation tissue.

B. Potential Role of rMF with Regard to IL-6 Production

Because IL-6 expression was detectable in rMF cultures, a cytokine regarded as one of the major mediators of the acute phase reaction in the liver (Baumann and Gauldie, 1994; Heinrich *et al.*, 1990), rMF might play additional biological roles apart from connective tissue metabolism. The significance of this finding is further underlined by the fact that rMF expressed IL-6 coding messengers even at considerable higher quantities than unstimulated Kupffer cells, which are regarded the principal cellular source of IL-6 in the liver. In conclusion, these data could propose that hepatic myofibroblasts might trigger the acute phase response of the liver through local IL-6 production in response to hepatic tissue injury. However, these results also implicate that fibroblastic cells present in extrahepatic organs could stimulate the acute phase reaction in the case of remote localized inflammation, a hypothesis already proposed by some investigators (Gauldie *et al.*, 1990; Okamoto *et al.*, 1997; Xing *et al.*, 1993).

C. Heterogeneity of Fibroblasts Outside of the Liver

To our knowledge, extended studies concerning the morphological and functional differences of cells of the fibroblast lineage defined by an *in vitro* culturing model in combination with an *in vivo* model of tissue injury have not been performed yet, although several reports illustrated the heterogeneity of the latter cell populations. In particular, distinct subpopulations of smooth muscle cells have been identified (Bochaton-Piallat *et al.*, 1996; Frid *et al.*, 1994; Giuriato *et al.*, 1995; Kacem *et al.*, 1996; Majesky and Schwartz, 1990; Sakata *et al.*, 1990; Shanahan *et al.*, 1993; Villaschi *et al.*, 1994), which were studied in detail showing different expression patterns, e.g., of cellular retinal-binding protein-1 (Neuville *et al.*, 1997), of PDGF-B chain (Lindner *et al.*, 1995), or proliferation capacity (Wohrley *et al.*, 1995). In addition to smooth muscle cells, fibroblast derived from spleen (Borrello and Phipps, 1996), skin (Desmouliere *et al.*, 1992; Jelaska *et al.*, 1996; Kahari *et al.*, 1988; Maxwell *et al.*, 1987; Needleman *et al.*, 1990), gingiva (Hassell and Stanek, 1983), muscle (Desmouliere *et al.*, 1992), or kidney (Rodemann *et al.*, 1991) also comprise phenotypically or functionally subsets of cells. Furthermore, it has been demonstrated that differentiation of cells of the fibroblast lineage is regulated by certain soluble mediators. TGF-β1 mostly induced SMA expression in these cells, whereas interferon-γ exhibited opposite effects, suggesting antifibrotic activity (Desmouliere *et al.*, 1992, 1993; Knittel *et al.*, 1997c; Peehl and Sellers, 1997; Schmitt-Graff *et al.*, 1994; Verbeek *et al.*, 1994). However, the exact mechanisms responsible for differentiation into a particular phenotype according to the physiological needs still remain to be clarified (Desmouliere, 1995).

VIII. OUTLOOK

In summary, data published so far suggest that morphologically and functionally different fibroblast populations are present within a tissue, which are located in distinct compartments of the fibrotic tissue, suggesting similar but not identical roles during tissue repair reactions. Apart from HSC, rMF have to be regarded as an essential cell population involved in liver fibrogenesis, and antifibrotic treatment strategies in the future should pay high attention to this liver cell population. Although some of the differences might reflect special conditions present only in liver tissue, it is tempting to speculate that other findings represent common features of tissue repair reactions characterized by granulation tissue formation. Furthermore, because HSC and rMF were different in their growth kinetics *in vitro*, this experimental setting offers the unique opportunity to study the mechanisms, that control proliferation and apoptosis in different cells of the fibroblast lineage.

REFERENCES

Andus, T., Ramadori, G., Heinrich, P. C., Knittel, T., and Meyer zum Büschenfelde, K.-H. (1987). Cultured Ito cells of rat liver express the a2-macroglobulin gene. *Eur. J. Biochem.* **168**, 641–646.

Bachem, M. G., Meyer, D., Schäfer, W., Riess, U., Melchior, R., Sell, K. M., and Gressner, A. M. (1993). The response of rat liver perisinusoidal lipocytes to polypeptide growth regulator changes with their transdifferentiation into myofibroblast-like cells in culture. *J. Hepatol.* **18**, 40–52.

Ballardini, G., Fallani, M., Biagini, G., Bianchi, F. B., and Pisi, E. (1988). Desmin and actin in the identification of Ito cells and in monitoring their evolution to myofibroblasts in experimental liver fibrosis. *Virch. Arch. B* **56**, 45–49.

Baumann, H., and Gauldie, J. (1994). The acute phase response. *Immunol. Today* **15**, 74–80.

Bhunchet, E., and Wake, K. (1992). Role of mesenchymal cell populations in porcine serum-induced rat liver fibrosis. *Hepatology* **16**, 1452–1473.

Bochaton- Piallat, M., Ropraz, P., Gabbiani, F., and Gabbiani, G. (1996). Phenotypic heterogeneity of rat arterial smooth muscle cell clones. Implications for the development of experimental intimal thickening. *Arterioscler. Thromb. Vasc. Biol.* **16**, 815–820.

Borrello, M. A., and Phipps, R. P. (1996). Differential Thy-1 expression by splenic fibroblasts defines functionally distinct subsets. *Cell. Immunol.* **173**, 198–206.

Burt, A. D., Robertson, J. L., Heir, J., and MacSween, R. N. (1986). Desmin-containing stellate cells in rat liver; distribution in normal animals and response to experimental acute liver injury. *J. Pathol.* **150**, 29–35.

De Leeuw, A. M., Barelds, R. J., De Zanger, R., and Knook, D. L. (1982). Primary culture of endothelial cells of the rat liver. *Cell Tissue Res.* **223**, 201–215.

De Leeuw, A. M., Brouwer, A., Barelds, R. J., and Knook, D. L. (1983). Maintenance cultures of Kupffer cells isolated from rats of various ages: Ultrastructure, enzyme cytochemistry, and endocytosis. *Hepatology* **3**, 497–506.

De Leeuw, A. M., McCarthy, S. P., Geerts, A., and Knook, D. L. (1984). Purified rat liver fat storing cells in culture divide and contain collagen. *Hepatology* **4**, 392–403.

Desmouliere, A. (1995). Factors influencing myofibroblast differentiation during wound healing and fibrosis. *Cell. Biol. Int.* **19**, 471–476.

Desmouliere, A., and Gabbiani, G. (1994). Modulation of fibroblastic cytoskeletal features during pathological situations: The role of extracellular matrix and cytokines. *Cell. Motil. Cytoskel.* **29**, 195–203.

Desmouliere, A., and Gabbiani, G. (1995). Myofibroblast differentiation during fibrosis. *Exp. Nephrol.* **3**, 134–139.

Desmouliere, A., Geinoz, A., Gabbiani , F., and Gabbiani, G. (1993). Transforming growth factor-beta 1 induces alpha-smooth muscle actin expression in granulation tissue myofibroblasts and in quiescent and growing cultured fibroblasts. *J. Cell Biol.* **122**, 103–111.

Desmouliere, A., Rubbia, B. L., Abdiu, A., Walz, T., Macieira, C. A., and Gabbiani, G. (1992). Alpha-smooth muscle actin is expressed in a subpopulation of cultured and cloned fibroblasts and is modulated by gamma-interferon. *Exp. Cell Res.* **201**, 64–73.

Enzan, H., Himeno, H., Iwamura, S., Onishi, S., Saibara, T., Yamamoto, Y., and Hara, H. (1994). Alpha-smooth muscle actin-positive perisinusoidal stromal cells in human hepatocellular carcinoma. *Hepatology* **19**, 895–903.

Frid, M. G., Moiseeva, E. P., and Stenmark, K. R. (1994). Multiple phenotypically distinct smooth muscle cell populations exist in the adult and developing bovine pulmonary arterial media in vivo. *Circ. Res.* **75**, 669–681.

Friedman, S. L. (1996). Hepatic stellate cells. *Prog. Liver Dis.* **14**, 101–130.

Gabbiani, G. (1996). The cellular derivation and the life span of the myofibroblast. *Pathol. Res. Pract.* **192**, 708–711.

Gauldie, J., Northemann, W., and Fey, G. H. (1990). IL-6 functions as an exocrine hormone in inflammation: Hepatocytes undergoing acute phase responses require exogenous IL-6. *J. Immunol.* **144**, 3804–3808.

Geerts, A., Lazou, J., De Bleser, P., and Wisse, E. (1991). Tissue distribution, quantitation and proliferation kinetics of fat-storing cells in carbon tetrachloride-injured rat liver. *Hepatology* **13**, 1193–1202.

Giuriato, L., Chiavegato, A., Pauletto, P., and Sartore, S. (1995). Correlation between the presence of an immature smooth muscle cell population in tunica media and the development of atherosclerotic lesion: A study on different-sized rabbit arteries from cholesterol-fed and Watanabe heritable hyperlipemic rabbits. *Atherosclerosis* **116**, 77–92.

Greenwel, P., Rubin, J., Schwartz, M., Hertzberg, E. L., and Rojkind, M. (1993). Liver fat-storing cell clones obtained from a CCl_4-cirrhotic rat are heterogeneous with regard to proliferation, expression of extracellular matrix components, interleukin-6, and connexin 43. *Lab. Invest.* **69**, 210–216.

Gressner, A. M. (1996). Mediators of hepatic fibrogenesis. *Hepato-Gastroenterol.* **43**, 92–103.

Gressner, A. M., and Bachem, M. G. (1995). Molecular mechanisms of liver fibrogenesis: A homage to the role of activated fat-storing cells. *Digestion* **56**, 335–346.

Hassell, T. M., and Stanek, E. D. (1983). Evidence that healthy human gingiva contains functionally heterogeneous fibroblast subpopulations. *Arch. Oral Biol.* **28**, 617–625.

Hautekeete, M. L., and Geerts, A. (1997). The hepatic stellate (Ito) cell: Its role in human liver disease. *Virch. Arch.* **430**, 195–207.

Heinrich, P., Castell, J., and Andus, T. (1990). Interleukin-6 and the acute phase response. *Biochem. J.* **265**, 621–636.

Herbst, H., Frey, A., Heinrichs, O., Milani, S., Bechstein, W. O., Neuhaus, P., and Schuppan, D. (1997). Heterogenity of liver cells expressing procollagen types I and IV in vivo. *Histochem. Cell Biol.* **107**, 399–409.

Jelaska, A., Arakawa, M., Broketa, G., and Korn, J. H. (1996). Heterogeneity of collagen synthesis in normal and systemic sclerosis skin fibroblasts: Increased proportion of high collagen-producing cells in systemic sclerosis fibroblasts. *Arthritis Rheum.* **39**, 1338–1346.

Johnson, S., Hines, J., and Burt, A. (1992). Phenotypic modulation of perisinusoidal cells following acute liver injury: A quantitative analysis. *Int. J. Exp. Pathol.* **73**, 765–772.

Kacem, K., Seylaz, J., and Aubineau, P. (1996). Differential processes of vascular smooth muscle cell differentiation within elastic and muscular arteries of rats and rabbits: An immunofluorescence study of desmin and vimentin distribution. *Histochem. J.* **28**, 53–61.

Kahari, V. M., Sandberg, M., Kalimo, H., Vuorio, T., and Vuorio, E. (1988). Identification of fibroblasts responsible for increased collagen production in localized scleroderma by in situ hybridization. *J. Invest. Dermatol.* **90**, 664–670.

Knittel, T., Armbrust, T., Neubauer, K., and Ramadori, G. (1995). Expression of von Willebrand factor in normal and diseased rat livers and in cultivated liver cells. *Hepatology* **21**, 470–476.

Knittel, T., Armbrust, T., Schwögler, S., Schuppan, D., Meyer zum Büschenfelde, K.-H., and Ramadori, G. (1992a). Distribution and cellular origin of undulin in the rat liver. *Lab. Invest.* **67**, 779–787.

Knittel, T., Aurisch, S., Neubauer, K., Eichhorst, S., and Ramadori, G. (1996a). Cell type specific expression of neural cell adhesion molecule (N-CAM) in Ito cells of rat liver: Up-regulation during in vitro activation and in hepatic tissue repair. *Am. J. Pathol.* **149**, 449–462.

Knittel, T., Dinter, C., Kobold, D., Neubauer, K., Mehde, M., Eichhorst, S., and Ramadori, G. (1999a). Expression and regulation of cell adhesion molecules by hepatic stellate cells (HSC) of rat liver: Involvement of HSC in the recruitment of inflammatory cells during tissue repair. *Am. J. Pathol.* **154**, 153–167.

Knittel, T., Fellmer, P., Müller, L., and Ramadori, G. (1997a). Bone morphogenetic protein-6 is expressed in non parenchymal liver cells and upregulated by transforming growth factor-β1. *Exp. Cell Res.* **232**, 263–269.

Knittel, T., Fellmer, P., Neubauer, K., Kawakami, M., Grundmann, A., and Ramadori, G. (1997b). The complement activating protease P100 is expressed by hepatocytes and is induced by IL-6 in vitro und during the acute phase reaction in vivo. *Lab. Invest.* 77, 221–230.

Knittel, T., Fellmer, P., and Ramadori, G. (1996b). Gene expression and regulation of plasminogen activator inhibitor type I in hepatic stellate cells of rat liver. *Gastroenterology* **111**, 745–754.

Knittel, T., Janneck, T., Müller, L., Fellmer, P., and Ramadori, G. (1996c). Transforming growth factor-β1 regulated gene expression of Ito cells. *Hepatology* **24**, 352–360.

Knittel, T., Kobold, D., Piscaglia, F., Saile, B., Neubauer, K., Mehde, M., Timpl, R., and Ramadori, G. (1999b). Localisation of liver myofibroblasts and hepatic stellate cells in normal and diseased rat livers: Distinct roles of (myo-) fibroblast-subpopulations in hepatic tissue repair. *Histochem. Cell Biol.* **112**, 387–401.

Knittel, T., Kobold, D., Saile, B., Grundmann, A., Neubauer, K., Piscaglia, F., and Ramadori, G. (1999c). Rat liver myofibroblasts and hepatic stellate cells: Different cell populations of the fibroblast lineage with fibrogenic potential. *Gastroenterology* **117**, 1205–1221.

Knittel, T., Müller, L., Saile, B., and Ramadori, G. (1997c). Effect of tumor necrosis factor-alpha on proliferation, activation and protein synthesis of rat hepatic stellate cells. *J. Hepatol.* **27**, 1067–1080.

Knittel, T., Odenthal, M., Schuppan, D., Schwögler, S., Just, M., Meyer zum Büschenfelde, K.-H., and Ramadori, G. (1992c). Synthesis of undulin by rat liver fat storing cells: Comparison with fibronectin and tenascin. *Exp. Cell Res.* **203**, 312–320.

Knittel, T., Schuppan, D., Meyer zum Büschenfelde, K.-H., and Ramadori, G. (1992b). Differential expression of collagen types I, III, and IV by fat-storing (Ito) cells in vitro. *Gastroenterology* **102**, 1724–1735.

Knook, D. L., Blansjaar, N., and Sleyster, E. C. (1977). Isolation and characterization of Kupffer and endothelial cells from the rat liver. *Exp. Cell Res.* **109**, 317–329.

Leyland, H., Gentry, J., Arthur, M. J., and Benyon, R. C. (1996). The plasminogen-activating system in hepatic stellate cells. *Hepatology* **24**, 1172–1178.

Lindner, V., Giachelli, C. M., Schwartz, S. M., and Reidy, M. A. (1995). A subpopulation of smooth muscle cells in injured rat arteries expresses platelet-derived growth factor-B chain mRNA. *Circ. Res.* **76**, 951–957.

Majesky, M. W., and Schwartz, S. M. (1990). Smooth muscle diversity in arterial wound repair. *Toxicol. Pathol.* **18**, 554–559.

Mathew, J., Geerts, A., and Burt, A. D. (1996). Pathobiology of hepatic stellate cells. *Hepato-Gastroenterol.* **43**, 72–91.

Maxwell, D. B., Grotendorst, C. A., Grotendorst, G. R., and LeRoy, E. C. (1987). Fibroblast heterogeneity in scleroderma: Clq studies. *J. Rheumatol.* **14**, 756–759.

Needleman, B. W., Ordonez, J. V., Taramelli, D., Alms, W., Gayer, K., and Choi, J. (1990). In vitro identification of a subpopulation of fibroblasts that produces high levels of collagen in scleroderma patients. *Arthritis. Rheum.* **33**, 842–852.

Neubauer, K., Knittel, T., Aurisch, S., Fellmer, P., and Ramadori, G. (1996). Glial fibrillary acidic protein: A cell type specific marker protein for Ito cells in vivo and in vitro. *J. Hepatol.* **24**, 719–730.

Neuville, P., Geinoz, A., Benzonana, G., Redard, M., Gabbiani, F., Ropraz, P., and Gabbiani, G. (1997). Cellular retinol-binding protein-1 is expressed by distinct subsets of rat arterial smooth muscle cells in vitro and in vivo. *Am. J. Pathol.* **150**, 509–521.

Niki, T., De Bleser, P. J., Xu, G., van den Berg, K., Wisse, E., and Geerts, A. (1996). Comparison of glial fibrillary acidic protein and desmin staining in normal and CCl_4-induced fibrotic rat livers. *Hepatology* **23**, 1538–1545.

Niki, T., Pekny, M., Hellemans, K., Bleser, P., Berg, K., Vaeyens, F., Quartier, E., Schuit, F., and Geerts, A. (1999). Class VI intermediate filament protein nestin is induced during activation of rat hepatic stellate cells. *Hepatology* **29**, 520–527.

Okamoto, H., Yamamura, M., Morita, Y., Harada, S., Makino, H., and Ota, Z. (1997). The synovial expression and serum levels of interleukin-6, interleukin-11, leukemia inhibitory factor, and oncostatin M in rheumatoid arthritis. *Arthritis. Rheum.* **40**, 1096–1105.

Pan, T., Sasaki, T., Zhang, R., Fassler, R., Timpl, R., and Chu, M. (1993). Structure and expression of fibulin-2, a novel extracellular matrix protein with multiple EGF-like repeats and consensus motifs for calcium binding. *J. Cell Biol.* **123**, 1269–1277.

Peehl, D. M., and Sellers, R. G. (1997). Induction of smooth muscle cell phenotype in cultured human prostatic stromal cells. *Exp. Cell Res.* **232**, 208–215.

Pinzani, M. (1995). Novel insights into the biology and physiology of the Ito cell. *Pharmacol. Ther.* **66**, 387–412.

Ramadori, G. (1991). The stellate cell (Ito-cell, fat-storing cell, lipocyte, perisinusoidal cell) of the liver: New insights into pathophysiology of an intriguing cell. *Virch. Arch. B* **61**, 147–158.

Ramadori, G., Knittel, T., Odenthal, M., Schwögler, S., Neubauer, K., and Meyer zum Büschenfelde, K.-H. (1992). Synthesis of cellular fibronectin by rat liver fat-storing (Ito) cells: Regulation by cytokines. *Gastroenterology* **103**, 1313–1321.

Ramadori, G., Schwögler, S., Veit, T., Rieder, H., Chiquet-Ehrismann, R., Mackie, E. J., and Meyer zum Büschenfelde, K.-H. (1991). Tenascin gene expression in rat liver and in rat liver cells. In vivo and in vitro studies. *Virch. Arch. B* **60**, 145–153.

Ramadori, G., Veit, T., Schwögler, S., Dienes, H. P., Knittel, T., Rieder, H., and Meyer zum Büschenfelde, K.-H. (1990). Expression of the gene of the alpha-smooth muscle actin isoform in rat liver and in rat fat storing (Ito) cells. *Virch. Arch. B* **59**, 349–357.

Rodemann, H. P., Muller, G. A., Knecht, A., Norman, J. T., and Fine, L. G. (1991). Fibroblasts of rabbit kidney in culture. I. Characterization and identification of cell-specific markers. *Am. J. Physiol.* **261**, F283–F291.

Saile, B., Knittel, T., Matthes, N., Schott, P., and Ramadori, G. (1997). CD95/CD95L-mediated apoptosis of the hepatic stellate cell: A mechanism terminating uncontrolled hepatic stellate cell proliferation during hepatic tissue repair. *Am. J. Pathol.* **151**, 1265–1272.

Sakata, N., Kawamura, K., Fujimitsu, K., Chiang, Y. Y., and Takebayashi, S. (1990). Immunocytochemistry of intermediate filaments in cultured arterial smooth muscle cells: Differences in desmin and vimentin expression related to cell of origin and/or plating time. *Exp. Mol. Pathol.* **53**, 126–139.

Sappino, A. P., Schurch, W., and Gabbiani, G. (1990). Differentiation repertoire of fibroblastic cells: Expression of cytoskeletal proteins as marker of phenotypic modulations. *Lab. Invest.* **63**, 144–161.

Sasaki, T., Wiedemann, H., Matzner, M., Chu, M., and Timpl, R. (1996). Expression of fibulin-2 by fibroblasts and deposition with fibronectin into a fibrillar matrix. *J. Cell Sci.* **109**, 2895–2904.

Schmitt-Graff, A., Chakroun, G., and Gabbiani, G. (1993). Modulation of perisinusoidal cell cytoskeletal features during experimental hepatic fibrosis. *Virch. Arch. A* **422**, 99–107.

Schmitt-Graff, A., Desmouliere, A., and Gabbiani, G. (1994). Heterogeneity of myofibroblast phenotypic features: An example of fibroblastic cell plasticity. *Virch. Arch.* **425**, 3–24.

Schwögler, S., Neubauer, K., Knittel, T., and Ramadori, G. (1994). Entactin gene expression in diseased rat livers and in cultivated liver cells. *Lab. Invest.* **4**, 525–536.

Schwögler, S., Odenthal, M., Meyer zum Büschenfelde, K.-H., and Ramadori, G. (1992). Alternative splicing products of the tenascin gene distinguish rat liver fat storing cells from arterial smooth muscle cells and skin fibroblasts. *Biochem. Bioph. Res. Commun.* **185**, 768–775.

Seifert, W., Roholl, P., Blauw, B., van der Ham, F., van Thiel-De Ruiter, C., Seifert-Bock, I., Bosma, A., Knook, D., and Brouwer, A. (1994). Fat-storing cells and myofibroblasts are involved in the initial phase of carbon tetrachloride-induced hepatic fibrosis in BN/BiRij rats. *Int. J. Exp. Pathol.* **75**, 131–146.

Shanahan, C. M., Weissberg, P. L., and Metcalfe, J. C. (1993). Isolation of gene markers of differentiated and proliferating vascular smooth muscle cells. *Circ. Res.* 73, 193–204.

Tang, L., Tanaka, Y., Marumo, F., and Sato, C. (1994). Phenotypic change in portal fibroblasts in biliary fibrosis. *Liver* **14**, 76–82.

Tiggelman, A. M., Linthorst, C., Boers, W., Brand, H. S., and Chamuleau, R. A. (1997). Transforming growth factor-β-induced collagen-synthesis by human liver myofibroblasts is inhibited by a2-macroglobulin. *J. Hepatol.* **26**, 1220–1228.

Tsutsumi, M., Takada, A., and Takase, S. (1987). Characterization of desmin-positive rat liver sinusoidal cells. *Hepatology* 7, 277–284.

Tuchweber, B., Desmouliere, A., Bochaton, P. M., Rubbia, B. L., and Gabbiani, G. (1996). Proliferation and phenotypic modulation of portal fibroblasts in the early stages of cholestatic fibrosis in the rat. *Lab. Invest.* **74**, 265–278.

Verbeek, M. M., Otte, H. I., Wesseling, P., Ruiter, D. J., and de Waal, R. (1994). Induction of alpha-smooth muscle actin expression in cultured human brain pericytes by transforming growth factor-beta 1. *Am. J. Pathol.* **144**, 372–382.

Villaschi, S., Nicosia, R. F., and Smith, M. R. (1994). Isolation of a morphologically and functionally distinct smooth muscle cell type from the intimal aspect of the normal rat aorta: Evidence for smooth muscle cell heterogeneity. *In Vitro Cell Dev. Anat.* **30**, 589–595.

Wohrley, J. D., Frid, M. G., Moiseeva, E. P., Orton, E. C., Belknap, J. K., and Stenmark, K. R. (1995). Hypoxia selectively induces proliferation in a specific subpopulation of smooth muscle cells in the bovine neonatal pulmonary arterial media. *J. Clin. Invest.* **96**, 273–281.

Xing, Z., Jordana, M., Braciak, T., Ohtoshi, T., and Gauldie, J. (1993). Lipopolysaccharide induces expression of granulocyte/macrophage colony-stimulating factor, interleukin-8, and interleukin-6 in human nasal, but not lung, fibroblasts: Evidence for heterogeneity within the respiratory tract. *Am. J. Res. Cell Mol.* 9, 255–263.

CHAPTER 8

Role of Sinusoidal Endothelial Cells in Liver Inflammation and Repair

GIULIANO RAMADORI AND KATRIN NEUBAUER

Department of Internal Medicine, Section of Gastroenterology and Endocrinology, University of Göttingen, Göttingen D-37075, Germany

Inflammatory processes are thought to be the consequence of tissue damage induced by different toxins and agents. However, while in some cases the damaging agent itself is able to induce cell death directly, in many cases, cell death probably takes place through recruited inflammatory cells. Inflammatory processes of the liver are supposed to be different from those of other organs because of the special structure of liver sinusoids, which do not possess a basement membrane, and because of the specialized liver sinusoidal endothelial cells (SECs), which are in contact with hepatic stellate cells (HSCs) and are separated from the hepatocytes by the space of Disse. We could show that intercellular adhesion molecule-1 (ICAM-1) gene expression in the liver sinusoid increases early after carbon tetrachloride (CCl_4) administration before leukocyte function antigen-1 (LFA-1) positive inflammatory cells are recruited into the pericentral area and before hepatocellular damage occurs. The increase of ICAM-1 expression is preceded by an increase of tumor necrosis factor (TNF)-α gene expression in the liver. The same cytokine is able to upregulate ICAM-1 expression in isolated rat SECs. The expression of another member of the immunoglobulin family of adhesion molecules, platelet endothelial cell adhesion molecule-1 (PECAM-1), is crucial for transmigration and is decreased in livers following CCl_4 administration,

Copyright 2003, Elsevier Science (USA). All rights reserved.

as well as in isolated SECs after TNF-α treatment. However, next to SECs, HSCs are also capable of responding to TNF-α treatment with increased ICAM-1 expression. The functional blockage of liver macrophages by the administration of gadolinium chloride (GD) prior to CCl_4 treatment leads to a reduction of the number of inflammatory cells and of hepatocellular damage, suggesting an important role of Kupffer cells (KC) as a source for TNF-α. In case of chronic persistence of the damaging agent, some sinusoids become involved in bridging necrosis and septa formation. The endothelial cells of those sinusoids modify their phenotype and possibly become involved in the so-called capillarization of the sinusoid. In fact, we could demonstrate that isolated SECs are able to synthesize several matrix proteins. The synthesis of matrix proteins by SECs is enhanced when cells are isolated from chronically CCl_4-treated livers and by the treatment of isolated SECs with the main fibrogenic mediator transforming growth factor (TGF)-β. In conclusion, sinusoidal endothelial liver cells represent an efficient barrier against corpusculate matter under normal conditions. Migration of inflammatory cells through the sinusoidal wall seems to happen only after activation of SECs and HSCs and probably after the involvement of Kupffer cells through messengers released by stressed hepatocytes. Once cellular debris has been cleared by recruited inflammatory cells, endothelial cells, together with HSCs, can participate in the repair process, as well as in scar formation.

I. INTRODUCTION

Inflammatory processes are characterized by the migration of inflammatory cells into areas affected by various noxious agents. Migration induced by chemokines, as e.g., interleukin II-8 and IP-10 (Masumoto *et al.*, 1998; Patzwahl *et al.*, 2001), released at sites of injury is mediated by the interaction of adhesion molecules on migrating inflammatory cells and adhesion molecules on resident cells (Butcher and Picker, 1996; Garcia-Barcina *et al.*, 1995; Springer, 1994). Four classes of adhesion molecules are known; adhesion molecules of the immunoglobulin superfamily, selectins, integrins, and cadherins. Members of the immunoglobulin (Ig) superfamily are characterized by one or more immunoglobulin domains. The immunoglobulin family comprises molecules such as ICAM, VCAM, NCAM, MadCAM, and PECAM. Selectins are preferentially involved in the first step of cell–cell interaction, the so-called "rolling," whereas VCAM and ICAM are important for later steps, the so-called "sticking" and "transmigration" (Butcher and Picker, 1996; Dunon *et al.*, 1996).

Inflammatory processes in the liver differ from those taking place in other organs. Liver SECs are, in many aspects, different from capillary or large vessel endothelium, e.g., they possess typical fenestrations clustered in sieve plates, which freely admit plasma constituents, but not formed blood elements, to the surface of hepatocytes. Whereas endothelium of most tissue rests on a basement membrane, sinusoids of the liver lack an electron-dense basement membrane. Moreover, SECs of the normal liver, in contrast to capillary endothelial cells, were suggested not to express certain adhesion molecules, such as most selectins, cadherins, CD34, and PECAM-1(CD31) (Couvelard *et al.*, 1993; Scoazec and Feldmann, 1991, 1994).

Due to the restricted set of adhesion molecules expressed by SECs, two pathways of cell–cell interaction during inflammatory processes in the liver were suggested (Garcia-Barcina *et al.*, 1995). One of them is the interaction between ICAM-1 of SECs and its ligands LFA-1 and MAC-1 expressed by inflammatory cells; the other pathway involves the VCAM-1 and VLA-4 interaction.

However, the first steps leading to the infiltration of the liver with inflammatory cells are still poorly understood. In a model of hepatitis, Ando *et al.* (1993) suggested that after viral infection, hepatocytes have a change of their surface antigen(s), which inflammatory cells recognize. Thereafter these inflammatory cells are capable of initiating a cascade of events that finally ends up in hepatocellular death. The induction of liver cell damage with the migration of inflammatory cells into the pericentral area of the liver lobule after a single administration of carbon tetrachloride (CCl_4) to the rat represents a good model for also studying early steps during liver inflammation. In fact, this model offers the almost unique possibility to isolate the different cellular participants of the inflammatory process at different time points before, during, and after inflammation and liver damage. Furthermore, these cells can be isolated from the liver of untreated animals and used to analyze the effects of mediators thought to be involved in the inflammatory process. In the CCl_4 model of liver injury used for our studies, we observed the following sequence of events.

II. INTERCELLULAR ADHESION MOLECULE-1 UPREGULATION ON SINUSOIDAL ENDOTHELIAL CELLS PRECEDES INFILTRATION WITH INFLAMMATORY CELLS AND DEVELOPMENT OF NECROTIC AREAS

In contrast to other endothelial cells, which only express ICAM-1 after activation, ICAM-1 positivity was detected along the sinusoids of normal rat livers as seen by

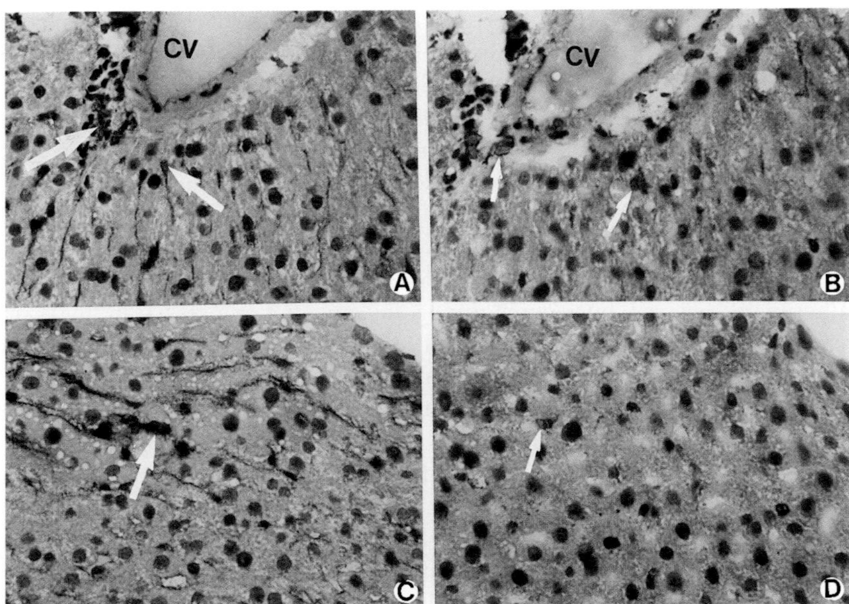

FIGURE 1 Indirect immunodetection of ICAM-1 and LFA-1-positive cells in sections of livers from control animals (A,B) or from rats 6 h after CCl_4 administration (C,D) using the APAAP staining method. Sections were stained with a monoclonal antibody directed against rat ICAM-1 (A,C) or with a monoclonal antibody directed against rat LFA-1 (B,D), followed by APAAP immunodetection (original magnification ×250). Big arrows indicate ICAM-1-positive nonparenchymal cells with small nuclei along the sinusoids and around the vessels. Vessel walls are ICAM-1 negative. Small arrows indicate LFA-1 immunoreactive cells located around the vessels and rarely distributed along the sinusoids. (C) Increased ICAM immunoreactivity along the sinusoids 6 h after treatment, whereas the pattern of LFA-1 staining remains unchanged. CV, central vein. Reproduced from Neubauer *et al.* (1998), with permission.

immunohistological staining of liver sections (Neubauer *et al.*, 1998). However, ICAM-1 positivity increased along the sinusoids 3–6 h after CCl_4 administration and finally accumulated in the necrotic areas (24–48 h after the administration) (Figs. 1 and 2). The ICAM-1 steady-state mRNA level in liver tissue increased 3–6 h after the CCl_4 treatment and returned to normal levels 48 h after the treatment (Fig. 3) as revealed by Northern blot analysis. Increased amounts of ICAM-1-specific transcripts could be observed in isolated SECs already 3–6 h after CCl_4 administration (Fig. 4). Immunohistology of normal rat livers revealed a few LFA-1-immunoreactive cells around the vessel walls. Not before 12 h after CCl_4 administration, the number of LFA-1-immunoreactive cells increased around the vessel walls and along the sinusoids and later accumulated in the necrotic area

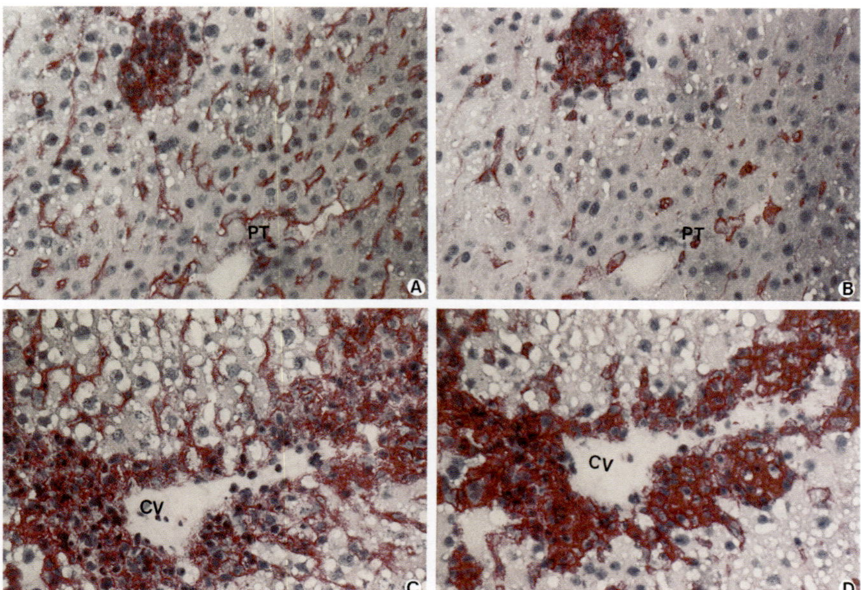

FIGURE 2 Indirect immunodetection of ICAM-1 and LFA-1-positive cells in sections of livers from rats 18 h after CCl_4 administration (A,B) or 48 h after CCl_4 administration (C,D). Sections were stained with a monoclonal antibody directed against rat ICAM-1(A,C) or with a monoclonal antibody directed against rat LFA-1 (B,D), followed by APAAP immunodetection. (A,B) Eighteen hours post-CCl_4 administration, the amount of LFA-1 immunoreactive cells increased around the vessels and along the sinusoids and formed accumulations (original magnification ×250). PT, portal tract; CV, central vein. Reproduced from Neubauer *et al.* (1998), with permission.

(Fig. 2), which suggested an early upregulation of ICAM-1 in liver cells prior to the accumulation of LFA-1 expressing cells, prior to the development of necrotic areas.

III. TUMOR NECROSIS FACTOR-α ENHANCES INTERCELLULAR ADHESION MOLECULE-1 TRANSCRIPT LEVEL IN CULTURE OF SINUSOIDAL ENDOTHELIAL CELLS

As TNF-α has been shown to be upregulated early after CCl_4 administration (Bruccoleri *et al.*, 1997; Knittel *et al.*, 1999), the influence of TNF-α on ICAM-1 gene expression was analyzed by Northern blot analysis. TNF-α treatment (100 U/ml for 8 h) of 1-day cultured rat SECs from normal liver increased the ICAM-1-specific transcript level (Fig. 5) (Neubauer *et al.*, 2000a).

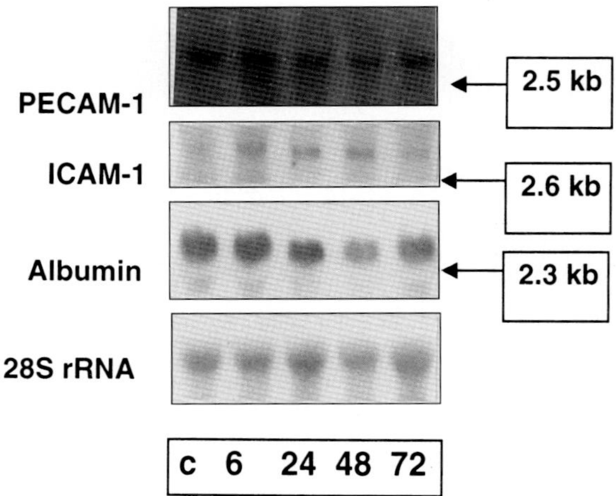

FIGURE 3 Northern blot analysis of total RNA extracted from livers from control animals (c) or from CCl_4-treated rats (6, 24, 48, and 72 h after a single administration of CCl_4). RNA was electrophoresed (10 μg of total RNA per lane), blotted, and subsequently hybridized with ^{32}P-dCTP-labeled cDNAs specific for PECAM-1, ICAM-1, and albumin. Reproduced from Neubauer *et al.* (2000b), with permission.

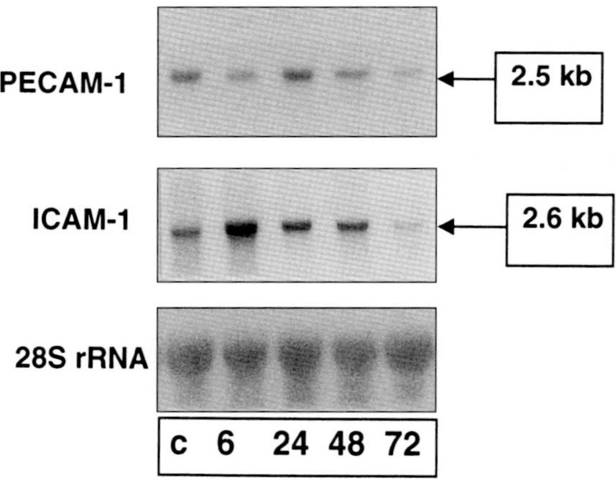

FIGURE 4 PECAM-1 gene expression in sinusoidal endothelial cells isolated from rat livers after a single CCl_4 administration. Northern blot analysis of total RNA extracted from sinusoidal endothelial cells isolated from normal rat livers (c), 6, 24, 48, or 72 h after a single CCl_4 administration. Three rats were treated at each time point. Five milligrams of total RNA was separated on agarose gel, blotted, and hybridized with a ^{32}P-dCTP-labeled cDNA probe specific for rat PECAM-1, ICAM-1. Normalization was performed with an oligonucleotide specific for the 28S rRNA. Reproduced from Neubauer *et al.* (2000b), with permission.

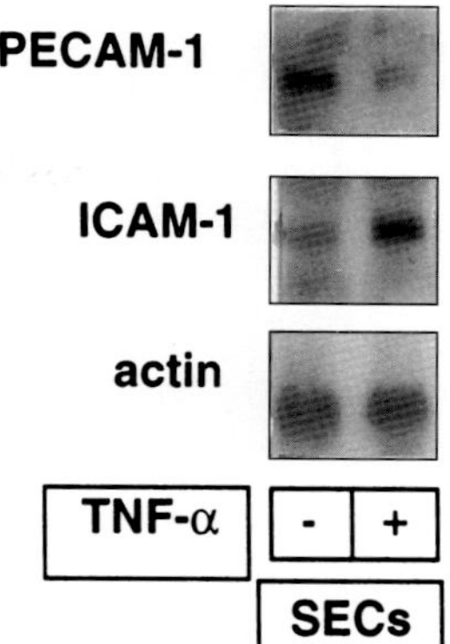

FIGURE 5 Sinusoidal endothelial cells at day 1 after isolation were cultured in the absence or presence of TNF-α (1000 U/well) for 8 h under serum-reduced conditions (0.3% fetal calf serum). TNF-α causes an increase in the amount of ICAM-1-specific transcripts and a decrease of the PECAM-1-specific transcript level. Normalization was performed with an oligonucleotide specific for the 28S rRNA. Reproduced from Neubauer et al. (2000a), with permission.

A. Kupffer Cells Seem to Be Responsible for Enhanced Synthesis of TNF-α Following CCl_4 Treatment, as Inactivation of Kupffer Cells by Pretreatment with Gadolinium Chloride Results in a Reduced TNF-α Level and a Reduction of Areas of Necrosis

Because Kupffer cells are suggested to be an important source of secreted TNF-α, we analyzed the TNF-α transcript level in CCl_4-treated livers with or without GD pretreatment by Northern blot analysis. These studies indicated that, indeed, when livers are treated with gadolinium chloride prior to CCl_4 treatment, there is a reduction of the areas of necrosis and there is a decrease in the TNF-α transcript level in liver tissue compared to solely CCl_4-treated livers (Fig. 6). These data let us take into consideration that CCl_4 induces a "disturbance" of hepatocytes but does not necessarily induce cell death. Hepatocellular necrosis may be a result of the interaction of "disturbed" hepatocytes with recruited inflammatory cells. Recruitment of inflammatory cells seems to take place after activation of SECs and HSCs and after enhanced TNF-α secretion by KC, which could be induced by messengers released by stressed hepatocytes. The involvement of Kupffer cells has been demonstrated in different T-cell-dependent liver injury in mice. When KC were depleted by clodronate liposomes, TNF RNA and protein production were strongly attenuated and liver damage was restricted to a few small necrotic areas (Schumann *et al.*, 2000).

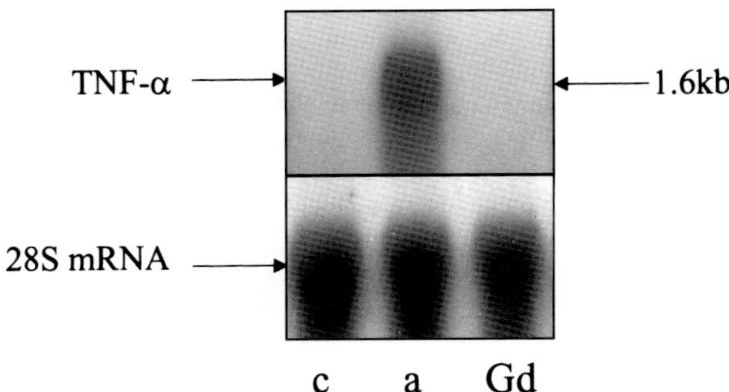

FIGURE 6 Northern blot analysis of total RNA extracted from control livers (c) or from acutely damaged livers (48 h after the administration of CCl_4) (a) or from gadolinium chloride-pretreated (24 h prior to CCl_4 treatment) CCl_4-treated livers (48 h after the administration of CCl_4) (Gd). RNA was electrophoresed (10 μg of total RNA per lane), blotted, and subsequently hybridized with a ^{32}P-dCTP-labeled cDNA specific for TNF-α and actin.

IV. TRANSMIGRATION

A. PECAM, which Is Crucial for Transmigration, Is Decreased Following CCl_4 Treatment of Rat Livers, as well as after Treatment of Isolated Sinusoidal Endothelial Cells by TNF-α

PECAM-1, another member of the Ig superfamily, plays a crucial role in transendothelial migration of leukocytes and monocytes, as transmigration and inflammation could be inhibited *in vivo* and *in vitro* by applying antibodies directed against PECAM-1 (DeLisser *et al.*, 1994; Liao *et al.*, 1997; Muller *et al.*, 1993).

PECAM-1 is a major constituent of endothelial cell intercellular junctions and therefore regulates endothelial cell–cell contact. It is known to be expressed at a lower level on the surface of circulating platelets, monocytes, and leukocytes (Newman, 1994, 1997).

Our data revealed PECAM-1 immunoreactivity along the sinusoids of normal rat livers in a pattern similar to ICAM-1 staining (Neubauer *et al.*, 2000b) (Fig. 7) and PECAM-1-specific transcripts were detected in freshly isolated SECs by Northern blot analysis) (Neubauer *et al.*, 2000b) (Fig. 4). After a single CCl_4 administration, PECAM-1 immunoreactivity did not increase in contrast to the increase of ICAM-1 (Fig. 8). Northern blot analysis of livers taken at different time points after the administration of CCl_4 indicated that

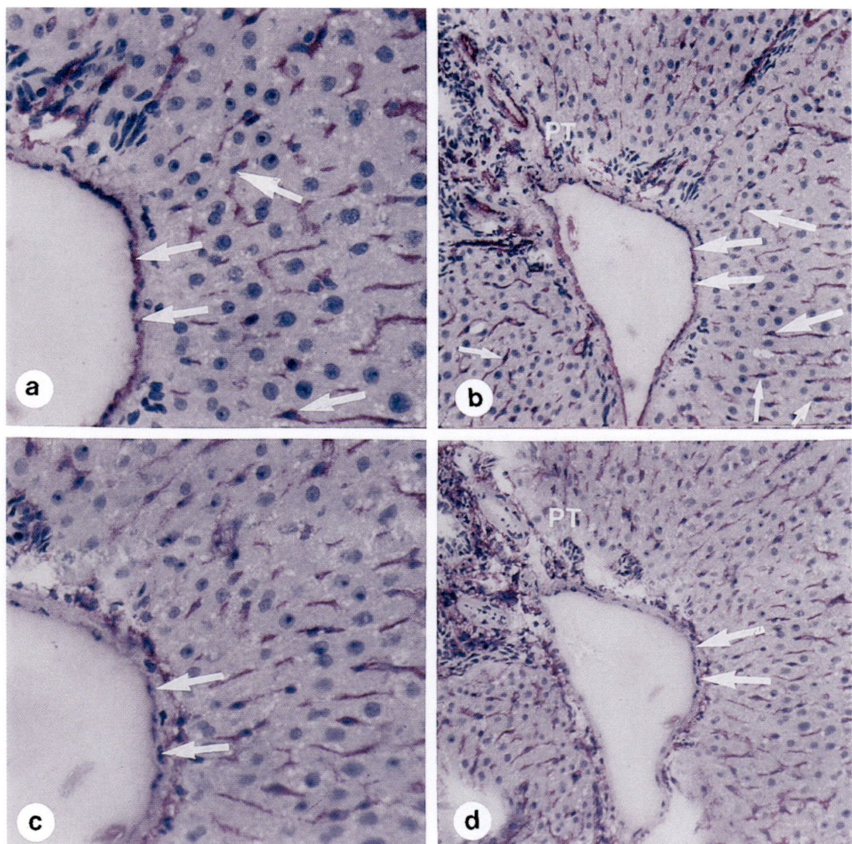

FIGURE 7 Indirect immunodetection of PECAM-1 and ICAM-1 in sections of normal rat livers using the APAAP method. Sections were stained with a monoclonal antibody directed against rat PECAM-1 (a,b) or with a monoclonal antibody directed against rat ICAM-1 (c,d), followed by APAAP immunodetection. Arrows indicate PECAM-1 positivity along the sinusoids and along the intimal lining of the vessels or ICAM-1 negativity of the intimal lining. PT, portal tract (a and c original magnification ×400, b and d original magnification ×250). Reproduced from Neubauer *et al.* (2000b), with permission.

PECAM-1 expression does not increase after a single administration of CCl_4, whereas the ICAM-1 steady-state level increased (Fig. 3). Furthermore, PECAM-1-specific transcript levels in SECs isolated at different time points after CCl_4 administration decreased in parallel to the increase of ICAM-1 transcript levels. The amount of PECAM-1-specific transcripts was also decreased in cultured SECs following the treatment of TNF-α as analyzed by Northern blot analysis (Fig. 5).

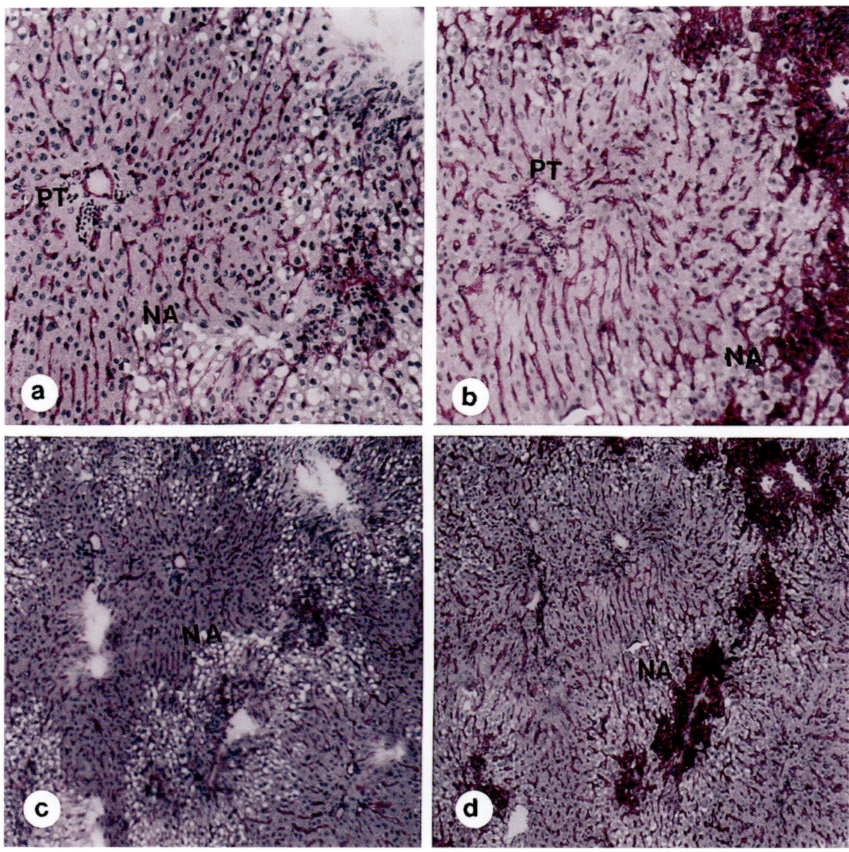

FIGURE 8 Indirect immunodetection of PECAM-1 and ICAM-1 in sections from acutely damaged rat livers (48 h after a single CCl_4 administration). Sections were stained with a monoclonal antibody directed against rat PECAM-1 (a,c) or with a monoclonal antibody directed against rat ICAM-1 (b,d), followed by APAAP immunodetection. PT, portal tract; NA, necrotic area. (a and b original magnification ×250, c and d original magnification ×100). Reproduced from Neubauer *et al.* (2000b), with permission.

Early production of TNF-α after liver injury could induce increased ICAM-1 expression and decreased PECAM-1 expression in SECs, which might be essential for the transmigration of inflammatory cells into the parenchyma. Our data, in combination with the finding that there is no reduced inflammation in the PECAM-1 knockout mouse (Duncan *et al.*, 1999), might indicate that PECAM-1 may be a member of a group of molecules that have to be inactivated rather than activated to allow infiltration during inflammation.

B. Hepatic Stellate Cells Are Capable of Expressing ICAM-1 and VCAM-1, which Are Inducible upon Stimulation with TNF-α and IFN-γ

So far, HSCs have been shown to be the primary source of matrix proteins. HSCs, also designated as Ito cells, fat-storing cells, or lipocytes, play a major role in vitamin A metabolism. HSCs are situated in the space of Disse between the sinusoidal endothelium and hepatocytes and exhibit long cytoplasmic processes that underlie the endothelium and embrace the sinusoid but also have contact with hepatocytes.

We tested if HSCs during early culture can contribute to the recruitment of inflammatory cells by the synthesis of ICAM-1 and VCAM-1 under conditions of inflammation (Knittel *et al.*, 1999). In addition, the regulation of neural cell adhesion molecule (NCAM, also termed CD56) was analyzed because NCAM, known to be expressed by HSC following activation, might be involved in the migration of CD56-positive lymphocytes or HSC and in the termination of HSC proliferation induced by tissue injury.

Northern blot analysis of HSCs during early culture, as well as several days after isolation, indicated that HSCs express adhesion molecules of the Ig family as ICAM-1, VCAM-1, and NCAM. Next to TNF-α, IFN-γ had also been shown to be involved in liver inflammation, the influence of both cytokines on the expression of ICAM-1 and VCAM-1, as well as NCAM in HSCs, was studied. TNF-α, as well as IFN-γ, upregulated the expression of ICAM-1 and VCAM-1, but decreased the NCAM-specific transcript level in HSCs (Figs. 9 and 10).

V. REGENERATION AND SCAR FORMATION

Liver fibrogenesis represents the uniform response of the liver to toxic, infectious or metabolic agents and is characterized by increased synthesis and altered deposition of newly formed extracellular matrix components. Deposition of collagen type IV, laminin, and entactin in the perisinusoidal space results in the formation of a complete basement membrane, a process called "collagenization" or "capillarization" of the sinusoids. Capillarization of the sinusoids is an early event in liver fibrogenesis and seems to be of crucial importance for the liver function, as the exchange of macromolecules between sinusoidal blood and hepatocytes is fundamentally affected (Jezequel *et al.*, 1990; Neubauer *et al.*, 2001; Schuppan, 1990).

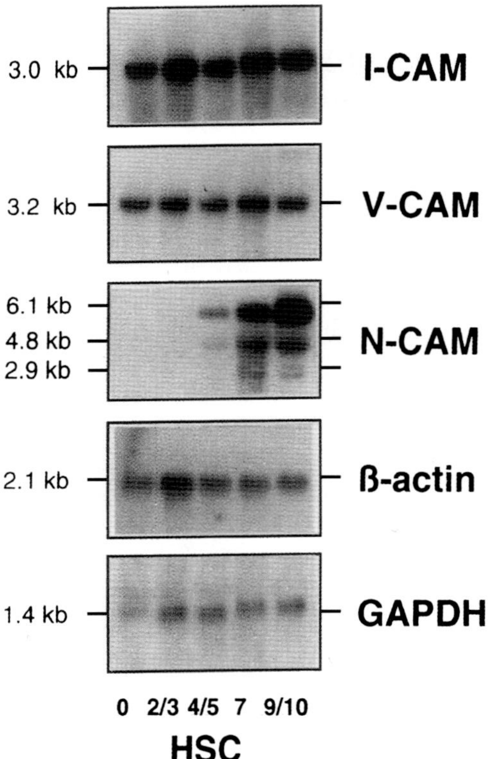

FIGURE 9 Expression of cell adhesion molecules by hepatic stellate cells *in vitro*: Expression of ICAM-1, VCAM-1, and NCAM as assessed by Northern blot analysis. Total RNA was purified from hepatic stellate cells at these time points: freshly isolated (0), day 2 or 3 (2/3), day 4 or 5 (4/5), day 7 (7), and day 9 or 10 (9/10) after plating. RNA (1.25 μg total) recovered from four different hepatic stellate cell isolations was pooled and size selected by 1% agarose gel electrophoresis followed by hybridization using specific cDNA probes. Reproduced from Knittel *et al.* (1999), with permission.

Next to HSCs, a possible contribution of SECs to "capillarization" by the synthesis of basement membrane proteins and to septa formation by the synthesis of extracellular matrix proteins has been hypothesized, as ultrastructural and immunocytochemical data indicate that SECs produce collagen type III, IV, and fibronectin (Clement *et al.*, 1986; Irving *et al.*, 1984). Because SECs are capable of endocytosing matrix proteins via the hyaluron, the collagen, and the P III-NP receptor (Smedsröd *et al.*, 1990), these results have been viewed with some criticism. Synthesis of cellular fibronectin in SECs was demonstrated by Rieder *et al.*, (1987).

As TGF-β1 seems to be the most effective mediator in the stimulation of matrix protein synthesis (Czaja *et al.*, 1989; Knittel *et al.*, 1996), we studied the modulation

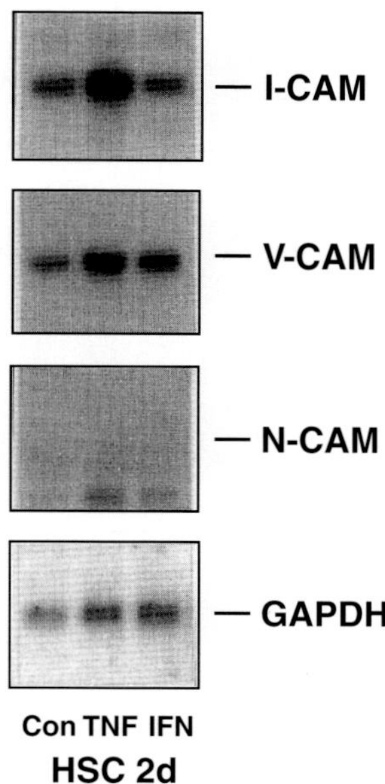

FIGURE 10 Regulation of ICAM-1, VCAM-1, and NCAM expression by TNF-α and IFN-γ in hepatic stellate cell (HSC) primary culture. HSC cultured for 2 days (2d) were incubated without (Con) or with TNF-α (TNF) or IFN-γ (IFN) for 20 h. RNA (5 μg total) was size selected by 1% agarose gel electrophoresis, and filters were then hybridized using a cDNA probe specific for ICAM-1, VCAM-1, NCAM, or GAPDH. Reproduced from Knittel *et al.* (1999), with permission.

of basement membrane proteins collagen type IV, laminin, and entactin expression by TGF-β1 in liver SECs (Neubauer *et al.*, 1999). Specific immunoprecipitates for the basement membrane (BM) proteins entactin, laminin, and collagen type IV and for extracellular matrix (ECM) proteins tenascin and fibronectin were detectable in freshly isolated or cultured SECs. The synthesis of all tested BM proteins and ECM proteins was stimulated at least threefold by TGF-β1 (Fig. 11). In SECs isolated from acutely or chronically CCl_4-damaged rat livers, there was an enhanced transcript level of matrix proteins (fibronectin, laminin, collagen type IV, and thrombospondin) as demonstrated by Northern blot analysis (Fig. 12). Stimulation of the synthesis of BM proteins by TGF-β1 *in vitro* and accumulation of transcripts for ECM proteins in SECs isolated from CCl_4-treated livers suggest that SECs are involved in the formation of a basement membrane during "capillarization" during liver disease. Although HSCs seem to be responsible for the synthesis of matrix proteins at later stages of disease, endothelial cells play a role under basal conditions, as well as during "capillarization" of the sinusoids at the beginning of fibrosis.

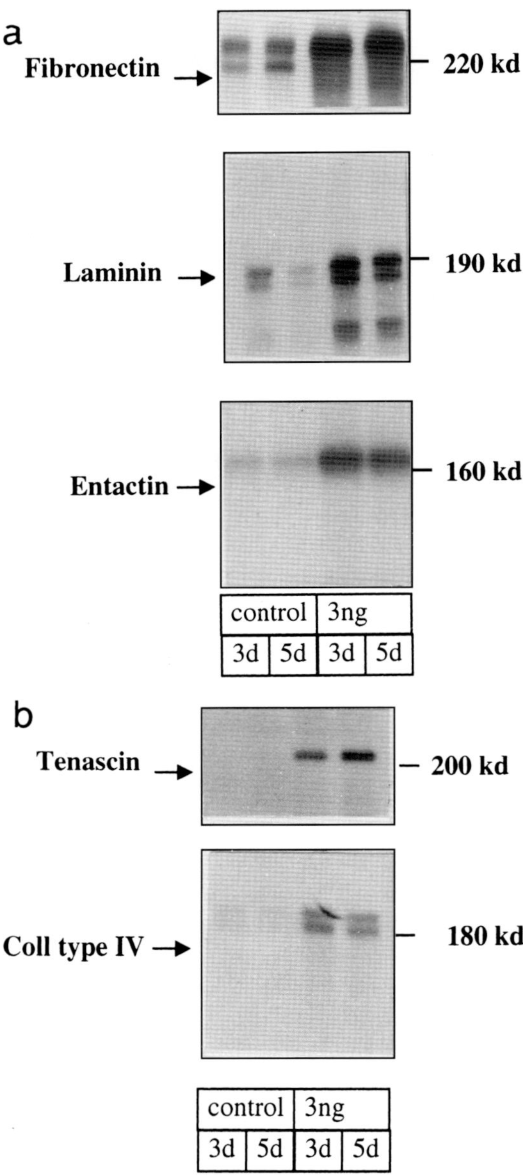

FIGURE 11 Stimulation of tenascin, entactin, fibronectin, laminin, and collagen type IV synthesis by TGF-β1 in primary cultured sinusoidal endothelial cells. On day 5 of primary culture, sinusoidal endothelial cells were pulse labeled with 100 mCi [^{35}S] methionine in the absence (control) or presence of 3 ng of TGF-β1 per well for 24 h. Aliquots of culture medium (3×10^5 cpm) containing the same amount of protein-bound radioactivity were immunoprecipitated with antibodies directed against fibronectin, laminin, entactin (a), tenascin, and collagen type IV (b). Samples were subjected to SDS–PAGE followed by autoradiography. Reproduced from Neubauer *et al.* (1999), with permission.

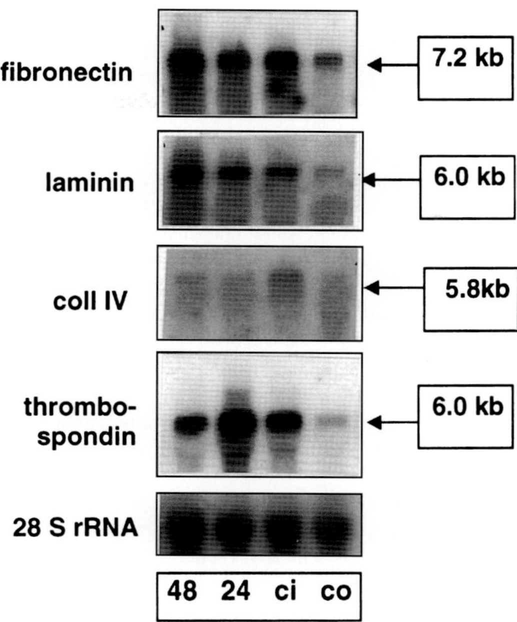

FIGURE 12 Matrix protein gene expression in sinusoidal endothelial cells isolated from rat livers after a single CCl_4 administration and after repeated administrations of CCl_4. Northern blot analysis of total RNA extracted from SECs isolated from normal rat livers (co), 24 h (24), and 48 h (48) after a single CCl_4 administration and 7 days after 10 weekly CCl_4 administrations (ci). Three rats were treated at each time point. Five milligrams of total RNA was separated on an agarose gel, blotted, and hybridized with a ^{32}P-dCTP-labeled cDNA probe specific for fibronectin, laminin, collagen type IV, and thrombospondin. Experiments were repeated with cells of three different rats. Reproduced from Neubauer *et al.* (1999), with permission.

REFERENCES

Ando, K., Moriyama, T., Guidotti, L. G., Wirth, S., Schreiber, R. D., Schlicht, H. J., Huang, S. N., and Chisari, F. V. (1993). Mechanisms of class I restricted immunopathology: A transgenic mouse model of fulminant hepatitis. *J. Exp. Med.* **178**, 1541.

Bruccoleri, A., Galluci, R., Germolec, D., Blackshear, P., Simeonova, P., Thurman, R., and Luster, M. (1997). Induction of early immediate genes by tumor necrosis factor a contribute to liver repair following chemical induced hepatotoxicity. *Hepatology* **25**, 133.

Butcher, E. C., and Picker, L. J. (1996). Lymphocyte homing and homeostasis. *Science* **272**, 60.

Clement, B., Grimaud, J. A., Campion, J. P., Deugnier, Y., and Guillouzo, A. (1986). Cell types involved in collagen and fibronectin production in normal and fibrotic human liver. *Hepatology* **6**, 225.

Couvelard, A., Scoazec, J. Y., and Feldmann, G. (1993). Expression of cell-cell and cell-matrix adhesion proteins by sinusoidal endothelial cells in the normal and cirrhotic human liver. *Am. J. Pathol.* **143**, 738.

Czaja, M. J., Weiner, F. R., Flanders, K. C., Giambrone, M. A., Wind, R., Biempica, L., and Zern, M. A. (1989). In vitro and in vivo association of transforming growth factor-β1 with hepatic fibrosis. *J. Cell Biol.* **108**, 2477.

DeLisser, H. M., Newman, P. J., and Albelda, S. M. (1994). Molecular and functional aspects of PECAM-1/CD31. *Immunol. Today* **15**, 490.

Duncan, G. S., Andrew, D. P., Takimoto, H., Kaufman, S. A., Yoshido, H., Spellberg, J., de la Pompa, J. L., Elia, A., Wakeham, A., Karan-Tami, B., and Muller, W. A. (1999). Genetic evidence for functional redundancy of platelet endothelial cell adhesion molecule-1 (PECAM-1): CD31-deficient mice reveal PECAM-1 dependent and PECAM-1 independent functions. *J. Immunol.* **162**, 3022.

Dunon, D., Piali, L., and Imhof, B. (1996). To stick or not to stick: The new leukocyte homing paradigm. *Curr. Opin. Cell Biol.* **8**, 714.

Garcia-Barcina, M., Lukomska, B., Gawron, W., Winnock, M., Vidal-Vanaclocha, F., Bioulac-Sage, P., Balabaud, C., and Olszewski, W. (1995). Expression of cell adhesion molecules on liver associated lymphocytes and their ligands on sinusoidal lining cells in patients with benign or malign liver disease. *Am. J. Pathol.* **146**, 1406.

Irving, M. G., Roll, J., Huang, S., and Bissell, D. (1984). Characterization and culture of sinusoidal endothelium from normal rat liver: Lipoprotein uptake and collagen phenotype. *Gastroenterology* **87**, 1233.

Jezequel, A. M., Ballardini, G., Mancini, R., Paolucci, F., Bianchi, F. B., and Orlandi, F. (1990). Modulation of extracellular matrix components during dimethylnitrosamine-induced cirrhosis. *J. Hepatol.* **11**, 206.

Knittel, T., Dinter, C., Kobold, D., Neubauer, K., Mehde, M., Eichhorst, S., and Ramadori, G. (1999). Expression and regulation of adhesion molecules by hepatic stellate cells of rat liver. *Am. J. Pathol.* **154**, 153.

Knittel, T., Janneck, T., Mueller, L., Fellmer, P., and Ramadori, G. (1996). Transforming growth factor-β1-regulated gene expression of Ito cells. *Hepatology* **24**, 352.

Liao, F., Ali, J., Greene, T., and Muller, W. A. (1997). Soluble domain 1 of platelet-endothelial cell adhesion molecule (PECAM) is sufficient to block transendothelial migration in vitro and in vivo. *J. Exp. Med.* **185**, 1349.

Masumoto, T., Ohkubo, K., Yamamoto, K., Ninomiya, T., Abe, M., Akbar, S. M., Michitaka, K., Horiike, N., and Onji, M. (1998). Serum IL-8 levels and localization of IL-8 in liver from patients with chronic viral hepatitis. *Hepatogastroenterology* **45**, 1630.

Muller, W. A., Weigl, S. A., Deng, X., and Phillips, D. M. (1993). PECAM-1 is required for transendothelial migration of leukocytes. *J. Exp. Med.* **178**, 449.

Neubauer, K., Eichhorst, S. T., Wilfling, T., Buchenau, M., Xia, L., and Ramadori, G. (1998). Sinusoidal ICAM-1 upregulation precedes the accumulation of LFA-1 positive cells and tissue necrosis in a model of CCl_4-induced acute rat liver injury. *Lab. Invest.* **78**, 185.

Neubauer, K., Krueger, M., Quondamatteo, F., Knittel, T., Saile, B., and Ramadori, G. (1999). Transforming growth factor-β1 stimulates the synthesis of basement membrane proteins laminin, collagen type IV and entactin in rat liver sinusoidal endothelial cells. *J. Hepatol.* **31**, 692.

Neubauer, K., Ritzel, A., Saile, B., and Ramadori, G. (2000a). Decrease of platelet-endothelial cell adhesion molecule 1: Gene-expression in inflammatory cells and in endothelial cells in the rat liver following CCl_4-administration and in vitro after treatment with TNF-alpha. *Immunol. Lett.* **74**, 153.

Neubauer, K., Saile, B., and Ramadori, G. (2001). Liver fibrosis and altered matrix synthesis. *Can. J. Gastroenterol.* **15**, 187.

Neubauer, K., Wilfling, T., Ritzel, A., and Ramadori, G. (2000b). PECAM-1 expression in sinusoidal endothelial cells of the liver. *J. Hepatol.* **32**, 921.

Newman, P. J. (1994). The role of PECAM-1 in vascular cell biology. *Ann. N. Y. Acad. Sci.* **714**, 165.

Newman, P. J. (1997). The biology of PECAM-1. *J. Clin. Invest.* **99**, 3.

Patzwahl, R., Meier, V., Ramadori, G., and Mihm, S. (2001). Enhanced expression of interferon-regulated genes in the liver of patients with chronic hepatitis C virus infection: Detection by suppression-subtractive hybridization. *J. Virol.* **75**, 1332.

Rieder, H., Ramadori, G., Dienes, H. P., and Meyer zum Bueschenfelde, K. H. (1987). Sinusoidal endothelial cells from guinea pig liver synthesize and secrete cellular fibronectin in vitro. *Hepatology* **7**, 856.

Schumann, J., Wolf, D., Pahl, A., Brune, K., Papadopoulos, T., van Rooijen, N., and Tiegs, G. (2000). Importance of Kupffer cells for T-cell-dependent liver injury in mice. *Am. J. Pathol.* **157**, 1671.

Schuppan, D. (1990). Structure of the extracellular matrix in normal and fibrotic liver: Collagens and glycoproteins. *Semin. Liver Dis.* **10**, 1.

Scoazec, J. Y., and Feldmann, G. (1991). In situ immunotyping study of endothelial cells of the human hepatic sinusoid: Result and functional implication. *Hepatolog* **14**, 789.

Scoazec, J. Y., and Feldmann, G. (1994). The cell adhesion molecules of hepatic sinusoidal cells. *J. Hepatol.* **20**, 296.

Smedsröd, B., Pertoft, H., Gustafsen, S., and Laurent, T. C. (1990). Scavenger functions of liver endothelial cell. *Biochem. J.* **266**, 313.

Springer, T. (1994). Traffic signals for lymphocyte recirculation and leukocyte emigration: The multi-step paradigm. *Cell* **76**, 301.

PART III

Activation Mechanism of Hepatic Stellate Cells and Signal Transduction

CHAPTER 9

Molecular Mechanism of Stellate Cell Activation and Extracellular Matrix Remodeling

DAN LI AND SCOTT L. FRIEDMAN

Division of Liver Diseases, Mount Sinai School of Medicine, New York, New York 10029

I. INTRODUCTION

In normal liver, hepatic stellate cells are perisinusoidal vitamin A-storing cells located in the subendothelial space of Disse. Following liver injury, they undergo a spectrum of phenotypic changes and transform into proliferative, fibrogenic, and contractile myofibroblasts—a processed collectively termed "activation." This chapter reviews the biological characteristics of hepatic stellate cells, addressing their functions in injured liver, with special emphasis on the molecular mechanisms of stellate cell activation.

A. FUNCTIONS OF HEPATIC STELLATE CELLS: MOLECULAR ASPECTS

In normal liver, stellate cells plays a major role in maintaining hepatic hemeostatsis. Their main functions include (1) production of cytokines and other mediators, (2) expression of membrane receptors, (3) cell matrix synthesis and degradation,

Copyright 2003, Elsevier Science (USA). All rights reserved.

(4) regulation of hepatic sinusoidal blood flow through their contractility, and (5) retinoid storage and metabolism.

1. Production of Cytokines

One of the main features of hepatic stellate cells is their versatility in producing a rich array of cytokines in the liver. Signal transduction mediated by binding of cytokines to their membrane receptors comprises the main mode of cell–cell interaction in both normal and injured liver (Friedman, 1999) (Table I).

TABLE 1 Products and Components of Hepatic Stellate Cells

1. Vitamin A-related components and lipids	
Retinoids	Retinol, retinyl esters, retroretinoids
Binding proteins	CRBP, CRABP, RBP (controversial)
Enzymes	Retinyl palmitate hydrolase, acetyl-CoA: retinyl acyltransferase
Nuclear retinoid receptors	RARα, RARγ, RXRα, RXRβ, PPARs (controversial)
Apolipoproteins	Apo E, apo A-I, and A-IV (controversial)
2. Cytoskeletal markers	
Vimentin, desmin, α-SMA, GFAP, nestin	
3. Extracellular matrix	
Collagens	Types I, III, IV, V, VI, XIV
Proteoglycans	Heparan, dermatan and chrodroitin sulfates, perlecan, syndecan-1, biglycan, decorin
Glyconproteins	Cellular fibronectin, laminin, merosin, tenascin, nidogen/entactin, undulin, hyaluronic acid
4. Proteases and inhibitors	
Matrix proteases	MMP-2, stromelysin-1 (transin), MMP-1, MT-MMP
Protease inhibitors	TIMP-1, TIMP-2, PAI-1
5. Cytokines, growth factors, and inflammatory mediators	
Prostanoids	Prostaglandin (PG)$F_2\alpha$, PGD2, PGI2, PGE2; LTC4, LTB4
Leukocyte mediators	M-CSF, MCP-1, PAF
Acute phase components	α_2-macroglobulin, IL-6
Mitogens	HGF, EGF, PDGF, SCF, IGF I and II, αFGF
Adhesion molecules	I-CAM-1, V-CAM-1, N-CAM
Vasoactive mediators	ET-1, NO
Fibrogenic compounds	TGF-β1, TGF-β2, TGF-β3, CTGF
Others	IL-10, CINC
6. Receptors	
Cytokine receptors	PDGF-R, TGFβ-R types I, II, and III, ET-R, EGF-R, VEGF-R
Others	Integrin, DDR, thrombin-R, mannose-6-phosphate-R, uPA-R
7. Signaling molecules and transcription factors	
Signaling components	Raf; raf and MAP kinase
Transcription factors	Sp1, NFκB, c-myb, Zf9/KLF6

Platelet-derived growth factor (PDGF) consists of A and B chains (as AA, AB, or BB dimers) and is the most potent stellate cell mitogen described to date (Pinzani *et al.*, 1989, 1991, 1992a, 1995). PDGF A chain mRNA has been detected in activated human stellate cells (Marra *et al.*, 1994; Win *et al.*, 1993). During liver injury, stellate cells produce more PDGF and upregulate PDGF receptors (Pinzani *et al.*, 1994b, 1996b, Wong *et al.*, 1994). Following ligand binding, the PDGF receptor recruits the signaling molecule Ras, followed by activation of the protein kinase mitogen-activated 1 (MAP) kinase pathway (Friedman, 1999). In addition, phosphoinositol 3 kinase (PI3K) and STAT-1 may mediate PDGF signaling in stellate cells (Kawada *et al.*, 1997; Marra *et al.*, 1997).

Stellate cells secrete transforming growth factor-α (TGF-α) and epidermal growth factor (EGF), two potent epithelial growth factors that play important roles in hepatocyte proliferation during liver regeneration (Bachem *et al.*, 1992; Meyer *et al.*, 1990; Mullhaupt *et al.*, 1994). TGF-α and EGF also stimulate mitosis in stellate cells (Meyer *et al.*, 1990; Win *et al.*, 1993), creating an autocrine loop for cellular proliferation. Hepatocyte growth factor (HGF) is a more potent hepatocyte mitogen produced by stellate cells (Maher, 1993; Schirmacher *et al.*, 1992); however, its production diminishes during acute liver injury (Maher, 1993). Stem cell factor (SCF) was also identified in stellate cells in rats undergoing liver regeneration induced by a partial hepatectomy combined with 2-acetoaminofluorene (Fujio *et al.*, 1994). In addition, insulin-like growth factors I and II (IGF-I and -II) and their receptors are also products of stellate cells (Pinzani *et al.*, 1990; Zindy *et al.*, 1992).

Acidic fibroblast growth factor (aFGF) is another mitogenic cytokine that has been identified in stellate cells *in situ* during liver injury, in late hepatic development, and during hepatic regeneration (Marsden *et al.*, 1992). Stellate cells also create an autocrine loop for basic FGF (bFGF), which is mitogenic toward culture-activated rodent and human stellate cells (Marsden *et al.*, 1992; Pinzani *et al.*, 1991; Rosenbaum *et al.*, 1995; Win *et al.*, 1993).

Stellate cells produce macrophage colony-stimulating factor (M-CSF) (Pinzani *et al.*, 1992a) and monocyte chemotactic peptide 1 (MCP-1) (Czaja *et al.*, 1994; Marra *et al.*, 1993), which regulate macrophage accumulation and growth. MCP-1 production is stimulated by thrombin, interleukin-lα, interferon γ, and tissue necrosis factor-α (TNF-α) (Marra *et al.*, 1993, 1995). It is blocked by H-7, an inhibitor of protein kinase C (PKC) (Marra *et al.*, 1993), suggesting the involvement of PKC in the signaling pathway leading to MCP production. M-CSF synthesis is stimulated by PDGF and bFGF(Pinzani *et al.*, 1992a). Secretion by stellate cells of these macrophage growth factors may amplify the inflammatory and fibrogenic responses during liver injury.

The neutrophil inflammatory response in injured liver is also amplified by stellate cells through the production of platelet-activating factor (PAF). PAF promotes chemotaxis of neutrophils and stimulates their activation (Pinzani *et al.*, 1994a). Its production is increased by thrombin, lipopolysaccharide, and calcium ionophores (Pinzani *et al.*, 1994a). In addition to PAF, rat stellate cells also produce cytokine-induced neutrophil chemoattractant (CINC), a rat form of human interleukin-8 (Maher *et al.*, 1998). The upregulation of CINC expression accompanies stellate cell activation both *in vivo* and in culture (Maher *et al.*, 1998).

Stellate cells secrete interleukin-6 (IL-6), thereby contributing to the acute-phase response (Greenwel *et al.*, 1993; Tiggelman *et al.*, 1995). Lipopolysaccharide, interleukin-1β and tumor necrosis factor-α are potent stimuli of interleukin-6 production (Tiggelman *et al.*, 1995). However, interleukin-10 (IL-10) is an anti-inflammatory cytokine also produced by stellate cells. Upregulation of IL-10 occurs in early stellate cell activation (Thompson *et al.*, 1998b; Wang *et al.*, 1998). It has prominent antifibrogenic activity by downregulating collagen type I expression while upregulating interstitial collagenase. IL-10 knockout mice develop more severe hepatic fibrosis following CCl_4 administration than wild-type mice, substantiating its antifibrotic activity *in vivo* (Louis *et al.*, 1998; Thompson *et al.*, 1998a). It is uncertain, however, whether the antifibrotic effect of IL-10 is direct or is mediated through immune downregulation. Clinical trials are currently evaluating the efficacy of IL-10 in chronic liver fibrosis.

Stellate cells also express several adhesion molecules, including intercellular adhesion molecule 1 (I-CAM-1) (Hellerbrand *et al.*, 1996), vascular adhesion molecule 1 (V-CAM-1) (Knittel *et al.*, 1999), and neural adhesion molecule (N-CAM) (Knittel *et al.*, 1996; Nakatani *et al.*, 1996). The expression of I-CAM-1 is increased following stellate cell activation and may regulate lymphocyte adherence to activated stellate cells (Hellerbrand *et al.*, 1996). *In situ*, both I-CAM-1 and V-CAM-1 are upregulated following CCl_4-induced liver injury (Knittel *et al.*, 1999). The peaks of immunoreactivity of these two molecules coincide with maximal cell infiltration, and the inflammatory cytokine TNF-α increases the transcripts of both CAMs (Knittel *et al.*, 1999). Therefore, it is likely that I-CAM and V-CAM are involved in modulating the recruitment of inflammatory cells during liver injury. Stellate cells expressing N-CAM lie in close proximity to nerve endings in the liver (Nakatani *et al.*, 1996). The function of this adhesion molecule in stellate cells is not known.

Transforming growth factor-β (TGF-β) is one of the most important stimuli of hepatic fibrosis. Stellate cells secrete latent TGF-β1 in response to injury, which, after its activation, exerts potent fibrogenic effects in both autocrine and paracrine patterns. Of these two sources, autocrine activity is the most important (Bissell *et al.*, 1995; Gressner, 1995). TGF-β1 expression is increased in experimental and human hepatic fibrosis (Castilla *et al.*, 1991; Gressner and Bachem, 1995). Upregulation of TGF-β1 in activated stellate cells occurs through multiple mechanisms. Factors including Sp1 (Ji *et al.*, 1997) and Zf9/KLF6 (Kim *et al.*, 1998) transactivate the TGF-β1 promoter through interactions with multiple "GC box" motifs. There are also several mechanisms mediating the activation of latent TGF-β1, including cell surface activation following binding to the cell surface mannose-6-phosphate/insulin-like growth factor II receptor (de Bleser *et al.*, 1995) and binding to a number of proteins secreted by stellate cells (Gressner and Bachem, 1990), including α_2-macroglobulin (Andus *et al.*, 1987), decorin, and biglycan (Meyer *et al.*, 1992). Local plasminogen activator (PA)/plasmin is particularly important in activating latent TGF-β1 (Imai *et al.*, 1997; Okuno *et al.*, 1997, 1999). This activity is regulated by metabolites of retinoic acid. The signaling of TGF-β1 in rat stellate cells involves the activation of Ras, Raf-1, MEK, and MAP kinase (Reimann *et al.*, 1997). A family of signaling molecules known as SMAD proteins are also involved in TGF-β signaling (Attisano and Wrana, 1998; Heldin *et al.*, 1997; Wrana and Attisano, 1996), and their role in stellate cells is under study (Dooley *et al.*, 2000).

Connective tissue growth factor (CTGF) is a cytokine that promotes fibrogenesis in skin, lung, and kidney (Igarashi *et al.*, 1996; Ito *et al.*, 1998; Lasky *et al.*, 1998). It is strongly expressed by stellate cells during hepatic fibrosis (Paradis *et al.*, 1999). Regulation of its expression in stellate cells is not defined, although it is a downstream target of TGF-β in other cellular systems (Duncan *et al.*, 1999; Grotendorst, 1997).

Endothelin-1 (ET-1) was originally identified as a potent vasoconstrictor produced mainly by endothelial cells (Yanagisawa *et al.*, 1988). More recent studies have identified stellate cells as both a major source and a target of this cytokine during liver injury (Pinzani *et al.*, 1996a; Rockey *et al.*, 1998). Interestingly, stellate cells also produce nitric oxide (NO), a physiological antagonist to ET-1 (Rockey and Chung, 1995) (see Section I,A,4).

A more recent finding is that activated hepatic stellate cells also secrete leptin, which is a 16-kDa peptide hormone critical for regulating fuel stores and energy expenditure (Marti *et al.*, 1999; Potter *et al.*, 1998). Leptin decreases appetite and thereby food intake, and increases energy expenditure. Expression of leptin was demonstrated in culture-activated hepatic stellate cells but not in quiescent stellate cells, and leptin mRNA expression was inhibited by retinoids (Potter *et al.*, 1998). Because leptin augments both inflammatory and profibrogenic responses in the liver caused by hepatotoxic chemicals (Ikejima *et al.*, 2001), it may play a role in enhancing fibrogenic activities of stellate cells following liver injury. Circulating leptin levels were higher in patients with alcoholic cirrhosis (McCullough *et al.*, 1998), but whether stellate cells are a major source of leptin in this setting is controversial (Henriksen *et al.*, 1999).

2. Expression of Membrane Receptors

Cytokines regulate stellate cell behavior through specific, high-affinity binding to their cognate membrane receptors. Thus far, several cytokine and extracellular matrix receptors have been identified in either quiescent or activated stellate cells.

The PDGF receptor was the first receptor identified in stellate cells. It is composed of α or β subunits as either homodimers or heterodimers. In rat stellate cells, the β subunit is the predominant isoform (Heldin *et al.*, 1991; Pinzani *et al.*, 1991, 1994b), whereas in human stellate cells, both α and β subunits are detectable (Pinzani *et al.*, 1995; Win *et al.*, 1993).

TGF-β receptors have also been characterized in rat stellate cells. All three forms of TGF-β receptors, types I, II, and III (betaglycan), are expressed. TGF-β1 binding and responsiveness are enhanced greatly during activation *in vivo* and *in vitro* (Friedman *et al.*, 1994; Roulot *et al.*, 1995).

The effects of ET-1 are mediated through two G-protein-coupled receptors. Receptor types A and B have been identified in both quiescent and activated stellate cells (Housset *et al.*, 1993; Kawada *et al.*, 1995). The relative prevalence of ETA and ETB receptors changes during stellate cells activation (Pinzani *et al.*, 1996a). The ETB receptor is the predominant mediator of stellate cell contraction and growth inhibition (Rockey, 1995). In contrast, the proliferative effect of ET-1 in quiescent stellate cells is mediated through ETA receptor (Pinzani *et al.*, 1996a).

The induction of receptors for vascular endothelial growth factor (VEGF) has been observed during stellate cell activation both *in vivo* and in culture (Ankoma-Sey *et al.*, 1998). VEGF receptor upregulation is associated with enhanced mitogenesis in response to VEGF, which is further synergized by bFGF. Because VEGF plays a critical role in angiogenesis, this finding suggests that stellate cells may be involved in typical "angiogenic" responses, broadening their potential roles in both wound healing and tumor formation.

Stellate cells express the receptor for thrombin, a serine protease derived from prothrombin (Marra *et al.*, 1995). The binding of thrombin to its receptor leads to cellular proliferation and increased production of MCP-1.

In addition to receptors for cytokines, membrane receptors for matrix molecules have been characterized in stellate cells. Integrins are a specialized type of membrane receptors that transduce signals from the extracellular matrix (ECM) to cells (Clark and Brugge, 1995; Iredale and Arthur, 1994; Scoazec, 1995, 1996). They are heterodimeric transmembrane proteins composed of α and β subunits whose ligands are matrix molecules rather than cytokines. Several integrins and their downstream effectors have been identified in stellate cells, including α1β1, α2β1, αvβ1, and α6β4 (Carloni *et al.*, 1996, 1997; Imai and Senoo, 1998; Jarnagin *et al.*, 1994; Racine-Samson *et al.*, 1997). In particular, integrin ligands contain an arginine (Arg)–glycine (Gly)–aspartate (Asp) tripeptide sequence. The common presence of Arg-Gly-Asp (RGD) within many integrin ligands has raised the possibility of using competitive RGD antagonists to block integrin-mediated pathways in fibrogenesis (Bruck *et al.*, 1997; Iwamoto *et al.*, 1998).

A special family of tyrosine kinase receptors called "discoidin domain receptors" (DDR) has been identified, whose ligands are fibrillar collagens rather than growth factors (Shrivastava *et al.*, 1997; Vogel *et al.*, 1997). The intriguing identification of DDR2 mRNA in activated stellate cells (Ankoma-Sey *et al.*, 1998) raises the possibility that this receptor may mediate interactions between stellate cells and the surrounding interstitial matrix during progressive liver injury (Olaso *et al.*, 1999).

3. Involvement in Extracellular Matrix Remodeling

The extracellular matrix ECM refers to the array of macromolecules that comprise the scaffolding of the liver. These macromolecules consist of three main families: collagens, glycoproteins, and proteoglycans (Timpl, 1996), with new members of matrix molecules being identified continually (Frizell *et al.*, 1995; Milani *et al.*, 1994a; Musso *et al.*, 1998; Takahara *et al.*, 1995). In normal liver, the subendothelial space of Disse contains a basement membrane-like matrix, which is composed of nonfibril-forming collagens, including types IV, VI, and XIV, glycoproteins, and proteoglycans (Rescan *et al.*, 1993; Roskams *et al.*, 1995). In contrast, the so-called interstitial ECM is largely confined to the capsule, around large vessels and in the portal areas. The main components are fibril-forming collagens (e.g., types I and III), cellular fibronectin, undulin (collagen XIV), and other glycoconjugates.

Studies in recent years have confirmed that hepatic stellate cells are the major cellular source of ECM in both normal and injured liver (Friedman *et al.*, 1985; Geerts *et al.*, 1989; Kawase *et al.*, 1986). Stellate cells produce a wide array of ECM components, including collagens I, III, IV, V, VI, XIV (Friedman *et al.*, 1985; Geerts *et al.*, 1989), and XVIII (Musso *et al.*, 1998); proteoglycans (Arenson *et al.*, 1988; Schafer *et al.*, 1987): heparan, dermatan, and chondroitin sulfates, perlecan, syndecan-1 (Kovalszky *et al.*, 1994), biglycan and decorin (Meyer *et al.*, 1992); glycoproteins: cellular fibronectin (Ramadori *et al.*, 1992), laminin (Maher *et al.*, 1988), tenascin (Ramadori *et al.*, 1991), undulin (Knittel *et al.*, 1992), and hyaluronic acid (Gressner and Haarmann, 1988). The list of matrix products of stellate cells is still expanding and includes virtually every ECM component of the normal and injured liver.

In addition to the synthesis of components of ECM in the liver, stellate cells are also the major participant in its degradation. This is carried out through the coordinated actions of a family of zinc-dependent enzymes named matrix-metalloproteinases (MMPs), their inhibitors (tissue inhibitor of metalloproteinases, TIMPs), and several converting enzymes (see Section II).

4. Contractility

Because of their unique location in the subendothelial space encircling hepatic sinusoids, stellate cells have long been proposed as tissue-specific pericytes, which regulate blood flow through their perivascular contractility (Sims, 1986; Wake, 1988; Wake *et al.*, 1992). Human and rat stellate cells contract in response to several vasoconstrictors, including thrombin and angiotensin II, thromboxane

A_2 and prostaglandin $F_{2\alpha}$ (Kawada *et al.*, 1993; Pinzani *et al.*, 1992b). Only activated stellate cells are fully contractile, whereas quiescent stellate cells are less so (Rockey *et al.*, 1993). The contractility of stellate cells appears concomitantly with the expression of α smooth muscle actin, but it is unclear whether this protein is required for the cell to contract or, alternatively, is simply a nonfunctional marker of a contractile phenotype.

It is uncertain whether quiescent stellate cells are contractile in normal liver. Freshly isolated stellate cells from normal rat liver do not contract in collagen gel contraction assays (Rockey *et al.*, 1993). However, based on *in vivo* microscopy, stellate cell-associated fluorescence colocalizes with sinusoidal contraction in response to ET-1 (Zhang *et al.*, 1994, 1995). However, this *in vivo* method is susceptible to artifact, making it difficult to distinguish between contractions of different cell populations, such as portal smooth muscle cells, stellate cells, and endothelial cells. Nonetheless, considering the strategic position of stellate cells, the abundance of endothelin receptors on their surface (see later), and close proximity to nerve endings containing vasoactive neurotransmitters (Bioulac-Sage *et al.*, 1990; Lafon *et al.*, 1989), it is quite possible that stellate cells play a role in regulating hepatic microcirculation under physiological as well as pathological conditions.

Among many factors that induce stellate cell contraction, ET-1 is the most potent agonist. Stellate cells express ETA and ETB receptors in both quiescent and activated states (Housset *et al.*, 1993; Kawada *et al.*, 1995). They express more ET-1 receptors than other liver cells, suggesting that this cell type is the main target of ET-1 (Housset *et al.*, 1993). Accompanying endothelin-1-mediated contraction is an increase in cytoplasmic calcium (Pinzani *et al.*, 1992b). Activated stellate cells express prepro-ET-1 mRNA (Housset *et al.*, 1993; Pinzani *et al.*, 1996a) and release ET-1 in the supernatant, indicating an autocrine activity of this cytokine, in addition to the paracrine action provided by endothelial cells. In cirrhotic liver, expression of ET-1 is increased markedly in both endothelial cells and stellate cells (Pinzani *et al.*, 1996a). The ET-1 antagonist, bosentan, reduces portal pressure when perfused into the cirrhotic rat liver with portal hypertension (Rockey and Weisiger, 1996). These studies strongly suggest that activated stellate cells regulate sinusoidal blood flow during liver injury, possibly via the action of ET-1. In cirrhotic liver, the contractility of stellate cells could lead to sinusoidal constriction and organ contraction, resulting in portal hypertension.

The vasoconstrictive effects of endothelin-1 may be counteracted by locally produced vasodilators. Prominent among these is nitric oxide (NO), which is produced from l-arginine by different forms of NO synthase (NOSs) (Moncada and Higgs, 1993). When stimulated with interferon-γ, TNF-α, and lipopolysaccharide, stellate cells produce NO rapidly (Helyar *et al.*, 1994; Rockey and Chung, 1995), largely as the product of the inducible form of NO synthase (Rockey and Chung, 1995). The endogenously produced NO strongly opposes contractility induced by ET-1 (Rockey and Chung, 1995) during liver injury, especially within regions of inflammation. In addition to NO, carbon monoxide, a product of heme cleavage, also mediates sinusoidal relaxation *in vivo* through its effects on stellate cells.

5. Retinoid Storage and Metabolism

In normal liver, stellate cells play a key role in the storage and transport of retinoids (vitamin A compounds). The biological role of retinoids in regulating stellate cell activation remains a puzzle. Although loss of retinoid is a prominent feature accompanying stellate cell activation both *in vivo* (Minato *et al.*, 1983) and in culture (Bachem *et al.*, 1992; Friedman *et al.*, 1993), it is unknown whether this process is a prerequisite for activation to occur. Reports describing effects of retinoids on stellate cells and fibrogenesis are contradictory (Geubel *et al.*, 1991; Leo and Lieber, 1983; Mizobuchi *et al.*, 1998; Seifert *et al.*, 1989, 1994; Senoo and Wake, 1985). In culture, both retinol and retinoic acid suppress stellate cells proliferation, with retinoic acid 1000 times more potent than retinol (Davis *et al.*, 1990). Studies in rat have documented the effects of 9-*cis*-RA and 9, 13-di-*cis*-RA, two metabolites of retinoic acid, in porcine serum-induced liver fibrosis (Okuno *et al.*, 1997, 1999). Both 9-*cis*-RA and 9, 13-di-*cis*-RA promote fibrosis by upregulating the plasminogen activator, which in turn increases the production and activation of TGF-β. This process is mediated by RARα. However, in another study using a rat fibrosis model induced by bile duct ligation, the increase in TGF-β production has been attributed to diminished RA signaling in stellate cells (Ohata *et al.*, 1997). It is likely, therefore, that retinoic acid affects stellate cell fibrogenesis through more than one pathway, which may depend on the mechanism of liver injury.

II. MOLECULAR MECHANISM OF STELLATE CELL ACTIVATION AND EXTRACELLULAR MATRIX REMODELING

The past two decades have witnessed the rapid elucidation of the role of stellate cells in liver injury and fibrosis. In normal liver, stellate cells are quiescent perisinusoidal vitamin A-storing cells in the subendothelial space of Disse. During injury, they undergo a gradient of phenotypic changes collectively called *activation* (sometimes referred to as "*transdifferentiation*"), through which they transform into proliferative, fibrogenic, and contractile myofibroblasts (Friedman, 1993, 2000). Regulation of their different phenotypes in normal and injured liver is the result of interactions with neighboring cells through paracrine and autocrine pathways, as well as the interactions between stellate cells and extracellular matrix (Fig. 1).

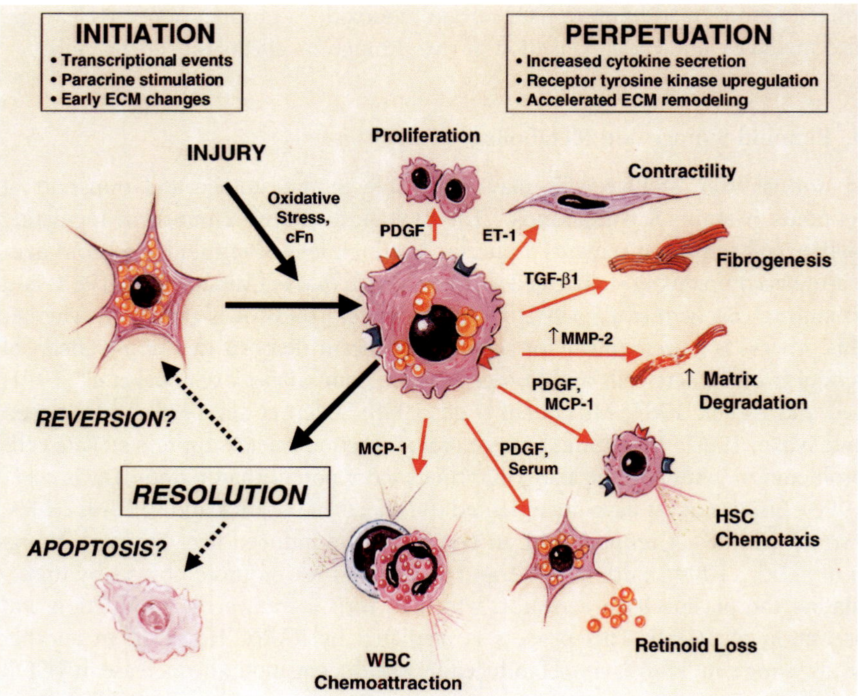

FIGURE 1 Sinusoidal events during fibrosing liver injury. Stellate cell activation leads to accumulation of the scar (fibril-forming) matrix, which in turn contributes to the loss of hepatocyte microvilli and sinusoidal endothelial fenestrae. Kupffer cell (macrophage) activation accompanies liver injury and contributes to the paracrine activation of stellate cells. From Friedman (2000), with permission.

The detailed pathways responsible for maintaining the quiescent phenotype of stellate cells in normal liver are unclear. Nonetheless, culture studies have demonstrated the importance of the basement membrane-like ECM in this process. When cultured on Engelbreth–Holm–Swarm (EHS) murine tumor-derived ECM, a gel matrix mimicking basement membrane-type ECM, stellate cells maintain the quiescent phenotype (Friedman *et al.*, 1989), in contrast to the activated phenotype when cultured on uncoated plastic. Individual components of the matrix (laminin, type IV collagen, and heparan sulfate proteoglycan) do not replicate the activity of the complete gel matrix, suggesting the requirement for complex matrix assembly in mediating this effect (Friedman *et al.*, 1989). When stellate cells are cultured on type I collagen, TGF-β stimulates the synthesis of collagen types I and III, whereas TGF-β has no such stimulating effects when cells are grown on type IV collagen (Davis, 1988).

The role of stellate cells in human liver disease has been clarified greatly. Features of stellate cell activation have been observed in viral hepatitis

(Enzan *et al.*, 1995; Inuzuka *et al.*, 1990), massive hepatic necrosis (Enzan *et al.*, 1995), and alcohol-induced liver disease (Hautekeete *et al.*, 1993; Horn *et al.*, 1986; Minato *et al.*, 1983; Schmitt-Graff *et al.*, 1993). In some diseases, such as those associated with iron overload, lipid peroxides stimulate stellate cell activation, leading to increased collagen production and liver fibrosis (Bedossa *et al.*, 1994; Houglum *et al.*, 1994; Kamimura *et al.*, 1992; Maher *et al.*, 1994; Parola *et al.*, 1993).

Activated stellate cells participate in tumor stroma accumulation in hepatocellular carcinoma (Enzan *et al.*, 1994; Torimura *et al.*, 1994). Stellate cell activation has also been implicated in a number of other human liver diseases, including biliary obstruction (Schmitt-Graff *et al.*, 1991), hematologic malignancy (Schmitt-Graff *et al.*, 1991), vascular disease (Schmitt-Graff *et al.*, 1991), mucopolysaccharidosis (Resnick *et al.*, 1994), acetaminophen overdose (Mathew *et al.*, 1994), leishmaniasis (el Hag *et al.*, 1994), and in drug abusers (Trigueiro de Araujo *et al.*, 1993).

A. A Model of Stellate Cell Activation

Stellate cell activation can be viewed conceptually as a two-stage process: initiation and perpetuation (Friedman, 1993, 2000). Initiation refers to early changes in gene expression and phenotype that render the cells responsive to other cytokines and stimuli, whereas perpetuation results from the effects of these stimuli on maintaining the activated phenotype and generating fibrosis. Initiation is largely due to paracrine stimulation, whereas perpetuation involves autocrine as well as paracrine loops (Fig. 2).

1. Initiation

Stellate cell activation is initiated by paracrine stimuli from injured neighboring cells, including hepatocytes, endothelial and Kupffer cells, and platelets, as well as infiltrating tumor cells in primary and metastatic cancer (Friedman *et al.*, 1989). Among the most well-studied paracrine stimuli are fibronectin (Jarnagin *et al.*, 1994), lipid peroxides (Paradis *et al.*, 1997), and cytokines, including PDGF, TGF-β1, and EGF. In recent years, increasing interest has focused on transcription factors involved in stellate cell activation, including Sp1, c-myb, NF-κB, c-jun/AP1, and STAT-1 (Kawada *et al.*, 1997) and Zf9/KLF6.

2. Perpetuation

After initiation, activated stellate cells undergo a series of phenotypic changes, which collectively lead to the accumulation of extracellular matrix. These changes include the following.

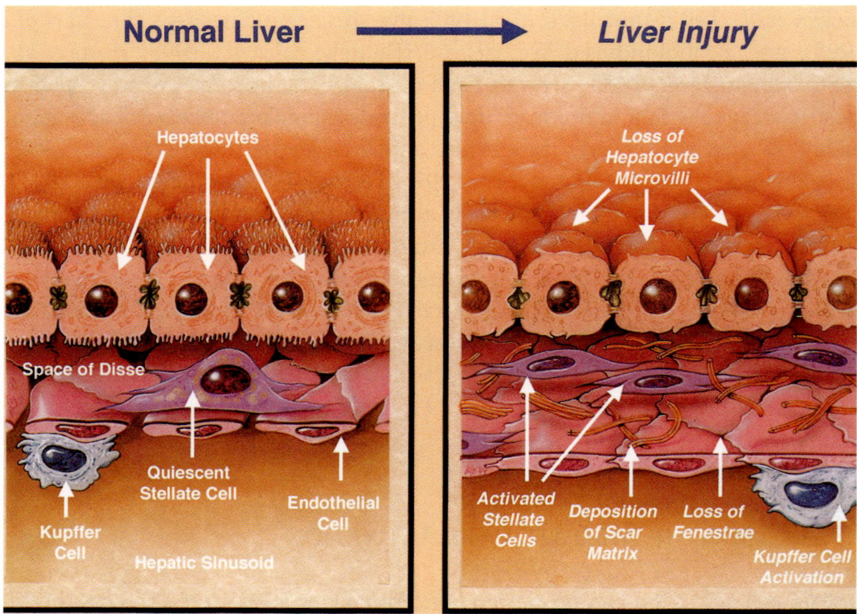

FIGURE 2 Features of hepatic stellate cell activation during liver injury and resolution. Following liver injury, hepatic stellate cells undergo "activation," a transition from quiescent vitamin A-rich cells into proliferative, fibrogenic, and contractile myofibroblasts. The major phenotypic changes after activation include proliferation, contractility, fibrogenesis, matrix degradation, chemotaxis, retinoid loss, and white blood cell (WBC) chemoattraction. Key mediators underlying these effects are shown. The fate of activated stellate cells during resolution of liver injury is uncertain, but may include reversion to a quiescent phenotype and/or selective clearance by apoptosis. ECM, extracellular matrix; ET-1, endothelin-1; MCP-1, monocyte chemotactic peptide 1; MMP, matrix-metalloproteinase; PDGF, platelet-derived growth factor; TGF-β1, transforming growth factor-β1. From Friedman (2000), with permission.

1. *Proliferation*. This is partly due to local proliferation in response to paracrine and autocrine mediators (Friedman, 1996), including PDGF (Pinzani *et al.*, 1994b; Wong *et al.*, 1994), ET-1 (Pinzani *et al.*, 1996a; Rockey *et al.*, 1998), thrombin (Marra *et al.*, 1995), FGF (Rosenbaum *et al.*, 1995), IGF (Pinzani *et al.*, 1990; Skrtic *et al.*, 1997), and TGF-β1 (Pinzani *et al.*, 1998), among others (Friedman, 1999; Pinzani *et al.*, 1998).

2. *Contractility*. Contraction by stellate cells may be a major determinant of early and late increases in portal resistance during liver fibrosis and cirrhosis. A key contractile stimulus toward stellate cells is ET-1 (Rockey, 1997). Other contractile agonists include arginine vasopressin, adrenomedullin, and eicosanoids.

3. *Fibrogenesis*. A large number of fibrogenic factors have been identified, including interleukin-1β, tumor necrosis factor, lipid peroxides, and acetaldehyde, but none is as potent as TGF-β1 (Friedman, 1996; Pietrangelo, 1996).

4. *Chemotaxis*. The accumulation of stellate cells during liver injury may be the result of both proliferation and directed migration into regions of injury. PDGF and the chemoattractant monocyte chemotactic peptide-1 (MCP-1) have been identified as stellate cell chemoattractants (Marra *et al.*, 1997).

5. *Matrix degradation*. Quantitative and qualitative changes in the activity of MMPs and their inhibitors play a vital role in extracellular matrix remodeling in liver fibrogenesis. The net effect is conversion of the low-density subendothelial matrix to one rich in interstitial collagens.

6. *Retinoid loss*. Activation of stellate cells is accompanied by the loss of their characteristic perinuclear retinoid droplets. It is still unknown whether retinoid loss is a requirement for stellate cell activation.

7. *Cytokine release*. Increased production of cytokines may be critical for the perpetuation of stellate cell activation. These cytokines include TGF-β1, PDGF, FGF, HGF, PAF, ET-1, MCP-1 (Marra *et al.*, 1998, 1993), and CINC (Maher *et al.*, 1998), among others (Friedman, 1999).

3. Fate of Activated Stellate Cells

Once activated, the ultimate outcome of these cells is still not clear. It is not certain whether activated stellate cells can revert to the quiescent state or are cleared selectively by apoptosis, although increasing evidence implicates apoptosis *in vivo*. During the resolution of liver injury, the percentage of stellate cells undergoing apoptosis is increased (Iredale *et al.*, 1998). Interestingly, these cells express high levels of TIMP (Iredale *et al.*, 1998). Therefore, their selective clearance would remove a key molecule preventing the degradation of fibrotic scar by interstitial collagenase, unleashing the capacity of the liver to resorb excess ECM. Activated stellate cells also display an increased susceptibility to apoptotic signals (such as soluble Fas ligand) and decreased expression of the antiapoptotic protein bcl-2 (Gong *et al.*, 1998). These findings are consistent with wound healing in other tissues such as the kidney and vascular wall (Desmouliere *et al.*, 1997).

Signals in the hepatic microenvironment responsible for modulating stellate cell apoptosis are not yet clarified. The extracellular matrix may play a role in regulating apoptosis by providing permissive or inhibitory signals (Shi *et al.*, 1998; Sugiyama *et al.*, 1998). Interruption of integrin-mediated interaction between stellate cells and ECM induces apoptosis (Iwamoto *et al.*, 1999) and similar effects occur in kidney (Sugiyama *et al.*, 1998). Interestingly, TGF-β and TNF-α inhibit apoptosis of activated stellate cells, suggesting that fibrogenic cytokines may promote the survival of activated, fibrogenic stellate cells (Saile *et al.*, 1999).

B. Extracellular Matrix Remodeling

During chronic liver injury, the matrix phenotype of stellate cells changes both qualitatively and quantitatively. Overall, its ECM production increases remarkably, accompanied by a shift in the type of ECM in subendothelial space from the normal low-density basement membrane-like matrix to interstitial type (Gressner and Bachem, 1990, 1995).

The accumulation of ECM, especially collagen, is a dynamic process reflecting the imbalance between matrix accumulation and degradation. The repertoire of factors responsible for ECM remodeling is continually being identified (Arthur, 1998). These include the MMPs family, their inhibitors (TIMPs), and several converting enzymes [e.g., MT1-MMP and stromelysin (Takahara *et al.*, 1997; Vyas *et al.*, 1995)]. Although cellular sources of matrix proteases and their regulators in liver have not yet been fully elucidated, stellate cells are a key source of MMP-2 and stromelysin (Arthur *et al.*, 1992; Milani *et al.*, 1994b; Vyas *et al.*, 1995). They also express TIMP-1 and -2 mRNA (Herbst *et al.*, 1997) and produce TIMP-1 and MT1-MMP (Theret *et al.*, 1997). MMP-9, which is a type IV collagenase, is locally secreted by Kupffer cells (Winwood *et al.*, 1995). The hepatic source of MMP-1 (interstitial collagenase, collagenase I), which plays a crucial role in degrading the excess interstitial matrix in advanced liver disease, is still uncertain, although it is known that upregulation of plasmin and stromelysin-1 can activate latent MMP1 (Arthur, 1995).

In human liver diseases, there is a downregulation of MMP1 and upregulation of MMP2 (gelatinase A) and MMP9 (gelatinase B) (Milani *et al.*, 1994a). Because these enzymes have different substrate specificities, the result is the increased degradation of the basement membrane collagen and decreased degradation of the interstitial type collagen. However, activated MMPs are regulated in part by their tissue inhibitors (TIMPs). TIMP1 and TIMP2 are upregulated relative to MMP1 in progressive experimental liver fibrosis, which may explain the decreased degradation of interstitial type matrix observed in experimental and human liver injury (Benyon *et al.*, 1996; Herbst *et al.*, 1997; Iredale, 1995, 1996, 1997). In contrast, during the recovery phase of experimental liver injury, the expression of TIMP1 and TIMP2 is decreased, whereas the collagenase mRNA level is unchanged (Iredale *et al.*, 1998), resulting in a net increase in collagenase activity and increased resorption of scar matrix. Increased MT1-MMP (membrane-type 1-matrix metalloproteinase) has also been described in fibrotic human livers (Takahara *et al.*, 1997). MT1-MMP activates latent MMP2 in other tissues and might have a similar role in stellate cells. The net result of ECM remodeling following chronic liver injury is the diminished basement membrane-type matrix and accumulation of interstitial type collagens, which ultimately evolve into thick scars seen in a cirrhotic liver.

III. CONCLUSIONS AND FUTURE PROSPECTS

Since the early 1980s, the development of modern technologies has enabled us to more fully characterize the biological features of hepatic stellate cells. However, mysteries remain. To date, the key genes that initiate stellate cell activation are still to be defined, and we have not found any genes that can be called "stellate cell specific." The fate of activated stellate cells during the resolution of liver injury needs to be clarified, and more importantly, the factors that determine the reversibility of liver fibrosis must be identified. Elucidation of these questions will greatly enrich our knowledge about this important cell type and ultimately translate into effective therapies for chronic liver disease and fibrosis.

REFERENCES

Andus, T., Ramadori, G., Heinrich, P. C., Knittel, T., and Meyer zum Buschenfelde, K. H. (1987). Cultured Ito cells of rat liver express the alpha 2-macroglobulin gene. *Eur. J. Biochem.* **168**, 641–646.

Ankoma-Sey, V., Matli, M., Chang, K. B., Lalazar, A., Donner, D. B., Wong, L., Warren, R. S., and Friedman, S. L. (1998). Coordinated induction of VEGF receptors in mesenchymal cell types during rat hepatic wound healing. *Oncogene* 17, 115–121.

Arenson, D. M., Friedman, S. L., and Bissell, D. M. (1988). Formation of extracellular matrix in normal rat liver: Lipocytes as a major source of proteoglycan. *Gastroenterology* **95**, 441–447.

Arthur, M. J., Stanley, A., Iredale, J. P., Rafferty, J. A., Hembry, R. M., and Friedman, S. L. (1992). Secretion of 72 kDa type IV collagenase/gelatinase by cultured human lipocytes: Analysis of gene expression, protein synthesis and proteinase activity. *Biochem. J.* **287**, 701–707.

Arthur, M. J. (1995). Collagenases and liver fibrosis. *J. Hepatol.* **22**, 43–48.

Arthur, M. J. (1998). Fibrosis and altered matrix degradation. *Digestion* **59**, 376–380.

Attisano, L., and Wrana, J. L. (1998). Mads and Smads in TGF beta signalling. *Curr. Opin. Cell Biol.* **10**, 188–194.

Bachem, M. G., Meyer, D., Melchior, R., Sell, K. M., and Gressner, A. M. (1992). Activation of rat liver perisinusoidal lipocytes by transforming growth factors derived from myofibroblastlike cells: A potential mechanism of self perpetuation in liver fibrogenesis. *J. Clin. Invest.* **89**, 19–27.

Bedossa, P., Houglum, K., Trautwein, C., Holstege, A., and Chojkier, M. (1994). Stimulation of collagen alpha 1(I) gene expression is associated with lipid peroxidation in hepatocellular injury: A link to tissue fibrosis? *Hepatology* **19**, 1262–1271.

Benyon, R. C., Iredale, J. P., Goddard, S., Winwood, P. J., and Arthur, M. J. (1996). Expression of tissue inhibitor of metalloproteinases 1 and 2 is increased in fibrotic human liver. *Gastroenterology* **110**, 821–831.

Bioulac-Sage, P., Lafon, M. E., Saric, J., and Balabaud, C. (1990). Nerves and perisinusoidal cells in human liver. *J. Hepatol.* **10**, 105–112.

Bissell, D. M., Wang, S. S., Jarnagin, W. R., and Roll, F. J. (1995). Cell-specific expression of transforming growth factor-beta in rat liver: Evidence for autocrine regulation of hepatocyte proliferation. *J. Clin. Invest.* **96**, 447–455.

Bruck, R., Shirin, H., Hershkoviz, R., Lider, O., Kenet, G., Aeed, H., Matas, Z., Zaidel, L., and Halpern, Z. (1997). Analysis of Arg-Gly-Asp mimetics and soluble receptor of tumour necrosis factor as therapeutic modalities for concanavalin A induced hepatitis in mice. *Gut* **40**, 133–138.

Carloni, V., Romanelli, R. G., Pinzani, M., Laffi, G., and Gentilini, P. (1996). Expression and function of integrin receptors for collagen and laminin in cultured human hepatic stellate cells. *Gastroenterology* **110**, 1127–1136.

Carloni, V., Romanelli, R. G., Pinzani, M., Laffi, G., and Gentilini, P. (1997). Focal adhesion kinase and phospholipase C gamma involvement in adhesion and migration of human hepatic stellate cells. *Gastroenterology* **112**, 522–531.

Castilla, A., Prieto, J., and Fausto, N. (1991). Transforming growth factors beta 1 and alpha in chronic liver disease: Effects of interferon alfa therapy. *N. Engl. J. Med.* **324**, 933–940.

Clark, E. A., and Brugge, J. S. (1995). Integrins and signal transduction pathways: The road taken. *Science* **268**, 233–239.

Czaja, M. J., Geerts, A., Xu, J., Schmiedeberg, P., and Ju, Y. (1994). Monocyte chemoattractant protein 1 (MCP-1) expression occurs in toxic rat liver injury and human liver disease. *J. Leuk. Biol.* **55**, 120–126.

Davis, B. H. (1988). Transforming growth factor beta responsiveness is modulated by the extracellular collagen matrix during hepatic ito cell culture. *J. Cell Physiol.* **136**, 547–553.

Davis, B. H., Kramer, R. T., and Davidson, N. O. (1990). Retinoic acid modulates rat Ito cell proliferation, collagen, and transforming growth factor beta production. *J. Clin. Invest.* **86**, 2062–2070.

de Bleser, P. J., Jannes, P., van Buul-Offers, S. C., Hoogerbrugge, C. M., van Schravendijk, C. F., Niki, T., Rogiers, V., van den Brande, J. L., Wisse, E., and Geerts, A. (1995). Insulinlike growth factor-II/mannose 6-phosphate receptor is expressed on CCl_4-exposed rat fat-storing cells and facilitates activation of latent transforming growth factor-beta in cocultures with sinusoidal endothelial cells. *Hepatology* **21**, 1429–1437.

Desmouliere, A., Badid, C., Bochaton-Piallat, M. L., and Gabbiani, G. (1997). Apoptosis during wound healing, fibrocontractive diseases and vascular wall injury. *Int. J. Biochem. Cell Biol.* **29**, 19–30.

Dooley, S., Delvoux, B., Lahme, B., Mangasser-Stephan, K., and Gressner, A. M. (2000). Modulation of transforming growth factor beta response and signaling during transdifferentiation of rat hepatic stellate cells to myofibroblasts. *Hepatology* **31**, 1094–1106.

Duncan, M. R., Frazier, K. S., Abramson, S., Williams, S., Klapper, H., Huang, X., and Grotendorst, G. R. (1999). Connective tissue growth factor mediates transforming growth factor beta-induced collagen synthesis: Down-regulation by cAMP. *FASEB J.* **13**, 1774–1786.

el Hag, I. A., Hashim, F. A., el Toum, I. A., Homeida, M., el Kalifa, M., el Hassan, A. M., Torimura, T., Ueno, T., Inuzuka, S., Kin, M., Ohira, H., Kimura, Y., Majima, Y., Sata, M., Abe, H., and Tanikawa, K. (1994). Liver morphology and function in visceral leishmaniasis (Kala-azar). *J. Clin. Pathol.* **47**, 547–551.

Enzan, H., Himeno, H., Iwamura, S., Onishi, S., Saibara, T., Yamamoto, Y., and Hara, H. (1994). Alpha-smooth muscle actin-positive perisinusoidal stromal cells in human hepatocellular carcinoma. *Hepatology* **19**, 895–903.

Enzan, H., Himeno, H., Iwamura, S., Saibara, T., Onishi, S., Yamamoto, Y., Miyazaki, E., and Hara, H. (1995). Sequential changes in human Ito cells and their relation to postnecrotic liver fibrosis in massive and submassive hepatic necrosis. *Virch. Arch.* **426**, 95–101.

Friedman, S. L. (1993). Seminars in medicine of the Beth Israel Hospital, Boston. The cellular basis of hepatic fibrosis. Mechanisms and treatment strategies. *N. Engl. J. Med.* **328**, 1828–1835.

Friedman, S. L. (1996). Hepatic stellate cells. *Progr. Liver Dis.* **14**, 101–130.

Friedman, S. L. (1999). Cytokines and fibrogenesis. *Semin. Liver Dis.* **19**, 129–140.

Friedman, S. L. (2000). Molecular regulation of hepatic fibrosis, an integrated cellular response to tissue injury. *J. Biol. Chem.* **275**, 2247–2250.

Friedman, S. L., Roll, F. J., Boyles, J., Arenson, D. M., and Bissell, D. M. (1989). Maintenance of differentiated phenotype of cultured rat hepatic lipocytes by basement membrane matrix. *J. Biol. Chem.* **264**, 10756–10762.

Friedman, S. L., Roll, F. J., Boyles, J., and Bissell, D. M. (1985). Hepatic lipocytes: The principal collagen-producing cells of normal rat liver. *Proc. Natl. Acad. Sci. USA* **82**, 8681–8685.

Friedman, S. L., Wei, S., and Blaner, W. S. (1993). Retinol release by activated rat hepatic lipocytes: Regulation by Kupffer cell-conditioned medium and PDGF. *Am. J. Physiol.* **264**, G947–G952.

Friedman, S. L., Yamasaki, G., and Wong, L. (1994). Modulation of transforming growth factor beta receptors of rat lipocytes during the hepatic wound healing response: Enhanced binding and reduced gene expression accompany cellular activation in culture and in vivo. *J. Biol. Chem.* **269**, 10551–10558.

Frizell, E., Liu, S. L., Abraham, A., Ozaki, I., Eghbali, M., Sage, E. H., and Zern, M. A. (1995). Expression of SPARC in normal and fibrotic livers. *Hepatology* **21**, 847–854.

Fujio, K., Evarts, R. P., Hu, Z., Marsden, E. R., and Thorgeirsson, S. S. (1994). Expression of stem cell factor and its receptor, c-kit, during liver regeneration from putative stem cells in adult rat. *Lab. Invest.* **70**, 511–516.

Geerts, A., Vrijsen, R., Rauterberg, J., Burt, A., Schellinck, P., and Wisse, E. (1989). In vitro differentiation of fat-storing cells parallels marked increase of collagen synthesis and secretion. *J. Hepatol.* **9**, 59–68.

Geubel, A. P., De Galocsy, C., Alves, N., Rahier, J., and Dive, C. (1991). Liver damage caused by therapeutic vitamin A administration: Estimate of dose-related toxicity in 41 cases. *Gastroenterology* **100**, 1701–1709.

Gong, W., Pecci, A., Roth, S., Lahme, B., Beato, M., and Gressner, A. M. (1998). Transformation-dependent susceptibility of rat hepatic stellate cells to apoptosis induced by soluble Fas ligand. *Hepatology* **28**, 492–502.

Greenwel, P., Rubin, J., Schwartz, M., Hertzberg, E. L., and Rojkind, M. (1993). Liver fat-storing cell clones obtained from a CCl_4 cirrhotic rat are heterogeneous with regard to proliferation, expression of extracellular matrix components, interleukin-6, and connexin 43. *Lab. Invest.* **69**, 210–216.

Gressner, A. M. (1995). Cytokines and cellular crosstalk involved in the activation of fat-storing cells. *J. Hepatol.* **22**, 28–36.

Gressner, A. M., and Bachem, M. G. (1990). Cellular sources of noncollagenous matrix proteins: Role of fat-storing cells in fibrogenesis. *Semin. Liver Dis.* **10**, 30–46.

Gressner, A. M., and Bachem, M. G. (1995). Molecular mechanisms of liver fibrogenesis: A homage to the role of activated fat-storing cells. *Digestion* **56**, 335–346.

Gressner, A. M., and Haarmann, R. (1988). Hyaluronic acid synthesis and secretion by rat liver fat storing cells (perisinusoidal lipocytes) in culture. *Biochem. Biophys. Res. Commun.* **151**, 222–229.

Grotendorst, G. R. (1997). Connective tissue growth factor: A mediator of TGF-beta action on fibroblasts. *Cytokine Growth Factor Rev.* **8**, 171–179.

Hautekeete, M. L., Geerts, A., Seynaeve, C., Lazou, J. M., Kloppel, G., and Wisse, E. (1993). Contributions of light and transmission electron microscopy to the study of the human fat-storing cell. *Eur. J. Morphol.* **31**, 72–76.

Heldin, C. H., Miyazono, K., and ten Dijke, P. (1997). TGF-beta signalling from cell membrane to nucleus through SMAD proteins. *Nature* **390**, 465–471.

Heldin, P., Pertoft, H., Nordlinder, H., Heldin, C. H., and Laurent, T. C. (1991). Differential expression of platelet-derived growth factor alpha and beta-receptors on fat-storing cells and endothelial cells of rat liver. *Exp. Cell Res.* **193**, 364–369.

Hellerbrand, Wang, S. C., Tsukamoto, H., Brenner, D. A., and Rippe, R. A. (1996). Expression of intracellular adhesion molecule 1 by activated hepatic stellate cells. *Hepatology* **24**, 670–676.

Helyar, L., Bundschuh, D. S., Laskin, J. D., and Laskin, D. L. (1994). Induction of hepatic Ito cell nitric oxide production after acute endotoxemia. *Hepatology* **20**, 1509–1515.

Henriksen, J. H., Holst, J. J., Moller, S., Brinch, K., and Bendtsen, F. (1999). Increased circulating leptin in alcoholic cirrhosis: Relation to release and disposal. *Hepatology* **29**, 1818–1824.

Herbst, H., Wege, T., Milani, S., Pellegrini, G., Orzechowski, H. D., Bechstein, W. O., Neuhaus, P., Gressner, A. M., and Schuppan, D. (1997). Tissue inhibitor of metalloproteinase-1 and -2 RNA expression in rat and human liver fibrosis. *Am. J. Pathol.* **150**, 1647–1659.

Horn, T., Junge, J., and Christoffersen, P. (1986). Early alcoholic liver injury: Activation of lipocytes in acinar zone 3 and correlation to degree of collagen formation in the Disse space. *J. Hepatol.* **3**, 333–340.

Houglum, K., Bedossa, P., and Chojkier, M. (1994). TGF-beta and collagen-alpha 1 (I) gene expression are increased in hepatic acinar zone 1 of rats with iron overload. *Am. J. Physiol.* **267**, G908–G913.

Housset, C., Rockey, D. C., and Bissell, D. M. (1993). Endothelin receptors in rat liver: Lipocytes as a contractile target for endothelin 1. *Proc. Natl. Acad. Sci. USA* **90**, 9266–9270.

Igarashi, A., Nashiro, K., Kikuchi, K., Sato, S., Ihn, H., Fujimoto, M., Grotendorst, G. R., and Takehara, K. (1996). Connective tissue growth factor gene expression in tissue sections from localized scleroderma, keloid, and other fibrotic skin disorders. *J. Invest. Dermatol.* **106**, 729–733.

Ikejima, K., Honda, H., Yoshikawa, M., Hirose, M., Kitamura, T., Takei, Y., and Sato, N. (2001). Leptin augments inflammatory and profibrogenic responses in the murine liver induced by hepatotoxic chemicals. *Hepatology* **34**, 288–297.

Imai, K., and Senoo, H. (1998). Morphology of sites of adhesion between hepatic stellate cells (vitamin A-storing cells) and a three-dimensional extracellular matrix. *Anat. Rec.* **250**, 430–437.

Imai, S., Okuno, M., Moriwaki, H., Muto, Y., Murakami, K., Shudo, K., Suzuki, Y., Kojima, S., Sato, T., Kitamoto, T., Kawada, N., Yoshimura, H., Onuki, K., Masushige, S., Friedman, S. L., and Kato, S. (1997). 9,13-di-cis-Retinoic acid induces the production of tPA and activation of latent TGF-beta via RAR alpha in a human liver stellate cell line, LI90. *FEBS Lett.* **411**, 102–106.

Inuzuka, S., Ueno, T., Torimura, T., Sata, M., Abe, H., and Tanikawa, K. (1990). Immunohistochemistry of the hepatic extracellular matrix in acute viral hepatitis. *Hepatology* **12**, 249–256.

Iredale, J. P. (1997). Tissue inhibitors of metalloproteinases in liver fibrosis. *Int. J. Biochem. Cell Biol.* **29**, 43–54.

Iredale, J. P., and Arthur, M. J. (1994). Hepatocyte-matrix interactions. *Gut* **35**, 729–732.

Iredale, J. P., Benyon, R. C., Arthur, M. J., Ferris, W. F., Alcolado, R., Winwood, P. J., Clark, N., and Murphy, G. (1996). Tissue inhibitor of metalloproteinase-1 messenger RNA expression is enhanced relative to interstitial collagenase messenger RNA in experimental liver injury and fibrosis. *Hepatology* **24**, 176–184.

Iredale, J. P., Benyon, R. C., Pickering, J., McCullen, M., Northrop, M., Pawley, S., Hovell, C., and Arthur, M. J. (1998). Mechanisms of spontaneous resolution of rat liver fibrosis. Hepatic stellate cell apoptosis and reduced hepatic expression of metalloproteinase inhibitors. *J. Clin. Invest.* **102**, 538–549.

Iredale, J. P., Goddard, S., Murphy, G., Benyon, R. C., and Arthur, M. J. (1995). Tissue inhibitor of metalloproteinase-I and interstitial collagenase expression in autoimmune chronic active hepatitis and activated human hepatic lipocytes. *Clin. Sci. (Colch.)* **89**, 75–81.

Ito, Y., Aten, J., Bende, R. J., Oemar, B. S., Rabelink, T. J., Weening, J. J., and Goldschmeding, R. (1998). Expression of connective tissue growth factor in human renal fibrosis. *Kidney Int.* **53**, 853–861.

Iwamoto, H., Sakai, H., and Nawata, H. (1998). Inhibition of integrin signaling with Arg-Gly-Asp motifs in rat hepatic stellate cells. *J. Hepatol.* **29**, 752–759.

Iwamoto, H., Sakai, H., Tada, S., Nakamuta, M., and Nawata, H. (1999). Induction of apoptosis in rat hepatic stellate cells by disruption of integrin-mediated cell adhesion. *J. Lab. Clin. Med.* **134**, 83–89.

Jarnagin, W. R., Rockey, D. C., Koteliansky, V. E., Wang, S. S., and Bissell, D. M. (1994). Expression of variant fibronectins in wound healing: Cellular source and biological activity of the EIIIA segment in rat hepatic fibrogenesis. *J. Cell Biol.* **127**, 2037–2048.

Ji, C., Casinghino, S., McCarthy, T. L., and Centrella, M. (1997). Multiple and essential Sp1 binding sites in the promoter for transforming growth factor-beta type I receptor. *J. Biol. Chem.* **272**, 21260–21267.

Kamimura, S., Gaal, K., Britton, R. S., Bacon, B. R., Triadafilopoulos, G., and Tsukamoto, H. (1992). Increased 4-hydroxynonenal levels in experimental alcoholic liver disease: Association of lipid peroxidation with liver fibrogenesis. *Hepatology* **16**, 448–453.

Kawada, N., Kuroki, T., Kobayashi, K., Inoue, M., Kaneda, K., and Decker, K. (1995). Action of endothelins on hepatic stellate cells. *J. Gastroenterol.* **30**, 731–738.

Kawada, N., Tran, T. T., Klein, H., and Decker, K. (1993). The contraction of hepatic stellate (Ito) cells stimulated with vasoactive substances: Possible involvement of endothelin 1 and nitric oxide in the regulation of the sinusoidal tonus. *Eur. J. Biochem.* **213**, 815–823.

Kawada, N., Uoya, M., Seki, S., Kuroki, T., and Kobayashi, K. (1997). Regulation by cAMP of STAT1 activation in hepatic stellate cells. *Biochem. Biophys. Res. Commun.* **233**, 464–469.

Kawase, T., Shiratori, Y., and Sugimoto, T. (1986). Collagen production by rat liver fat-storing cells in primary culture. *Exp. Cell Biol.* **54**, 183–192.

Kim, Y., Ratziu, V., Choi, S. G., Lalazar, A., Theiss, G., Dang, Q., Kim, S. J., and Friedman, S. L. (1998). Transcriptional activation of transforming growth factor beta1 and its receptors by the Kruppel-like factor Zf9/core promoter-binding protein and Sp1: Potential mechanisms for autocrine fibrogenesis in response to injury. *J. Biol. Chem.* **273**, 33750–33758.

Knittel, T., Armbrust, T., Schwogler, S., Schuppan, D., and Ramadori, G. (1992). Distribution and cellular origin of undulin in rat liver. *Lab. Invest.* **67**, 779–787.

Knittel, T., Aurisch, S., Neubauer, K., Eichhorst, S., and Ramadori, G. (1996). Cell-type-specific expression of neural cell adhesion molecule (N-CAM) in Ito cells of rat liver: Up-regulation during in vitro activation and in hepatic tissue repair. *Am. J. Pathol.* **149**, 449–462.

Knittel, T., Dinter, C., Kobold, D., Neubauer, K., Mehde, M., Eichhorst, S., and Ramadori, G. (1999). Expression and regulation of cell adhesion molecules by hepatic stellate cells (HSC) of rat liver: Involvement of HSC in recruitment of inflammatory cells during hepatic tissue repair. *Am. J. Pathol.* **154**, 153–167.

Kovalszky, H., Gallai, M., Armbrust, T., and Ramadori, G. (1994). Syndecan-1 gene expression in isolated rat liver cells (hepatocytes, Kupffer cells, endothelial and Ito cells). *Biochem. Biophys. Res. Commun.* **204**, 944–949.

Lafon, M. E., Bioulac-Sage, P., Le Bail, B., Saric, J., Quinton, A., and Balabaud, C. (1989). Nerves and perisinusoidal cells in human liver. 230–234.

Lasky, J. A., Ortiz, L. A., Tonthat, B., Hoyle, G. W., Corti, M., Athas, G., Lungarella, G., Brody, A., and Friedman, M. (1998). Connective tissue growth factor mRNA expression is upregulated in bleomycin-induced lung fibrosis. *Am. J. Physiol.* **275**, L365–L371.

Leo, M. A., and Lieber, C. S. (1983). Hepatic fibrosis after long-term administration of ethanol and moderate vitamin A supplementation in the rat. *Hepatology* **3**, 1–11.

Louis, H., Van Laethem, J. L., Wu, W., Quertinmont, E., Degraef, C., Van den Berg, K., Demols, A., Goldman, M., Le Moine, O., Geerts, A., and Deviere, J. (1998). Interleukin-10 controls neutrophilic infiltration, hepatocyte proliferation, and liver fibrosis induced by carbon tetrachloride in mice. *Hepatology* **28**, 1607–1615.

Maher, J. J. (1993). Cell-specific expression of hepatocyte growth factor in liver: Upregulation in sinusoidal endothelial cells after carbon tetrachloride. *J. Clin. Invest.* **91**, 2244–2252.

Maher, J. J., Friedman, S. L., Roll, F. J., and Bissell, D. M. (1988). Immunolocalization of laminin in normal rat liver and biosynthesis of laminin by hepatic lipocytes in primary culture. *Gastroenterology* **94**, 1053–1062.

Maher, J. J., Lozier, J. S., and Scott, M. K. (1998). Rat hepatic stellate cells produce cytokine-induced neutrophil chemoattractant in culture and in vivo. *Am. J. Physiol.* **275**, G847–G853.

Maher, J. J., Tzagarakis, C., and Gimenez, A. (1994). Malondialdehyde stimulates collagen production by hepatic lipocytes only upon activation in primary culture. *Alcohol Alcohol.* **29**, 605–610.

Marra, F., Choudhury, G. G., Pinzani, M., and Abboud, H. E. (1994). Regulation of platelet-derived growth factor secretion and gene expression in human liver fat-storing cells. *Gastroenterology* **107**, 1110–1117.

Marra, F., DeFranco, R., Grappone, C., Milani, S., Pastacaldi, S., Pinzani, M., Romanelli, R. G., Laffi, G., and Gentilini, P. (1998). Increased expression of monocyte chemotactic protein-1 during active hepatic fibrogenesis: Correlation with monocyte infiltration. *Am. J. Pathol.* **152**, 423–430.

Marra, F., Grandaliano, G., Valente, A. J., and Abboud, H. E. (1995). Thrombin stimulates proliferation of liver fat-storing cells and expression of monocyte chemotactic protein-1: Potential role in liver injury. *Hepatology* **22**, 780–787.

Marra, F., Gentilini, A., Pinzani, M., Choudhury, G. G., Parola, M., Herbst, H., Dianzani, M. U., Laffi, G., Abboud, H. E., and Gentilini, P. (1997). Phosphatidylinositol 3-kinase is required for platelet-derived growth factor's actions on hepatic stellate cells. *Gastroenterology* **112**, 1297–1306.

Marra, F., Valente, A. J., Pinzani, M., and Abboud, H. E. (1993). Cultured human liver fat-storing cells produce monocyte chemotactic protein-1: Regulation by proinflammatory cytokines. *J. Clin. Invest.* **92**, 1674–1680.

Marsden, E. R., Hu, Z., Fujio, K., Nakatsukasa, H., Thorgeirsson, S. S., and Evarts, R. P. (1992). Expression of acidic fibroblast growth factor in regenerating liver and during hepatic differentiation. *Lab. Invest.* **67**, 427–433.

Marti, A., Berraondo, B., and Martinez, J. A. (1999). Leptin: Physiological actions. *J. Physiol. Biochem.* **55**, 43–49.

Mathew, J., Hines, J. E., James, O. F., and Burt, A. D. (1994). Non-parenchymal cell responses in paracetamol (acetaminophen)-induced liver injury. *J. Hepatol.* **20**, 537–541.

McCullough, A. J., Bugianesi, E., Marchesini, G., and Kalhan, S. C. (1998). Gender-dependent alterations in serum leptin in alcoholic cirrhosis. *Gastroenterology* **115**, 947–953.

Meyer, D. H., Bachem, M. G., and Gressner, A. M. (1990). Modulation of hepatic lipocyte proteoglycan synthesis and proliferation by Kupffer cell-derived transforming growth factors type beta 1 and type alpha. *Biochem. Biophys. Res. Commun.* **171**, 1122–1129.

Meyer, D. H., Krull, N., Dreher, K. L., and Gressner, A. M. (1992). Biglycan and decorin gene expression in normal and fibrotic rat liver: Cellular localization and regulatory factors. *Hepatology* **16**, 204–216.

Milani, S., Grappone, C., Pellegrini, G., Schuppan, D., Herbst, H., Calabro, A., Casini, A., Pinzani, M., and Surrenti, C. (1994a). Undulin RNA and protein expression in normal and fibrotic human liver. *Hepatology* **20**, 908–916.

Milani, S., Herbst, H., Schuppan, D., Grappone, C., Pellegrini, G., Pinzani, M., Casini, A., Calabro, A., Ciancio, G., Stefanini, F., *et al.* (1994b). Differential expression of matrix-metalloproteinase-1 and -2 genes in normal and fibrotic human liver. *Am. J. Pathol.* **144**, 528–537.

Minato, Y., Hasumura, Y., and Takeuchi, J. (1983). The role of fat-storing cells in Disse space fibrogenesis in alcoholic liver disease. *Hepatology* **3**, 559–566.

Mizobuchi, Y., Shimizu, I., Yasuda, M., Hori, H., Shono, M., and Ito, S. (1998). Retinyl palmitate reduces hepatic fibrosis in rats induced by dimethylnitrosamine or pig serum. *J. Hepatol.* **29**, 933–943.

Moncada, S., and Higgs, A. (1993). The L-arginine-nitric oxide pathway. *N. Engl. J. Med.* **329**, 2002–2012.

Mullhaupt, B., Feren, A., Fodor, E., and Jones, A. (1994). Liver expression of epidermal growth factor RNA. Rapid increases in immediate-early phase of liver regeneration. *J. Biol. Chem.* **269**, 19667–19670.

Musso, O., Rehn, M., Saarela, J., Theret, N., Lietard, J., Hintikka, Lotrian, D., Campion, J. P., Pihlajaniemi, T., and Clement, B. (1998). Collagen XVIII is localized in sinusoids and basement membrane zones and expressed by hepatocytes and activated stellate cells in fibrotic human liver. *Hepatology* **28**, 98–107.

Nakatani, K., Seki, S., Kawada, N., Kobayashi, K., and Kaneda, K. (1996). Expression of neural cell adhesion molecule (N-CAM) in perisinusoidal stellate cells of the human liver. *Cell Tissue Res.* **283**, 159–165.

Ohata, M., Lin, M., Satre, M., and Tsukamoto, H. (1997). Diminished retinoic acid signaling in hepatic stellate cells in cholestatic liver fibrosis. *Am. J. Physiol.* **272**, G589–G596.

Okuno, M., Moriwaki, H., Imai, S., Muto, Y., Kawada, N., Suzuki, Y., and Kojima, S. (1997). Retinoids exacerbate rat liver fibrosis by inducing the activation of latent TGF-beta in liver stellate cells. *Hepatology* **26**, 913–921.

Okuno, M., Sato, T., Kitamoto, T., Imai, S., Kawada, N., Suzuki, Y., Yoshimura, H., Moriwaki, H., Onuki, K., Masushige, S., Muto, Y., Friedman, S. L., Kato, S., and Kojima, S. (1999). Increased 9,13-di-cis-retinoic acid in rat hepatic fibrosis: Implication for a potential link between retinoid loss and TGF-beta mediated fibrogenesis in vivo. *J. Hepatol.* **30**, 1073–1080.

Olaso, E., Eng, F., Lin, C., Yancopoulous, G., and Friedman, S. L. (1999). The discoidin domain receptor 2 (DDR2) is induced during stellate cell activation, mediates cell growth and MMP2 expression, and is regulated by extracellular matrix. *Hepatology* **30**, 413A.

Paradis, V., Kollinger, M., Fabre, M., Holstege, A., Poynard, T., and Bedossa, P. (1997). In situ detection of lipid peroxidation by-products in chronic liver diseases. *Hepatology* **26**, 135–142.

Paradis, V., Dargere, D., Vidaud, M., De Gouville, A. C., Huet, S., Martinez, V., Gauthier, J. M., Ba, N., Sobesky, R., Ratziu, V., and Bedossa, P. (1999). Expression of connective tissue growth factor in experimental rat and human liver fibrosis. *Hepatology* **30**, 968–976.

Parola, M., Pinzani, M., Casini, A., Albano, E., Poli, G., Gentilini, A., Gentilini, P., and Dianzani, M. U. (1993). Stimulation of lipid peroxidation or 4-hydroxynonenal treatment increases procollagen alpha 1 (I) gene expression in human liver fat-storing cells. *Biochem. Biophys. Res. Commun.* **194**, 1044–1050.

Pietrangelo, A. (1996). Metals, oxidative stress, and hepatic fibrogenesis. *Semin. Liver Dis.* **16**, 13–30.

Pinzani, M., Abboud, H. E., and Aron, D. C. (1990). Secretion of insulin-like growth factor-I and binding proteins by rat liver fat-storing cells: Regulatory role of platelet-derived growth factor. *Endocrinology* **127**, 2343–2349.

Pinzani, M., Abboud, H. E., Gesualdo, L., and Abboud, S. L. (1992a). Regulation of macrophage colony-stimulating factor in liver fat-storing cells by peptide growth factors. *Am. J. Physiol.* **262**, C876–C881.

Pinzani, M., Carloni, V., Marra, F., Riccardi, D., Laffi, G., and Gentilini, P. (1994a). Biosynthesis of platelet-activating factor and its 1O-acyl analogue by liver fat-storing cells. *Gastroenterology* **106**, 1301–1311.

Pinzani, M., Failli, P., Ruocco, C., Casini, A., Milani, S., Baldi, E., Giotti, A., and Gentilini, P. (1992b). Fat-storing cells as liver-specific pericytes: Spatial dynamics of agonist-stimulated intracellular calcium transients. *J. Clin. Invest.* **90**, 642–646.

Pinzani, M., Gentilini, A., Caligiuri, A., De Franco, R., Pellegrini, G., Milani, S., Marra, F., and Gentilini, P. (1995). Transforming growth factor-beta 1 regulates platelet-derived growth factor receptor beta subunit in human liver fat-storing cells. *Hepatology* **21**, 232–239.

Pinzani, M., Gesualdo, L., Sabbah, G. M., and Abboud, H. E. (1989). Effects of platelet-derived growth factor and other polypeptide mitogens on DNA synthesis and growth of cultured rat liver fat-storing cells. *J. Clin. Invest.* **84**, 1786–1793.

Pinzani, M., Knauss, T. C., Pierce, G. F., Hsieh, P., Kenney, W., Dubyak, G. R., and Abboud, H. E. (1991). Mitogenic signals for platelet-derived growth factor isoforms in liver fat-storing cells. *Am. J. Physiol.* **260**, C485–C491.

Pinzani, M., Marra, F., and Carloni, V. (1998). Signal transduction in hepatic stellate cells. *Liver* **18**, 2–13.

Pinzani, M., Milani, S., De Franco, R., Grappone, C., Caligiuri, A., Gentilini, A., Tosti-Guerra, C., Maggi, M., Failli, P., Ruocco, C., and Gentilini, P. (1996a). Endothelin 1 is overexpressed in human cirrhotic liver and exerts multiple effects on activated hepatic stellate cells. *Gastroenterology* **110**, 534–548.

Pinzani, M., Milani, S., Grappone, C., Weber, F. J., Gentilini, P., and Abboud, H. E. (1994b). Expression of platelet-derived growth factor in a model of acute liver injury. *Hepatology* **19**, 701–707.

Pinzani, M., Milani, S., Herbst, H., DeFranco, R., Grappone, C., Gentilini, A., Caligiuri, A., Pellegrini, G., Ngo, D. V., Romanelli, R. G., and Gentilini, P. (1996b). Expression of platelet-derived growth factor and its receptors in normal human liver and during active hepatic fibrogenesis. *Am. J. Pathol.* **148**, 785–800.

Potter, J. J., Womack, L., Mezey, E., and Anania, F. A. (1998). Transdifferentiation of rat hepatic stellate cells results in leptin expression. *Biochem. Biophys. Res. Commun.* **244**, 178–182.

Racine-Samson, L., Rockey, D. C., and Bissell, D. M. (1997). The role of alpha1beta1 integrin in wound contraction: A quantitative analysis of liver myofibroblasts in vivo and in primary culture. *J. Biol. Chem.* **272**, 30911–30917.

Ramadori, G., Knittel, T., Odenthal, M., Schwogler, S., Neubauer, K., and Meyer zum Buschenfelde, K. H. (1992). Synthesis of cellular fibronectin by rat liver fat-storing (Ito) cells: Regulation by cytokines. *Gastroenterology* **103**, 1313–1321.

Ramadori, G., Schwogler, S., Veit, T., Rieder, H., Chiquet-Ehrismann, R., Mackie, E. J., and Meyer zum Buschenfelde, K. H. (1991). Tenascin gene expression in rat liver and in rat liver cells. In vivo and in vitro studies. *Virch. Arch. B Cell Pathol. Incl. Mol. Pathol.* **60**, 145–153.

Reimann, T., Hempel, U., Krautwald, S., Axmann, A., Scheibe, R., Seidel, D., Wenzel, K. W., Kawada, N., Suzuki, Y., Yoshimura, H., Moriwaki, H., Onuki, K., Masushige, S., Muto, Y., Friedman, S. L., Kato, S., and Kojima, S. (1997). Transforming growth factor-beta1 induces activation of Ras, Raf-1, MEK and MAPK in rat hepatic stellate cells. *FEBS Lett.* **403**, 57–60.

Rescan, P. Y., Loreal, O., Hassell, J. R., Yamada, Y., Guillouzo, A., and Clement, B. (1993). Distribution and origin of the basement membrane component perlecan in rat liver and primary hepatocyte culture. *Am. J. Pathol.* **142**, 199–208.

Resnick, J. M., Whitley, C. B., Leonard, A. S., Krivit, W., and Snover, D. C. (1994). Light and electron microscopic features of the liver in mucopolysaccharidosis. *Hum. Pathol.* **25**, 276–286.

Rockey, D. C. (1995). Characterization of endothelin receptors mediating rat hepatic stellate cell contraction. *Biochem. Biophys. Res. Commun.* **207**, 725–731.

Rockey, D. C. (1997). The cellular pathogenesis of portal hypertension: Stellate cell contractility, endothelin, and nitric oxide. *Hepatology* **25**, 2–5.

Rockey, D. C., and Chung, J. J. (1995). Inducible nitric oxide synthase in rat hepatic lipocytes and the effect of nitric oxide on lipocyte contractility. *J. Clin. Invest.* **95**, 1199–1206.

Rockey, D. C., Fouassier, L., Chung, J. J., Carayon, A., Vallee, P., Rey, C., and Housset, C. (1998). Cellular localization of endothelin-1 and increased production in liver injury in the rat: Potential for autocrine and paracrine effects on stellate cells. *Hepatology* **27**, 472–480.

Rockey, D. C., Housset, C. N., and Friedman, S. L. (1993). Activation-dependent contractility of rat hepatic lipocytes in culture and in vivo. *J. Clin. Invest.* **92**, 1795–1804.

Rockey, D. C., and Weisiger, R. A. (1996). Endothelin induced contractility of stellate cells from normal and cirrhotic rat liver: Implications for regulation of portal pressure and resistance. *Hepatology* **24**, 233–240.

Rosenbaum, J., Blazejewski, S., Preaux, A. M., Mallat, A., Dhumeaux, D., and Mavier, P. (1995). Fibroblast growth factor 2 and transforming growth factor beta 1 interactions in human liver myofibroblasts. *Gastroenterology* **109**, 1986–1996.

Roskams, T., Moshage, H., De Vos, R., Guido, D., Yap, P., and Desmet, V. (1995). Heparan sulfate proteoglycan expression in normal human liver. *Hepatology* **21**, 950–958.

Roulot, D., Durand, H., Coste, T., Rautureau, J., Strosberg, A. D., Benarous, R., and Marullo, S. (1995). Quantitative analysis of transforming growth factor beta 1 messenger RNA in the liver of patients with chronic hepatitis C: Absence of correlation between high levels and severity of disease. *Hepatology* **21**, 298–304.

Saile, B., Matthes, N., Knittel, T., and Ramadori, G. (1999). Transforming growth factor beta and tumor necrosis factor alpha inhibit both apoptosis and proliferation of activated rat hepatic stellate cells. *Hepatology* **30**, 196–202.

Schafer, S., Zerbe, O., and Gressner, A. M. (1987). The synthesis of proteoglycans in fat-storing cells of rat liver. *Hepatology* 7, 680–687.

Schirmacher, P., Geerts, A., Pietrangelo, A., Dienes, H. P., and Rogler, C. E. (1992). Hepatocyte growth factor/hepatopoietin A is expressed in fat-storing cells from rat liver but not myofibroblast-like cells derived from fat-storing cells. *Hepatology* **15**, 5–11.

Schmitt-Graff, A., Chakroun, G., and Gabbiani, G. (1993). Modulation of perisinusoidal cell cytoskeletal features during experimental hepatic fibrosis. *Virch. Arch. A Pathol. Anat. Histopathol.* **422**, 99–107.

Schmitt-Graff, A., Kruger, S., Bochard, F., Gabbiani, G., and Denk, H. (1991). Modulation of alpha smooth muscle actin and desmin expression in perisinusoidal cells of normal and diseased human livers. *Am. J. Pathol.* **138**, 1233–1242.

Scoazec, J. Y. (1995). Expression of cell-matrix adhesion molecules in the liver and their modulation during fibrosis. *J. Hepatol.* **22**, 20–27.

Scoazec, J. Y. (1996). Adhesion molecules in normal human liver. *Hepatogastroenterology* **43**, 1103–1105.

Seifert, W. F., Bosma, A., Brouwer, A., Hendriks, H. F., Roholl, P. J., van Leeuwen, R. E., van Thiel-de Ruiter, G. C., Seifert-Bock, I., and Knook, D. L. (1994). Vitamin A deficiency potentiates carbon tetrachloride-induced liver fibrosis in rats. *Hepatology* **19**, 193–201.

Seifert, W. F., Bosma, A., and Hendriks, H. F. J. (1989). Dual role of vitamin A in experimentally induced liver fibrosis. 43–48.

Senoo, H., and Wake, K. (1985). Suppression of experimental hepatic fibrosis by administration of vitamin A. *Lab. Invest.* **52**, 182–194.

Shi, Y. B., Li, Q., Damjanovski, S., Amano, T., and Ishizuya-Oka, A. (1998). Regulation of apoptosis during development: Input from the extracellular matrix. *Int. J. Mol. Med.* **2**, 273–282.

Shrivastava, A., Radziejewski, C., Campbell, E., Kovac, L., McGlynn, M., Ryan, T. E., Davis, S., Goldfarb, M. P., Glass, D. J., Lemke, G., and Yancopoulos, G. D. (1997). An orphan receptor tyrosine kinase family whose members serve as nonintegrin collagen receptors. *Mol. Cell* **1**, 25–34.

Sims, D. E. (1986). The pericyte: A review. *Tissue Cell* **18**, 153–174.

Skrtic, S., Wallenius, V., Ekberg, S., Brenzel, A., Gressner, A. M., and Jansson, J. O. (1997). Insulin-like growth factors stimulate expression of hepatocyte growth factor but not transforming growth factor beta1 in cultured hepatic stellate cells. *Endocrinology* **138**, 4683–4689.

Sugiyama, H., Kashihara, N., Maeshima, Y., Okamoto, K., Kanao, K., Sekikawa, T., and Makino, H. (1998). Regulation of survival and death of mesangial cells by extracellular matrix. *Kidney Int.* **54**, 1188–1196.

Takahara, T., Furui, K., Yata, Y., Jin, B., Zhang, L. P., Nambu, S., Sato, H., Seiki, M., and Watanabe, A. (1997). Dual expression of matrix metalloproteinase-2 and membrane-type 1-matrix metalloproteinase in fibrotic human livers. *Hepatology* **26**, 1521–1529.

Takahara, T., Sollberg, S., Muona, P., and Uitto, J. (1995). Type VI collagen gene expression in experimental liver fibrosis: Quantitation and spatial distribution of mRNAs, and immunodetection of the protein. *Liver* **15**, 78–86.

Theret, N., Musso, O., L'Helgoualc'h, A., and Clement, B. (1997). Activation of matrix metalloproteinase-2 from hepatic stellate cells requires interactions with hepatocytes. *Am. J. Pathol.* **150**, 51–58.

Thompson, K., Maltby, J., Fallowfield, J., McAulay, M., Millward-Sadler, H., and Sheron, N. (1998a). Interleukin-10 expression and function in experimental murine liver inflammation and fibrosis. *Hepatology* **28**, 1597–1606.

Thompson, K. C., Trowern, A., Fowell, A., Marathe, M., Haycock, C., Arthur, M. J., and Sheron, N. (1998b). Primary rat and mouse hepatic stellate cells express the macrophage inhibitor cytokine interleukin-10 during the course of activation In vitro. *Hepatology* **28**, 1518–1524.

Tiggelman, A. M., Boers, W., Linthorst, C., Brand, H. S., Sala, M., and Chamuleau, R. A. (1995). Interleukin-6 production by human liver (myo)fibroblasts in culture: Evidence for a regulatory role of LPS, IL-1 beta and TNF alpha. *J. Hepatol.* **23**, 295–306.

Timpl, R. (1996). Macromolecular organization of basement membranes. *Curr. Opin. Cell Biol.* **8**, 618–624.

Torimura, T., Ueno, T., Inuzuka, S., Kin, M., Ohira, H., Kimura, Y., Majima, Y., Sata, M., Abe, H., and Tanikawa, K. (1994). The extracellular matrix in hepatocellular carcinoma shows different localization patterns depending on the differentiation and the histological pattern of tumors: Immunohistochemical analysis. *J. Hepatol.* **21**, 37–46.

Trigueiro de Araujo, M. S., Gerard, F., Chossegros, P., Guerret, S., Barlet, P., Adeleine, P., Grimaud, J. A., Sata, M., Abe, H., and Tanikawa, K. (1993). Cellular and matrix changes in drug abuser liver sinusoids: A semiquantitative and morphometric ultrastructural study. *Virch. Arch. A Pathol. Anat. Histopathol.* **422**, 145–152.

Vogel, W., Gish, G. D., Alves, F., and Pawson, T. (1997). The discoidin domain receptor tyrosine kinases are activated by collagen. *Mol. Cell* **1**, 13–23.

Vyas, S. K., Leyland, H., Gentry, J., and Arthur, M. J. (1995). Rat hepatic lipocytes synthesize and secrete transin (stromelysin) in early primary culture. *Gastroenterology* **109**, 889–898.

Wake, K. (1988). Liver perivascular cells revealed by gold and silver impregnation methods and electron microscopy. 23–36.

Wake, K., Kishiye, T., Yamamoto, H., Kaneda, K., and Sato, C. (1992). Sinusoidal cell function life under the microscope. 45–51.

Wang, S. C., Tsukamoto, H., Rippe, R. A., Schrum, L., and Ohata, M. (1998). Expression of interleukin-10 by in vitro and in vivo activated hepatic stellate cells. *J. Biol. Chem.* **273**, 302–308.

Win, K. M., Charlotte, F., Mallat, A., Cherqui, D., Martin, N., Mavier, P., Preaux, A. M., Dhumeaux, D., and Rosenbaum, J. (1993). Mitogenic effect of transforming growth factor-beta 1 on human Ito cells in culture: Evidence for mediation by endogenous platelet-derived growth factor. *Hepatology* **18**, 137–145.

Winwood, P. J., Schuppan, D., Iredale, J. P., Kawser, C. A., Docherty, A. J., and Arthur, M. J. (1995). Kupffer cell-derived 95-kd type IV collagenase/gelatinase B: Characterization and expression in cultured cells. *Hepatology* **22**, 304–315.

Wong, L., Yamasaki, G., Johnson, R. J., and Friedman, S. L. (1994). Induction of beta-platelet-derived growth factor receptor in rat hepatic lipocytes during cellular activation in vivo and in culture. *J. Clin. Invest.* **94**, 1563–1569.

Wrana, J. L., and Attisano, L. (1996). MAD-related proteins in TGF-beta signalling. *Trends Genet.* **12**, 493–496.

Yanagisawa, M., Kurihara, H., Kimura, S., Tomobe, Y., Kobayashi, M., Mitsui, Y., Yazaki, Y., Goto, K., Masaki, T., Duncan, M. R., Frazier, K. S., Abramson, S., Williams, S., Klapper, H., Huang, X., and Grotendorst, G. R. (1988). A novel potent vasoconstrictor peptide produced by vascular endothelial cells. *Nature* **332**, 411–415.

Zhang, J. X., Bauer, M., and Clemens, M. G. (1995). Vessel- and target cell-specific actions of endothelin-1 and endothelin-3 in rat liver. *Am. J. Physiol.* **269**, G269–G277.

Zhang, J. X., Pegoli, W., Jr., and Clemens, M. G. (1994). Endothelin-1 induces direct constriction of hepatic sinusoids. *Am. J. Physiol.* **266**, G624–G632.

Zindy, F., Lamas, E., Schmidt, S., Kirn, A., and Brechot, C. (1992). Expression of insulin-like growth factor II (IGF-II) and IGF-II, IGF-I and insulin receptors mRNAs in isolated non-parenchymal rat liver cells. *J. Hepatol.* **14**, 30–34.

CHAPTER 10

Peroxisome Proliferator-Activated Receptor γ and Hepatic Stellate Cell Activation

TAKEO MIYAHARA, SASWATI HAZRA, SHIGANG XIONG, KENTA MOTOMURA, HONGYUN SHE, AND HIDEKAZU TSUKAMOTO

USC-UCLA Research Center for Alcoholic Liver and Pancreatic Diseases, USC Research Center for Liver Diseases, Keck School of Medicine of the University of Southern California, Los Angeles, California 90033 and VA Greater Los Angeles Healthcare System, California 90073

I. DEFICIENCY OF RXR AND 9-*cis* RA IN ACTIVATED HEPATIC STELLATE CELLS

One salient feature of hepatic stellate cell (HSC) activation is the loss of intracellular vitamin A. This occurs both *in vitro* (Friedman *et al.*, 1993) and *in vivo* (Tsukamoto *et al.*, 1996). A mechanistic link among vitamin A depletion, HSC activation, and liver fibrosis is supported by *in vivo* findings that treatment of rats with retinol ameliorates liver fibrosis induced by carbon tetrachloride (CCl_4) (Senoo and Wake, 1985), whereas this experimental liver fibrosis is exacerbated by vitamin A deficiency (Seifert *et al.*, 1994). In contrast, vitamin A supplementation aggravates liver injury and fibrosis caused by CCl_4 (elSisi *et al.*, 1993) or alcohol (Leo and Lieber, 1983). However, these latter effects likely reflect enhanced oxidative injury of hepatocytes via metabolism of retinoids by induced microsomal cytochromes. Treatment of cultured HSC with all-*trans* retinoic acid (RA) causes suppression of cell proliferation, collagen, and transforming growth factor (TGF)-β production (Davis *et al.*, 1990). Spontaneous activation of cultured HSC is associated with the release of retinol into the medium, the process that is

Copyright 2003, Elsevier Science (USA). All rights reserved.

enhanced by Kupffer cell-conditioned medium (Friedman *et al.*, 1993). The first demonstration of depletion of all-*trans* and 9-*cis* RA in *in vivo*-activated HSC was made by Ohata *et al.* (1997) using cells isolated from rats with cholestatic liver fibrosis. These changes were accompanied by decreased mRNA expression of nuclear receptors for RA (RARβ and RXRα) and suppressed binding of nuclear proteins to retinoic acid response element (RARE). Because both RARβ and RXRα are regulated positively by RARE, these changes are considered molecular consequences of RA deficiency at the nuclear level in activated HSC. RAR and RXR are known to antagonize AP-1 promoters in a RA-dependent manner either through direct cross-coupling (Ohata *et al.*, 1997) or dephosphorylation of c-jun by the induction of mitogen-activated protein kinase (MAPK) phosphatase-1 (Lee *et al.*, 1999). Thus diminished RA signaling demonstrated in activated HSC may result in reduced negative regulation on AP-1 via these mechanisms and upregulate AP-1 containing genes such as TGF-β1 and α1(I) procollagen, which are implicated directly in fibrogenic activation of HSC. However, 9-*cis* RA (Okuno *et al.*, 1997) and 9,13-di-*cis* RA (Imai *et al.*, 1997) are shown to activate latent TGF-β via upregulation of the tissue plasminogen activator. These opposing effects of different RA isomers on HSC may reflect complex and multiple regulatory pathways rendered by RA (Hellemans *et al.*, 1999).

II. PPAR AND HSC

RXR is a promiscuous nuclear receptor that heterodimerizes with members of the steroid/thyroid hormone receptor superfamily, such as vitamin D receptor, thyroid hormone receptor, peroxisome proliferator- activated receptor (PPAR), liver X receptor (LXR), and pregnane X receptor (PXR). It is important to note that RXR serves as a ligand-dependent, active partner for PPAR and LXR. This means that ligand-activated RXR itself is capable of activating PPAR-RXR or LXR-RXR, and the reduced levels of 9-*cis* RA and RXR seen in activated HSC may limit activities of these heterodimeric transcription factors. In particular, we became interested in RXR-PPARγ activity in activated HSC for the following reasons. First, culture-activated HSC are shown to express leptin (Potter *et al.*, 1998), a 16-kDa *ob* gene product, which is expressed predominantly by adipose tissue to suppress food intake, increase metabolic rate, and reduce fat depot size (Spiegelman and Flier, 1996). The deficiency of leptin or its receptor is associated with obesity in mice and humans. The finding that leptin is expressed in culture-activated HSC suggests the possibility that HSC may share the adipocyte phenotype. In fact, quiescent HSC are characterized by the ample intracellular storage of lipids, including fat-soluble vitamin A, and activation of HSC causes depletion of lipid droplets and transdifferentiation to myofibroblastic cells. These changes resemble transdifferentiation of 3T3-L1 cells that possess the potential to differentiate into

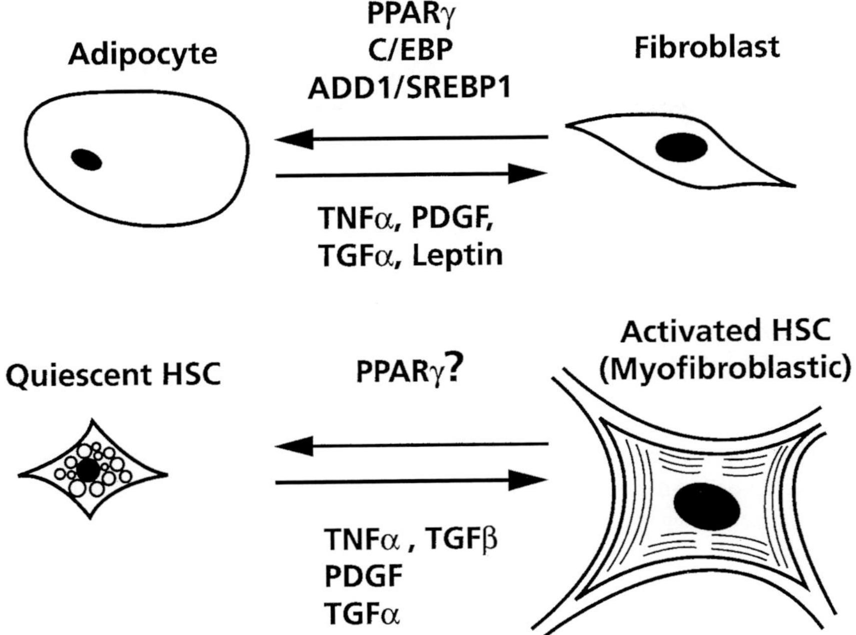

FIGURE 1 A schematic diagram depicting an analogy between adipocyte–fibroblast transdifferentiation and HSC activation.

the phenotype of either fibroblasts or adipocytes (Fig. 1). Under appropriate culture conditions, 3T3-L1 cells undergo adipocyte differentiation driven by two key transcription factors, C/EBP and PPARγ, which induce adipocyte-specific genes (Spiegelman and Flier, 1996). Conversely, suppressed activities of these transcription factors result in the loss of fat storage capacity and fibroblastic differentiation similar to HSC activation (Fig. 1). Cytokines [tumor necrosis factor (TNF)-α] and growth factors (platelate-derived growth factor, PDGF; epidermal growth factor, EGF; TGFβ) are known to inhibit PPARγ activity and adipocyte differentiation (Spiegelman and Flier, 1996; Zhang *et al.*, 1996), and the same mediators are also implicated in the activation of HSC, again supporting similar transdifferentiation mechanisms (Fig. 1). Thus, a natural question from this analogy is whether diminished PPARγ activity is involved in the activation of HSC as in fibroblastic transdifferentiation of adipocytes. This hypothesis was tested by Miyahara *et al.* (2000) and is discussed later.

PPARγ is composed of at least three isoforms (γ1, γ2, and γ3) due to alternative splicing. PPARγ2 is expressed abundantly in adipose tissue, whereas low levels of PPARγ1 are found in many other tissues. PPARγ2 constitutes a "master" regulator that facilitates the entire program of adipocyte differentiation via the transcriptional induction of adipocyte-specific genes such as fatty acid-binding protein (aP2), acyl-CoA synthase, and phosphoendpyruvate carboxykinase (PEPCK). This is highlighted experimentally by the transdifferentiation of fibroblasts into adipocytes via ectopic expression of PPARγ2 using a viral vector and activation of the receptor by ligands (Tontonoz *et al.*, 1994).

Nonadipogenic effects of PPARγ ligands have received much attention. One such effect is anti-inflammatory effects on macrophages characterized by inhibition of the expression of proinflammatory cytokines and inducible nitric oxide synthase (iNOS) via the mechanisms that in part involve the antagonism of transcriptional factors, AP-1, SAT, and NF-κB (Jiang *et al.*, 1998; Ricote *et al.*, 1998). Macrophages also appear to rely on the activity of 12/15-lipoxygenase for the generation of 13-hydroxyoctadecadienoic acid (13-HODE) and 15-hydroxyeicosatetraenoic acid (15-HETE) as endogenous PPARγ ligands in response to interleukin (IL)-4, providing potential feedback regulation involving PPARγ (Huang *et al.*, 1999). In contrast, macrophage differentiation is also promoted by PPARγ ligands such as 15-deoxy-$\Delta^{12,14}$-prostaglandin J2 (15dPGJ2), which induces the expression of macrophagespecific genes, including CD36 (Tontonoz *et al.*, 1998). This paradoxical effect seems evident when undifferentiated monocytes or monocytic cell lines are exposed to PPARγ ligands (Jarrous and Kaempfer, 1994; Tontonoz *et al.*, 1998). In fact, Chawla *et al.* (2001), who used embryonic stem cells with a null mutation in PPARγ, demonstrated that neither development of the macrophage lineage nor anti-inflammatory effects of thiazolidinediones (TZD), synthetic PPARγ ligands, are dependent of PPARγ but that PPARγ is an important regulator of CD36 expression. Thus, pleiotropic effects of PPARγ ligands need to be examined carefully as to whether they are really mediated via PPARγ.

III. PPARγ IS REDUCED IN ACTIVATED HSC

Is PPARγ activity suppressed in activated HSC analogous to transdifferentiation of adipocytes to fibroblasts? This central hypothesis was tested by Miyahara *et al.* (2000). First, we analyzed the expression of PPARγ isoforms in HSC by a RNase protection assay that is capable of detecting both γ1 and γ2 isoforms. This revealed weak expression of mRNA for the PPARγ1 isoform but not that for the γ2 isoform in HSC freshly isolated from sham-operated normal rats (HSC-sham, Fig. 2). In contrast, HSC from cholestatic liver fibrosis induced by bile duct ligation (HSC-BDL) showed no expression of either isoforms (Fig. 2). This result suggested that HSC may not share the adipocyte phenotype because they don't express the γ2 isoform but supported our hypothesis that PPARγ expression may indeed be suppressed in activated HSC. To confirm this observation, reverse transcriptase-polymerase chain reaction (RT-PCR) was performed, which showed decreased mRNA levels of PPARγ and expected induction of α1(I) procollagen and α-smooth muscle actin in HSC-BDL. To directly assess binding of PPARγ by nuclear

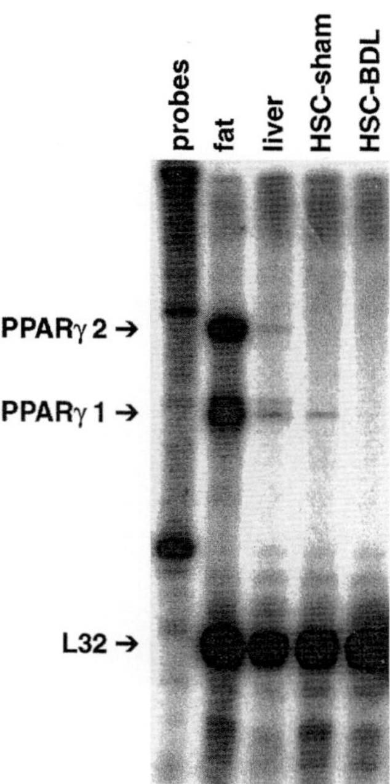

FIGURE 2 RNase protection assay for PPARγ1 and PPARγ2 mRNA in HSC. Note abundant expression of both isoforms in the fat tissue. PPARγ1 expression in this sample is probably derived from mesenchymal cells in the fat tissue. Even though the level of expression is much less, the liver also expresses both forms. However, only PPARγ1 mRNA but not PPARγ2 mRNA is expressed in HSC isolated from a sham-operated normal rat (HSC-sham), and this expression is diminished in HSC isolated from a bile duct-ligated rat (HSC-BDL). From Miyahara *et al.* (2000).

extracts, electrophoretic mobility shift assay (EMSA) was performed using a ARE7 probe (a PPRE of aP2 gene), which preferentially binds PPARγ over PPARα (Juge-Aubry *et al.*, 1997). This analysis showed clearly decreased PPRE binding of nuclear extracts from HSC-BDL as compared to the HSC sham. The specificity of the binding was supported by competition with ×500 excess cold probe and a supershift assay using antibodies against PPARγ, which demonstrated a diminution of the DNA binding and the appearance of a supershifted band (arrow, last lane, Fig. 3A). At the same time, the same HSC-BDL nuclear extracts were shown to have increased NF-κB and AP-1 binding (Figs. 3B and 3C) as predicted from activated HSC. Thus these results demonstrated that activated HSC from cholestatic liver fibrosis have reduced PPARγ expression and DNA-binding activity. Then we tested culture-activated HSC to determine whether they also exhibit the same defect. Indeed, culture activation of HSC for 3 to 7 days caused a progressive reduction in the mRNA level of PPARγ. This finding was also supported by a study by Galli *et al.* (2000), who demonstrated a similar progressive inhibition in PPRE promoter activation in culture-activated human HSC.

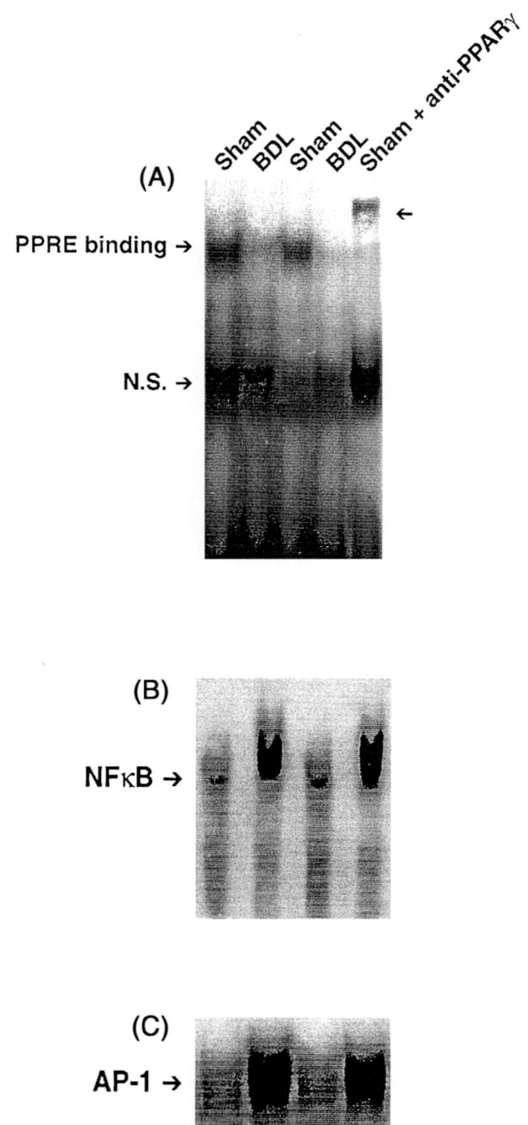

FIGURE 3 PPRE binding is diminished in *in vivo*-activated HSC. (A) Note that PPRE binding is decreased in nuclear extracts from HSC isolated from cholestatic liver fibrosis (BDL) as compared to those from sham-operated rats (Sham). The supershift assay is shown in the last lane. The same nuclear extracts from BDL show increased NF-κB (B) and AP-1 (C) binding. From Miyahara *et al.* (2000).

IV. PPARγ LIGANDS SUPPRESS ACTIVATION OF HSC

To further assess the cause-and-effect relationship between PPARγ and activation of HSC, we have treated culture-activated HSC with various PPARγ ligands to determine whether they were capable of inhibiting HSC activation markers. First, the effects of 15dPGJ2 and BRL49653 (rosiglitazone) on DNA synthesis were examined. 15dPGJ2 and BRL49653, both at 10 μ*M*, suppressed DNA synthesis by 40–50%. We also tested WY14643, which inhibited this parameter at 100 and 250 μ*M*, concentrations known to activate both PPARα and PPARγ. Around the same time, Galli *et al.* (2000) and Marra *et al.* (2000) also demonstrated independently that PDGF-stimulated proliferation of human HSC was inhibited by 15dPGJ2 (Galli *et al.*, 2000; Marra *et al.*, 2000), ciglitazone (Galli *et al.*, 2000), and troglitazone (Marra *et al.*, 2000). In addition, Marra *et al.* (2000) demonstrated that PDGF-induced HSC migration was also inhibited by the ligands. The next parameter we tested was collagen production, which was also inhibited by all three ligands. To address the specificity of PPARγ ligands, particularly 15PGJ2, we used a PPAR antagonist (GW9662). The use of this antagonist blocked 15dPGJ2-mediated inhibition of HSC collagen synthesis by 70%, suggesting that most of the effect by this ligand is mediated by PPAR. Next, we examined the effects of 15dPGJ2 on mRNA levels of HSC activation markers. The 15dPGJ2 treatment decreased the levels of α1(I) procollagen, α-smooth muscle actin, and MCP-1. These effects were not due to systematic inhibition of HSC functions. The treatment induced MMP-3 and CD36 expression. The latter encodes a scavenger receptor/fatty acid translocase, whose promoter contains PPRE (Tontonoz *et al.*, 1998). These results demonstrated collectively that PPARγ ligands suppress multiple functional markers of HSC activation in culture and support the importance of PPARγ in the maintenance of the quiescent HSC phenotype. Inhibition of α-smooth muscle actin and MCP-1 expression by PPARγ ligands was also shown in cultured human HSC by Galli *et al.* (2000) and Marra *et al.* (2000), respectively.

V. PPARγ LIGANDS AND COLLAGEN GENE EXPRESSION

We have further examined the mechanisms of PPARγ ligand-mediated suppression of collagen expression by cultured HSC. Reduced mRNA levels for α1(I) procollagen suggested the pretranslational effect. A nuclear run-on assay was performed to determine whether this effect was transcriptional. Our results demonstrated that this was the case (Xiong *et al.*, 2000). We then examined whether the effect was due to suppressed promoter activity by cotransfection of HSC with an α1(I) procollagen promoter (–2.2 kb/116 b)-luciferase and a PPARα or PPARγ expression vector. 15dPGJ2 inhibited the promoter activity in a PPARγ-dependent manner (Miyahara *et al.*, 2000).

VI. REMAINING QUESTIONS CONCERNING PPARγ DEPENDENCE OF ANTIFIBROGENIC EFFECTS OF TZD

In light of the finding of Chawla *et al.* (2001) on PPARγ-independent effects of TZD on macrophages, we also need to be aware of the possibility that the effects observed on HSC with the ligands still may not be receptor dependent. To address this question, we are currently overexpressing PPARγ or a dominant-negative form of PPARγ in cultured HSC. Results from this study should yield insightful information concerning this critical question. The observed anti-inflammatory or antifibrogenic effects of TZD on a variety of cell types have led to a possibility that these agents may inhibit fibroproliferative pathology *in vivo*. In fact, several studies have already demonstrated that this was the case. Examples include inhibition of atherosclerosis (Marra *et al.*, 2000), gromerulosclerosis (Xiong *et al.*, 2000), and arthritis (Buckingham *et al.*, 1998) in animal models. Even though PPARγ-mediated improvements in the metabolism of lipids and glucose may have these beneficial secondary effects, it is tempting to speculate that TZD treatment has direct antifibrotic effects on fibrogenic effector cells, such as smooth muscle cells, mesangial cells, and synoviocytes as demonstrated so *in vitro* (Chen *et al.*, 2001; Kawahito *et al.*, 2000; Marx *et al.*, 1998; Okura *et al.*, 2000). In the same vein, results from Miyahara *et al.* (2000) and other studies (Galli *et al.*, 2000; Marra *et al.*, 2000) strongly support direct inhibitory effects on activation of HSC and suggest that PPARγ ligands may be beneficial for liver fibrosis. When testing this hypothesis, it is important to differentiate the potential effects of the ligands on multiple cell types in the liver. In particular, anti-inflammatory effects of TZD on macrophages and endothelial cells are firmly established and may serve as primary effects in the liver, rendering secondary effects on HSC. For this reason, testing PPARγ ligands in liver fibrosis models of different etiology may be important in fully understanding the mechanisms of the effects. This ultimate question was effectively approached by the most recent study by Galli *et al.* (2002) who demonstrated that TZD treatment ameliorated liver fibrosis in not only dimethylnitrosamine and carbon tetrachloride but also bile duct ligation models. Confirmation and further extension of this important finding should advance our understanding of the therapeutic efficacy of TZD. With respect to the mechanisms of the action, critical questions still remain as to whether all the TZD's effects are really mediated by PPARγ and whether PPARγ exerts anti-fibrogenic effects in a ligand dependent or independent manner. Lastly, it is still unknown why PPARγ becomes depleted in HSC during activation.

ACKNOWLEDGMENTS

Supported by NIH Grants R37AA06603, P50AA11999 (USC-UCLA Research Center for Alcoholic Liver and Pancreatic Diseases), P30DK48522 (USC Research Center for Liver Diseases), and Medical Research Service of Department of Veterans Affairs.

REFERENCES

Buckingham, R. E., Al Barazanji, K. A., Toseland, C. D., Slaughter, M., Connor, S. C., West, A., Bond, B., Urner, N. C., and Clapham, J. C. (1998). Peroxisome proliferator-activated receptor-gamma agonist, rosiglitazone, protects against nephropathy and pancreatic islet abnormalities in Zucker fatty rats. *Diabetes* **47**, 1326–1334.

Chawla, A., Barak, Y., Nagy, L., Liao, D., Tontonoz, P., and Evans, R. M. (2001). PPAR-gamma dependent and independent effects on macrophage-gene expression in lipid metabolism and inflammation. *Nature Med.* **7**, 48–52.

Chen, Z., Ishibashi, S., Perrey, S., Osuga, J., Gotoda, T., Kitamine, T., Tamura, Y., Okazaki, H., Yahagi, N., Iizuka, Y., Shionoiri, F., Ohashi, K., Harada, K., Shimano, H., Nagai, R., and Yamada, N. (2001). Troglitazone inhibits atherosclerosis in apolipoprotein E-knockout mice: Pleiotropic effects on CD36 expression and HDL. *Arterioscler. Thromb. Vasc. Biol.* **21**, 372–377.

Davis, B. H., Kramer, R. T., and Davidson, N. O. (1990). Retinoic acid modulates rat Ito cell proliferation, collagen, and transforming growth factor beta production. *J. Clin. Invest.* **86**, 2062–2070.

elSisi, A. E., Hall, P., Sim, W. L., Earnest, D. L., and Sipes, I. G. (1993). Characterization of vitamin A potentiation of carbon tetrachloride-induced liver injury. *Toxicol. Appl. Pharmacol.* **119**, 280–288.

Friedman, S. L., Wei, S., and Blaner, W. S. (1993). Retinol release by activated rat hepatic lipocytes: Regulation by Kupffer cell-conditioned medium and PDGF. *Am. J. Physiol.* **264**, G947–G952.

Galli, A., Crabb, D., Price, D., Ceni, E., Salzano, R., Surrenti, C., and Casini, A. (2000). Peroxisome proliferator-activated receptor gamma transcriptional regulation is involved in platelet-derived growth factor-induced proliferation of human hepatic stellate cells. *Hepatology* **31**, 101–108.

Galli, A., Crabb, D. W., Ceni, E., Salzano, R., Mello, T., Svegliati-Baroni, G., Ridolfi, F., Trozzi, L., Surrenti, C., and Casini, A. (2002). Antidiabetic thiazolidinediones inhibit collagen synthesis and hepatic stellate cell activation in vivo and in vitro. *Gastroenterology* **122**, 1924–1940.

Hellemans, K., Grinko, I., Rombouts, K., Schuppan, D., and Geerts, A. (1999). All-trans and 9-cis retinoic acid alter rat hepatic stellate cell phenotype differentially. *Gut* **45**, 134–142.

Huang, J. T., Welch, J. S., Ricote, M., Binder, C. J., Willson, T. M., Kelly, C., Witztum, J. L., Funk, C. D., Conrad, D., and Glass, C. K. (1999). Interleukin-4-dependent production of PPAR-gamma ligands in macrophages by 12/15-lipoxygenase. *Nature* **400**, 378–382.

Imai, S., Okuno, M., Moriwaki, H., Muto, Y., Murakami, K., Shudo, K., Suzuki, Y., and Kojima, S. (1997). 9,13-di-cis-Retinoic acid induces the production of tPA and activation of latent TGF-beta via RAR alpha in a human liver stellate cell line, LI90. *FEBS Lett.* **411**, 102–106.

Jarrous, N., and Kaempfer, R. (1994). Induction of human interleukin-1 gene expression by retinoic acid and its regulation at processing of precursor transcripts. *J. Biol. Chem.* **269**, 23141–23149.

Jiang, C., Ting, A. T., and Seed, B. (1998). PPAR-gamma agonists inhibit production of monocyte inflammatory cytokines. *Nature* **391**, 82–86.

Juge-Aubry, C., Pernin, A., Favez, T., Burger, A. G., Wahli, W., Meier, C. A., and Desvergne, B. (1997). DNA binding properties of peroxisome proliferator-activated receptor subtypes on various natural peroxisome proliferator response elements: Importance of the 5'-flanking region. *J. Biol. Chem.* **272**, 25252–25259.

Kawahito, Y., Kondo, M., Tsubouchi, Y., Hashiramoto, A., Bishop-Bailey, D., Inoue, K., Kohno, M., Yamada, R., Hla, T., and Sano, H. (2000). 15-Deoxy-delta(12,14)-PGJ(2) induces synoviocyte apoptosis and suppresses adjuvant-induced arthritis in rats. *J. Clin. Invest.* **106**, 189–197.

Lee, H. Y., Sueoka, N., Hong, W. K., Mangelsdorf, D. J., Claret, F. X., and Kurie, J. M. (1999). All-trans-retinoic acid inhibits Jun N-terminal kinase by increasing dual-specificity phosphatase activity. *Mol. Cell. Biol.* **19**, 1973–1980.

Leo, M. A., and Lieber, C. S. (1983). Hepatic fibrosis after long-term administration of ethanol and moderate vitamin A supplementation in the rat. *Hepatology* **3**, 1–11.

Marra, F., Efsen, E., Romanelli, R. G., Caligiuri, A., Pastacaldi, S., Batignani, G., Bonacchi, A., Caporale, R., Laffi, G., and Pinzani, M. E. A. (2000). Ligands of peroxisome proliferator-activated receptor gamma modulate profibrogenic and proinflammatory actions in hepatic stellate cells. *Gastroenterology* **119**, 466–478.

Marx, N., Schonbeck, U., Lazar, M. A., Libby, P., and Plutzky, J. (1998). Peroxisome proliferator-activated receptor gamma activators inhibit gene expression and migration in human vascular smooth muscle cells. *Circ. Res.* **83**, 1097–1103.

Miyahara, T., Schrum, L., Rippe, R., Xiong, S., Yee, H. F., Jr., Motomura, K., Anania, F. A., Willson, T. M., and Tsukamoto, H. (2000). Peroxisome proliferator-activated receptors and hepatic stellate cell activation. *J. Biol. Chem.* **275**, 35715–35722.

Ohata, M., Lin, M., Satre, M., and Tsukamoto, H. (1997). Diminished retinoic acid signaling in hepatic stellate cells in cholestatic liver fibrosis. *Am. J. Physiol.* **272**, G589–G596.

Okuno, M., Moriwaki, H., Imai, S., Muto, Y., Kawada, N., Suzuki, Y., and Kojima, S. (1997). Retinoids exacerbate rat liver fibrosis by inducing the activation of latent TGF-beta in liver stellate cells. *Hepatology* **26**, 913–921.

Okura, T., Nakamura, M., Takata, Y., Watanabe, S., Kitami, Y., and Hiwada, K. (2000). Troglitazone induces apoptosis via the p53 and Gadd45 pathway in vascular smooth muscle cells. *Eur. J. Pharmacol.* **407**, 227–235.

Potter, J. J., Womack, L., Mezey, E., and Anania, F. A. (1998). Transdifferentiation of rat hepatic stellate cells results in leptin expression. *Biochem. Biophys. Res. Commun.* **244**, 178–182.

Ricote, M., Li, A. C., Willson, T. M., Kelly, C. J., and Glass, C. K. (1998). The peroxisome proliferator-activated receptor-gamma is a negative regulator of macrophage activation. *Nature* **391**, 79–82.

Seifert, W. F., Bosma, A., Brouwer, A., Hendriks, H. F., Roholl, P. J., van Leeuwen, R. E., van Thiel-de Ruiter, G. C., Seifert-Bock, I., and Knook, D. L. (1994). Vitamin A deficiency potentiates carbon tetrachloride-induced liver fibrosis in rats. *Hepatology* **19**, 193–201.

Senoo, H., and Wake, K. (1985). Suppression of experimental hepatic fibrosis by administration of vitamin A. *Lab. Invest.* **52**, 182–194.

Spiegelman, B. M., and Flier, J. S. (1996). Adipogenesis and obesity: Rounding out the big picture. *Cell* **87**, 377–389.

Tontonoz, P., Hu, E., and Spiegelman, B. M. (1994). Stimulation of adipogenesis in fibroblasts by PPAR gamma 2, a lipid-activated transcription factor. *Cell* **79**, 1147–1156.

Tontonoz, P., Nagy, L., Alvarez, J. G., Thomazy, V. A., and Evans, R. M. (1998). PPARgamma promotes monocyte/macrophage differentiation and uptake of oxidized LDL. *Cell* **93**, 241–252.

Tsukamoto, H., Cheng, S., and Blaner, W. S. (1996). Effects of dietary polyunsaturated fat on ethanol-induced Ito cell activation. *Am. J. Physiol.* **270**, G581–G586.

Xiong, S., Miyahara, T., Schrum, L., Rippe, R., Yee, H. F., Jr., Yavrom, S. M., She, H., Willson, T. M., and Tsukamoto, H. (2000). PPAR-gamma inhibits hepatic stellate cell collagen gene expression. *Hepatology* **32**, 89. [Abstract]

Zhang, B., Berger, J., Hu, E., Szalkowski, D., White-Carrington, S., Spiegelman, B. M., and Moller, D. E. (1996). Negative regulation of peroxisome proliferator-activated receptor-gamma gene expression contributes to the antiadipogenic effects of tumor necrosis factor-alpha. *Mol. Endocrinol.* **10**, 1457–1466.

CHAPTER 11

Role of Histone Deacetylases in Transcriptional Control of the Hepatic Stellate Cell Phenotype

KRISTA ROMBOUTS,* TOSHIRO NIKI,[†] MINURA YOSHIDA,[‡]
AND ALBERT GEERTS*,[§]

*Laboratory for Molecular Liver Cell Biology, Free University of Brussels (VUB), 1090 Brussels, Belgium; Departments of [†]Pathology and [‡]Biotechnology, University of Tokyo, Tokyo 113-8655, Japan; and [§]Department of Medical Cell Biology and Centre for Liver Research, University of Newcastle-upon-Tyne, United Kingdom

Eukaryotic gene expression has mainly been studied in the context of *trans*-acting transcription factors and their interaction with regulatory *cis* elements. A role for histone modifications in transcription processes and the remodeling of chromatin structure has been established.

Hepatic stellate cells (HSC) are the major cellular sources of extracellular matrix (ECM) synthesis in chronic liver diseases leading to fibrosis. We explored the antifibrogenic effect of TSA, a histone deacetylase inhibitor, on HSC *in vitro*. Primary and fully activated HSC were exposed to 10^{-7}–10^{-9} *M* TSA. Collagens type I and III and smooth muscle α-actin (α-SMA) were investigated on the protein and mRNA steady-state level by performing Northern hybridization and *de novo* immunoprecipitation. The antiproliferative effect was examined by [^{3}H]thymidine incorporation and cell counting. Differential mRNA display, Northern hybridization, and Western blotting were performed to identify TSA-sensitive genes.

TSA (10^{-7} *M*) was shown to be strongly antifibrogenic and antiproliferative in primary HSC when compared to fully activated HSC. TSA affected collagen type III and α-SMA protein and the mRNA steady-state level, whereas collagen type I

Copyright 2003, Elsevier Science (USA). All rights reserved.

synthesis was influenced at the posttranscriptional level. These biochemical changes were paralleled by an inhibitory effect on the differentiation of primary HSC into fully activated HSC. In addition, we identified several cytoskeleton-related proteins that were affected by TSA. TSA influenced novel actin filament formation by downregulation of two nucleating proteins, Arp2 and Arp3, and by upregulation of ADDL70 and gelsolin, two capping proteins. TSA increased the steady-state mRNA level of SSeCKS, a cytoskeletal protein with tumor suppressor activity. RhoA, a key mediator in the development of the actin cytoskeleton, decreased following TSA exposure. In conclusion, TSA inhibits activation of primary HSC into fully activated HSC by interfering with the level of hyperacetylation of histones. This coincides with the specific inhibition of ECM components and a strong effect on actin cytoskeleton-related proteins.

I. INTRODUCTION

During the last decade much effort has been made to find therapeutic agents to prevent or treat liver fibrosis and cirrhosis (Li and Friedman, 1999; Lieber, 1999; Wu and Zern, 2000; Rockey, 2000). The ultimate goal of fibrosis research is the development of a rational basis for effective antifibrotic therapy. Previous research has focused on several cellular molecular mechanisms, which lead to fibrosis, whereby different signaling transduction pathways have been partially elucidated (Pinzani and Gentilini, 1999; Friedman, 1999, 2000).

A. Hepatic Stellate Cells and Fibrogenesis

In liver, the major fibrogenic cell type is the HSC (previously called lipocyte, Ito cell, fat-storing, or perisinusoidal cell). These cells are resident nonparenchymal (nonhepatocyte) cells found in the subendothelial space between hepatocytes and sinusoidal endothelial cells. These cells undergo activation during liver injury, and this process is associated with the induction of key genes that can be influenced at transcriptional and posttranslational levels. Several groups have worked at identifying transcription factors, such as c-myb (Lee *et al.*, 1995), GC receptor (Raddatz *et al.*, 1996), Stat1 (Kawada *et al.*, 1997), Kruppel-like factor 6 (KLF6) (Ratziu *et al.*, 1998), Ets-1 (Knittel *et al.*, 1999), PPAR-γ (Galli *et al.*, 2000), PPAR-β (Hellemans *et al.*, unpublished data), RAR/RXR (Weiner *et al.*, 1992; Ohata *et al.*, 1997), c-jun/AP1 (Armendariz-Borunda *et al.*, 1994), junD (Smart *et al.*, 2001), Sp1 (Rippe *et al.*, 1995), AP2 (Chen *et al.*, 1996), NF-κB (Lee *et al.*, 1995; Hellerbrand *et al.*, 1998; Elsharkawy *et al.*, 1999), CREB (Houglum *et al.*, 1997), E-box BP (Weiner *et al.*, 1998), and others involved in the transdifferentiation of stellate cells (Britton and Bacon, 1999; Rippe, 1999).

HSC activation refers to transition from a quiescent vitamin A-rich, stellate-shaped cell to a proliferative, fibrogenic, and contractile myofibroblast-like cell with a reduced vitamin A content (Friedman, 2000; Geerts, 2001). This transition is gradual and occurs through an intermediate stage defined as "transdifferentiation" (Fig. 1). Activation of HSC can be subdivided in two phases: initiation and perpetuation. The first phase, *initiation*, refers to the earliest changes in phenotype that render the HSC more responsive to paracrine growth factors (Friedman, 2000). This process of initiation is followed by *perpetuation*, whereby the cells (i) proliferate; (ii) synthesize large quantities of extracellular matrix; (iii) increase their secretion of matrix metalloproteinases (MMPs), paralleled by an even stronger increase of secretion of tissue inhibitors of metalloproteinases (TIMPs) (Knittel *et al.*, 1999; Arthur *et al.*, 1999); (iv) secretion of a number of growth factors and cytokines, including transforming growth factor-β (TGF-β) (Weiner *et al.*, 1992; Kerstin Mangasser and Gressner, 1999), platelet-derived growth factor (PDGF) (Pinzani *et al.*, 1996), endothelin-1 (ET-1) (Shao *et al.*, 1999), and monocyte chemoattractant protein-1 (MCP-1) (Marra *et al.*, 1999); and (v) the development of contractile properties (Thimgan *et al.*, 1999).

HSC transdifferentiation also occurs *in vitro* by culturing the cells after isolation on plastic substratum in the presence of fetal calf serum. The first days after isolation HSC show the characteristics of quiescent HSC, whereas later these cells transdifferentiate into so-called "transitional cells," which further differentiate into fully activated HSC, also called myofibroblast-like cells. Thus, isolation and

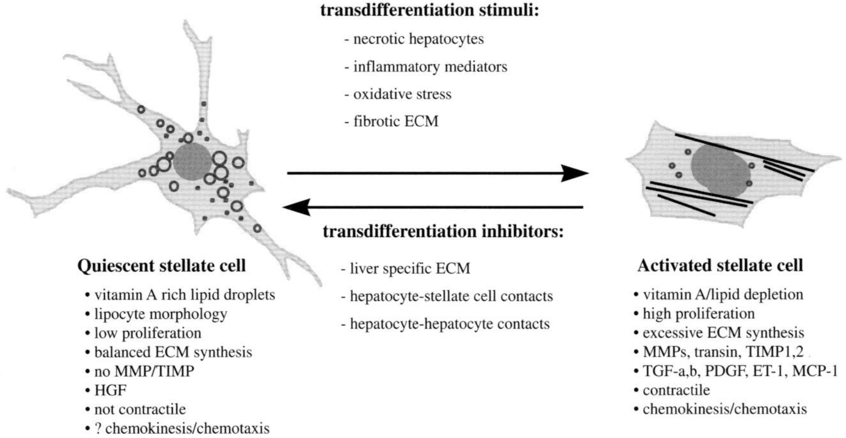

FIGURE 1 Transdifferentiation of HSC is regulated by three types of local control mechanisms: cell–cell contacts, cell–matrix contacts, and soluble factors secreted in the microenvironment.

cultivation of HSC constitute a model system to investigate the process of cell differentiation *in vitro*.

B. Chromatin and Modification of Histone Tails

The DNA template is embedded within a matrix of histone and nonhistone proteins termed chromatin, which serves to achieve the enormous compaction of the eukaryotic genome (Wade and Wolffe, 1997; Grunstein, 1997). Transcription of genes is regulated by chromatin-remodeling events, which can render the DNA either more or less accessible to transcriptional factors. To achieve access to the DNA while complexed in chromatin, cells have at their disposal modifying enzyme complexes, which serve to locally alter the structure of chromatin, facilitating gene-specific release from nucleosomal repression (Felsenfeld, 1992; Kornberg and Lorch, 1995). Certain enzymes and protein complexes are now known to bring about changes in the state of chromatin by numerous mechanisms, with resultant effects on gene expression. One major process by which the chromatin structure can be modulated concerns posttranslational modifications of the core histones, the building blocks of the fundamental structural unit of chromatin called the nucleosome. Core histones are susceptible to a wide range of posttranslational modifications, including acetylation, phosphorylation, methylation, ubiquitination, and ADP-ribosylation. Acetylation occurs at the ε amino groups of evolutionarily conserved lysine residues located in N termini of the core histones (Davie and Spencer, 1999). Steady-state levels of histone acetylation in core histones result from the balance of the antagonistic activities of histone acetyltransferases (HATs) and histone deacetylases (HDACs) (Csordas, 1990; Vidal and Gaber, 1991).

Experimentally, one can interfere with this mechanism of transcriptional regulation by using specific histone deacetylase inhibitors. Specific pharmacological inhibitors of histone deacetylases can be used to further investigate the possible working mechanism of HATs and HDACs on influencing the transcriptional regulation of specific genes.

C. Histone Deacetylases

During the last decade, many investigators have focused on the identification of HATs (reviewed in Sterner and Berger, 2000; Gregory *et al.*, 2001), as well as HDACs and their inhibitors. At this moment, three classes of HDACs have been identified (Table I). Class I HDACs are also called RPD3-like HDACs because of the strong homology of all members with the RPD3 (HDACs) isolated from yeast.

TABLE I Different Classes of Histone Deacetylases

Class
Class I
HDAC1 (Taunton *et al.*, 1996)
HDAC2 (Yang *et al.*, 1996)
HDAC3 (Yang *et al.*, 1996)
HDAC8 (Hu *et al.*, 2000; Buggy *et al.*, 2000; Vandenwyngaert *et al.*, 2000)
Class II
HDAC4 (Wang *et al.*, 1999; Miska *et al.*, 1999; Grozinger *et al.*, 1999)
HDAC5 (Grozinger *et al.*, 1999; Fischle *et al.*, 1999)
HDAC6 (Wang *et al.*, 1999; Miska *et al.*, 1999)
HDAC7 (Fischle *et al.*, 1999; Kao *et al.*, 2000)
Putative HDAC (Dunham *et al.*, 1999)
Putative HDAC-C (Fischle *et al.*, 1999; Wang *et al.*, 1999; Miska *et al.*, 1999)
Class III
SIRT1 (Frye, 1999)
SIRT2 (Afshar and Murnane, 1999; Frye, 1999)
SIRT3 (Frye, 1999)
SIRT4 (Frye, 1999)
SIRT5 (Frye, 1999)
SIRT6 (Frye, 2000)
SIRT7 (Frye, 2000)

These HDACs associate with two core complexes: Sin3 corepressor complexes and Mi-2/Nurd complexes. In addition, HDACs of class I are found to associate with a number of other complexes, including TGIF/Smad (Wotton *et al.*, 1999), glucocorticoid receptors (Ito *et al.*, 2000), and other important proteins, including Blimp-1 (Yu *et al.*, 2000), β-catenin (Billin *et al.*, 2000), and Sp1 (Doetzlhofer *et al.*, 1999). All four members of this class are sensitive to TSA. Class II HDACs or HDA1-like HDACs possess a strong homology with the yeast HDA1. Members of class II HDACs form large multiprotein complexes (Grozinger *et al.*, 1999; Fischle *et al.*, 1999; Huang *et al.*, 2000). This class of HDACs plays a major role in cellular proliferation/differentiation and is TSA sensitive. The histone deacetylase activity of Sir2-like HDACs or class III of HDACs is shown to be NAD^+ dependent. In contrast to class I and class II, members of class III are TSA insensitive (Imai *et al.*, 2000).

In order to identify HDACs in HSC, RNA was extracted from quiescent HSC and fully activated HSC. Data from our group showed the existence of HDAC1, 2, 3, 5, 6, and 7, whereas HDAC8 gene expression was absent. By using RTQ-PCR, we could quantify the expression of different HDACs in quiescent HSC *versus* fully activated HSC. All HDACs remained constant during the process of transdifferentiation of the HSC except the HDAC3 transcript, which was downregulated significantly during HSC activation (Geerts *et al.*, unpublished data).

D. TSA, a Histone Deacetylase Inhibitor

In 1987, Yoshida and co-workers extracted fungistatic antibiotics trichostatins A and C from culture broth of *Steptomyces platensis* No. 145. These trichostatins were found to be potent inducers of differentiation of murine erythroleukemia cells. Later, Yoshida and colleagues (1990) demonstrated that R-trichostatin A not only induced Friend erythroleukemia cell differentiation, but also inhibited cell cycle proliferation in normal rat fibroblasts. These effects were paralleled with an accumulation of acetylated histone species caused by the inhibition of histone deacetylases (Yoshida *et al.*, 1990). Many other investigators identified and synthesized other histone deacetylase inhibitors. At present, different groups of inhibitors are available (Table II).

II. EFFECT OF TSA ON HEPATIC STELLATE CELLS

In this study we investigated whether TSA affected those proteins that play a major role in the process of stellate cell differentiation. Several parameters of toxicity were investigated *in vitro* and *in vivo*. Furthermore, TSA-sensitive genes were identified, the translation products which were shown to retain the cytoskeleton of the primary HSC.

TABLE II Different Structural Classes of Histone Deacetylase Inhibitors

Short chain fatty acids: Butyrate and derivatives (Newmark *et al.*, 1994; Walczak *et al.*, 2001)
Hydroxamic acids
Trichostatin A (Yoshida *et al.*, 1990)
SAHA (Richon *et al.*, 1998)
Oxamflatin (Kim *et al.*, 1999)
Cyclic tetrapeptides containing an AOE moiety
Trapoxin A (Kijima *et al.*, 1993)
CHAP1 (Furumai *et al.*, 2001)
Cyclic tetrapeptides not containing an AOE moiety
FR901228 (Nakajima *et al.*, 1998)
Apicidin (Darkin Rattray *et al.*, 1996)
Benzamides: MS-27-275 (Saito *et al.*, 1999)
Benzoylamino alkano hydroxamates (Geerts *et al.*, unpublished data)
Miscellaneous
Depudecin (Kwon *et al.*, 1998)
DMBA (Jung *et al.*, 1997)
Diallyl disulfides (Lea *et al.*, 1999)

A. Influence of TSA on *de Novo* Protein Synthesis of Primary Hepatic Stellate Cells

We first examined the effect of TSA on the synthesis of procollagens type I and III, the major fibril-forming collagens that predominate in fibrotic livers (Schuppan *et al.*, 1990). TSA at 10^{-7} *M* strongly suppressed the synthesis of collagens type I and III by 62 and 70%, respectively (Table III).

Inhibition of histone deacetylase activity also suppressed the synthesis of α-SMA, an established marker for myofibroblast differentiation (Friedman *et al.*, 1989). As shown, synthesis of this molecule was inhibited by 84, 46, and 12% of the control levels by 10^{-7}, 10^{-8} and 10^{-9} *M* TSA (Table III). Suppression of α-SMA expression suggested that TSA blocked the synthesis of collagens type I and III at least partially by preventing the differentiation of primary HSC into myofibroblasts. This assumption was supported by the observation that TSA was less effective when tested on cells at day 14, when transition into myofibroblast-like cells had already occurred (Niki *et al.*, 1999).

B. Influence of TSA on mRNA Synthesis of Primary Hepatic Stellate Cells

To explore at which level of collagen synthesis TSA exerted its effects, we performed Northern hybridization analysis on RNA extracted from primary HSC. TSA (10^{-7} *M*) suppressed collagen α_1(III) mRNA levels by 61% and α-SMA mRNA levels by 75%, which was in keeping with the extent of suppression at the protein level (–70%) (Table III). In contrast, collagen α_1(I) mRNA levels were modestly altered by TSA (–21.5%). This suggested that the suppressive effect of TSA on collagen type I synthesis occurred mainly at the posttranscriptional level. The inhibitory action of TSA was selective for collagens type I, III, and α-SMA, as mRNA levels of collagen α_1(IV) and the housekeeping gene GAPDH were not altered (Niki *et al.*, 1999).

TABLE III Effect of TSA on *de Novo* Protein Synthesis, mRNA Steady-State Level, and Cell Proliferation

	10^{-7} *M* TSA			10^{-8} *M* TSA	10^{-9} *M* TSA
	mRNA		Protein	Protein	Protein
Collagen α_1(I)	25%		62%	31%	4%
Collagen α_1(III)	61%		70%	25%	24%
α-SMA	75%		84%	46%	12%
Proliferation		89%		1%	2%
Cell counting		72%		4%	2%

C. Effect of TSA on Proliferation of Primary Hepatic Stellate Cells

In addition to showing a strong antifibrogenic effect of TSA in primary HSC, we also examined the effect of TSA on the proliferation rate of HSC. TSA at 10^{-7} *M* showed a strong suppressive effect on [^{3}H]thymidine incorporation (−89%); lower concentrations were not inhibitory. The suppressive effect of 10^{-7} *M* TSA on cell proliferation observed by [^{3}H]thymidine incorporation was confirmed by counting cells (Table III).

D. Toxicity of TSA

When considering clinical usage, the question whether HDAC inhibitors are toxic is important. Skin fibroblasts, HSC, or hepatocytes *in vitro* did not show any morphological sign of toxicity when exposed to 100 n*M* TSA for up to 2–24 h. In primary hepatocyte cultures, we found no significant changes of albumin synthesis and no increased LDH leakage, epoxide hydrolase activity, or ethoxycoumarine *O*-deethylase activity (Niki *et al.*, 1999).

No toxicity was observed in an *in vivo* model of CCl_4-induced fibrosis in which inbred BALB/c mice were given TSA continuously at a release rate of 0.6 μg/g body weight/day for 2 weeks. In this *in vivo* model, the extent of fibrosis was reduced by 65.7% and the number of fully activated HSC by 43% (Geerts *et al.*, unpublished data).

E. Identification of TSA-Sensitive Genes

Because the antifibrotic effect of TSA was paralleled by maintaining the morphology of primary HSC (Fig. 2), we set out to further identify TSA-sensitive genes.

By performing RAP-PCR mRNA differential display, we identified TSA-sensitive genes, which were investigated further by performing Northern hybridization and Western blot analysis. One group of genes that was affected by short exposure to TSA was a group of actin cytoskeleton-related proteins (Fig. 3).

Our data show that TSA influences certain building blocks of the actin cytoskeleton. The net effect of these changes is that transdifferentiating HSC retain the actin cytoskeleton of primary cells. As a consequence, TSA inhibits the development of a strong cytoskeleton, which is associated with the fully activated HSC.

One important complex that plays a major role in the polymerization of actin filaments is the Arp2/Arp3 complex. The purified Arp2/3 complex binds pointed ends, nucleates the formation of actin filaments with free barbed ends, and binds preexisting actin filaments, thereby forming a branching network of actin

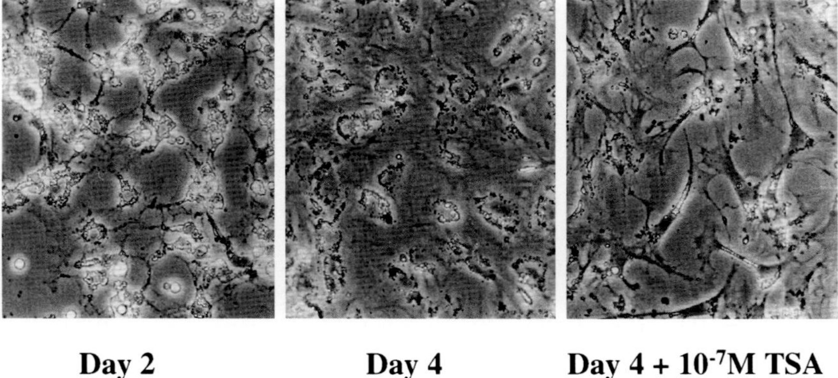

FIGURE 2 Morphology of HSC *in vitro*. Untreated control cells at day 4 showed intermediate morphology between quiescent and activated stellate cells. Cells treated with trichostatin A (10^{-7} *M*) at day 4 were slender shaped and retained cytoplasmic extensions. Magnification: 125×.

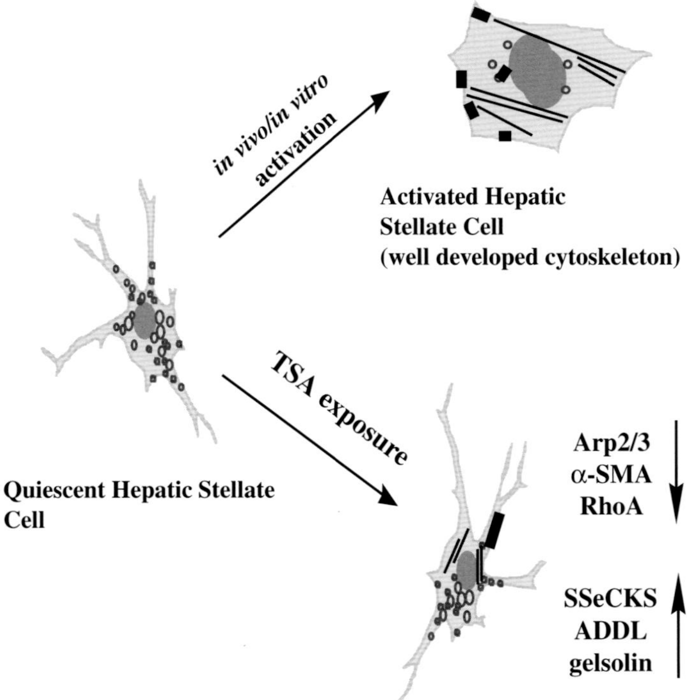

FIGURE 3 Schematic presentation on the identification of TSA-sensitive genes. TSA affects actin cytoskeleton-related proteins, Arp2/3 complex, α-SMA, RhoA, SSeCKS, ADDL70, and gelsolin, and therefore retains the morphology of the primary HSC.

filaments (Machesky *et al.*, 1998; Mullins *et al.*, 1998). A change in gene expression was observed for two other capping proteins: ADDL70 and gelsolin. It is known that adducin blocks elongation and polymerization of actin filaments by its capping capacity at the barbed ends of actin filaments (Kuhlman *et al.*, 1996). In contrast, gelsolin blocks actin monomer exchange at the barbed ends of actin filaments (Burtnick *et al.*, 1997). SSeCKS, known as *Src*-suppressed C kinase substrate (pronounced essex), is a negative mitogenic regulator that modulates the actin cytoskeleton. SSeCKS is localized in focal contact sites known to be enriched for PKCα and actin-binding proteins. Because SSeCKS is thought to be involved in anchoring actin filaments to focal adhesions through GTPase proteins such as Rho and Rac (Lin *et al.*, 1995, 1996), we further investigated the role of RhoA known to be important in the formation of actin stress fibers and focal adhesions (Sastry and Burridge, 2000; Schmitz *et al.*, 2000; Schoenwaelder, 1999; Zohn *et al.*, 1998). The inhibitory effect of TSA on RhoA protein expression emphasized again the suppressive role of TSA on actin cytoskeleton formation. Further evidence shows that by influencing the actin cytoskeleton of primary HSC by TSA, functions such as migration are impaired, whereas cell adhesion and contraction are unaffected by TSA (Rombouts *et al.*, unpublished data).

III. DISCUSSION

We have shown that hyperacetylation of core histone H4 has a potent antifibrogenic effect on HSC. These cells are the major cellular sources of extracellular matrix proteins in fibrotic livers (Pinzani and Gentilini, 1999; Friedman, 2000). To our knowledge, this is the first evidence that hyperacetylation of histones has a suppressive action on collagen synthesis and other cellular features of fibrogenesis.

At present we use the following hypothetical working model to explain how the hyperacetylation of histones could bring about the observed antifibrogenic effects (Fig. 4). In healthy liver, the quiescent HSC express a fairly constant array of genes. When homeostasis is disrupted, either by liver injury or by isolating and subculturing cells, paracrine and autocrine factors (cytokines, growth factors, reactive oxygen species, eicosanoids) induce a new type of repressor complex that contains one or more HDACs. This repressor complex binds to promoters of genes that are characteristic for the quiescent phenotype. Incubation of cells with a selective inhibitor of HDAC will lead to alleviating gene repression and to reverting activated into quiescent cells. This model is consistent with available experimental data and is also the leading working hypothesis to explain the differentiating effect of TSA on promyelocytic leukemia cells (He *et al.*, 1998). Whether TSA has other effects in addition to the inhibition of histone deacetylation cannot be entirely excluded but is unlikely. Tumor cell lines with mutant HDAC are resistant to the antiproliferative effect of TSA; if TSA would have other

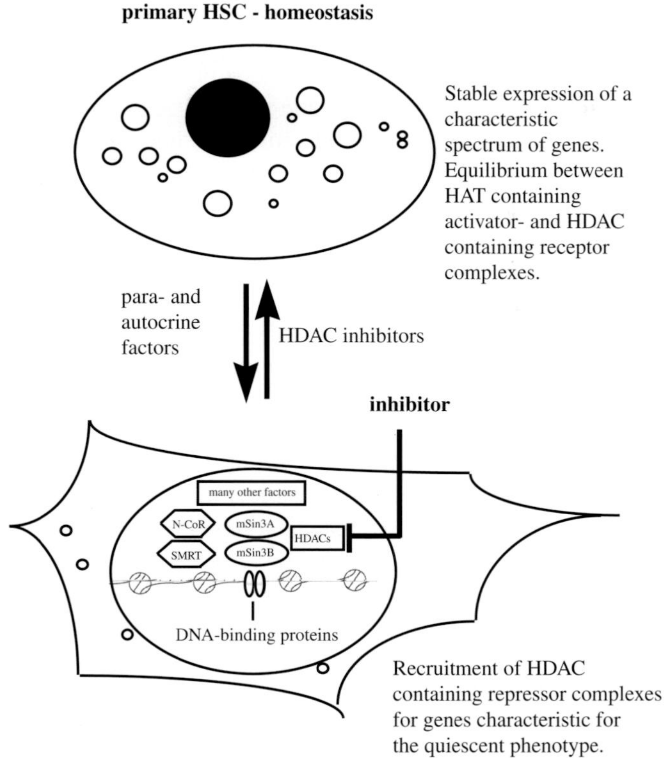

FIGURE 4 Hypothetical working model to explain why inhibition of HDACs, in primary HSC, keeps the cells in a quiescent phenotype.

effects, they should show up in these mutant cells, which is not the case (Yoshida *et al.*, 1987).

In view of the generally assumed role of histone acetylation in gene regulation, how selectivity is achieved needs clarification. In our study, we found that TSA affects the expression of collagens type I and III and α-SMA strongly, but does not alter the gene expression of collagen type IV and GAPDH. A study using differential mRNA display analysis has shown that in lymphoid cell lines, TSA affects the expression of only 2% of genes analyzed (Van Lint *et al.*, 1996). When applying differential mRNA display to primary HSC exposed to TSA, comparable low numbers of genes were affected (Rombouts *et al.*, unpublished observations). The selectivity of TSA is brought about by different mechanisms. First, this may be explained by the presence of several HDACs that show different sensitivity to TSA. Preliminary experiments in our laboratory showed the presence of HDAC classes I and II in HSC (Geerts *et al.*, unpublished data). Neither HATs nor HDACs appear to work alone, as both exist in multiprotein complexes. Recruitment of these complexes to specific regions of the genome plays a major part in creating the spectrum of expressed and silenced genes that characterizes a cell type. Second, promoter regions of different genes are organized in different ways in nucleosomes. Third, acetylation of nucleosomes varies topographically within the nucleus and with time. In general, approximately 10% of nucleosomes are acetylated; 90% are not. The latter nucleosomes are located in transcriptionally silenced heterochromatin. The degree of acetylation of each individual nucleosome is the result of local action of HATs that increase the level of acetylation and of HDACs that decrease the level. Fifth, selectivity in the regulation of transcription is brought about by the composition of the protein complexes in which HDACs are incorporated. Coimmunoprecipitation experiments have shown that these enzymes are present in large transcriptional complexes, the composition of which is dependent on the cell type and state of differentiation (Hassig *et al.*, 1997; Roopra *et al.*, 2000; Yu *et al.*, 2000; Park *et al.*, 2000). The latter finding supports the results of our study that histone acetylation can be targeted to specific promoters by gene-specific activator/repressor protein core complexes. Two core complexes have been identified: the Sin3 corepressor complex and Mi-2/Nurd complex. HDACs associate with these complexes and direct gene-specific transcriptional repression. The composition of these repressor complexes depends on the cell type, the target gene, and the state of cellular differentiation (Struhl and Moqtaderi, 1998; Dagond *et al.*, 1998; Knoepfler and Eisenman, 1999; Ng and Bird, 2000; Cress and Seto, 2000). Therefore, identification of the HDAC/corepressor complex(es) at the different gene promoters under investigation can provide additional information about which HDAC is sensitive to TSA and thus responsible for the observed effects (Richon *et al.*, 2000; Maeda *et al.*, 2000; Yu *et al.*, 2000). Moreover, Huang *et al.* (2000) have demonstrated the existence of multiple HDAC complexes in one cell type; each of these complexes has different effects on gene transcription.

In summary, we found that submicromolar concentrations of TSA inhibited strongly (1) synthesis of collagens type I and III, (2) cellular proliferation, and (3) expression of α-SMA in primary cultures of rat HSC. These biochemical changes were paralleled by inhibition of the morphological changes characteristic for activation of these cells. Thus, the histone deacetylase inhibitor TSA may be a promising lead compound in the development of novel agents to treat fibroproliferative diseases.

REFERENCES

Afshar, G., and Murnane, J. P. (1999). Characterization of a human gene with sequence homology to *Saccharomyces cerevisiae* SIR2. *Gene* **234**, 161–168.

Armendariz-Borunda, J., Simkevich, C. P., Roy, N., Raghow, R., Kang, A. H., and Seyer, J. M. (1994). Activation of Ito cells involves regulation of AP-1 binding proteins and induction of type I collagen gene expression. *Biochem. J.* **304**, 817–824.

Arthur, M. J., Iredale, J. P., and Mann, D. A. (1999). Tissue inhibitors of metalloproteinases: Role in liver fibrosis and alcoholic liver disease. *Alcohol Clin. Exp. Res.* **23**, 940–943.

Billin, A. N., Thirlwell, H., and Ayer, D. E. (2000). Beta-catenin-histone deacetylase interactions regulate the transition of LEF1 from a transcriptional repressor to an activator. *Mol. Cell Biol.* **20**, 6882–6890.

Britton, R. S., and Rippe, R. A. (1999). Intracellular signaling pathways in stellate cell activation. *Alcohol Clin. Exp. Res.* **23**, 922–925.

Buggy, J. J., Sideris, M. L., Mak, P., Lorimer, D. D., McIntosh, B., and Clark, J. M. (2000). Cloning and characterization of a novel human histone deacetylase, HDAC8. *Biochem. J.* **350**, 199–205.

Burtnick, L. D., Koepf, E. K., Grimes, J., Jones, E. Y., Stuart, D. I., McLaughlin, P. J., and Robinson, P. J. (1997). The crystal structure of plasma gelsolin; implications for actin severing, capping, and nucleation. *Cell* **90**, 661–670.

Chen, H. Y., Sun, J. M., Hendzel, M. J., Rattner, J. B., and Davie, J. R. (1996). Changes in the nuclear matrix of chicken erythrocytes that accompany maturation. *Biochem. J.* **320**, 257–265.

Cress, W. D., and Seto, E. (2000). Histone deacetylases, transcriptional control, and cancer. *J. Cell Physiol.* **184**, 1–16.

Csordas, A. (1990). On the biological role of histone acetylation. *Biochem. J.* **265**, 23–38.

Dagond, F., Hafler, D. A., Tong, J. K., Randall, J., Kojima, R., Utku, N., and Gullans, S. R. (1998). Differential display cloning of a novel human histone deacetylase (HDAC3) cDNA from PHA-activated immune cells. *Biochem. Biophys. Res. Commun.* **242**, 648–652.

Darkin Rattray, S. J., Gurnett, A. M., Myers, R. W., Dulski, P. M., Crumley, T. M., Allocco, J. J., Cannova, C., Meinke, P. T., Colletti, S. L., Bednarek, M. A., Singh, S. B., Goetz, M. A., Dombrowski, A. W., Polishook, J. D., and Schmatz, D. M. (1996). Apicidin: A novel anti-protozoal agent that inhibits parasite histone deacetylase. *Proc. Natl. Acad. Sci. USA* **93**, 13143–13147.

Davie, J. R., and Spencer, R. (1999). Control of histone modifications. *J. Cell Biochem. Suppl.* **32–33**, 141–148.

Doetzlhofer, A., Rotheneder, H., Lagger, G., Koranda, M., Kurtev, V., Brosch, G., Wintersberger, E., and Seiser, C. (1999). Histone deacetylase 1 can repress transcription by binding to Sp1. *Mol. Cell Biol.* **19**, 5504–5511.

Dunham, I., Shimizu, N., Roe, B. A., Chissoe, S., Hunt, A. R., Collins, J. E., Bruskiewich, R., Beare, D. M., Clamp, M., Smink, L. J., Ainscough, R., Almeida, J. P., Babbage, A., Bagguley, C., Bailey, J., Barlow, K., Bates, K. N., Beasley, O., Bird, C. P., Blakey, S., Bridgeman, A. M., Buck, D., Burgess, J., Burrill, W. D. P., and O'Brien, K. (1999). The DNA sequence of human chromosome 22. *Nature* **402**, 489–495.

Elsharkawy, A. M., Wright, M. C., Hay, R. T., Arthur, M. J., Hughes, T., Bahr, M. J., Degitz, K., and Mann, D. A. (1999). Persistent activation of nuclear factor-kappaB in cultured rat hepatic stellate cells involves the induction of potential novel Rel-like factors and prolonged changes in the expression of Ikappa-B family proteins. *Hepatology* **30**, 761–769.

Felsenfeld, G. (1992). Chromatin as an essential part of the transcriptional mechanism. *Nature* **355**, 219–224.

Fischle, W., Emiliani, S., Hendzel, M. J., Nagase, T., Nomura, N., Voelter, W., and Verdin, E. (1999). A new family of human histone deacetylases related to *Saccharomyces cerevisiae* HDA1p. *J. Biol. Chem.* **274**, 11713–11720.

Friedman, S. L. (1999). Cytokines and fibrogenesis. *Semin. Liver Dis.* **19**, 129–140.

Friedman, S. L. (2000). Molecular regulation of hepatic fibrosis, an integrated cellular response to tissue injury. *J. Biol. Chem.* **275**, 2247–2250.

Friedman, S. L., Roll, J., Arenson, D., and Bissell, D. M. (1989). Maintenance of differentiated phenotype of cultured rat hepatic lipocytes by basement membrane matrix. *J. Biol. Chem.* **264**, 10756–10762.

Frye, R. A. (1999). Characterization of five human cDNAs with homology to the yeast SIR2 gene: Sir2-like proteins (sirtuins) metabolize NAD and may have protein ADP-ribosyltransferase activity. *Biochem. Biophys. Res. Commun.* **260**, 273–279.

Frye, R. A. (2000). Phylogenetic classification of prokaryotic and eukaryotic Sir2-like proteins. *Biochem. Biophys. Res. Commun.* **273**, 793–798.

Furumai, R., Komatsu, Y., Khochbin, S., Yoshida, M., and Horinouchi, S. (2001). Potent histone deacetylase inhibitors built from trichostatin A and cyclic tetrapeptide antibiotics including trapoxin. *Proc. Natl. Acad. Sci. USA* **98**, 87–92.

Galli, A., Crabb, D., Price, D., Ceni, E., Salzano, R., Surrenti, C., and Casini, A. (2000). Peroxisome proliferator-activated receptor gamma transcriptional regulation is involved in platelet-derived growth factor - induced proliferation of human hepatic stellate cells. *Hepatology* **31**, 101–108.

Geerts, A. (2001). History, heterogeneity, developmental biology and functions of quiescent hepatic stellate cells. *Semin. Liver Dis.* **21**, 311–335.

Gregory, P., Wagner, K., and Hörz, W. (2001). Histone acetylation and chromatin remodeling. *Exp. Cell Res.* **265**, 195–202.

Grozinger, C., Hassig, C. A., and Schreiber, S. L. (1999). Three define a class of human histone deacetylase related to yeast Hda1p. *Proc. Natl. Acad. Sci. USA* **96**, 4868–4873.

Grunstein, M. (1997). Histone acetylation in chromatin structure and transcription. *Nature* **389**, 349–352.

Hassig, C. A., Fleischer, T. C., Billin, A. N., Schreiber, S. L., and Ayer, D. E. (1997). Histone deacetylase activity is required for full transcriptional repression by mSin3A. *Cell* **89**, 341–347.

He, L. Z., Guidez, F., Tribioli, C., Peruzzi, D., Ruthardt, M., Zelent, A., and Pandolfi, P. P. (1998). Distinct interactions of PML-RAR alpha and PLZF-RAR alpha with co-repressors determine differential responses to RA in APL. *Nature Genet.* **18**, 126–135.

Hellerbrand, C., Jobin, C., Iimuro, L., Sartor, R. B., and Brenner, D. A. (1998). Inhibition of NFkappaB in activated rat hepatic stellate cells by proteasome inhibitors and an IkappaB super-repressor. *Hepatology* **27**, 1285–1295.

Houglum, K., Lee, K. S., and Chojkier, M. (1997). Proliferation of hepatic stellate cells is inhibited by phosphorylation of CREB on serine 133. *J. Clin. Invest.* **99**, 1322–1328.

Hu, E., Chen, Z., Fredrickson, T., Zhu, Y., Kirkpatrick, R., Zhang, G. F., Johanson, K., Sung, C. M., Liu, R., and Winkler, J. (2000). Cloning and characterization of a novel human class I histone deacetylase that functions as a transcriptional repressor. *J. Biol. Chem.* **275**, 15254–15264.

Huang, E. Y., Zhang, J., Miska, E. A., Guenther, M. G., Kouzarides, T., and Lazar, M. A. (2000). Nuclear receptor corepressors partner with class II histone deacetylases in a Sin3-independent repression pathway. *Genes Dev.* **14**, 45–54.

Imai, S., Armstrong, C., Kaeberlein, M., and Guarente, L. (2000). Transcriptional silencing and longevity protein Sir-2 is an NAD-dependent histone deacetylase. *Nature* **403**, 795–800.

Ito, K., Barnes, P., and Adcock, I. M. (2000). Glucocorticoid receptor recruitment of histone deacetylase 2 inhibits interleukin-1beta-induced histone H4 acetylation on lysines 8 and 12. *Mol. Cell Biol.* **20**, 6891–6903.

Jung, M., Hoffmann, K., Brosch, G., and Loidl, P. (1997). Analogues of trichostatin A and trapoxin B as histone deacetylase inhibitors. *Bioorgan. Med. Chem. Lett.* **7**, 1655–1658.

Kao, H. Y., Downes, M., Ordentlich, P., and Evans, R. M. (2000). Isolation of a novel histone deacetylase reveals that class I and class II deacetylases promote SMRT-mediated repression. *Genes Dev.* **14**, 55–66.

Kawada, N., Uoya, M., Seki, S., Kuroki, T., and Kobayashi, K. (1997). Regulation by cAMP of STAT1 activation in hepatic stellate cells. *Biochem. Biophys. Res. Commun.* **233**, 464–469.

Kerstin Mangasser, S., and Gressner, A. M. (1999). Molecular and functional aspects of latent transforming growth factor-beta binding protein: Just a masking protein? *Cell Tissue Res.* **297**, 363–370.

Kijima, M., Yoshida, M., Sugita, K., Horinouchi, S., and Beppu, T. (1993). Trapoxin, an antitumor cyclic tetrapeptide, is an irreversible inhibitor of mammalian histone deacetylase. *J. Biol. Chem.* **268**, 22429–22435.

Kim, Y. B., Lee, J. H., Sugita, K., Yoshida, M., and Horinouchi, S. (1999). Oxamflatin is a novel antitumor compound that inhibits mammalian histone deacetylase. *Oncogene* **15**, 2461–2470.

Knittel, T., Mehde, M., Kobold, D., Saile, B., Dinter, C., and Ramadori, G. (1999). Expression patterns of matrix metalloproteinases and their inhibitors in parenchymal and non-parenchymal cells of rat liver: Regulation by TNF-alpha and TGF-beta-1. *J. Hepatol.* **30**, 48–60.

Knoepfler, P. S., and Eisenman, R. N. (1999). Sin meets NuRD and other tails of repression. *Cell* **99**, 447–450.

Kornberg, R. D., and Lorch, Y. (1995). Interplay between chromatin structure and transcription. *Curr. Opin. Cell Biol.* **7**, 371–375.

Kuhlman, P. A., Hughes, C. A., Bennett, V., and Fowler, V. M. (1996). A new function for adducin: Calcium/calmodulin-regulated capping of the barbed ends of actin filaments. *J. Biol. Chem.* **271**, 7986–7991.

Kwon, H. J., Owa, T., Hassig, C. A., Shimada, J., and Schreiber, S. L. (1998). Depudecin induces morphological reversion of transformed fibroblasts via the inhibition of histone deacetylase. *Proc. Natl. Acad. Sci. USA* **95**, 3356–3361.

Lea, M. A., Randolph, V. M., and Hodge, S. K. (1999). Induction of histone acetylation and growth regulation in erythroleukemia cells by 4-phenylbutyrate and structural analogs. *Anticancer Res.* **19**, 1971–1976.

Lee, K. S., Buck, M., Houglum, K., and Chojkier, M. (1995). Activation of hepatic stellate cells by TGF alpha and collagen type I is mediated by oxidative stress through c-myb expression. *J. Clin. Invest.* **95**, 2461–2468.

Li, D., and Friedman, S. L. (1999). Liver fibrogenesis and the role of hepatic stellate cells: new insights and prospects for therapy. *J. Gastroenterol. Hepatol.* **14**, 618–633.

Lieber, C. S. (1999). Prevention and treatment of liver fibrosis based on pathogenesis. *Alcohol Clin. Exp. Res.* **23**, 944–949.

Lin, X., Nelson, P., Frankfort, B., Tombler, E., Johnson, R., and Gelman, I. H. (1995). Isolation and characterization of a novel mitogenic regulatory gene, 322, which is transcriptionally suppressed in cells transformed by src and ras. *Mol. Cell Biol.* **15**, 2754–2762.

Lin, X., Tombler, E., Nelson, P. J., Ross, M., and Gelman, I. H. (1996). A novel src- and ras-suppressed protein kinase C substrate associated with cytoskeletal architecture. *J. Biol. Chem.* **271**, 28430–28438.

Machesky, L. M., and Insall, R. H. (1998). Scar1 and the related Wiskott-Aldrich syndrome protein, WASP, regulate the actin cytoskeleton through the Arp2/3 complex. *Curr. Biol.* **8**, 1347–1356.

Maeda, T., Towatari, M., Kosugi, H., and Saito, H. (2000). Up-regulation of costimulatory/adhesion molecules by histone deacetylase inhibitors in acute myeloid leukemia cells. *Blood* **96**, 3847–3856.

Marra, F., DeFranco, R., Grappone, C., Parola, M., Milani, S., Leonarduzzi, G., Pastacaldi, S., Wenzel, U. O., Pinzani, M., Dianzani, M. U., Laffi, G., and Gentilini, P. (1999). Expression of monocyte chemotactic protein-1 precedes monocyte recruitment in a rat model of acute liver injury, and is modulated by vitamin E. *J. Invest. Med.* **47**, 66–75.

Miska, E. A., Karlsson, C., Langley, E., Nielsen, S. J., Pines, J., and Kouzarides, T. (1999). HDAC4 deacetylase associated with and represses the MEF2 transcription factor. *EMBO J.* **18**, 5099–5107.

Mullins, R. D., Heuser, J. A., and Pollard, T. D. (1998). The interaction of Arp2/3 complex with actin: nucleation, high affinity pointed end capping, and formation of branching networks of filaments. *Proc. Natl. Acad. Sci. USA* **95**, 6181–6186.

Nakajima, H., Kim, Y. B., Terano, H., Yoshida, M., and Horinouchi, S. (1998). FR901228, a potent antitumor antibiotic, is a novel histone deacetylase inhibitor. *Exp. Cell Res.* **241**, 126–133.

Newmark, H. L., Lupton, J. R., and Young, C. M. (1994). Butyrate as a differentiating agent: Pharmacokinetics, analogues and current status. *Cancer Lett.* **78**, 1–5.

Ng, H. H., and Bird, A. (2000). Histone deacetylases: silencers for hire. *Trends Biochem. Sci.* **25**, 121–126.

Niki, T., Rombouts, K., De Bleser, P., Rogiers, V., De Smet, K., Schuppan, D., Yoshida, M., Gabbiani, G., and Geerts, A. (1999). A histone deacetylase inhibitor, trichostatin A, suppresses transdifferentiation of cultured rat hepatic stellate cells. *Hepatology* **29**, 859–867.

Ohata, M., Lin, M., Satre, M., and Tsukamoto, H. (1997). Diminished retinoic acid signaling in hepatic stellate cells in cholestatic liver fibrosis. *Am. J. Physiol.* **272**, G589–G596.

Park, J. S., Kim, E. J., Kwon, H. J., Hwang, E. S., Namkoong, S. E., and Um, S. J. (2000). Inactivation of interferon regulatory factor-1 tumor suppressor protein by HPV E7 oncoprotein: Implication for E7-mediated immune evasion mechanism in cervical carcinogenesis. *J. Biol. Chem.* **275**, 6764–6769.

Pinzani, M., and Gentilini, P. (1999). Biology of hepatic stellate cells and their possible relevance in the pathogenesis of portal hypertension in cirrhosis. *Semin. Liver Dis.* **19**, 397–410.

Pinzani, M., Milani, S., Herbst, H., DeFranco, R., Grappone, C., Gentilini, A., Caligiuri, A., Pellegrini, G., Ngo, D. V., Romanelli, R. G., and Gentilini, P. (1996). Expression of platelet-derived growth factor and its receptors in normal human liver and during active hepatic fibrogenesis. *Am. J. Pathol.* **148**, 785–800.

Raddatz, D., Henneken, M., Armbrust, T., and Ramadori, G. (1996). Subcellular distribution of glucocorticoid receptor in cultured rat and human liver-derived cells and cell lines: Influence of dexamethasone. *Hepatology* **24**, 928–933.

Ratziu, V., Lalazar, A., Wong, L., Dang, Q., Collins, C., Shaulian, E., Jensen, S., and Friedman, S. L. (1998). Zf9, a Kruppel-like transcription factor up-regulated in vivo during early hepatic fibrosis. *Proc. Natl. Acad. Sci. USA* **95**, 9500–9505.

Richon, V., Sandhoff, T. W., Rifkind, R. A., and Marks, P. A. (2000). Histone deacetylase inhibitor selectively induces p21WAF1 expression and gene-associated histone acetylation. *Proc. Natl. Acad. Sci. USA* **97**, 10014–10019.

Richon, V. M., Emiliani, S., Verdin, E., Webb, Y., Breslow, R., Rifkind, R. A., and Marks, P. A. (1998). A class of hybrid polar inducers of transformed cell differentiation inhibits histone deacetylases. *Proc. Natl. Acad. Sci. USA* **95**, 3003–3007.

Rippe, R. A. (1999). Role of transcriptional factors in stellate cell activation. *Alcohol Clin. Exp. Res.* **23**, 926–929.

Rippe, R. A., Almounajed, G., and Brenner, D. A. (1995). Sp1 binding activity increases in activated Ito cells. *Hepatology* **22**, 241–251.

Rockey, D. C. (2000). The cell and molecular biology of hepatic fibrogenesis: Clinical and therapeutic implication. *Clin. Liver Dis.* **4**, 319–355.

Roopra, A., Sharling, L., Wood, I. C., Briggs, T., Bachfisher, U., Paquette, A. J., and Bucley, N. J. (2000). Transcriptional repression by neuron-restrictive silencer factor is mediated via the Sin3- histone deacetylase complex. *Mol. Cell Biol.* **20**, 2147–2157.

Saito, A., Yamashita, T., Nosaka, Y., Tsuchiya, K., Ando, T., Suzuki, T., Tsuruo, T., and Nakanishi, O. (1999). A synthetic inhibitor of histone deacetylase, MS-27-275, with marked in vivo antitumor activity against human tumors. *Proc. Natl. Acad. Sci. USA* **96**, 4592–4597.

Sastry, S. K., and Burridge, K. (2000). Focal adhesion: A nexus for intracellular signaling and cytoskeletal dynamics. *Exp. Cell Res.* **261**, 25–36.

Schmitz, A. A. P., Govek, E. E., Böttner, B., and VanAelst, L. (2000). Rho GTPases: Signaling, migration, and invasion. *Exp. Cell Res.* **261**, 1–12.

Schoenwaelder, S. M., and Burridge, K. (1999). Bidirectional signaling between the cytoskeleton and integrins. *Curr. Opin. Cell Biol.* **11**, 274–286.

Schuppan, D. (1990). Structure of the extracellular matrix in normal and fibrotic liver: Collagens and glycoproteins. *Semin. Liver Dis.* **10**, 1–10.

Shao, R., Yan, W., and Rockey, D. (1999). Regulation of endothelin-1 synthesis by endothelin-converting enzyme-1 during wound healing. *J. Biol. Chem.* **274**, 3228–3234.

Smart, D. E., Vincent, K. J., Arthur, M. J., Eickelberg, O., Castellazzi, M., Mann, J., and Mann, D. A. (2001). JunD regulates transcription of the tissue inhibitor of metalloproteinases-1 and interleukin-6 in activated hepatic stellate cells. *J Biol. Chem.* **276**, 24414–24421.

Sterner, D. E., and Berger, S. L. (2000). Acetylation of histones and transcription-related factors. *Microbiol. Mol. Biol. Rev.* **64**, 435–459.

Struhl, K., and Moqtaderi, Z. (1998). The TAFs in the HAT. *Cell* **94**, 1–4.

Taunton, J., Hassig, C. A., and Schreiber, S. L. (1996). A mammalian histone deacetylase related to the yeast transcriptional regulator Rpd3p. *Science* **272**, 408–411.

Thimgan, M., and Yee, H. J. (1999). Quantitation of rat hepatic stellate cell contraction: Stellate cells' contribution to sinusoidal resistance. *Am. J. Physiol.* **277**, G137–G143.

Van Den Wyngaert, I., de Vries, W., Kremer, A., Neefs, J. M., Verhasselt, P., Luyten, W., and Kass, S. U. (2000). Cloning and characterization of human histone deacetylase 8. *FEBS Lett.* **478**, 77–83.

Van Lint, C., Emiliani, S., and Verdin, E. (1996). The expression of a small fraction of cellular genes is changed in response to histone hyperacetylation. *Gene Expr.* **5**, 245–253.

Vidal, M., and Gaber, R. F. (1991). RPD3 encodes a second factor required to achieve maximum positive and negative transcriptional states in *Saccharomyces cerevisiae*. *Mol. Cell Biol.* **11**, 6317–6327.

Wade, P. A., and Wolffe, A. P. (1997). Histone acetyltransferases in control. *Curr. Biol.* **7**, R82–R84.

Walczak, J., Wood, H., Wilding, G., Williams, T., Jr., Bishop, C. W., and Carducci, M. (2001). Prostate cancer prevention strategies using antiproliferative or differentiating agents. *Urology* **57**, 81–85.

Wang, A. H., Bertos, N. R., Vezmar, M., Pelletier, N., Crosato, M., Heng, H. H., Th'ng, J., Han, J., and Yang, X. J. (1999). HDAC4, a human histone deacetylase related to yeast HDA1, is a transcriptional corepressor. *Mol. Cell Biol.* **19**, 7816–7827.

Weiner, F., Blaner, W. S., Czaja, M., Shah, A., and Geerts, A. (1992). Ito cell expression of a nuclear retinoic acid receptor. *Hepatology* **15**, 336–342.

Weiner, J. A., Chen, A., and Davis, B. H. (1998). E-box-binding repressor is down-regulated in hepatic stellate cells during up-regulation of mannose 6-phosphate/insulin-like growth factor II receptor expression in early hepatic fibrogenesis. *J. Biol. Chem.* **273**, 15913–15919.

Wotton, D., Lo, R. S., Swaby, L. C., and Massagué, J. (1999). Multiple modes of repression by the Smad transcriptional corepressor TGIF. *J. Biol. Chem.* **274**, 37105–37110.

Wu, J., and Zern, M. A. (2000). Hepatic stellate cells: A target for the treatment of liver fibrosis. *J. Gastroenterol.* **35**, 665–672.

Yang, W. M., Inouye, C. J., Zeng, Y., Bearss, D., and Seto, E. (1996). Transcriptional repression by YY1 is mediated by interaction with a mammalian homologue of the yeast global regulator RPD3. *Proc. Natl. Acad. Sci. USA* **93**, 12845–12850.

Yoshida, M., Hoshikawa, Y., Koseki, K., Mori, K., and Beppu, T. (1990). Structural specificity for biological activity of trichostatin A, a specific inhibitor of mammalian cell cycle with potent differentiation-inducing activity in Friend leukemia cells. *J. Antibiot. Tokyo.* **43**, 1101–1106.

Yoshida, M., Nomura, S., and Beppu, T. (1987). Effects of Trichostatins on differentiation of murine erythroleucemia cells. *Cancer Res.* **47**, 3688–3691.

Yu, J., Angelin-Duclos, C., Greenwood, J., Liao, J., and Calame, K. (2000). Transcriptional repression by blimp - 1 (PRDI-BF1) involves recruitment of histone deacetylase. *Mol. Cell Biol.* **20**, 2592–2603.

Zohn, I. M., Campbell, S. L., Khosravi-Far, R., Rossman, K. L., and Der, C. J. (1998). Rho family proteins and Ras transformation: The RHOad less traveled gets congested. *Oncogene* **17**, 1415–1438.

CHAPTER 12

Profibrogenic Actions of Hepatic Stellate Cells: Major Intracellular Signaling Pathways

MASSIMO PINZANI AND FABIO MARRA

Dipartimento di Medicina Interna, Università degli Studi di Firenze, I-50134, Firenze, Italy

Following acute or chronic liver tissue damage, hepatic stellate cells (HSC) undergo a process of activation toward a phenotype characterized by increased proliferation, motility, contractility, and synthesis of extracellular matrix (ECM) components. In addition to these changes in the biology of HSC, several factors have been shown to play a key role in the promotion of the full-blown picture of activated HSC. These include extensive changes in the composition and organization of the ECM, the secretion of several growth factors, cytokines, chemokines, products of oxidative stress, and other soluble factors. Indeed, in the presence of chronic liver tissue damage/inflammation, clusters of soluble factors, specifically directed at different cell targets, are simultaneously active in the tissue and are, at least in part, responsible for the fibrogenic outcome of the wound-healing process. It is likely that a complex network of interactions occurs between these mediators, their targets, and the extracellular matrix. Different groups of profibrogenic soluble factors could be classified according to their prevalent biological effect: factors promoting HSC proliferation and/or migration [i.e., platelet-derived growth factor (PDGF), basic fibroblast growth factor (bFGF), insulin-like growth factor-1 (IGF-1)]; factors promoting fibrillar ECM accumulation, particularly transforming growth factor (TGF)-β1 (induction of procollagen I/III gene expression, induction of TIMP-1 gene expression, inhibition of MMP-1, and increase

Copyright 2003, Elsevier Science (USA). All rights reserved.

of MMP-2); factors with a prevalent contractile effect on HSC, such as thrombin and ET-1, although both agents may also promote HSC proliferation; chemokines, such as MCP-1, one of the most potent chemoattractant for leukocytes; cytokines with a prominent anti-inflammatory/antifibrogenic activity, such as interleukin (IL)-10 (produced also by HSC) and interferon (IFN)-γ. Additional important issues to be addressed in this context include the autocrine/paracrine action of many of these factors synthesized by activated HSC; the relationship occurring between these substances and the rapidly evolving ECM microenvironment, where they bind and can be stored; and the biological effects of oxidative stress-related molecules (ROI and reactive aldehydes). In the past decade a major effort has been made in order to elucidate the major intracellular signaling pathways elicited by these factors on HSC. As also shown in other cell types showing similar features and profibrogenic roles in extrahepatic tissues, signaling pathways such as Ras/ERK and PI 3-K have been demonstrated to be indispensable to transduce signals leading to cell proliferation and motility in HSC. Similarly, the involvement of SMAD proteins has been shown to transduce the biologic effects of TGF-β1 in these cells, and several other examples are available. In addition to the specific signaling elicited by the interaction of a given cytokine with this membrane receptor, a central key issue concerns the cooperation between cytokine signaling and signals originated from the interaction of integrin receptors with specific ECM components. This relationship appears indeed to be highly relevant for the effective action of several cytokines. Another emerging key issue is the influence of cytoskeletal structures, particularly their assembly and tension, on the intracellular signaling of growth factors and cytokines. In conclusion, an increasing body of evidence is becoming available for a more complete understanding of the profibrogenic role of several soluble factors acting on HSC. It is hoped that these acquisitions will contribute to design pharmacological and biotechnological strategies able to modulate the fibrogenic progression of chronic liver diseases.

I. INTRODUCTION

Growth factors, cytokines, chemokines, and oxidative stress products play a role in the activation of hepatic stellate cells. In the presence of chronic liver tissue damage/inflammation, these factors directed at specific cell targets are simultaneously active in the tissue and are, at least in part, responsible for the fibrogenic outcome of the wound-healing process (Friedman, 2000; Pinzani, 2000). It is important to stress that they do not work alone, rather a complex network of interactions occurs between these mediators, their targets, and the ECM. Thus, the response to single agonists on cultured HSC does not completely reflect the complexity of the *in vivo* situation. Nevertheless, these approaches are quite informative in unearthing key signaling events.

The different groups of cytokines can be grouped according to their class of receptors, which tend to generate similar intracellular signals within each group: factors promoting HSC proliferation, migration, and survival (polypeptide growth factor receptors); factors promoting fibrillar ECM accumulation, particularly TGF-β1 (TGF-β receptor superfamily); factors with a prevalent contractile effect on HSC, such as endothelins, angiotensin II, vasopressin, and thrombin, although all these agents may also promote HSC proliferation (seven transmembrane domain receptors); and receptors for chemokines and proinflammatory cytokines. An important determinant of the biologic response to these ligands is the interaction between these substances and the rapidly evolving ECM microenvironment, where they bind and can be stored. In this context, the relationship between cytokine- and integrin-mediated intracellular signaling is increasingly relevant.

II. POLYPEPTIDE GROWTH FACTOR RECEPTORS

Platelet-derived growth factor (PDGF), a dimer of two polypeptide chains referred to as A and B chains, is the most potent mitogen for cultured HSC isolated from rat, mouse, or human liver (Pinzani *et al.*, 1989; Pinzani *et al.*, 1992a; Pinzani *et al.*, 1995). Of the three possible dimeric forms of PDGF (-AA, -AB, and -BB), PDGF-BB is most potent in stimulating HSC growth and intracellular signaling, in agreement with a predominant expression of PDGF-receptor β (or type B) subunits compared to PDGF-receptor α (or type A) subunits in activated HSC (Pinzani *et al.*, 1995). Codistribution of PDGF with cells expressing PDGF receptor subunits has been demonstrated following both acute and chronic liver tissue damage (Pinzani *et al.*, 1994b; Pinzani *et al.*, 1996b), thereby confirming an active role of this growth factor in liver repair and fibrosis. Figure 1 illustrates the major signaling pathways and biologic effects elicited by the interaction of PDGF with its membrane receptors, i.e., cell proliferation and migration. PDGF receptors, which have intrinsic tyrosine kinase activity, dimerize and become autophosphorylated on tyrosine residues upon binding to their ligand (Claesson-Welsh, 1994). Phosphotyrosines on the activated receptor operate as high-affinity-binding sites for several molecules involved in the downstream propagation of the signal, which bind through src-homology-2 (SH-2) domains or phosphotyrosine-binding (PTB) domains (Cohen *et al.*, 1995; Pawson, 1995). Association of the PDGF receptor with the adapter protein Grb2 leads to recruitment of the exchange factor mSos with the consequent activation of Ras. This event is followed by the sequential activation of Raf-1, MEK, and extracellular signal-regulated kinase (ERK) (Marshall, 1995). Nuclear translocation of ERK is associated to the phosphorylation of several transcription factors, including Elk-1 and SAP, and represents an absolute requirement for triggering a proliferative response (Pages *et al.*, 1993). In cultured human HSC, there is activation of the ERK pathway followed by increased expression of *c-fos* in response to PDGF (Marra *et al.*, 1995b, 1996, 1999).

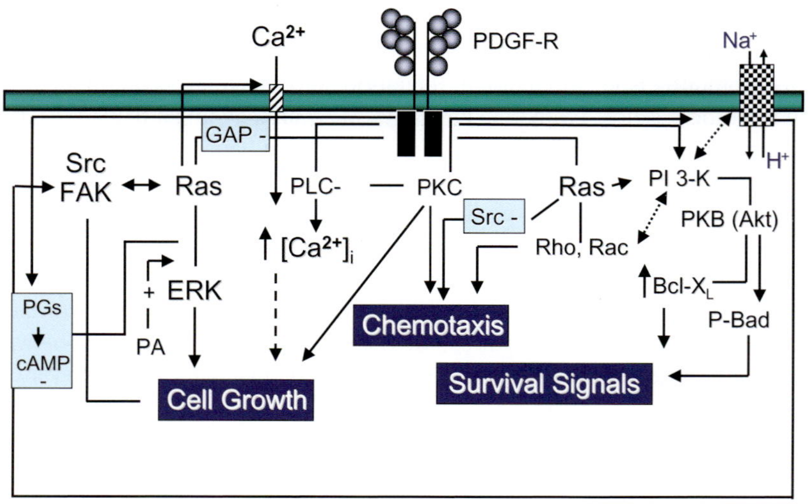

FIGURE 1 PDGF signaling. PDGF binds to a receptor with intrinsic tyrosine kinase activity. Receptor dimerization leads to autophosphorylation with formation of high-affinity-binding sites for signaling proteins with SH-2 or PTB domains. The downstream pathways are differentially implicated in the regulation of biologic activities of PDGF. For a detailed discussion, see text.

The activation of this pathway is necessary for PDGF-induced cell proliferation and, accordingly, the pharmacological blockade of signaling molecules upstream of ERK (e.g., MEK) leads to a dose-dependent inhibition of cell growth. This observation is supported by the reduction in the downstream activation of the protooncogene *c-fos* and of the AP-1 complex binding activity that follows the inhibition of PDGF-induced ERK activation (Marra *et al.*, 1999). ERK activation in rat HSC occurs following *in vivo* liver injury induced by the acute administration of CCl_4 (Marra *et al.*, 1999). In this model, increased ERK activity temporally precedes HSC proliferation and peaks 48 h after the administration of the toxin. Remarkably, this time point is associated with maximal availability of PDGF in acutely injured liver tissue (Pinzani *et al.*, 1994b) and precedes HSC proliferation, which begins at 48 h and peaks 72 h after liver damage.

Phosphatidylinositol 3-kinase (PI 3-K), another molecule that is recruited by the activated PDGF receptor, is composed of a 85-kDa regulatory subunit, equipped with two SH-2 domains, and a catalytic 110-kDa subunit (Parker and Waterfield, 1992). PDGF stimulation leads to the association of PI 3-K with the activated receptor and to tyrosine phosphorylation of p85 but not of p110. The downstream effectors of PI 3-K activation are only partially known and include protein kinase Cζ, ribosomal S6 kinase, and protein kinase B (c-Akt). Nevertheless, this pathway is sufficient to transduce PDGF-dependent mitogenic signals (Valius and Kazlaukas, 1993) and to be necessary for cell chemotaxis (Kundra *et al.*, 1993). Thus, in human HSC cultures, PI 3-K activation is necessary for both mitogenesis and chemotaxis induced by PDGF (Marra *et al.*, 1997). The *in vivo* relevance of this finding is suggested by the recruitment of the p85 subunits by the PDGF receptor and activation of PI 3-K following acute CCl_4-induced liver damage in the rat. Wortmannin, a fungal metabolite that binds and inhibits PI 3-K noncompetitively, induces a dose-dependent inhibition of PDGF-BB-induced PI 3-K activation in HSC with a maximal effect at 100 n*M*. This concentration, which does not affect either PDGF receptor autophosphorylation or the physical association between the PI 3-K p85 subunit and the receptor, virtually abolishes PDGF-induced mitogenesis and chemotaxis in HSC, indicating a functional involvement of this pathway. Similar observations have been made with other PI 3-K inhibitors, such as LY294002 (Gentilini *et al.*, 2000). In addition, PI 3-K is involved in the activation of the Ras-ERK pathway in human HSC, although it is not strictly necessary, as both wortmannin and LY294002 inhibit ERK activation only by 40–50% (Gentilini *et al.*, 2000; Marra *et al.*, 1995b). Therefore, in HSC, PI 3-K regulates PDGF-related mitogenesis and cell migration by pathways that are at least in part independent of ERK activation.

In addition to the involvement in cell growth and migration, growth factor-induced PI 3-K activation may contribute to the downstream signaling that regulates cell survival. PDGF and insulin-like growth factor-I (IGF-I) provide an example of two contrasting paradigms of action. In human HSC, PDGF induces a 10-fold increase in DNA synthesis and cell migration, whereas the effect of IGF-I is in general one-fifth of that of PDGF. Regardless, PDGF and IGF-I are equipotent in the activation of the Ras/ERK and the PI 3-k pathways, at least at early time points (10–15 min) after stimulation (Gentilini *et al.*, 2000). In addition, IGF-I acts as a survival rather than a mitogenic growth factor in this cell type (Issa *et al.*, 2001). Current evidence suggests that the "survival" or antiapoptotic action of PI 3-K is mediated by the activation of c-Akt (also referred to as protein kinase B – PKB), a signaling protein whose activity is regulated by several upstream events, and particularly the generation of phoshoinositides by PI 3-K (Datta *et al.*, 1999). Gentilini and co-workers (2001) have shown that in human activated HSC, IGF-I can activate the c-Akt pathway and its downstream targets regulating cell survival and to reduce nerve growth factor (NGF)-induced apoptosis. Importantly, in these experiments, activation of the c-Akt pathway is a PI 3-K-dependent event that is reversed by PI 3-K inhibitors.

In addition to specific intracellular signaling pathways that involve protein phosphorylation, PDGF signaling also relies on changes in $[Ca^{2+}]_i$ and pH. In particular, in HSC and other cells, sustained changes in $[Ca^{2+}]_i$ and intracellular pH are necessary for the correct articulation of pathways involving protein phosphorylation. The mitogenic potential of different PDGF dimeric forms is proportional to their effects on $[Ca^{2+}]_i$ in activated rat and human HSC (Pinzani *et al.*, 1991, 1995). The increase in $[Ca^{2+}]_i$ induced by PDGF in HSC is characterized by two main components: (1) a consistent and transient increase (peak increase), due to calcium release from intracellular stores following the activation of PLCg and the consequent PIP_2 hydrolysis, and (2) a lower but longer lasting increase (plateau phase) due to an influx from the external medium. Accumulated evidence indicates that the induction of replicative competence by PDGF is dependent on the maintenance of sustained increase in $[Ca^{2+}]_i$ due to calcium entry rather than from the release from intracellular stores (Kondo *et al.*, 1993; Wang *et al.*, 1993). Accordingly, stimulation of human HSC with PDGF in the virtual absence of extracellular calcium results in an almost complete abrogation of the mitogenic effect of this growth factor and supports the view that the plateau phase of the increase in $[Ca^{2+}]_i$ is essential for eliciting full PDGF-induced replicative competence in this cell type (Failli *et al.*, 1995).

Extracellular calcium entry induced by PDGF was originally ascribed to the opening of low threshold voltage-gated calcium channels consistent with a "T" type designation (Wang *et al.*, 1993). Subsequently, this channel has been better characterized and defined as a PDGF receptor-operated nonselective cation channel controlled by the tyrosine kinase activity of the PDGF-R and, particularly, by the activation of Ras through Grb2-Sos (Ma *et al.*, 1996). The existence of this PDGF receptor-operated channel in activated human HSC is suggested by the functional uncoupling between PDGF-R and this calcium channel caused by the inhibition of Ras processing following incubation of HSC with GGTI-298, an inhibitor of protein geranylgeranylation (Carloni *et al.*, 2000).

Stimulation with PDGF increases the activity of the Na^+/H^+ exchanger in rat or human HSC with consequent sustained changes in intracellular pH (Di Sario *et al.*, 1997, 1999, 2001). This increased activity appears to occur through calcium–calmodulin and protein kinase C-dependent pathways (Di Sario *et al.*, 1999).

Inhibition of the activity of the Na^+/H^+ exchanger by pretreatment with amiloride inhibits PDGF-induced mitogenesis, thus indicating that changes in intracellular pH induced by this growth factor are essential for its full biologic activity (Benedetti *et al.*, 2001). Data suggest that PDGF-induced Na^+/H^+ exchanger activity is linked to the activation of PI 3-K and is blocked by preincubation with PI 3-K inhibitors. Furthermore, inhibition of the Na^+/H^+ exchanger leads to the interruption of downstream signaling events essential for growth factor-mediated cytoskeletal reorganization such as PDGF-induced focal adhesion kinase (FAK) phosphorylation (Caligiuri *et al.*, 2001).

III. TRANSFORMING GROWTH FACTOR-β RECEPTOR SUPERFAMILY

TGF-β are pleiotropic cytokines that play a pivotal role in the development of fibrosis in the liver and in other organs, including the lung and the kidney

(Branton and Kopp, 1999). Expression of TGF-β is increased markedly in animal models of liver fibrosis and in patients with chronic liver disease (Castilla *et al.*, 1991; Nakatsukasa *et al.*, 1990). Overexpression of TGF-β is associated with an increased deposition of matrix in the target tissue, and neutralization of the biologic activity of this cytokine ameliorates experimental liver fibrosis (George *et al.*, 1999; Sanderson *et al.*, 1995). The biologic actions of TGF-β on HSC are in great part related to the profibrogenic role of this factor. TGF-β is the most potent stimulus for the production of fibrillar and nonfibrillar matrix by HSC and it also induces qualitative changes in the matrix by differentially stimulating its components (Friedman, 1999). In addition, TGF-β also has effects on matrix degradation, which are characterized by mixed actions on matrix metalloproteinases, and by inhibition of tissue inhibitors of metalloproteinases (TIMPs) and plasminogen activator inhibitor (PAI). Although TGF-β is a growth inhibitor in several cell types, conflicting results have been reported in HSC, where TGF-β has no effect (Pinzani *et al.*, 1989), inhibits (Bachem *et al.*, 1992; Saile *et al.*, 1999), or induces HSC proliferation via production of PDGF-AA (Win *et al.*, 1993). In addition, prolonged exposure of HSC to this cytokine may increase PDGF-dependent DNA synthesis via PDGF receptor upregulation (Pinzani *et al.*, 1995). Finally, TGF-β behaves as a survival factor for activated HSC, where spontaneous apoptosis was downregulated by this cytokine (Saile *et al.*, 1999).

The receptors for TGF-β are serinethreonine kinases belonging to a superfamily that includes the receptors for activins, inhibins, bone morphogenetic proteins, and other multifunctional cytokines (Piek *et al.*, 1999). Signal propagation occurs via the formation of heteromeric complexes comprising type I and type II TGF-β receptors. TGF-β and other proteins of this superfamily bind to TGF-β type II receptor (TβRII), which subsequently recruits TGF-β type I receptor (TβRI), leading to the formation of a heterotrimeric complex (Fig. 2) (Piek *et al.*, 1999). TβRII has a constitutively active serine/threonine kinase, which phosphorylates TβRI, which in turn is responsible for the propagation of downstream signals through the phosphorylation of Smads. Betaglycan, or TGF-β receptor type III, also binds TGF-β with high affinity. This receptor, however, has no role in TGF-β signaling, but it may be relevant for the presentation of TGF-β to TβRII, and this interaction appears to be particularly important for TGF-β2 (Piek *et al.*, 1999). The signal downstream of receptors of the TGF-β superfamily is transduced by the Smad proteins (Massague, 2000). Smads undergo phosphorylation by type I receptors of the TGF-β superfamily and can be divided in three categories depending on the ability to be activated by the receptor (R-Smads), to function as common partners (Co-Smads), or to inhibit signal (anti-Smads). Following formation of the TGF-β receptor complex, Smad2 and Smad3 are activated by the TβRI and thus function as R-Smads for this cytokine (Fig. 2). Upon activation by phosphorylation, Smad2 or Smad3 heterodimerizes with the common partner Smad4 and migrates to the nucleus where it regulates transcription.

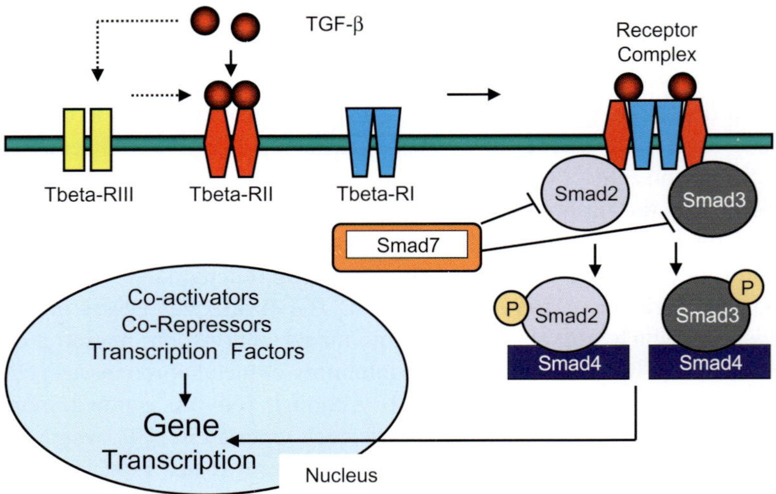

FIGURE 2 TGF-β signaling. TGF-β binds to TβRII, and this binding may be enhanced by the presence of TβIII. After binding to TGF-β, TβRII recruits and phosphorylates TβRI, leading to activation of Smad2 and Smad3 (R-Smads). This process is inhibited by Smad7 (anti-Smad). Activated Smad2 and Smad3 heterodimerize with Smad4 (co-Smad) and migrate to the nucleus where they regulate gene transcription. Smad heterodimers may cooperate with other transcription factors, coactivators, or corepressors.

Transcriptional control occurs not only by direct interaction with a target DNA element (Smad-binding element), but also by cooperation with other transcription factors, e.g., AP-1, or interaction with coactivators (e.g., CREB-binding protein) or corepressors (Massague, 2000). Smad7 is an inhibitory Smad for TGF-β signaling. It binds TβRI efficiently, but lacks the C-terminal phosphorylation motif and therefore behaves as an endogenous dominant negative Smad (Piek *et al.*, 1999). Interestingly, a Smad-binding element has been found in the Smad7 promoter, indicating that TGF-β-induced activation of this molecule plays a role in the modulation of TGF-β signals. Smad2 and Smad3 are likely to mediate different sets of biologic actions, as indicated by the fact that they may mediate antagonistic actions and by the different phenotype of knockout mice. While Smad2-deficient animals show embryonic lethality, Smad3 knockout mice survive to adulthood and may actually display enhanced wound healing (Ashcroft *et al.*, 1999). In addition, direct DNA binding of Smad2 has not yet been demonstrated (Piek *et al.*, 1999).

Although signaling generated by TGF-β has been explored extensively, a limited amount of information has been obtained on the pathways that regulate matrix production, its activity most relevant to the development of liver fibrosis. Signals originating from TβRI and leading to growth inhibition may be differentiated from those that regulate extracellular matrix synthesis (Saitoh *et al.*, 1996).

Quiescent HSC are poorly responsive to TGF-β when maintained in suspension, a condition that prevents transition to an activated state (Friedman *et al.*, 1994). Accordingly, affinity labeling studies have demonstrated small amounts of TβRIII in quiescent cells, whereas all three TGF-β receptors are present after activation. Nevertheless, in nonactivated cells, TβRII expression could be demonstrated by Western blotting, and mRNA transcripts for TβRII and TβRIII (betaglycan) were actually greater than after activation on uncoated plastic (Friedman *et al.*, 1994). Decreased mRNA levels for TβRII in activated HSC have been confirmed by Roulot *et al.* (1999), who suggest that the ratio between type II and type I receptors may be critical to differentially mediate the biologic effects of TGF-β, including matrix synthesis or inhibition of cell proliferation. This hypothesis is supported by data obtained in the *Xenopus laevis*, where small changes in the amount of signaling can dramatically affect the nature of the cellular response to the TGF-β–like factor, activin (Massague, 1998). The biological significance of reduced expression of TβRIII in activated versus quiescent HSC remains to be established. Similar to other TGF-β responsive cells, HSC show phosphorylation of Smad2 and Smad3 upon exposure to this cytokine (Dooley *et al.*, 2000).

Interestingly, compared to early cultured HSC, cells with a fully activated, myofibroblast-like phenotype show reduced Smad activation, and this finding is associated with lower efficacy of the cytokine in inducing biologic actions in fully activated cells (Dooley *et al.*, 2000). It has been hypothesized that altered compartmentalization of TGF-β receptors, with a higher proportion of receptor in the intracellular than in the membrane compartment, may explain these findings (Dooley *et al.*, 2000). However, these changes are likely to be limited to fully activated, myofibroblast-like cells because binding of TGF-β to its receptors, as evaluated by affinity binding, is not reduced in HSC isolated after 48 or 72 h after acute CCl_4 intoxication in the rat (Inoue and Thomas, 2000). Interaction between activated Smad proteins and other factors may be relevant for the cell specificity of TGF-β actions. Sp1 binding to the promoter of the α2(I) collagen gene mediates the increased expression by TGF-β in HSC (Inagaki *et al.*, 1994), whereas in hepatocytes, Sp3 binds the same element, but with little transactivating activity. In HSC, but not in hepatocytes, activated Smad3 physically interacts with Sp1, thus providing a link between TGF-β signaling and matrix upregulation in specific cell types (Inagaki *et al.*, 2001b). Moreover, constitutive phosphorylation and nuclear localization of Smad3 were found in a clone of HSC exhibiting high levels of collagen and PAI-1 expression together with a poor response to TGF-β, confirming the relevance of this pathway for the upregulation of extracellular matrix production in HSC (Inagaki *et al.*, 2001a).

On addition to Smads, other signaling pathways are activated by TGF-β in HSC, with potential involvement in extracellular matrix synthesis and other activities. TGF-β1 induces activation of the ERK pathway through the sequential activation of Ras, Raf-1, and MEK (Reimann *et al.*, 1997). Interestingly, in the absence of TGF-β, members of this pathway have divergent actions on collagen gene expression. While dominant-negative ERK, which inhibits ERK signaling, reduced procollagen α1(I) transcription, dominant-negative *raf* increased collagen reporter gene expression. These results indicate that activation of *ras* exerts negative signals on collagen transcription via a Raf-dependent, ERK-independent pathway, whereas ERK positively modulates expression of the collagen gene (Davis *et al.*, 1996). In addition, the action of ERK was dependent on Sp-1- and NF-1-binding sites of the proximal 5′-untranslated region of the procollagen α1(I) promoter, whereas the Raf inhibitory signal was mapped upstream (Davis *et al.*, 1996). Other Smad-independent signaling pathways contribute to the biologic effects of TGF-β in HSC, in particular the exposure of the cells to reactive oxygen intermediates. TGF-β induction of procollagen α1(I) mRNA is mimicked by the addition of H_2O_2, and is prevented by the addition of the antioxidant PDTC, or of catalase to cultured HSC. DeBleser and co-workers (1999) have proposed a model according to which glutathione levels on HSC enable the cells to discriminate between exogenously produced H_2O_2, and therefore oxidative stress, and H_2O_2 generated as a signaling molecule, such as in response to TGF-β. In particular, exposure to TGF-β would be associated with high intracellular levels of H_2O_2, which could be removed efficiently by catalase, yielding high levels of glutathione and allowing autocrine secretion of TGF-β. On the contrary, exogenously added H_2O_2, simulating a condition of oxidative stress, generates low intracellular levels of H_2O_2, which is catabolized by glutathione peroxidase, resulting in low glutathione levels and interruption of a TGF-β autocrine pathway (De Bleser *et al.*, 1999).

IV. SEVEN TRANSMEMBRANE DOMAIN RECEPTORS

Several factors acting through receptors belonging to the "seven transmembrane domain" family are active in HSC. Due to similarities in biologic actions and signaling, these receptors may be divided, for the purpose of this review, into two subgroups: receptors for "vasoconstrictors" and chemokine receptors.

Endothelins, angiotensin II, vasopressin, and thrombin, although generally referred to as "vasoconstrictors," promote profibrogenic actions and are considered pleiotropic cytokines when viewed in the context of the chronic wound-healing process. Endothelin (ET)-1, a potent vasoactive 21 amino acid peptide secreted by endothelial, as well as other cell types, exerts a multifunctional role in a variety of tissues and cells (Simonson, 1993; Simonson and Dunn, 1991; Yanagisawa and Masaki, 1989). Endothelins (ET-1, ET-2, and ET-3) bind to G-protein-coupled receptors, termed ET_A, ET_B, and to a still not well-characterized ET_C receptor. The ET_A receptor binds ET-1 with a higher affinity than ET-3, the ET_B receptor displays similar affinity for both peptides, and the ET_C receptor exhibits a higher affinity for ET-3 than ET-1. Activation of ET receptors originates intracellular signals leading to differentiation, proliferation, growth inhibition, cell contraction, and a variety of metabolic effects. However, it is increasingly clear that activation of a given ET receptor does not necessarily lead to identical biologic effects in different cell types. This intriguing and peculiar feature is currently explained in terms of receptor structure, as the ET receptor is a typical heptahelical G-protein-coupled receptor. In ligand binding to the heptahelical receptor, the receptor has two functions, i.e., "address" (*address domain:* regulates the affinity for the ligand) and "message" (*message domain:* regulates the activation of different G proteins and their downstream effectors). A different part of the ligand structure also corresponds to each domain of the receptor. Although ET receptors are currently classified according to their affinity for the three known forms of ET, it has been proposed that they should be classified according to their *message* domain. Indeed, each Ga protein acts on different target molecules, resulting in different responses. In addition, the activation of each Ga protein presumably depends on its intracellular level. Therefore, although the same ET-receptor is activated by the same ligand, the resulting final response may be different from cell to cell (Masaki *et al.*, 1999). This complex organization of ET-related intracellular signaling may explain at least some of the discrepancies in the biologic effects of this class of peptides reported in the literature.

ET expression in liver tissue is increased markedly during chronic hepatic fibrogenic disorders both in humans (Alam *et al.*, 2000; Pinzani *et al.*, 1996a) and in animal models (Tieche *et al.*, 2001). Morphological studies have clearly indicated that ET-1 (both at mRNA and protein levels) is overexpressed in different cellular elements present within cirrhotic liver tissue (Pinzani *et al.*, 1996a; Rockey *et al.*, 1998). Furthermore, expression of ET-1 and its receptors in liver tissue correlates with the severity of the disease and its complications in cirrhotic patients (Alam *et al.*, 2000; Leivas *et al.*, 1998). *In vitro* studies have confirmed that both sinusoidal endothelial cells and activated HSC are able to synthesize ET-1 (Housset *et al.*, 1993; Pinzani *et al.*, 1996a; Rieder *et al.*, 1991). In activated HSC, ET-1 synthesis and release are promoted by agonists such as angiotensin II, PDGF, TGF-β, reactive oxygen species, and ET-1 itself (Gabriel *et al.*, 1998; Housset *et al.*, 1993; Leivas *et al.*, 1998; Pinzani *et al.*, 1996a; Rieder *et al.*, 1991; Rockey *et al.*, 1998). Moreover, ET-induced ET-1 synthesis in HSC is regulated through the modulation of endothelin-converting enzyme-1 (ECE-1) rather than by modulation of the precursor pre-proET-1 (Shao *et al.*, 1999).

Overall, it is increasingly evident that the process of HSC activation and phenotypical modulation is characterized by a close and complex modulation of the ET system. The ability to synthesize and release ET-1 is associated with a progressive shift in the relative predominance of ET_A and ET_B receptors observed during serial subculture: ET_A are predominant in the early phases of activation, whereas ET_B become increasingly more abundant in "myofibroblast-like" cells

(Pinzani *et al.*, 1996a; Reinehr *et al.*, 1998). This shift in the relative receptor densities may be directed at differentiating the possible paracrine and autocrine effects of ET-1 on HSC during the activation process. Indeed, when HSC are provided with a majority of ET_A receptors (early phases of activation), stimulation with ET-1 causes a dose-dependent increase in cell growth, ERK activity, and expression of c-*fos*. These effects, likely related to the activation of the Ras-ERK pathway, are completely blocked by pretreatment with BQ-123, a specific ET_A receptor antagonist (Pinzani *et al.*, 1996a), and are in agreement with studies performed in other vascular pericytes, such as glomerular mesangial cells (Wang *et al.*, 1994). Conversely, in later stages of activation, when the number of ET_B receptors increases, ET-1 induces a prevalent antiproliferative effect linked to the activation of this receptor subtype (Mallat *et al.*, 1995b). In this setting, activation of the ET_B receptor stimulates the production of PGs, leading to an increase in intracellular cAMP, which in turn reduces the activation of both ERK and JNK (Mallat *et al.*, 1996). In addition, both cAMP and PGs upregulate ET_B-binding sites, thus suggesting the possibility of a positive feedback regulatory loop.

Further studies have clarified that ET-1-induced PGs synthesis in human activated HSC is due to an increased enzymatic activity of cycloxygenase (COX)-2 and that this increase is dependent on the activation of NF-κB by ET-1 (Gallois *et al.*, 1998). Additional qualitative and quantitative changes in the expression of ET-receptors may be due to the action of proinflammatory and profibrogenic stimuli present in the hepatic tissue microenvironment, including TGF-β1, reactive oxygen species, and endotoxins (Gabriel *et al.*, 1998, 1999; Gandhi *et al.*, 2000).

Because ET-1 induces a dose-dependent increase in intracellular-free calcium coupled with cell contraction in activated HSC (Pinzani *et al.*, 1992b, 1996a), an autocrine action of this peptide has been postulated, leading to the contraction of scar tissue and portal hypertension. At least in human HSC, the ET-1-induced $[Ca^{2+}]_i$ increase, coupled with cell contraction, occurs at any stage of cellular activation (Pinzani *et al.*, 1996a). Because HSC contraction is always blocked by ET_A receptor antagonists and never reproduced by selective ET_B agonists, it is conceivable that the signaling pathways regulating HSC contraction require the activation of a small number of ET_A receptors and/or are somehow divergent from those regulating cell growth. As described in other contractile cell types, it is conceivable that ET-1-induced HSC contraction occurs through phosphorylation of the myosin light chain by Ca^{2+}/calmodulin-dependent myosin light chain kinase. Additionally, a portion of ET-1-induced contraction is partially mediated by the calcium-independent activation of the small G protein RhoA and of a downstream target, Rho-kinase. Accordingly, activation of Rho-kinase by ET-1 has been reported in rat HSC (Tangkijvanich *et al.*, 2001).

Circulating levels of angiotensin II (A-II), a powerful vasoconstrictor, are frequently increased in cirrhotic patients and have been implicated in the circulatory disturbances typical of this clinical condition. However, A-II, in addition to its action as a vasoconstrictor, is provided with biologic properties potentially relevant for the progression of chronic fibrogenic disorders. These include an increase in cell proliferation and cell hypertrophy; accordingly, A-II may also be considered a pleiotropic cytokine. Bataller and co-workers (2000) reported that activated human HSC express A-II receptors of the AT1 subtype and that an increased expression of this type of receptor may represent a feature of HSC activation. Stimulation with A-II elicits a marked dose-dependent increase in $[Ca^{2+}]_i$ concentration associated with rapid cell contraction, in agreement with a previous report (Pinzani *et al.*, 1992b). Moreover, A-II stimulates DNA synthesis and cell growth. The involvement of AT1 receptors in these effects of A-II is confirmed by their complete abrogation following preincubation with the AT1 receptor antagonist losartan. Although the antifibrogenic effects of pharmacological AT1 receptor blockade appear promising in other conditions, i.e., myocardiosclerosis (Lim *et al.*, 2001), no information is currently available for hepatic fibrogenesis. Analogous effects on HSC biology have been described for arginine vasopressin (AVP). Human-activated HSC express V1 receptors, and stimulation with AVP elicits a dose-dependent increase in intracellular $[Ca^{2+}]_i$ coupled with cell contraction. Moreover, AVP increases ERK activity, DNA synthesis, and cell growth (Bataller *et al.*, 1997).

The serine protease thrombin (THR) regulates platelet aggregation, endothelial cell activation, and other important responses in vascular biology and in acute and chronic wound repair. Although THR is a protease, it acts as a traditional hormone or as a pleiotropic cytokine based on the nature of its receptors, protease-activated receptors or PARs. PARs are G-protein-coupled receptors that use a fascinating mechanism to convert an extracellular proteolytic cleavage event into a transmembrane signal: these receptors carry their own ligands, which remain cryptic until unmasked by receptor cleavage (Coughlin, 2000). Four PARs are known in mouse and human: human PAR1, PAR3, and PAR4 can be activated by THR, whereas PAR2 is activated by trypsin and tryptase, as well as by coagulation factors VIIa and Xa, but not by THR. Studies by Marra and co-workers have shown that the expression of PAR1 is increased markedly in chronic fibrogenic disorders involving liver. In addition, human HSC express PAR1, and this expression increases during HSC activation (Marra *et al.*, 1995a, 1998). Stimulation of human HSC with THR induces cell contraction (Pinzani *et al.*, 1992b), proliferation (Racine-Samson *et al.*, 1997), synthesis, and release of chemokines such as MCP-1 (Marra *et al.*, 1998) or platelet-activating factor (Pinzani *et al.*, 1994a). The signaling mechanisms specifically regulating THR action in HSC have not been reported thus far.

The chemokine system is a major modulator of many critical functions, both in physiologic and pathologic conditions, including inflammation, development, leukocyte trafficking, angiogenesis, and cancer (Rossi and Zlotnik, 2000). In HSC, secretion of several chemokines belonging to different subclasses regulates the recruitment of inflammatory cells to sites of damage (Marra, 1999). These findings demonstrate a first line of interaction between the inflammatory and the reparative phases of the wound-healing response.

However, chemokines also directly regulate the behavior of cells involved in tissue repair. In fact, glomerular mesangial cells, or smooth muscle cells, express chemokine receptors, which mediate functional responses that are relevant for tissue fibrosis (Gerard and Rollins, 2001; Romagnani *et al.*, 1999). In agreement with these observations, the CC chemokine monocyte chemoattractant protein-1 (MCP-1) induces concentration-dependent chemotaxis of HSCs (Marra, 1999). Interestingly, HSC do not express the receptor CCR2, which mediates the biologic responses of this chemokine in leukocytes. Nevertheless, incubation of stellate cells with MCP-1 is associated with a rise in intracellular calcium concentration and activation of intracellular signaling pathways, including protein tyrosine phosphorylation, and activation of PI 3-K, which are necessary to stimulate cell migration. The receptor(s) responsible for these effects of MCP-1 has not been identified yet, but a possible candidate could be CCR11, which can bind MCP-1 with high affinity (Schweickart *et al.*, 2000). The chemokine receptor CXCR3, which binds the ligands IP-10, Mig, and I-TAC, is the first chemokine receptor to be identified in HSC. Its expression increases after a few days in culture on uncoated plastic, and receptor activation is associated with increased cell migration, establishing that CXCR3 is functional in HSC.

Interaction of CXCR3 with its ligands leads to activation of the Ras/ERK cascade through a Src-dependent pathway and to activation of PI 3-K and its downstream kinase Akt. Interestingly, CXCR3 activation increases cell proliferation in glomerular mesangial cells but not in HSC. This discrepancy in the biologic actions elicited by this receptor is accompanied by a different time course of ERK activation in the two cell types. Indeed, in HSC, ERK is activated only transiently, whereas in mesangial cells, a biphasic activation occurs, including a late peak at 15–24 h (Schweickart *et al.*, 2000). The molecular mechanisms underlying this different behavior in the two cell types have not yet been elucidated. We have obtained evidence for expression of the CCR7 chemokine receptor by activated HSC (A. Bonacchi and F. Marra, unpublished observations). The biologic effects and downstream signaling triggered by activation of this receptor are currently being studied.

V. TUMOR NECROSIS FACTOR RECEPTOR SUPERFAMILY

The receptors for tumor necrosis factor (TNF) belong to a superfamily that includes several transmembrane molecules that bind cytokines or other ligands and orphan receptors and are characterized by similar, cysteine-rich extracellular domains (Ashkenazi and Dixit, 1998). This family may be divided into two large subgroups depending on the presence or absence of a "death domain," which allows physical association with molecules containing the same domain. TNFR1, Fas, and the p75 receptor for the nerve growth factor belong to the group of "death domain" receptors, whereas TNFR2 and CD40 lack this domain (Ledgerwood *et al.*, 1999). TNFR1 plays a major role in mediating the biologic actions of TNF because animals lacking this receptor, unlike those lacking TNFR2, are unable to mount inflammatory reactions in response to TNF (Rothe *et al.*, 1993). TNF activates pathways that regulate gene transcription and inflammation, and others leading to cell death. Binding of TNF induces homotrimerization of TNFR-1, which binds to the death domain containing protein TRADD (Ledgerwood *et al.*, 1999). Association of TRADD with RIP and TRAF2 is responsible for the downstream activation of NF-κB and JNK, respectively, leading to the regulation of gene transcription. Activation of other members of the MAPK family, including ERK1/2 and p38MAPK, is also elicited by TNF. In contrast, cell death is mediated by the interaction of TRADD with FADD, which initiates a cascade of proteases, known as caspases, leading to the apoptotic effect. It is important to note that most cells become sensitive to TNF-induced apoptosis only when protein synthesis or RNA transcription is inhibited.

TNF has many important effects on HSC relevant to the pathophysiology of liver fibrosis. TNF participates in the activation process (Knittel *et al.*, 1997), but has an inhibitory effect on the expression and synthesis of type I collagen (Armendariz-Borunda *et al.*, 1992) and on the proliferation of HSC (Gallois *et al.*, 1998; Knittel *et al.*, 1997). Remarkably, TNF is a critical factor for the "proinflammatory" role of HSC because it upregulates the expression and secretion of several cytokines and chemokines (for a review, see Gallois *et al.*, 1998).

Unlike other cytokine receptors, quiescent HSC express mRNA transcripts for TNFR1, and TNF efficiently binds to the cell surface (Hellerbrand *et al.*, 1998). However, the receptor expressed by quiescent cells seems to be only partially functional because exposure of nonactivated cells to the ligand does not result in activation of NF-κB due to the inability to degrade the inhibitory protein I-κBα in quiescent cells. Accordingly, increased expression of NF-κB-regulated proteins, such as the adhesion molecule ICAM-1, is observed only in activated cells. Interestingly, the JNK pathway may be activated by TNF in both quiescent and activated HSC, indicating that the block in NF-κB activation in quiescent cells occurs at a postreceptor level. Activation of NF-κB plays a pivotal role in mediating the proinflammatory effects of TNF on HSC. Interference with NF-κB activation by proteasome degradation inhibitors or an I-κB super-repressor blocks the expression of several cytokines, chemokines, and adhesion molecules, including interleukin-6, MCP-1, CINC, MIP-2, and ICAM-1 (Efsen *et al.*, 2001; Marra, 1999).

Interestingly, novel Rel-like proteins may contribute to the NF-κB DNA-binding complex observed in activated HSC and upregulated by exposure to TNF (Elsharkawy *et al.*, 1999). NF-κB activation is also required to induce cyclooxygenase 2 (COX-2), which mediates the growth inhibitory effect of TNF-α in these cells and contributes to chemokine expression (Efsen *et al.*, 2001; Hellerbrand *et al.*, 1998). NF-κB is also an important mediator of cell survival (Beg *et al.*, 1995), and exposure of HSC to TNF, together with inhibition of NF-κB activation, resulted in apoptosis (Lang *et al.*, 2000). However, when used alone, TNF actually protects HSC from apoptosis via the reduction of Fas-ligand expression (Saile *et al.*, 1999). However, activation of stress-activated protein kinases, such as JNK or p38, may be involved in the phenotypic transition from quiescent to activated HSC (Reeves *et al.*, 2000) and in the expression of matrix metalloproteinases (Poulos *et al.*, 1997). The inhibition of type I collagen expression is mediated by a complex mechanism. Preincubation of HSC with pertussis toxin abolishes the inhibitory effects of TNF on procollagen α1(I) mRNA expression, and ceramide mimicks the effects of this cytokine, indicating the involvement of a pathway requiring a G protein and sphingomyelin/ceramide (Hernandez-Munoz *et al.*, 1997). Moreover, several transcription factors and regulatory elements are implicated in the inhibitory effects of TNF, including a tissue-specific regulatory region, increased binding of p20C/EBPβ and C/EBPδ, and reduced binding of Sp1 (Hernandez *et al.*, 2000; Houglum *et al.*, 1998; Iraburu *et al.*, 2000).

VI. OTHER CYTOKINE RECEPTORS

The actions of interleukin-1 on HSC are remarkably similar to those elicited by TNF regarding its effects on proinflammatory molecules, and these actions are mediated by the activation of NF-κB (Marra, 1999). Similarly, quiescent HSC express IL-1 receptors, but the activation of NF-κB is restricted to activated cells (Hellerbrand *et al.*, 1998). IL-1 also has inhibitory effects on procollagen α1(I) expression, which occurs at a posttranscriptional level (Armendariz-Borunda *et al.*, 1992), and stimulates HSC proliferation (Matsuoka *et al.*, 1989). However, the net effects of this cytokine are profibrogenic because administration of the IL-1 receptor antagonist reduces matrix deposition (Mancini *et al.*, 1994).

Receptors for interferon-γ (IFN-γ) are ubiquitous, although no studies have examined their expression during HSC activation. A critical step in IFN-γ signaling is activation of the latent transcription factor STAT1, which upon dimerization migrates to the nucleus where it regulates transcription (Darnell, 1997). IFN-γ downregulates activation, matrix synthesis, and proliferation of HSC *in vitro* and *in vivo* (Darnell, 1997; Mallat *et al.*, 1995a; Ramadori *et al.*, 1992; Rockey and Chung, 1994, 1995). In addition, particularly if used in combination with other cytokines or LPS, IFN-γ stimulates the expression of inducible NO synthase (Rockey and Chung, 1994). However, the molecular mechanisms underlying these effects have not been studied. IFN-γ upregulates the expression of chemokines such as MCP-1 in HSC (Marra, 1999), and this effect is differentially regulated as compared to other cytokines (Efsen *et al.*, 2001). Surprisingly, when human HSC are incubated with IFN-γ for prolonged periods and exposed to mitogens such as serum or PDGF, an increase in cell proliferation is observed, which is dependent on the synergistic activation of STAT1α (Gressner and Althaus, 1990; Marra *et al.*, 1996).

HSC also respond to other cytokines, including IL-10, IL-6, and oncostatin M (Greenwel *et al.*, 1995; Levy *et al.*, 2000; Wang *et al.*, 1998). IL-10 inhibits procollagen α(1) expression at the transcriptional level and may have important antifibrogenic properties (Wang *et al.*, 1998). Specific information on intracellular signals generated by these cytokines in HSC has not been reported.

VII. COOPERATION BETWEEN GROWTH FACTOR RECEPTOR AND INTEGRIN SIGNALING

In recent years, studies have begun to address the multiple potential interactions of cells with the microenvironment. The major advances reflect elucidation of the fine-tuning occurring upon cell adhesion and the consequent cytoskeletal organization. Key elements in these responses are (a) the specificity of integrin receptors and their downstream signaling, (b) the cross talk between integrin and cytokine signaling, and (c) the relationship between the aforementioned mechanisms and the organization and tension of the cytoskeleton.

Binding of cell to ECM is mediated, at least in part, by cell surface receptors belonging to the integrin family. Integrins are heterodimeric transmembrane proteins that consist of an α subunit and a β subunit. Activated HSC express several integrin β_1-associated α subunits. A particularly high expression has been demonstrated for $\alpha_1\beta_1$ and $\alpha_2\beta_1$ (Carloni *et al.*, 1996; Racine-Samson *et al.*, 1997). Protein phosphorylation is one of the earliest events detected in response to integrin stimulation, and in particular tyrosine phosphorylation is a common response to integrin engagement in many cell types (Burridge *et al.*, 1992; Schaller and Parsons, 1993). Several protein tyrosine kinases have been implicated in integrin-related signaling events. Focal adhesion kinase (FAK) plays a central role in integrin-mediated signal transduction (Schaller *et al.*, 1995). In addition to its activation by integrins, FAK is also activated by several growth factors. FAK phosphorylation is stimulated by mitogenic neuropeptides such as bombesin and vasopressin and by PDGF and ET-1. Indeed, many of the signaling proteins regulated by integrins are also involved in signal transduction pathways activated by growth factor receptors, indicating that synergistic interactions between growth factor and integrin signaling pathways are involved in the regulation of cell proliferation, adhesion, and migration. For example, PI 3-K coimmunoprecipitates with tyrosine-phosphorylated FAK in response to cell adhesion. Tyrosine-phosphorylated PLCγ may be a transducer molecule in integrin-mediated signaling pathways. Adhesion of human HSC to ECM proteins does not result in PLCγ tyrosine phosphorylation (Carloni *et al.*, 1997). Nevertheless, adhesion of HSC induces interactions between PLCγ and cellular proteins undergoing tyrosine phosphorylation, one of which has been identified as FAK, suggesting that adhesion of HSC is followed by the recruitment of PLCγ to phosphorylated FAK. Because PLCγ also physically associates with the PDGF receptor, there may be cross talk between this receptor and proteins of the focal adhesion complex. Indeed, stimulation of HSC with PDGF-BB leads to clustering of PDGF-β receptor subunits in areas possibly corresponding to focal adhesion complexes. Along these lines on autophosphorylation, PDGF receptors are codistributed with FAK, thus suggesting a potential functional cross talk between these signaling molecules (Carloni *et al.*, 2000). In addition, experimental evidence indicates that Ras plays a key role in the cross talk between the PDGF receptor and FAK in human HSC (Carloni *et al.*, 2000). Therefore, it appears that multiple receptor systems can synergize with integrins to regulate biological phenomena such as cell proliferation and cell motility, and the signaling proteins activated by these synergistic agents are common to different receptor pathways. However, more recent advances indicate that cell adhesion and the consequent signaling events are necessary but not sufficient to provide a complete control of cell functions, and in particular their response to growth factors and cytokines. Indeed, a modern concept, defined in its complexity with the term "cellular tensegrity architecture" (Huang and Ingber, 1999), implies that in the presence of growth factor stimulation and adequate adhesion to the substratum, the cell is unable to progress into the cell cycle if restricted in its spreading. This view is supported by several observations indicating that lack of cytoskeletal tension, obtainable only with full cell spreading, is equivalent to cytoskeletal disruption and leads to the inability to enter the cell cycle. This general key mechanism appears relevant for the regulation of the molecular signaling cascades operating in the context of the structural and mechanical complexity of living tissue and their pathophysiological alterations, including hepatic fibrogenesis.

REFERENCES

Alam, I., Bass, N. M., Bacchetti, P., and Rockey, D. C. (2000). Hepatic tissue endothelin-1 levels in chronic liver disease correlate with disease severity and ascites. *Am. J. Gastroenterol.* **95**, 199–203.

Armendariz-Borunda, J., Katayama, K., and Seyer, J. M. (1992). Transcriptional mechanisms of type I collagen gene expression are differentially regulated by interleukin-1 beta, tumor necrosis factor alpha, and transforming growth factor beta in Ito cells. *J. Biol. Chem.* **267**, 14316–14321.

Ashcroft, G. S., Yang, X., Glick, A. B., Weinstein, M., Letterio, J. L., Mizel, D. E., Anzano, M., Greenwell-Wild, T., Wahl, S. M., Deng, C., and Roberts, A. B. (1999). Mice lacking Smad3 show accelerated wound healing and an impaired local inflammatory response. *Nature Cell Biol.* **1**, 260–266.

Ashkenazi, A., and Dixit, V. M. (1998). Death receptors: Signaling and modulation. *Science* **281**, 1305–1308.

Bachem, M. G., Meyer, D., Melchior, R., Sell, K. M., and Gressner, A. M. (1992). Activation of rat liver perisinusoidal lipocytes by transforming growth factors derived from myofibroblast-like cells: A potential mechanism of self perpetuation in liver fibrogenesis. *J. Clin. Invest.* **89**, 19–27.

Bataller, R., Nicolas, J. M., Gines, P., Esteve, A., Nieves Gorbig, M., Garcia-Ramallo, E., Pinzani, M., Ros, J., Jimenez, W., Thomas, A. P., Arroyo, V., and Rodes, J. (1997). Arginine vasopressin induces contraction and stimulates growth of cultured human hepatic stellate cells. *Gastroenterology* **113**, 615–624.

Bataller, R., Gines, P., Nicolas, J. M., Gorbig, M. N., Garcia-Ramallo, E., Gasull, X., Bosch, J., Arroyo, V., and Rodes, J. (2000). Angiotensin II induces contraction and proliferation of human hepatic stellate cells. *Gastroenterology* **118**, 1149–1156.

Beg, A., Sha, W., Bronson, R., Ghosh, S., and Baltimore, D. (1995). Embryonic lethality and liver degeneration in mice lacking the RelA component of NF-κB. *Nature* **376**, 167–170.

Benedetti, A., Di Sario, A., Casini, A., Ridolfi, F., Bendia, E., Pigini, P., Tonnini, C., D'Ambrosio, L., Feliciangeli, G., Macarri, G., and Svegliati Baroni, G. (2001). Inhibition of the Na+/H+ exchanger reduces rat hepatic stellate cell activity and liver fibrosis: an in vitro and in vivo study. *Gastroenterology* **120**, 545–556.

Branton, M. H., and Kopp, J. B. (1999). TGF-β and fibrosis. *Microbes Infect.* **1**, 1349–1365.

Burridge, K., Turner, C. E., and Romer, L. H. (1992). Tyrosine phosphorylation of paxillin and pp125 FAK accompanies cell adhesion to extracellular matrix: A role in cytoskeletal assembly. *J. Cell Biol.* **119**, 893–903.

Caligiuri, A., Marra, F., Failli, P., DeFranco, R. M. S., Romanelli, R. G., Gentilini, P., and Pinzani, M. (2001). Cross-talk between phosphatidylinositol 3-kinase (PI3-K) and the Na+/H+ exchanger in hepatic stellate cells (HSC): Effect of canrenone. *J. Hepatol.* **34**(Suppl. 1), 84.

Carloni, V., Pinzani, M., Giusti, S., Romanelli, R. G., Parola, M., Bellomo, G., Failli, P., Hamilton, A. D., Sebti, S. M., Laffi, G., and Gentilini, P. (2000). Tyrosine phosphorylation of focal adhesion kinase by PDGF is dependent on Ras in human hepatic stellate cells. *Hepatology* **31**, 131–140.

Carloni, V., Romanelli, R. G., Pinzani, M., Laffi, G., and Gentilini, P. (1996). Expression and function of integrins receptors for collagen and laminin in cultured human hepatic stellate cells. *Gastroenterology* **110**, 1127–1136.

Carloni, V., Romanelli, R. G., Pinzani, M., Laffi, G., and Gentilini, P. (1997). Focal adhesion kinase and phospholipase Cg involvement in adhesion and migration of human hepatic stellate cells. *Gastroenterology* **112**, 522–531.

Castilla, A., Prieto, J., and Fausto, N. (1991). Transforming growth factors beta 1 and alpha in chronic liver disease: Effects of interferon alfa therapy. *N. Engl. J. Med.* **324**, 933–940.

Claesson-Welsh, L. (1994). Platelet-derived receptor signals. *J. Biol. Chem.* **269**, 32023–32026.

Cohen, G. B., Ren, R., and Baltimore, D. (1995). Modular binding domain in signal transduction proteins. *Cell* **80**, 237–248.

Coughlin, S. R. (2000). Thrombin signaling and protease-activated receptors. *Nature* **407**, 258–264.

Darnell, J. E. J. (1997). STATs and gene regulation. *Science* **277**, 1630–1633.

Datta, S. R., Brunet, A., and Greenberg, M. E. (1999). Cellular survival: A play in three Akts. *Genes Dev* **13**, 2905–2927.

Davis, B. H., Chen, A. P., and Beno, D. W. A. (1996). Raf and mitogen-activated protein kinase regulate stellate cell collagen gene expression. *J. Biol. Chem.* **271**, 11039–11042.

De Bleser, P. J., Xu, G., Rombouts, K., Rogiers, V., and Geerts, A. (1999). Glutathione levels discriminate between oxidative stress and transforming growth factor-beta signaling in activated rat hepatic stellate cells. *J. Biol. Chem.* **274**, 33881–33887.

Di Sario, A., Bendia, E., Svegliati Baroni, G., Ridolfi, F., Bolognini, L., Feliciangeli, G., Jezequel, A. M., Orlandi, F., and Benedetti, A. (1999). Intracellular pathways mediating Na+/H+ exchange activation by platelet-derived growth factor in rat hepatic stellate cells. *Gastroenterology* **116**, 1155–1166.

Di Sario, A., Svegliati Baroni, G., Bendia, E., D'Ambrosio, L., Ridolfi, F., Marileo, J. R., Jezequel, A. M., and Benedetti, A. (1997). Characterization of ion transport mechanisms regulating intracellular pH in hepatic stellate cells. *Am. J. Physiol.* **273**, G39–G48.

Di Sario, A., Svegliati Baroni, G., Bendia, E., Ridolfi, F., Saccomanno, S., Ugili, L., Trozzi, L., Marzioni, M., Jezequel, A. M., Macarri, G., and Benedetti, A. (2001). Intracellular pH regulation and Na+/H+ exchange activity in human hepatic stellate cells: Effects of platelet-derived growth factor, insulin-like growth factor 1 and insulin. *J. Hepatol.* **34**, 378–385.

Dooley, S., Delvoux, B., Lahme, B., Mangasser-Stephan, K., and Gressner, A. M. (2000). Modulation of transforming growth factor beta response and signaling during transdifferentiation of rat hepatic stellate cells to myofibroblasts. *Hepatology* **31**, 1094–1106.

Efsen, E., Pastacaldi, S., Bonacchi, A., Valente, A. J., Wenzel, U. O., Tosti-Guerra, C., Pinzani, M., Laffi, G., Abboud, H. E., Gentilini, P., and Marra, F. (2001). Agonist-specific regulation of monocyte chemoattractant protein-1 expression by cyclooxygenase metabolites in hepatic stellate cells. *Hepatology* **33**, 713–721.

Elsharkawy, A. M., Wright, M. C., Hay, R. T., Arthur, M. J., Hughes, T., Bahr, M. J., Degitz, K., and Mann, D. A. (1999). Persistent activation of nuclear factor-kappaB in cultured rat hepatic stellate cells involves the induction of potentially novel Rel-like factors and prolonged changes in the expression of IkappaB family proteins. *Hepatology* **30**, 761–769.

Failli, P., Ruocco, C., DeFranco, R., Caligiuri, A., Gentilini, A., Gentilini, P., Giotti, A., and Pinzani, M. (1995). The mitogenic effect of platelet-derived growth factor in human hepatic stellate cells requires calcium influx. *Am. J. Physiol.* **269**, C1133–C1139.

Friedman, S. L., Yamasaki, G., and Wong, L. (1994). Modulation of transforming growth factor beta receptors of rat lipocytes during the hepatic wound healing response: Enhanced binding and reduced gene expression accompany cellular activation in culture and in vivo. *J. Biol. Chem.* **269**, 10551–10558.

Friedman, S. L. (1999). Cytokines and fibrogenesis. *Semin. Liver Dis.* **19**(2), 129–140.

Friedman, S. L. (2000). Molecular regulation of hepatic fibrosis, an integrated cellular response to tissue injury. *J. Biol. Chem.* **275**, 2247–2250.

Gabriel, A., Kuddus, R. H., Rao, A. S., and Gandhi, C. R. (1999). Down-regulation of endothelin receptors by transforming growth factor beta1 in hepatic stellate cells. *J. Hepatol.* **30**, 440–450.

Gabriel, A., Kuddus, R. H., Rao, A. S., Watkins, W. D., and Gandhi, C. R. (1998). Superoxide-induced changes in endothelin (ET) receptors in hepatic stellate cells. *J. Hepatol.* **29**, 614–627.

Gallois, C., Habib, A., Tao, J., Moulin, S., Maclouf, J., Mallat, A., and Lotersztajn, S. (1998). Role of NF-kappaB in the antiproliferative effect of endothelin-1 and tumor necrosis factor-alpha in human hepatic stellate cells. Involvement of cyclooxygenase-2. *J. Biol. Chem.* **273**, 23183–23190.

Gandhi, C. R., Uemura, T., and Kuddus, R. (2000). Endotoxin causes up-regulation of endothelin receptors in cultured hepatic stellate cells via nitric oxide-dependent and -independent mechanisms. *Br. J. Pharmacol.* **131**, 319–327.

Gentilini, A., Marra, F., Gentilini, P., and Pinzani, M. (2000). Phosphatidylinositol-3 kinase and extracellular signal-regulated kinase mediate the chemotactic and mitogenic effects of insulin-like growth factor-I in human hepatic stellate cells. *J. Hepatol.* **32**, 227–234.

Gentilini, A., Romanelli, R. G., Marra, F., Gentilini, P., and Pinzani, M. (2001). IGF-I is a survival factor for human hepatic stellate cells (HSC): Involvement of PI 3-K and c-Akt. *J. Hepatol.* **34**(Suppl. 1), 7.

George, J., Roulot, D., Koteliansky, V. E., and Bissell, D. M. (1999). In vivo inhibition of rat stellate cell activation by soluble transforming growth factor beta type II receptor: A potential new therapy for hepatic fibrosis. *Proc. Natl. Acad. Sci. USA* **96**, 12719–12724.

Gerard, C., and Rollins, B. J. (2001). Chemokines and disease. *Nature Immunol.* **2**, 108–115.

Greenwel, P., Iraburu, M. J., Reyes-Romero, M., Meraz-Cruz, N., Casado, E., Solis-Herruzo, J. A., and Rojkind, M. (1995). Induction of an acute phase response in rats stimulates the expression of alpha1(I) procollagen messenger ribonucleic acid in their livers: Possible role of interleukin-6. *Lab. Invest.* **72**, 83–91.

Gressner, A. M., and Althaus, M. (1990). Effects of murine recombinant interferon-gamma on rat liver fat storing cell proliferation, cluster formation and proteoglycan synthesis. *Biochem. Pharmacol.* **40**, 1953–1962.

Hellerbrand, C., Jobin, C., Licato, L. L., Sartor, R. B., and Brenner, D. A. (1998). Cytokines induce NF-kappaB in activated but not in quiescent rat hepatic stellate cells. *Am. J. Physiol.* **275**, G269–G278.

Hernandez, I., de la Torre, P., Rey-Campos, J., Garcia, I., Sanchez, J. A., Munoz, R., Rippe, R. A., Munoz-Yague, T., and Solis-Herruzo, J. A. (2000). Collagen alpha1(I) gene contains an element responsive to tumor necrosis factor-alpha located in the 5' untranslated region of its first exon. *DNA Cell Biol.* **19**(6), 341–352.

Hernandez-Munoz, I., de la Torre, P., Sanchez-Alcazar, J. A., Garcia, I., Santiago, E., Munoz-Yague, M. T., and Solis-Herruzo, J. A. (1997). Tumor necrosis factor alpha inhibits collagen alpha 1(I) gene expression in rat hepatic stellate cells through a G protein. *Gastroenterology* **113**, 625–640.

Houglum, K., Buck, M., Kim, D. J., and Chojkier, M. (1998). TNF-alpha inhibits liver collagen-alpha 1(I) gene expression through a tissue-specific regulatory region. *Am. J. Physiol.* **274**, G840–G847.

Housset, C. N., Rockey, D. C., and Bissel, D. M. (1993). Endotelin receptors in rat liver: Lipocytes as a contractile target for endothelin 1. *Proc. Natl. Acad. Sci. USA* **90**, 9266–9270.

Huang, S., and Ingber, D. E. (1999). The structural and mechanical complexity of cell-growth control. *Nature Cell Biol.* **1**, E131–E138.

Inagaki, Y., Mamura, M., Kanamaru, Y., Greenwel, P., Nemoto, T., Takehara, K., Ten Dijke, P., and Nakao, A. (2001a). Constitutive phosphorylation and nuclear localization of Smad3 are correlated with increased collagen gene transcription in activated hepatic stellate cell. *J. Cell Physiol.* **187**, 117–123.

Inagaki, Y., Nemoto, T., Nakao, A., ten Dijk, P., Kobayashi, K., Takehara, K., and Greenwel, P. (2001b). Interaction between GC box binding factors and Smad proteins modulates cell lineage-specific alpha 2(I) collagen gene transcription. *J. Biol. Chem.* **276**, 16573–16579.

Inagaki, Y., Truter, S., and Ramirez, F. (1994). Transforming growth factor-beta stimulates alpha 2(I) collagen gene expression through a cis-acting element that contains an Sp1-binding site. *J. Biol. Chem.* **269**, 14828–14834.

Inoue, T., and Thomas, J. H. (2000). Suppressors of transforming growth factor-beta pathway mutants in the *Caenorhabditis elegans* dauer formation pathway. *Genetics* **156**, 1035–1046.

Iraburu, M. J., Dominguez-Rosales, J. A., Fontana, L., Auster, A., Garcia-Trevijano, E. R., Covarrubias-Pinedo, A., Rivas-Estilla, A. M., Greenwel, P., and Rojkind, M. (2000). Tumor necrosis factor alpha down-regulates expression of the alpha1(I) collagen gene in rat hepatic stellate cells through a p20C/EBPbeta- and C/EBPdelta-dependent mechanism. *Hepatology* **31**, 1086–1093.

Issa, R., Williams, E., Trim, N., Kendall, T., Arthur, M. J. P., Reichen, J., Benyon, R. C., and Iredale, J. P. (2001). Apoptosis of hepatic stellate cells: Involvement in resolution of biliary fibrosis and regulation by soluble growth factors. *Gut* **48**, 548–557.

Knittel, T., Muller, L., Saile, B., and Ramadori, G. (1997). Effect of tumour necrosis factor-alpha on proliferation, activation and protein synthesis of rat hepatic stellate cells. *J. Hepatol.* **27**, 1067–1080.

Kondo, T., Konishi, F., Inui, H., and Inagami, T. (1993). Differing signal transduction elicited by three isoforms of platelet-derived growth factor in vascular smooth muscle cells. *J. Biol. Chem.* **268**, 4458–4464.

Kundra, V., Escobedo, J. A., and Kazlaukas, A. (1993). Regulation of chemotaxis by platelet-derived growth factor receptor beta. *Nature* **367**, 474–476.

Lang, A., Schoonhoven, R., Tuvia, S., Brenner, D. A., and Rippe, R. A. (2000). Nuclear factor kappaB in proliferation, activation, and apoptosis in rat hepatic stellate cells. *J. Hepatol.* **33**, 49–58.

Ledgerwood, E. C., Pober, J. S., and Bradley, J. R. (1999). Recent advances in the molecular basis of TNF signal tranduction. *Lab. Invest.* **79**, 1041–1050.

Leivas, A., Jimenez, W., Bruix, J., Boix, L., Bosch, J., Arroyo, V., Rivera, F., and Rodes, J. (1998). Gene expression of endothelin-1 and ET(A) and ET(B) receptors in human cirrhosis: relationship with hepatic hemodynamics. *J. Vasc. Res.* **35**, 186–193.

Levy, M. T., Trojanowska, M., and Reuben, A. (2000). Oncostatin M: A cytokine upregulated in human cirrhosis, increases collagen production by human hepatic stellate cells. *J. Hepatol.* **32**, 218–226.

Lim, D. S., Lutucuta, S., Bachireddy, P., Youker, K., Evans, A., Entman, M., Roberts, R., and Marian, A. J. (2001). Angiotensin II blockade reverses myocardial fibrosis in a transgenic mouse model of human hypertrophic cardiomyopathy. *Circulation* **103**, 789–791.

Ma, H., Matsunaga, H., Li, B., Schieffer, B., Marrero, M. B., and Ling, B. N. (1996). Ca^{2+} channel activation by platelet-derived growth factor-induced tyrosine phosphorylation and Ras guanine triphosphate-binding proteins in rat glomerular mesangial cells. *J. Clin. Invest.* **97**, 2332–2341.

Mallat, A., Fouassier, F., Preaux, A. M., Serradeil-Le Gal, C., Raufaste, D., Rosembaum, J., Dhumeaux, D., Journeaux, C., Mavier, P., and Lotersztajn, S. (1995a). Growth inhibitory properties of endothelin-1 in human hepatic myofibroblastic Ito cells: An endothelin B receptor-mediated pathway. *J. Clin. Invest.* **96**, 42–49.

Mallat, A., Preaux, A. M., Blazejewski, S., Rosenbaum, J., Dhumeaux, D., and Mavier, P. (1995b). Interferon alfa and gamma inhibit proliferation and collagen synthesis of human Ito cells in culture. *Hepatology* **21**, 1003–1010.

Mallat, A., Preaux, A. M., Serradeil-Le Gal, C., Raufaste, D., Gallois, C., Brenner, D. A., Bradham, C., Maclouf, J., Iourgenko, V., Fouassier, L., Dhumeaux, D., Mavier, P., and Lotersztajn, S. (1996). Growth inhibitory properties of endothelin-1 in activated human hepatic stellate cells: A cyclic adenosine monophosphate-mediated pathway. Inhibition of both extracellular signal-regulated kinase and c-Jun kinase and upregulation of endothelin B receptors. *J. Clin. Invest.* **98**, 2771–2778.

Mancini, R., Benedetti, A., and Jezequel, A. M. (1994). An interleukin-1 receptor antagonist decreases fibrosis induced by dimethylnitrosamine in rat liver. *Virch. Arch.* **424**, 25–31.

Marra, F. (1999). Hepatic stellate cells and the regulation of liver inflammation. *J. Hepatol.* **31**, 1120–1130.

Marra, F., Arrighi, M. C., Fazi, M., Caligiuri, A., Pinzani, M., Romanelli, R. G., Efsen, E., Laffi, G., and Gentilini, P. (1999). ERK activation differentially regulates PDGF's actions in hepatic stellate cells, and is induced by in vivo liver injury. *Hepatology* **30**, 951–958.

Marra, F., DeFranco, R., Grappone, C., Milani, S., Pinzani, M., Pellegrini, G., Laffi, G., and Gentilini, P. (1998). Expression of the thrombin receptor in human liver: Upregulation during acute and chronic injury. *Hepatology* **27**, 462–471.

Marra, F., Gentilini, A., Pinzani, M., Ghosh Choudhury, G., Parola, M., Herbst, H., Laffi, G., Abboud, H. E., and Gentilini, P. (1997). Phosphatidylinositol 3-kinase is required for platelet-derived growth factor's actions on hepatic stellate cells. *Gastroenterology* **112**, 1297–1306.

Marra, F., Ghosh Choudhury, G., and Abboud, H. E. (1996). Interferon-γ-mediated activation of STAT1a regulates growth factor-induced mitogenesis. *J. Clin. Invest.* **98**, 1218–1230.

Marra, F., Grandaliano, G., Valente, A. J., and Abboud, H. E. (1995a). Thrombin stimulates proliferation of liver fat-storing cells and expression of monocyte chemotactic protein-1: Potential role in liver injury. *Hepatology* **22**, 780–787.

Marra, F., Pinzani, M., DeFranco, R., Laffi, G., and Gentilini, P. (1995b). Involvement of phoshatidylinositol 3-kinase in the activation of extracellular signal-regulated kinase by PDGF in hepatic stellate cells. *FEBS Lett.* **376**, 141–145.

Marshall, C. J. (1995). Specificity of receptor tyrosine kinase signaling : Transient versus sustained extracellular signal-regulated kinase activation. *Cell* **80**, 179–185.

Masaki, T., Ninomiya, H., Sakamoto, A., and Okamoto, Y. (1999). Structural basis of the function of endothelin receptor. *Mol. Cell Biochem.* **190**, 153–156.

Massague, J. (1998). TGF-β signal transduction. *Annu. Rev. Biochem.* **67**, 753–759.

Massague, J. (2000). How cells read TGF-beta signals. *Nature Rev. Mol. Cell Biol.* 169–178.

Matsuoka, M., Pham, N. T., and Tsukamoto, H. (1989). Differential effects of interleukin-1 alpha, tumor necrosis factor alpha, and transforming growth factor beta 1 on cell proliferation and collagen formation by cultured fat-storing cells. *Liver* **9**, 71–78.

Nakatsukasa, H., Nagy, P., Evarts, R. P., Hsia, C. C., Marsden, E., and Thorgeirsson, S. S. (1990). Cellular distribution of transforming growth factor-beta 1 and procollagen types I, III, and IV transcripts in carbon tetrachloride-induced rat liver fibrosis. *J. Clin. Invest.* **85**, 1833–1843.

Pages, G., Lenormand, P., and L'Allemain, G. (1993). Mitogen-activated protein kinases p42MAPK and p44MAPK are required for fibroblast proliferation. *Proc. Natl. Acad. Sci. USA* **90**, 8319–8323.

Parker, P. J., and Waterfield, M. D. (1992). Phosphatidylinositol 3-kinase: A novel effector. *Cell Growth Differ.* **3**, 747–752.

Pawson, T. (1995). Protein modules and signaling networks. *Nature* **373**, 573–580.

Piek, E., Heldin, C. H., and Ten Dijke, P. (1999). Specificity, diversity, and regulation in TGF-beta superfamily signaling. *FASEB J.* **13**, 2105–2124.

Pinzani, M. (2000). Liver fibrosis. *Springer Semin. Immunopathol.* **21**, 475–490.

Pinzani, M., Abboud, H. E., Gesualdo, L., and Abboud, S. L. (1992a). Regulation of macrophage-colony stimulating factor in liver fat-storing cells by peptide growth factors. *Am. J. Physiol. Cell Physiol.* **262**, C876–C881.

Pinzani, M., Carloni, V., Marra, F., Riccardi, D., Laffi, G., and Gentilini, P. (1994a). Biosynthesis of platelet-activating factor and its 1O-acyl analogue by liver fat-storing cells. *Gastroenterology* **106**, 1301–1311.

Pinzani, M., Failli, P., Ruocco, C., Casini, A., Milani, S., Baldi, E., Giotti, A., and Gentilini, P. (1992b). Fat-storing cells as liver-specific pericytes: Spatial dynamics of agonist-stimulated intracellular calcium transients. *J. Clin. Invest.* **90**, 642–646.

Pinzani, M., Gentilini, A., Caligiuri, A., DeFranco, R., Pellegrini, G., Milani, S., Marra, F., and Gentilini, P. (1995). Transforming growth factor-w1 regulates platelet-derived growth factor subunit in human liver fat-storing cells. *Hepatology* **21**, 232–239.

Pinzani, M., Gesualdo, L., Sabbah, G. M., and Abboud, H. E. (1989). Effects of platelet-derived growth factor and other polypeptide mitogens on DNA synthesis and growth of cultured rat liver fat-storing cells. *J. Clin. Invest.* **84**, 1786–1793.

Pinzani, M., Knauss, T. C., Pierce, G. F., Hsieh, P., Kenney, W., Dubiak, G. R., and Abboud, H. E. (1991). Mitogenic signals for platelet-derived growth factor isoforms in liver fat-storing cells. *Am. J. Physiol.* **260**, C485–C491.

Pinzani, M., Milani, S., DeFranco, R., Grappone, C., Caligiuri, A., Gentilini, A., Tosti-Guerra, C., Maggi, M., Failli, P., Ruocco, C., and Gentilini, P. (1996a). Endothelin 1 is overexpressed in human cirrhotic liver and exerts multiple effects on activated hepatic stellate cells. *Gastroenterology* **110**, 534–548.

Pinzani, M., Milani, S., DeFranco, R., Grappone, C., Gentilini, A., Caligiuri, A., Pellegrini, G., Ngo, D.-V., Romanelli, R. G., and Gentilini, P. (1996b). Expression of platelet-derived growth factor and its receptors in normal human liver and during active hepatic fibrogenesis. *Am. J. Pathol.* **148**, 785–800.

Pinzani, M., Milani, S., Grappone, C., Weber, F. L. J., Gentilini, P., and Abboud, H. E. (1994b). Expression of platelet-derived growth factor in a model of acute liver injury. *Hepatology* **19**, 701–707.

Poulos, J. E., Weber, J. D., Bellezzo, J. M., Di Bisceglie, A. M., Britton, R. S., Bacon, B. R., and Baldassare, J. J. (1997). Fibronectin and cytokines increase JNK, ERK, AP-1 activity, and transin gene expression in rat hepatic stellate cells. *Am. J. Physiol.* **273**, G804–G811.

Racine-Samson, L., Rockey, D. C., and Bissell, D. M. (1997). The role of alpha1beta1 integrin in wound contraction: A quantitative analysis of liver myofibroblasts in vivo and in primary culture. *J. Biol. Chem.* **272**, 30911–30917.

Ramadori, G., Knittel, T., Odenthal, M., Schwogler, S., Neubauer, K., and Meyer zum Büschenfelde, K. H. (1992). Synthesis of cellular fibronectin by rat liver fat-storing (Ito) cells: Regulation by cytokines. *Gastroenterology* **103**(4), 1313–1321.

Reeves, H. L., Dack, C. L., Peak, M., Burt, A. D., and Day, C. P. (2000). Stress-activated protein kinases in the activation of rat hepatic stellate cells in culture. *J. Hepatol.* **32**, 465–472.

Reimann, T., Hempel, U., Krautwald, S., Axmann, A., Scheibe, R., Seidel, D., and Wenzel, K. W. (1997). Transforming growth factor-beta1 induces activation of Ras, Raf-1, MEK and MAPK in rat hepatic stellate cells. *FEBS Lett.* **403**, 57–60.

Reinehr, R. M., Kubitz, R., Peters-Regehr, T., Bode, J. G., and Haussinger, D. (1998). Activation of rat hepatic stellate cells in culture is associated with increased sensitivity to endothelin 1. *Hepatology* **28**, 1566–1577.

Rieder, H., Ramadori, G., and Meyer zum Büschenfelde, K. H. (1991). Sinusoidal endothelial liver cells in vitro release endothelin: Augmentation by transforming growth factor β and Kupffer cell-conditioned media. *Klin. Wochenschr.* **69**, 387–391.

Rockey, D. C., and Chung, J. J. (1994). Interferon gamma inhibits lipocyte activation and extracellular matrix mRNA expression during experimental liver injury: Implications for treatment of hepatic fibrosis. *J. Invest. Med.* **42**, 660–670.

Rockey, D. C., and Chung, J. J. (1995). Inducible nitric oxide synthase in rat hepatic lipocytes and the effect of nitric oxide on lipocyte contractility. *J. Clin. Invest.* **95**, 1199–1206.

Rockey, D. C., Fouassier, L., Chung, J. J., Vallee, P., Rey, C., and Housset, C. (1998). Cellular localization of endothelin-1 and increased production in liver injury in the rat: Potential for autocrine and paracrine effects on stellate cells. *Hepatology* **27**, 472–480.

Romagnani, P., Beltrame, C., Annunziato, F., Lasagni, L., Luconi, M., Galli, G., Cosmi, L., Maggi, E., Salvadori, M., Pupilli, C., and Serio, M. (1999). Role for interactions between IP-10/Mig and CXCR3 in proliferative glomerulonephritis. *J. Am. Soc. Nephrol.* **10**, 2518–2526.

Rossi, D., and Zlotnik, A. (2000). The biology of chemokines and their receptors. *Annu. Rev. Immunol.* **18**, 217–242.

Rothe, J., Lesslauer, W., Loetscher, H., Lang, Y., Koebel, P., Koentgen, F., Althage, A., Zinkernagel, R., Steinmetz, M., and Bluethmann, H. (1993). Mice lacking the tumour necrosis factor receptor-1 are resistant to TNF-mediated toxicity but highly susceptible to infection by Listeria monocytogenes. *Nature* **364**, 798–802.

Roulot, D., Sevcsik, A. M., Coste, T., Strosber, A. D., and Marullo, S. (1999). Role of transforming growth factor beta type II receptor in hepatic fibrosis: Studies of human chronic hepatitis C and experimental fibrosis in rats. *Hepatology* **29**, 1730–1738.

Saile, B., Matthes, N., Knittel, T., and Ramadori, G. (1999). Transforming growth factor beta and tumor necrosis factor alpha inhibit both apoptosis and proliferation of activated rat hepatic stellate cells. *Hepatology* **30**, 196–202.

Saitoh, M., Nishitoh, H., Amagasa, T., Miyazono, K., Takagi, M., and Ichijo, H. (1996). Identification of important regions in the cytoplasmic juxtamembrane domain of type I receptor that separate signaling pathways of transforming growth factor-beta. *J. Biol. Chem.* **271**, 2769–2775.

Sanderson, N., Factor, V., Nagy, P., Kopp, J., Kondaiah, P., Wakefield, L., Roberts, A. B., Sporn, M. B., and Thorgeirsson, S. S. (1995). Hepatic expression of mature transforming growth factor beta 1 in transgenic mice results in multiple tissue lesions. *Proc. Natl. Acad. Sci. USA* **92**, 2572–2576.

Schaller, M. D., Otey, C., Hildebrand, J. D., and Parsons, J. T. (1995). Focal adhesion kinase and paxillin bind to peptides mimicking β integrin cytoplasmic domains. *J. Cell Biol.* **130**, 1181–1187.

Schaller, M. D., and Parsons, J. T. (1993). Focal adhesion kinase: An integrin-linked protein tyrosine kinase. *Trends Cell Biol.* **3**, 258–262.

Schweickart, V. L., Epp, A., Raport, C. J., and Gray, P. W. (2000). CCR11 is a functional receptor for the monocyte chemoattractant protein family of chemokines. *J. Biol. Chem.* **275**, 9550–9556.

Shao, R., Yan, W., and Rockey, D. C. (1999). Regulation of endothelin-1 synthesis by endothelin-converting enzyme-1 during wound healing. *J. Biol. Chem.* **274**, 3228–3234.

Simonson, M. S. (1993). Endothelins: Multifunctional renal peptides. *Physiol. Rev.* **73**, 375–411.

Simonson, M. S., and Dunn, M. J. (1991). Endothelins: A family of regulatory peptides. *Hypertension* **17**, 856–863.

Tangkijvanich, P., Tam, S. P., and Yee, H. F. (2001). Wound-induced migration of rat hepatic stellate cells is modulated by endothelin-1 through rho-kinase-mediated alterations in the acto-myosin cytoskeleton. *Hepatology* **33**, 74–80.

Tieche, S., De Gottardi, A., Kappeler, A., Shaw, S., Sagesser, H., Zimmermann, A., and Reichen, J. (2001). Overexpression of endothelin-1 in bile duct ligated rats: Correlation with activation of hepatic stellate cells and portal pressure. *J. Hepatol.* **34**, 38–45.

Valius, M., and Kazlaukas, A. (1993). Phospholipase C-gamma and phosphatidylinositol 3-kinase are the downstream mediators of the PDGF receptor's mitogenic signal. *Cell* **73**, 321–334.

Wang, S. C., Ohata, M., Schrum, L., Rippe, R. A., and Tsukamoto, H. (1998). Expression of interleukin-10 by in vitro and in vivo activated hepatic stellate cells. *J. Biol. Chem.* **273**, 302–308.

Wang, Y. Z., Pouyssegur, J., and Dunn, M. J. (1994). Endothelin stimulates mitogen-activated protein kinase activity in mesangial cells through ET(A). *J. Am. Soc. Nephrol.* **5**, 1074–1080.

Wang, Z., Estacion, M., and Mordan, L. J. (1993). Ca^{2+} influx via T-type channels modulates PDGF-induced replication of mouse fibroblasts. *Am. J. Physiol.* **265**, C1239–C1246.

Win, K. M., Charlotte, F., Mallat, A., Cherqui, D., Martin, N., Mavier, P., Preaux, A. M., Dhumeaux, D., and Rosenbaum, J. (1993). Mitogenic effect of transforming growth factor-beta 1 on human Ito cells in culture: Evidence for mediation by endogenous platelet-derived growth factor. *Hepatology* **18**, 137–145.

Yanagisawa, M., and Masaki, T. (1989). Endothelin, a novel endothelium-derived peptide. *Biochem. Pharmacol.* **38**, 1877–1883.

CHAPTER 13

Transcriptional Activation of Type I Collagen Gene during Hepatic Fibrogenesis

YUTAKA INAGAKI
Department of Internal Medicine and Division of Clinical Research, National Kanazawa Hospital, Kanazawa 920-8650, Japan

TOMOYUKI NEMOTO
Department of Functional Biology, Kyoto University Graduate School of Biostudies, Kyoto 606-8502, Japan

ATSUHITO NAKAO
Allergy Research Center, Juntendo University School of Medicine, Tokyo 113-8421, Japan

Increased production of type I collagen is a common hallmark of fibrotic diseases in various organs, including the liver. This increase is exerted mainly by transcriptional upregulation of the genes coding for the α1 and α2 chains of type I collagen, and transforming growth factor-β (TGF-β) is a key player to stimulate type I collagen gene transcription. We have shown previously that the −313 to −183 upstream sequence of the α2(I) collagen gene (COL1A2) is essential for basal and TGF-β-stimulated transcription in skin fibroblasts and hepatic stellate cells (HSC). We therefore designated this region the TGF-β responsive element (TbRE) and revealed that a ubiquitous *trans*-activator Sp1 and unknown cofactor(s) bind to this region to mediate the stimulatory effect of TGF-β. Smad3, an intracellular mediator of TGF-β signal transduction, has been shown to bind to the TbRE, and its interaction with Sp1 has been implicated in TGF-β-elicited COL1A2 stimulation. This chapter provides a brief summary of the cell type-specific activation of the COL1A2 promoter during hepatic fibrogenesis and reveals molecular mechanisms responsible for pathologically accelerated collagen gene transcription in activated HSC.

Copyright 2003, Elsevier Science (USA). All rights reserved.

I. INTRODUCTION

Collagens represent a family of proteins involved not only in the maintenance of organ architecture and tissue integrity, but also in various physiological conditions, such as developmental program, tissue repair process, and wound healing. There is a dynamic balance between production and degradation of collagen, and a disruption of this equilibrium results in either excessive collagen deposition or inadequate tissue integrity. Irrespective of the initial stimuli, fibrosis in different organs is caused commonly by a chronic and uncontrolled inflammatory and repair process leading to excessive deposition of collagen and other components of extracellular matrix (ECM) (Diegelmann *et al.*, 1988). The liver is one of those organs undergoing progressive fibrosis as a result of chronic repeating inflammation.

Type I collagen, the major component of extracellular matrix in fibrotic tissues, is a heterotrimer composed of two α1 chains and one α2 chain. They are coordinately expressed but encoded by the distinct genes, COL1A1 and COL1A2, respectively. Transforming growth factor-β1 (henceforth referred to as TGF-β) plays critical roles in stimulating type I collagen gene expression mainly at the levels of transcription (Ramirez and Di Liberto, 1990).

During the last decade, we have been working on the transcriptional regulation of COL1A2 expression, especially on the molecular mechanisms responsible for TGF-β-elicited COL1A2 stimulation and its pathological roles during the fibrogenic process in the liver. We first identified the *cis*-acting DNA elements and *trans*-acting nuclear factors essential for basal and TGF-β-stimulated COL1A2 transcription in skin fibroblasts. Then we used activated hepatic stellate cells, the major source of collagen in the liver, and revealed that similar transcriptional mechanisms are utilized for COL1A2 transcription in skin fibroblasts and HSC. More recently, we have shown that constitutive phosphorylation and nuclear localization of Smad3, an intracellular mediator of TGF-β signal transduction, are correlated with increased COL1A2 transcription in activated HSC. Experiments using transgenic mice harboring the COL1A2 upstream sequence indicated that the COL1A2 promoter is activated in a cell type-specific manner during hepatic fibrogenesis *in vivo*. We have revealed that interactions between GC box-binding factors (Sp1/Sp3) and Smad proteins modulate cell type-specific COL1A2 transcription in the liver.

II. COL1A2 TRANSCRIPTION IN SKIN FIBROBLASTS

Our initial studies using primary culture of fetal skin fibroblasts have mapped the upstream sequence essential for COL1A2 transcription (Boast *et al.*, 1990; Inagaki *et al.*, 1994). Cell transfection experiments revealed that the sequence spanning

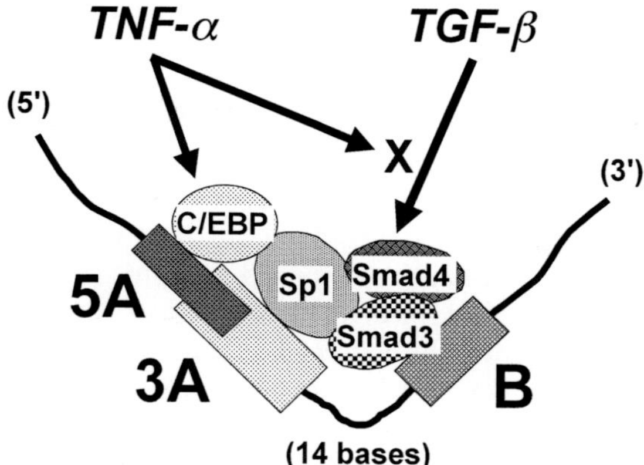

FIGURE 1 Schematic representation of *cis*-acting DNA elements and *trans*-acting nuclear factors mediating the effects of TGF-β and TNF-α on COL1A2 transcription in a primary culture of skin fibroblasts. Box 5A- and box B-bound unknown factors have been subsequently shown to be CAAT/enhancer-binding proteins (C/EBP) and Smad proteins, respectively (see text).

from –378 to –183 relative to the transcription start site is necessary for the high level of COL1A2 transcription. DNase I footprinting analyses identified at least two sites of DNA–protein interaction within this segment: box A and box B (Fig. 1). Gel mobility shift assays further divided box A into two overlapping subregions: box 5A and box 3A. Box 5A was found to be the binding site of an unknown repressor, whereas a ubiquitous *trans*-activator Sp1 binds to the 3A region and interacts with the box B-bound unknown factor(s) (Fig. 1).

Intensive functional assays have indicated that both box 3A-bound Sp1 and box B-bound cofactor(s) are necessary to mediate the stimulatory effects of TGF-β on COL1A2 transcription (Inagaki *et al.*, 1994; Greenwel *et al.*, 1997). However, box 5A does not contribute to the promoter response to TGF-β. There is a 14-bp distance between box 3A and box B. When box 3A was replaced 60 bp upstream from box B, the 3A+B region did not show TGF-β responsiveness, suggesting a functional interaction between box 3A and box B in mediating TGF-β stimulation of COL1A2 transcription (Fig. 1). We therefore designated the 3A+B region the TGF-β responsive element, TbRE (Inagaki *et al.*, 1994). One of the box B-bound cofactors has been subsequently shown to be Smad3, an intracellular mediator of TGF-β signal transduction, which is described in detail later in this chapter.

TGF-β treatment of skin fibroblasts increased the binding of both Sp1 and box B-bound cofactors when using the TbRE as a gel-shift probe (Inagaki *et al.*, 1994). However, when using the 3A oligonucleotide or Sp1 consensus sequence as a probe, there was no increase in Sp1 binding following TGF-β treatment. It is therefore suggested that TGF-β stimulates COL1A2 transcription, not by increasing the amount of Sp1, but by modifying the interaction between Sp1 and the B-bound cofactors, which in turn increases the binding affinity of Sp1 to the 3A region (Fig. 1).

Our study has also revealed that the counterrepression of COL1A2 transcription by tumor necrosis factor-α (TNF-α) is mediated through the same TbRE as well as by increasing the amount of a repressor protein bound to the immediately upstream 5A region (Inagaki *et al.*, 1995b) (Fig. 1). It has been shown that CAAT/enhancer-binding protein β is the major component of the box 5A-bound complex and mediates TNF-α-elicited COL1A2 repression (Greenwel *et al.*, 2000). Thus, the convergence of the TGF-β and TNF-α signals on the same COL1A2 promoter sequence represents an example of combinatorial gene regulation achieved through composite responsive elements (Inagaki *et al.*, 1995b) (Fig. 1).

III. COL1A2 TRANSCRIPTION IN HEPATIC STELLATE CELLS

HSC are sinusoidal cells localized within the space of Disse and are considered to be the main producers of type I collagen in both normal and fibrotic liver (Friedman, 1990). During the development of hepatic fibrosis, they undergo an activation process, which is characterized by a decrease in vitamin A-containing fat droplets, increased expression of α-smooth muscle actin, and enhanced production of collagen and other ECM components (Mak *et al.*, 1984). However, the molecular events responsible for the pathologic activation of collagen gene expression in HSC during hepatic fibrogenesis have not been fully understood.

Based on results obtained with skin fibroblasts, we next examined COL1A2 transcription in cultured HSC. We used two distinct HSC clones, CFSC-2G and CFSC-5H cells, derived from a single cirrhotic liver induced by carbon tetrachloride injection. These cells express desmin and type I collagen and maintain their phenotypes of activated myofibroblasts over a hundred passages (Greenwel *et al.*, 1993). Cell transfection experiments indicated that the upstream sequence between nucleotide −378 and −183 is essential for COL1A2 transcription in both HSC clones, as well as in the primary culture of skin fibroblasts (Inagaki *et al.*, 1995a). It was also shown that box 5A acts as a negatively *cis*-acting element, as deletion of this region resulted in an increase in transcriptional activity. In contrast, substitution of the Sp1-bound box 3A sequence with an unrelated polylinker sequence reduced COL1A2 transcription, indicating that Sp1 contributes to stimulation of gene transcription. It is therefore indicated that similar regulatory mechanisms control COL1A2 transcription in two different cell types of mesenchymal origin: skin fibroblasts and HSC (Inagaki *et al.*, 1995a).

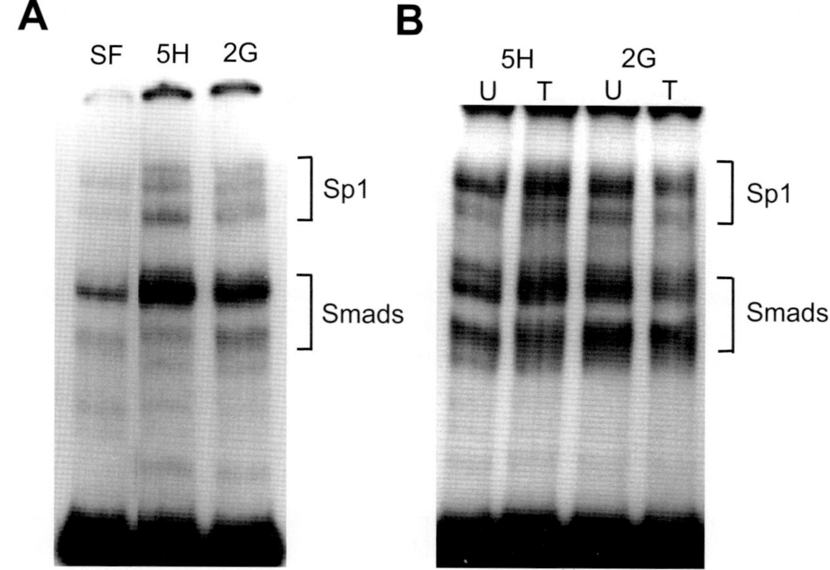

FIGURE 2 Gel mobility shift analyses of protein complexes bound to the TbRE. (**A**) The same amounts of nuclear proteins were prepared from a primary culture of skin fibroblasts (SF) and activated HSC clones, CFSC-2G (2G) and CFSC-5H (5H), and were subjected to polyacrylamide gel electrophoresis using the 3A+B fragment as a probe. (**B**) Nuclear proteins were prepared from untreated (U) or TGF-β-treated (T) CFSC-2G and CFSC-5H cells and were subjected to gel electrophoresis. The identities of nuclear proteins bound to box 3A and box B are indicated on the right side of each autoradiogram.

Nuclear extracts prepared from untreated CFSC-2G and CFSC-5H cells contained larger amounts of box 3A-bound Sp1 and the box B-bound cofactor(s) than those from a primary culture of skin fibroblasts (Fig. 2). While TGF-β treatment of skin fibroblasts significantly increased binding of both Sp1 and cofactor(s) as described earlier, it did not affect protein binding to the TbRE in activated HSC (Fig. 2). These results therefore suggested that COL1A2 transcription in activated HSC derived from a cirrhotic liver is upregulated via stimulation of the TbRE in the promoter (Inagaki *et al.*, 1995a).

IV. Smad PROTEINS REGULATING COL1A2 TRANSCRIPTION IN ACTIVATED HEPATIC STELLATE CELLS

How is collagen gene transcription stimulated pathologically in activated HSC? Smad proteins are a family of newly identified factors that play important roles in the intracellular signal transduction pathways of the TGF-β superfamily members (Heldin *et al.*, 1997). TGF-β first binds to the TGF-β type II receptor on the cell surface. It subsequently recruits the TGF-β type I receptor, thus forming a heteromeric complex between these two types of receptors. Once the type I receptor is phosphorylated by the type II receptor kinase, it in turn phosphorylates Smad2 and/or Smad3, which form heterooligomers with Smad4. They translocate from the cytoplasm to the nucleus, where they stimulate or repress gene transcription (Nakao *et al.*, 1997b) (Fig. 3A). In contrast to those signal transducing Smads, Smad7 is known as an inhibitory Smad: it inhibits TGF-β signaling by interfering with the phosphorylation of Smad2 and Smad3 by the type I receptor kinase (Nakao *et al.*, 1997a).

Smad3 has been shown to bind to box B of the TbRE and stimulate COL1A2 transcription when overexpressed in skin fibroblasts (Chen *et al.*, 1999, 2000). More importantly, we have demonstrated that synergistic cooperation between Sp1 and Smad3/Smad4, both of which bind to the TbRE, is critical for mediating TGF-β-stimulated COL1A2 transcription (Zhang *et al.*, 2000). However, pathological roles of Smad proteins and abnormalities in their signaling pathway were not revealed in the process of organ fibrosis, where increased production of TGF-β and ECM components is observed. We therefore examined the role of Smad proteins in regulating COL1A2 transcription in activated HSC.

We first compared the basal levels of COL1A2 transcription and response to TGF-β between an activated HSC clone (CFSC-2G) and a primary culture of skin fibroblasts (Inagaki *et al.*, 2001a). The basal levels of COL1A2 transcription in CFSC-2G cells were 10 times higher than those in the primary culture of skin fibroblasts, but less responded to TGF-β (Inagaki *et al.*, 2001a). Then we examined the effects of overexpression of Smad proteins on COL1A2 transcription. Among the Smad expression plasmids tested, cotransfection with a Smad3 plasmid exhibited the greatest effect on the basal COL1A2 transcription in primary culture of skin fibroblasts. It increased COL1A2 promoter activity 20-fold (Fig. 4A). In contrast to skin fibroblasts, and consistent with a low response to ligand stimulation, COL1A2 transcription in CFSC-2G cells increased only 2.4-fold after cotransfection with a Smad3 expression plasmid (Fig. 4A). However, overexpression of Smad2 did not affect the basal COL1A2 transcription in either skin fibroblasts or activated HSC (Fig. 4A). Increased COL1A2 transcription in CFSC-2G cells was not affected by overexpression of inhibitory Smad7 that interferes with the Smad3 phosphorylation (Fig. 4A), thus indicating that the increase occurred at the level downstream of Smad3 activation. Similar results were obtained when we transfected a primary culture of skin fibroblasts and CFSC-2G cells with a plasminogen activator inhibitor-1 reporter construct, another TGF-β/Smad responsive gene (Fig. 4B).

Consistent with the results of transfection assays, Western blot analyses indicated constitutive phosphorylation of Smad3 in CFSC-2G cells (Inagaki *et al.*, 2001a).

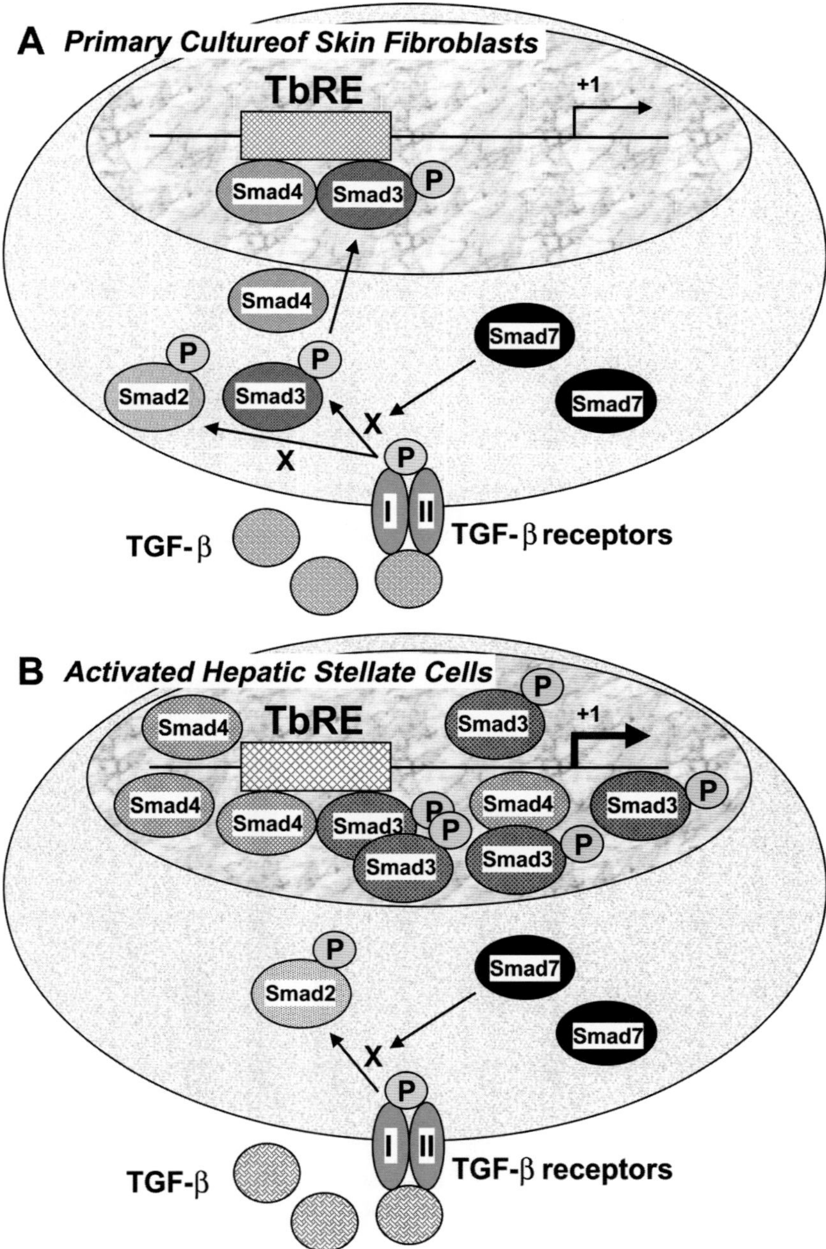

FIGURE 3 Schematic representation of the intracellular signal transduction pathways of TGF-β and COL1A2 transcription in a primary culture of skin fibroblasts (A) and activated HSC (B). +1, transcription start site.

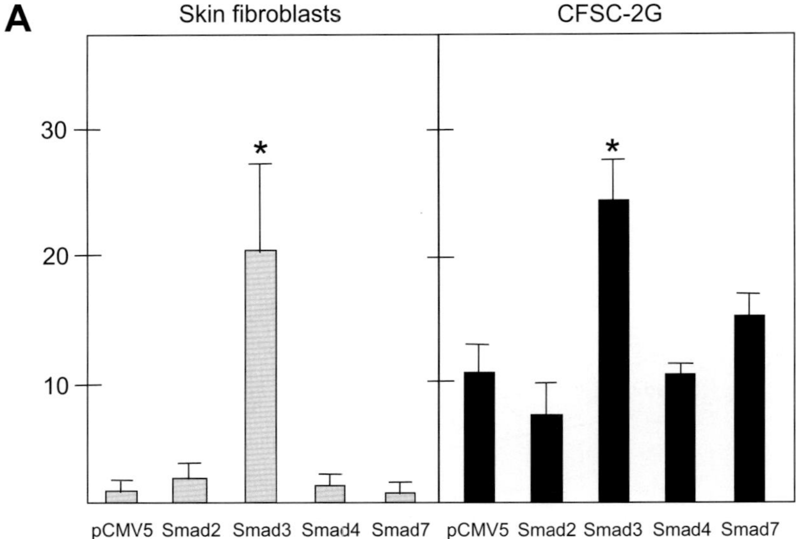

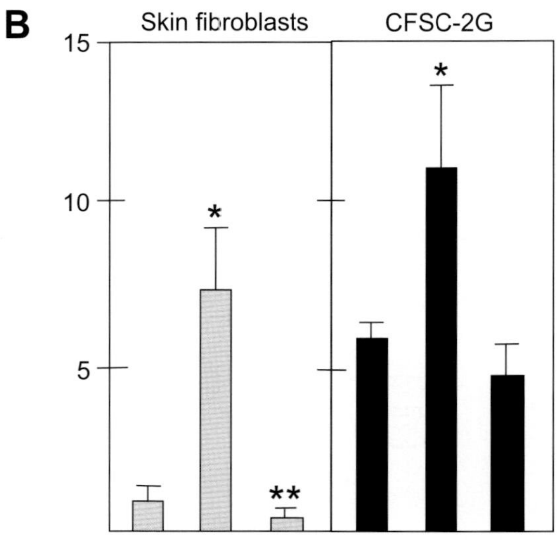

FIGURE 4 Effects of overexpression of Smad proteins on COL1A2 and PAI-1 transcription. A primary culture of skin fibroblasts and an activated HSC clone (CFSC-2G) were cotransfected with either the –378COL1A2 promoter/reporter construct (A) or the –800PAI-1 promoter/reporter construct (B), together with cytomegalovirus expression plasmids encoding the indicated Smad proteins. Transcriptional activities in each transfectants were expressed relative to the activity in skin fibroblasts transfected with the control empty vector (pCMV5). Asterisks indicate that the values are significantly higher (*) or lower (**) than that in pCMV5 transfectants.

Immunofluorescence studies revealed that, in contrast to Smad2 that translocated from the cytoplasm to the nucleus upon TGF-β treatment, Smad3 and Smad4 were always present in the nucleus irrespective of ligand stimulation (Inagaki *et al.*, 2001a). In contrast, both Smad3 and Smad4, as well as Smad2, were detected predominantly in the cytoplasm of untreated primary culture of skin fibroblasts and translocated into the nucleus following TGF-β treatment (Inagaki *et al.*, 2000a).

Altogether, the results correlated an alteration in TGF-β/Smad3 signaling with pathologically accelerated collagen gene transcription in activated HSC (Fig. 3B). They also indicated that increased COL1A2 transcription in CFSC-2G cells is attributed, at least in part, to constitutive activation of Smad3 and increased binding of Sp1 and Smad3 to the TbRE (Fig. 3B).

Increased collagen gene expression and insensitivity to TGF-β were mimicked during the activation process of primary culture of HSC transforming to myofibroblasts (Dooley *et al.*, 2000). In addition, it was reported previously that COL1A2 transcription levels in cultured scleroderma fibroblasts are higher than those in normal skin fibroblasts obtained from the same patients and that administration of exogenous TGF-β into culture media does not increase COL1A2 transcription in the disease cells (Kikuchi *et al.*, 1992). From these findings, it appears that increased collagen gene transcription, which does not respond to exogenous TGF-β, is the common feature of pathologically activated collagen-producing cells. It would thus be interesting to determine whether ligand-independent phosphorylation of Smad3 was commonly observed during fibrogenic processes of various organs such as liver, skin, lung, and kidney.

V. CELL TYPE-SPECIFIC ACTIVATION OF COL1A2 PROMOTER DURING HEPATIC FIBROGENESIS *IN VIVO*

The results of cell transfection and DNA-binding assays using cultured HSC may not be fully applicable to hepatic fibrogenesis *in vivo*. Because of the obvious limitations of these *in vitro* experiments, we sought further confirmation in transgenic mice harboring COL1A2 upstream sequences containing the TbRE.

Transgenic mouse lines were generated using the mouse or human α2(I) collagen promoter sequence linked to either a luciferase or a β-galactosidase reporter gene (Bou-Gharios *et al.*, 1996; Inagaki *et al.*, 1998). One of the two constructs contains the −17kb to +54 sequence of the mouse α2(I) collagen promoter linked to a firefly luciferase gene. Strong enhancer activity has been found between −13.5 and −17 kb of the mouse promoter sequence (Bou-Gharios *et al.*, 1996). The other construct contains the −313 to +58 upstream sequence of the human α2(I) collagen promoter linked to a bacterial β-galactosidase gene

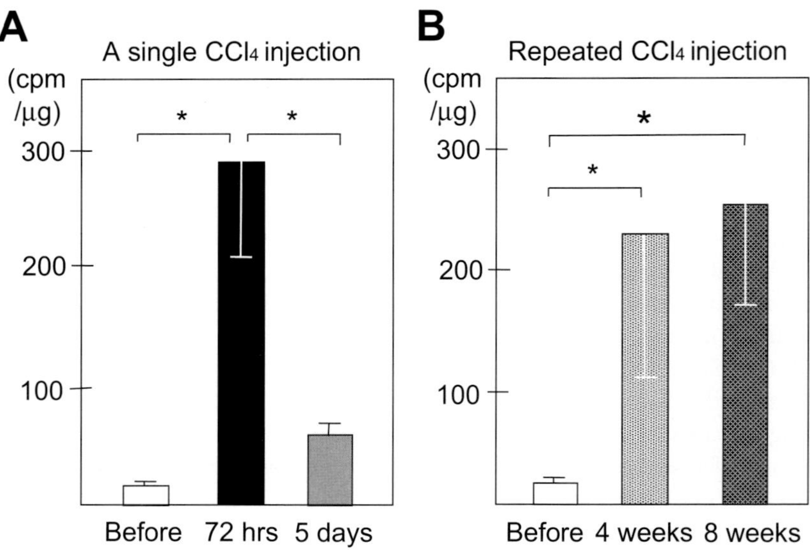

FIGURE 5 Activation of COL1A2 promoter following carbon tetrachloride administration into transgenic mice. Liver tissues of transgenic mice harboring the –17-kb COL1A2 upstream sequence/luciferase transgene were obtained after a single (A) or repeated (B) carbon tetrachloride (CCl_4) injections and subjected to luciferase assays. The asterisk indicates that there is a significant difference between the groups.

(*LacZ*). Acute and chronic liver injury was introduced by injecting once a week 0.1 ml/kg body weight of carbon tetrachloride (CCl_4) intraperitoneally.

A single or repeated intraperitoneal CCl_4 administration increased the –17-kb COL1A2 promoter activity more than 10-fold (Fig. 5). The activity went down in parallel with recovery from acute liver injury following a single CCl_4 injection. In contrast, high level of luciferase enzyme activity was constantly observed in chronically CCl_4-treated mouse liver (Fig. 5). Several transgenic mouse studies have demonstrated activation of the COL1A1 promoter after CCl_4 and ethanol administration (Brenner *et al.*, 1993; Houglum *et al.*, 1995; Walton *et al.*, 1996). It is therefore suggested that continuous activation of both α1(I) and α2(I) collagen promoters plays a critical role in the development of hepatic fibrosis (Inagaki *et al.*, 1998).

We next examined activation of the −313 COL1A2 promoter, which had been implicated in *in vitro* COL1A2 stimulation, in liver tissues following CCl_4 administration. X-Gal staining of liver sections indicated that the promoter seemed to be activated mainly in the necrotic areas and around fibrous septa. However, the *LacZ* expression driven by the −313 COL1A2 promoter was rather weak, and it was difficult to clearly indicate cellular localization. Thus, the identity of *LacZ*-expressing cells was confirmed by isolating parenchymal and nonparenchymal

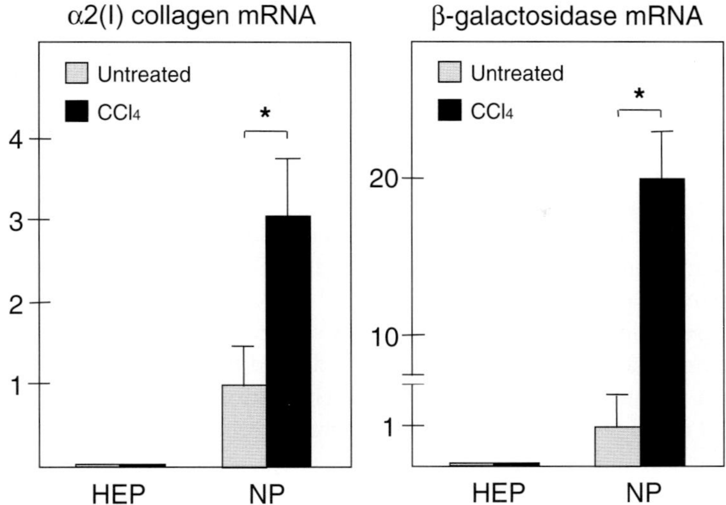

FIGURE 6 Expression of endogenous α2(I) collagen gene and β-galactosidase transgene in parenchymal hepatocytes (HEP) and nonparenchymal cells (NP) freshly isolated from untreated or CCl_4-treated (CCl_4) transgenic mice. Total RNA was prepared from liver tissues of transgenic mice harboring the –313COL1A2 promoter/β-galactosidase transgene and subjected to Northern blot analysis using α2(I) collagen or β-galactosidase cDNA as a probe. The expression levels were semiquantified using a densitometer and expressed relative to those in nonparenchymal cells from untreated mouse liver. The asterisk indicates that there is a significant difference between the groups.

cell fractions and by culturing HSC (Inagaki *et al.*, 1998). Results indicated that the −313 COL1A2 promoter was utilized only in nonparenchymal cells and that the promoter activity was significantly higher in cells from CCl_4-treated mice than in those from untreated animals (Fig. 6). Interestingly, TGF-β treatment increased the promoter activity and the amount of endogenous COL1A2 mRNA in HSC from untreated, but not CCl_4-treated mice (Inagaki *et al.*, 1998), supporting the results of our *in vitro* study showing insensitivity of activated HSC to TGF-β. From these findings, we conclude that the −313 COL1A2 promoter is activated in a cell type-specific manner to stimulate gene transcription during hepatic fibrogenesis *in vivo*.

VI. MOLECULAR MECHANISMS RESPONSIBLE FOR CELL TYPE-SPECIFIC ACTIVATION OF COL1A2 PROMOTER

Type I collagen is produced predominantly in mesenchymal cells such as osteoblasts, fibroblasts, and myofibroblasts. In the liver, activated HSC showing the morphological and functional features of myofibroblasts are the main producers of type I collagen, whereas parenchymal hepatocytes produce little, if any, of this protein during hepatic fibrogenesis (Friedman, 1990). Our transgenic mouse study described earlier further confirmed the minor contribution of parenchymal hepatocytes to collagen production (Inagaki *et al.*, 1998). However, very little was known regarding the molecular mechanisms determining differential type I collagen gene expression in activated HSC and parenchymal hepatocytes. We have demonstrated, at the molecular level, that different regulatory mechanisms control COL1A2 transcription in activated HSC and parenchymal hepatocytes (Inagaki *et al.*, 2001b).

We first compared the functional activity of the TbRE between activated HSC and parenchymal hepatocytes. The results indicated that, unlike in mesenchymal cells such as skin fibroblasts and activated HSC, strong enhancer activity of the TbRE and response to TGF-β were not observed in primary culture of hepatocytes (Inagaki *et al.*, 2001b). Gel mobility shift assays using the TbRE as a probe revealed that while Sp1 was the major box 3A-bound factor in activated HSC, Sp3 bound predominantly to this GC-rich sequence in parenchymal hepatocytes (Inagaki *et al.*, 2001b).

Transfection of activated HSC with an Sp3 expression plasmid abolished the COL1A2 response to TGF-β (Fig. 7A). In contrast, overexpression of Sp1 in parenchymal hepatocytes increased the basal level of COL1A2 transcription and conferred TGF-β responsiveness (Fig. 7A). Then we examined the effects of cotransfection of either the Sp1 or the Sp3 expression vector together with an expression plasmid encoding Smad3. Overexpression of Smad3 in CFSC-2G cells increased COL1A2 transcription significantly (Fig. 7B). While cotransfection with an Sp1 expression plasmid resulted in a further increase in COL1A2 transcription, overexpression of Sp3 did not affect the Smad3-stimulated transcription levels (Fig. 7B). Functional and physical interactions between Sp1 and Smad3, but not between Sp3 and Smad3, were further demonstrated using the bacterial GAL4 fusion protein system and immunoprecipitation–Western blot analyses, respectively (Inagaki *et al.*, 2001b).

Sp1 and Sp3 are closely related proteins with very similar structural features (Hagen *et al.*, 1994). They bind to the common GC-rich sequence named GC box with the same specificity and affinity and regulate gene transcription. However, Sp3 often acts as a transcriptional repressor by binding competitively to the Sp1-bound GC box sequences (Hagen *et al.*, 1994). It is now recognized that Sp3 can either activate or repress transcription of target genes depending on the cell type, the context of DNA-binding sites, and interactions with other nuclear factors (Majello *et al.*, 1997). Our results indicated that the two members of the GC box-binding factor family, Sp1 and Sp3, participate in the regulation of COL1A2 transcription through differential interaction with Smad3. They also suggested that predominant binding of Sp3, rather than Sp1, to the box 3A sequence and a lack of interaction with

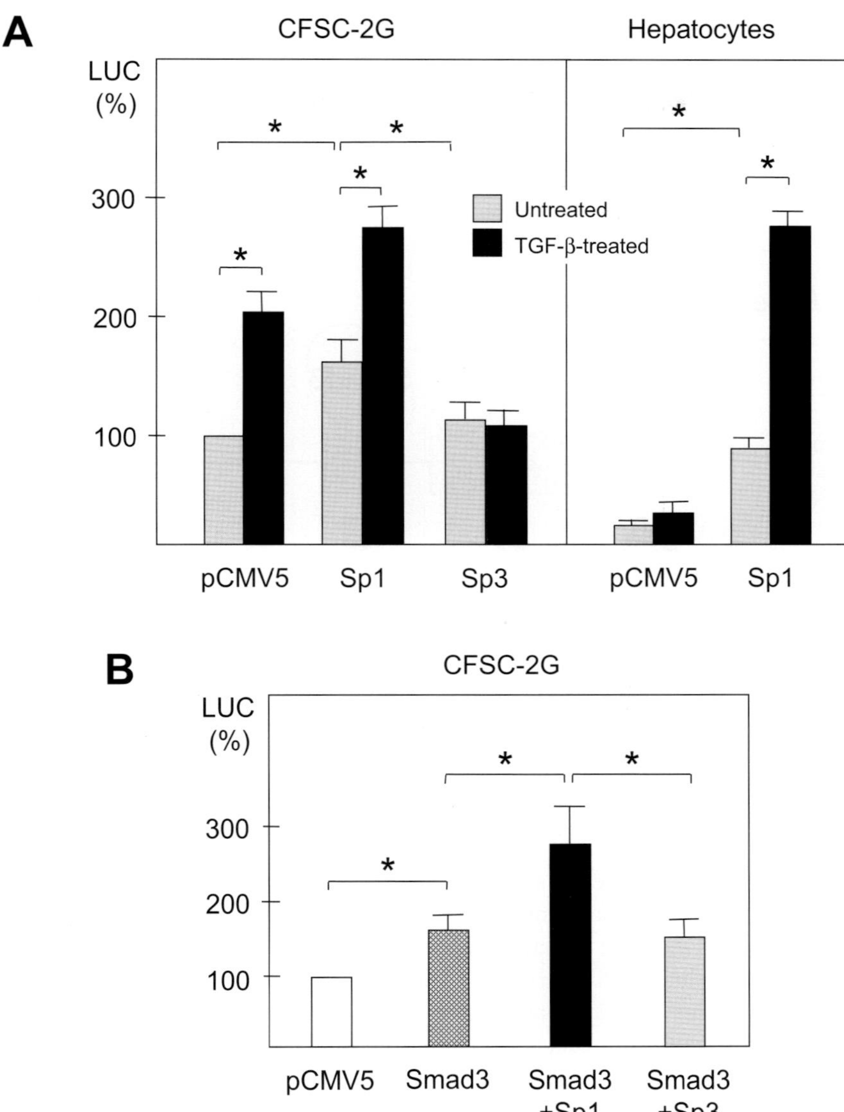

FIGURE 7 Functional cooperation between GC box-binding factors and Smad3. (A) An activated stellate clone (CFSC-2G cells) or a primary culture of hepatocytes was cotransfected with the −378COL1A2 promoter/reporter construct and either the Sp1 or the Sp3 expression plasmid. (B) Synergistic stimulatory effects of Smad3 and Sp1/Sp3 on COL1A2 transcription were examined by cotransfecting CFSC-2G cells with a Smad3 expression vector and either the Sp1 or the Sp3 expression plasmid. Transcriptional activities in each transfectants were expressed relative to the activity in CFSC-2G cells cotransfected with the control empty vector (pCMV5). The asterisk indicates that there is a significant difference between the groups.

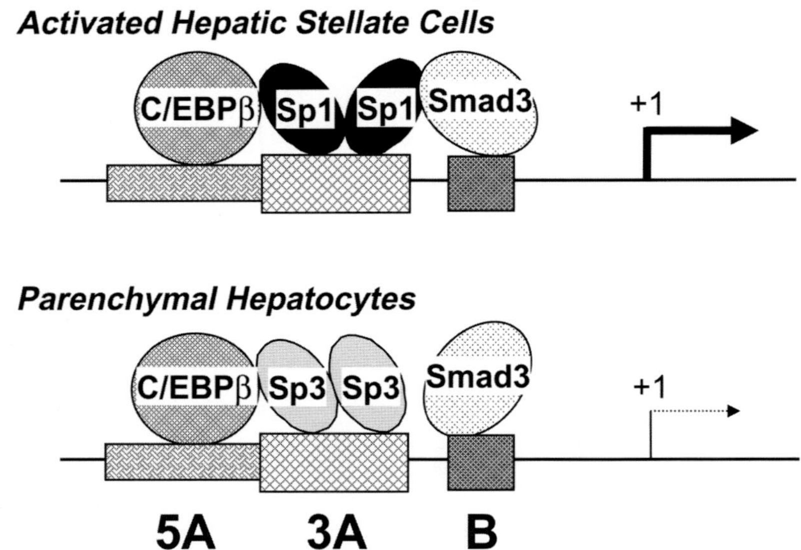

FIGURE 8 Schematic representation of the molecular mechanisms responsible for differential COL1A2 transcription and TGF-β responsiveness in activated HSC and parenchymal hepatocytes.

Smad3 may account, at least in part, for relatively low levels of COL1A2 transcription and loss of TGF-β responsiveness in parenchymal hepatocytes (Fig. 8).

VII. CONCLUSION

Altogether, our experimental evidence indicates that the upstream sequence of the COL1A2 promoter containing the TbRE is responsible for the transcriptional upregulation of the gene in activated HSC and is activated in a cell type-specific manner during hepatic fibrogenesis *in vivo*. Our original observation that Smad3 plays an important role in collagen gene transcription in activated HSC has been supported by a study utilizing Smad3 knockout mice (Schnabl *et al.*, 2001). These results lead not only to the better understanding of molecular pathogenesis underlying hepatic fibrogenesis but also to the development of therapeutic means that prevent hepatic fibrosis by suppressing pathologically activated collagen gene transcription in HSC.

ACKNOWLEDGMENTS

The following people have contributed greatly to the work: Drs. Patricia Greenwel and Sharada Truter at Mount Sinai School of Medicine in New York and Drs. Mizuko Mamura and Yutaka

Kanamaru at Juntendo University School of Medicine in Tokyo. The authors thank Dr. Francesco Ramirez at Mount Sinai School of Medicine in New York for his kind cooperation and continuous encouragement throughout the work. This work was supported in part by a research grant from the Scleroderma Research Committee of the Ministry of Health and Welfare, Japan.

REFERENCES

Boast, S., Su, M.-W., Ramirez, F., Sanchez, M., and Avvedimentro, E. V. (1990). Functional analysis of *cis*-acting DNA sequences controlling transcription of the human type I collagen genes. *J. Biol. Chem.* **265**, 13351–13356.

Bou-Gharios, G., Garrett, L. A., Rossert, J., Niederreither, K., Eberspaecher, H., Smith, C., Black, C., and de Crombrugghe, B. (1996). A potent far-upstream enhancer in the mouse proα2(I) collagen gene regulates expression of reporter genes in transgenic mice. *J. Cell Biol.* **134**, 1333–1344.

Brenner, D. A., Veloz, L., Jaenisch, R., and Alcorn, J. M. (1993). Stimulation of the collagen α1(I) endogenous gene and transgene in carbon tetrachloride-induced hepatic fibrosis. *Hepatology* **17**, 287–292.

Chen, S.-J., Yuan, W., Lo, S., Trojanowska, M., and Varga, J. (2000). Interaction of Smad3 with a proximal Smad-binding element of the human α2(I) procollagen gene promoter required for transcriptional activation by TGF-β. *J. Cell. Physiol.* **183**, 381–392.

Chen, S.-J., Yuan, W., Mori, Y., Levenson, A., Trojanowska, M., and Varga, J. (1999). Stimulation of type I collagen transcription in human skin fibroblasts by transforming growth factor-β: Involvement of Smad3. *J. Invest. Dermatol.* **112**, 49–57.

Diegelmann, R. F., Lindblad, W. J., and Cohen, I. K. (1988). Fibrogenic processes during tissue repair. *In* "Collagen: Biochemistry and Biomechanics" (M. E. Nimni, ed.), Vol. 2, pp. 113–131, CRC Press, Boca Raton.

Dooley, S., Delvoux, B., Lahme, B., Mangasser-Stephan, K., and Gressner, A. M. (2000). Modulation of transforming growth factor-β response and signaling during transdifferentiation of rat hepatic stellate cells to myofibroblasts. *Hepatology* **31**, 1094–1106.

Friedman, S. L. (1990). Cellular sources of collagen and regulation of collagen production in liver. *Semin. Liver Dis.* **10**, 20–29.

Greenwel, P., Inagaki, Y., Hu, W., Walsh, M., and Ramirez, F. (1997). Sp1 is required for the early response of α2(I) collagen to transforming factor-β1. *J. Biol. Chem.* **272**, 19738–19745.

Greenwel, P., Rubin, J., Schwartz, M., Hertzberg, E. L., and Rojkind, M. (1993). Liver fat-storing cell clones obtained from a CCl_4-cirrhotic rat are heterogeneous with regard to proliferation, expression of extracellular matrix components, interleukin-6 and connexin 43. *Lab. Invest.* **69**, 210–217.

Greenwel, P., Tanaka, S., Penkov, D., Zhang, W., Olive, M., Moll, J., Vinson, C., Di Liberto, M., and Ramirez, F. (2000). Tumor necrosis factor alpha inhibits type I collagen synthesis through repressive CCAAT/enhancer-binding proteins. *Mol. Cell. Biol.* **20**, 912–918.

Hagen, G., Muller, S., Beato, M., and Suske, G. (1994). Sp1-mediated transcriptional activation is repressed by Sp3. *EMBO J.* **13**, 3843–3851.

Heldin, C.-H., Miyazono, K., and ten Dijke, P. (1997). TGF-β signaling from cell membrane to nucleus via Smad proteins. *Nature* **390**, 465–471.

Houglum, K., Buck, M., Alcorn, J., Contreras, S., Bornstein, P., and Chojkier, M. (1995). Two different *cis*-acting regulatory regions direct cell-specific transcription of the collagen α1(I) gene in hepatic stellate cells and in skin and tendon fibroblasts. *J. Clin. Invest.* **96**, 2269–2276.

Inagaki, Y., Mamura, M., Kanamaru, Y., Greenwel, P., Nemoto, T., Takehara, K., ten Dijke, P., and Nakao, A. (2001a). Constitutive phosphorylation and nuclear localization of Smad3 are correlated with increased collagen gene transcription in activated hepatic stellate cells. *J. Cell. Physiol.* **187**, 117–123.

Inagaki, Y., Nemoto, T., Nakao, A., ten Dijke, P., Kobayashi, K., Takehara, K., and Greenwel, P. (2001b). Interaction between GC box binding factors and Smad proteins modulates cell lineage-specific α2(I) collagen gene transcription. *J. Biol. Chem.* **276**, 16573–16579.

Inagaki, Y., Truter, S., Bou-Gharios, G., Garrett, L.-A., de Crombrugghe, B., Nemoto, T., and Greenwel, P. (1998). Activation of proα2(I) collagen promoter during hepatic fibrogenesis in transgenic mice. *Biochem. Biophys. Res. Commun.* **250**, 606–611.

Inagaki, Y., Truter, S., Greenwel, P., Rojkind, M., Unoura, M., Kobayashi, K., and Ramirez, F. (1995a). Regulation of the α2(I) collagen gene transcription in fat-storing cells derived from a cirrhotic liver. *Hepatology* **22**, 573–579.

Inagaki, Y., Truter, S., and Ramirez, F. (1994). Transforming growth factor-β stimulates α2(I) collagen gene expression through a *cis*-acting element that contains an Sp1 binding site. *J. Biol. Chem.* **269**, 14828–14834.

Inagaki, Y., Truter, S., Tanaka, S., Di Liberto, M., and Ramirez, F. (1995b). Overlapping pathways mediate the opposing actions of tumor necrosis factor-α and transforming growth factor-β on α2(I) collagen gene transcription. *J. Biol. Chem.* **270**, 3353–3358.

Kikuchi, K., Hartl, C. W., Smith, E. A., LeRoy, E. C., and Trojanowska, M. (1992). Direct demonstration of transcriptional activation of collagen gene expression in systemic sclerosis fibroblasts: Insensitivity to TGF-β1 stimulation. *Biochem. Biophys. Res. Commun.* **187**, 45–50.

Majello, B., De Luca, P., and Lania, L. (1997). Sp3 is a bifunctional transcription regulator with modular independent activation and repression domains. *J. Biol. Chem.* **272**, 4021–4026.

Mak, K. M., Leo, M. A., and Lieber, C. S. (1984). Alcoholic liver injury in baboons: Transformation of lipocytes to transitional cells. *Gastroenterology* **87**, 188–200.

Nakao, A., Afrakhte, M., Moren, A., Nakayama, T., Christian, J. L., Heuchel, R., Itoh, S., Kawabata, M., Heldin, N.-E., Heldin, C.-H., and ten Dijke, P. (1997a). Identification of Smad7, a TGFβ-inducible antagonist of TGF-β signaling. *Nature* **389**, 631–635.

Nakao, A., Imamura, T., Souchenlnytskyi, S., Kawabata, M., Ishisaki, A., Oeda, E., Tamaki, K., Hanai, J., Heldin, C.-H., Miyazono, K., and ten Dijke, P. (1997b). TGF-β receptor-mediated signaling through Smad2, Smad3 and Smad4. *EMBO J.* **16**, 5353–5362.

Ramirez, F., and Di Liberto, M. (1990). Complex and diversified regulatory programs control the expression of vertebrate collagen genes. *FASEB J.* **4**, 1616–1623.

Schnabl, B., Kweon, Y. O., Frederick, J. P., Wang, X.-F., Rippe, R. A., and Brenner, D. A. (2001). The role of Smad3 in mediating mouse hepatic stellate cell activation. *Hepatology* **34**, 89–100.

Walton, C. M., Wu, G. Y., Petruff, C. A., Clark, S. H., Lichtler, A. C., and Wu, C. H. (1996). A collagen enhancer-promoter construct in transgenic mice is markedly stimulated by ethanol administration. *Hepatology* **23**, 310–315.

Zhang, W., Oul, J., Inagaki, Y., Greenwel, P., and Ramirez, F. (2000). Synergistic cooperation between Sp1 and Smad3/Smad4 mediates transforming growth factor β1 stimulation of α2(I)-collagen (COL1A2) transcription. *J. Biol. Chem.* **275**, 39237–39245.

PART IV

Basic Science of Matrix Metalloproteinases and Tissue Inhibitor Metalloproteinases

CHAPTER 14

Advances in Matrix Metalloproteinases (MMPs), Membrane-Type MMPs, and a Disintegrin and Metalloproteinase and Their Roles in Cellular Interaction and Migration

KAZUKI NABESHIMA,* TERUHIKO INOUE,* YOSHIYA SHIMAO,* AND TETSURO SAMESHIMA[†]

Departments of Pathology and Neurosurgery,[†] Miyazaki Medical College, Miyazaki 889-1692, Japan*

It is now generally thought that proteolytic cleavage of the extracellular matrix (ECM) and cell surface molecules is not simply remodeling of the pericellular structure, but is itself a component of cellular interactions that have inside-out and outside-in signals. From this point of view, this chapter provides advances in mechanisms for cell surface localization and activation of matrix metalloproteinases (MMPs), as well as their role in cell migration and cell–cell interaction-dependent induction of MMPs. Although the function of a disintegrin and metalloproteinase (ADAM) and a disintegrin and metalloproteinase with thrombospondin motif (ADAMTS) gene products is unknown, they can potentially provide both attachment and proteolysis in close topical proximity based on their domain structure. Their possible involvement in cell–cell and cell–ECM interactions is also discussed.

I. INTRODUCTION

The extracellular matrix provides cells with not only mechanical support, but also various cellular functions through cell–ECM interactions. Thus, proteolysis of

Extracellular Matrix and the Liver
Copyright 2003, Elsevier Science (USA). All rights reserved.

ECM affects both structural remodeling associated with development or repair caused by disease processes and diverse physiological events such as cellular differentiation, growth, programmed cell death, migration, and invasion in the tissue. Cell surface and ECM proteinases play pivotal roles in such interactions (Streuli, 1999).

Matrix metalloproteinases and ADAMs are involved in cell–ECM interactions. MMP and ADAM proteases fall within the matrixin and adamalysin/reprolysin subfamily of the metzincin superfamily of Zn-dependent metalloproteinases, respectively (Stocker *et al.*, 1995). MMPs share two well-conserved domains—the prodomain and the catalytic domain—and act on a broad spectrum of ECM components (Nagase and Woessner, 1999). The catalytic domain contains an active site Zn^{2+} that binds three conserved histidines in the sequence HEXXHXXGXXH(S/T) (Lohi *et al.*, 2001). Mammalian MMPs are classified into a soluble (secreted) type and a membrane type based on their domain structures (Fig. 1) (Seiki, 1999). MMPs are all synthesized as prepro enzymes, and soluble-type MMPs are secreted as inactive pro-MMPs in most cases, whereas membrane-type MMPs (MT-MMPs) are activated intracellularly and are expressed on the cell surface as an active enzyme (Nagase and Woessner, 1999; Seiki, 1999). Enzymatic activities of both soluble-type MMPs and MT-MMPs are inhibited by an endogenous MMP inhibitor, a tissue inhibitor of metalloproteinase (TIMP), which contains four members (TIMP-1–TIMP-4). ADAMs are a family of multidomain glycoproteins that contain both a disintegrin and a metalloproteinase domain and therefore combine features of both cell surface adhesion molecules and proteinases (Black and White, 1998; Kaushal and Shah, 2000; Tang, 2001). The ADAM family also contains soluble and membrane-type molecules. In typical

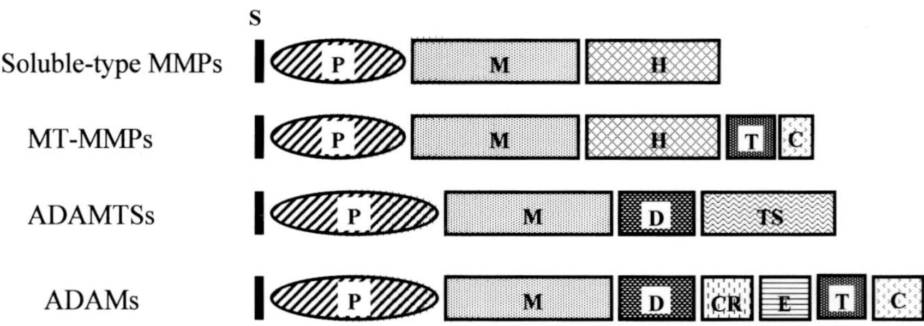

FIGURE 1 Schematic representation of the domain structure of MMP and ADAM families in mammals. Soluble-type MMPs and ADAMTSs do not contain transmembrane domains and are secreted, whereas MT-MMPs and ADAMs are transmembrane proteins that localize to the cell surface. S, signal peptide; P, prodomain; M, catalytic domain; H, hemopexin-like domain; T, transmembrane domain; C, cytoplasmic domain; D, disintegrin domain; TS, thrombospondin domain; CR, cysteine-rich domain; E, EGF-like domain.

ADAMs, an N-terminal signal peptide and a common prodomain are followed by metalloproteinase, disintegrin, cysteine-rich, EGF-like, transmembrane, and cytoplasmic tail domains (Fig. 1). New members of the ADAM family, known as ADAMTSs, lack the cysteine-rich, EGF-like, transmembrane, and cytoplasmic tail domains of ADAMs, but possess a unique thrombospondin domain that is responsible for ECM binding (Kaushal and Shah, 2000). ADAMTSs do not contain transmembrane domains and are therefore secreted, whereas ADAMs are transmembrane proteins that localize to the cell surface. Soluble MMPs and ADAMTs may show broader accessibility into ECM and play a role in tissue remodeling events. However, MT-MMPs and ADAMs are possibly involved in the regulation of pericellular environments closely associated with cellular functions, including migration and invasion.

II. SOLUBLE AND MEMBRANE-TYPE MMPs AND THEIR ACTIVATION

To date, 23 different MMPs have been cloned and additional members continue to be identified (Nagase and Woessner, 1999; Lohi *et al.*, 2001). Soluble MMPs can be further classified into collagenases, stromelysins, gelatinases, matrilysins, and others based on substrate specificity (Table I) (Chambers and Matrisian, 1997). The second group is made of 6 MT-MMPs and CA (cysteine-array)-MMP (MMP-23), all of which are expressed as integral membrane zymogens (Seiki, 1999; Pei *et al.*, 2000). Based on the membrane-anchoring mechanisms, these membrane-associated MMPs can be further divided into type I transmembrane MMPs (MT1-3, 5-MMPs), glycosylphosphatidylinositol (GPI)-anchored MMPs (MT4, 6-MMPs), and a type II transmembrane MMP (MMP-23) (Pei *et al.*, 2000). MT-MMPs contain a basic motif that can be recognized by proprotein convertases (PCs) between pro- and catalytic domains. Therefore, they are supposed to be activated intracellularly by PCs and appear on the cell surface as an active form (Seiki, 1999). MMP-23 also has a PC motif between its pro- and catalytic domains. Because of its topology with an N-terminal signal anchor, however, a proteolytic cleavage by PCs would not only activate MMP-23 but also release the processed enzyme into the lumen of the secretory pathway for secretion (see domain structure of MMP-23 in Table I) (Pei *et al.*, 2000). Thus, although all these MMPs contain transmembrane or membrane-anchoring domains, 6 MT-MMPs exert their activities on the cell surface, whereas MMP-23 may act like a soluble MMP. In this light, MMP-23 may be similar to stromelysin 3 (MMP-11) and epilysin (MMP-28), which also contain a PC motif, but not the transmembrane domain, and are presumably activated intracellularly and secreted (Pei and Weiss, 1995; Lohi *et al.*, 2001). Two sets of basic motifs have been identified in the propeptide region of MT1-MMP, and both furin (one of PCs)-dependent and

TABLE I The MMP Gene Family

Protein	MMP	Molecular mass (kDa)	Domain structure[a]
Soluble-type MMPs			
Collagenase			
Collagenase 1	MMP-1	52/61	S+P+M+H
Collagenase 2	MMP-8	85/64	S+P+M+H
Collagenase 3	MMP-13	65/55	S+P+M+H
Stromelysin			
Stromelysin 1	MMP-3	57/45, 28	S+P+M+H
Stromelysin 2	MMP-10	56/47, 24	S+P+M+H
Gelatinase			
Gelatinase A	MMP-2	72/67	S+P+M(Fn)+H
Gelatinase B	MMP-9	92/67	S+P+M(Fn)+H
Matrilysin			
Matrilysin 1	MMP-7	28/19	S+P+M
Matrilysin 2	MMP-26	29/19	S+P+M
Others			
Stromelysin 3	MMP-11	58/28	S+P(bm)+M+H
Epilysin	MMP-28	56/45	S+P(bm)+M+H
No trivial name	MMP-19	57	S+P+M+H
Metalloelastase	MMP-12	54/45, 22	S+P+M+H
Enamelysin	MMP-20	54/43	S+P+M+H
Membrane-type MMPs			
Type I transmembrane type			
MT1-MMP	MMP-14	66/60	S+P(bm)+M+H+TM+C
MT2-MMP	MMP-15	68/62	S+P(bm)+M+H+TM+C
MT3-MMP	MMP-16	64/55	S+P(bm)+M+H+TM+C
MT5-MMP	MMP-24	73/64	S+P(bm)+M+H+TM+C
GPI type			
MT4-MMP	MMP-17	71/67	S+P(bm)+M+H+GPI
MT6-MMP	MMP-25	62/58	S+P(bm)+M+H+GPI
Type II transmembrane type			
CA-MMP	MMP-23	~66	SA+TM+P(bm)+M+CA+Ig

[a]S, signal peptide; P, prodomain; M, catalytic domain; H, hemopexin-like domain; Fn, fibronectin-type II domain; bm, basic motif (RRKR or RRRR) that can be recognized by proprotein convertases; SA, signal anchor; CA, cysteine array; Ig, Ig-like domain; GPI, glycosyl-phosphatidyl-inositol anchor signal.

furin-independent MT1-MMP processing pathways are demonstrated that require tethering of the metalloproteinase to the cell surface (Yana and Weiss, 2000).

A. Soluble-Type MMPs

Most soluble-type MMPs are secreted from the cell as inactive zymogen with the prodomain. The prodomain of approximately 80 amino acid residues has a

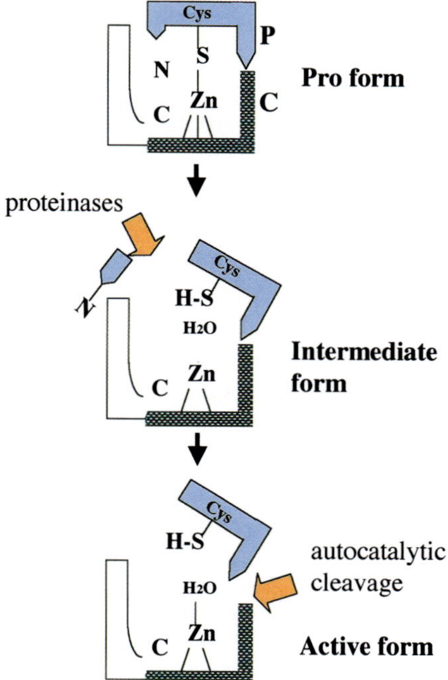

FIGURE 2 Schematic presentation of activation of pro-MMPs. The activation process consists of a two-step reaction: initial cleavage at the propeptide bait region, leading to removal of the N-terminal polypeptide, and an autoproteolytic reaction that generates the stable active enzyme.

conserved unique PRCG(V/N)PD sequence. The cysteine residue within this sequence (the "cysteine switch") ligates the catalytic zinc to maintain the latency of pro-MMPs (Van Wart and Birkedal-Hansen, 1990; Nagase and Woessner, 1999). Activation requires the disruption of the cysteine–zinc (cysteine switch) interaction possibly by tissue or plasma proteinases, and this process often consists of a two-step reaction (Fig. 2) (Nagase *et al.*, 1990; Nabeshima *et al.*, 1999b). An initial cleavage at the propeptide bait region leads to removal of the N-terminal polypeptide, which is followed by the second step, an intra- and inter-molecular autoproteolytic reactions that generate the stable active enzyme (Grant *et al.*, 1987). Pathophysiological significance of the urokinase-type plasminogen activator (uPA)/plasmin system as an activator of pro-MMPs has been suggested using transgenic mice deficient in uPA (Carmeliet *et al.*, 1997). This activation of pro-MMPs, including collagenases, stromelysins, and matrilysins, can occur distant from cells in ECM or at the cell surface through the uPA/PA receptor (PAR)/plasminogen cascade for plasmin generation (Murphy *et al.*, 1992).

B. MT-MMPs

The discovery of the first MT-MMP (MT1-MMP) by Sato *et al.* (1994) and subsequent five MT-MMPs (MT2-6-MMP) has strengthened the concept of pericellular activation of MMPs. Four MT-MMPs (MT1-3, 5-MMP) have been shown to activate progelatinase A (pro-MMP-2), although MT1-MMP activates it most efficiently (Seiki, 1999). This progelatinase A is known to have a propeptide that is not susceptible to proteolytic initiation of activation by serine proteinases (Okada *et al.*, 1990). Moreover, gelatinase A is believed to be particularly important for tumor invasion of the basement membrane because (i) it degrades type IV collagen, (ii) it is activated in a tumor-specific manner, and (iii) its activation correlates with tumor spread and poor prognosis (Seiki, 1999). Thus, elucidation of progelatinase A activation mechanisms has provided great progress in understanding cell migration and invasion that are closely related to the pericellular rearrangement of ECM. This progelatinase A activation by MT1-MMP expressed on the cell surface involves a two-step activation mechanism like the activation of other MMPs. MT1-MMP-mediated proteolysis causes an initial cleavage of the N-terminal propeptide sequence of progelatinase A, resulting in secondary autoproteolytic cleavage and generation of fully active gelatinase A (Fig. 3) (Will *et al.*, 1996).

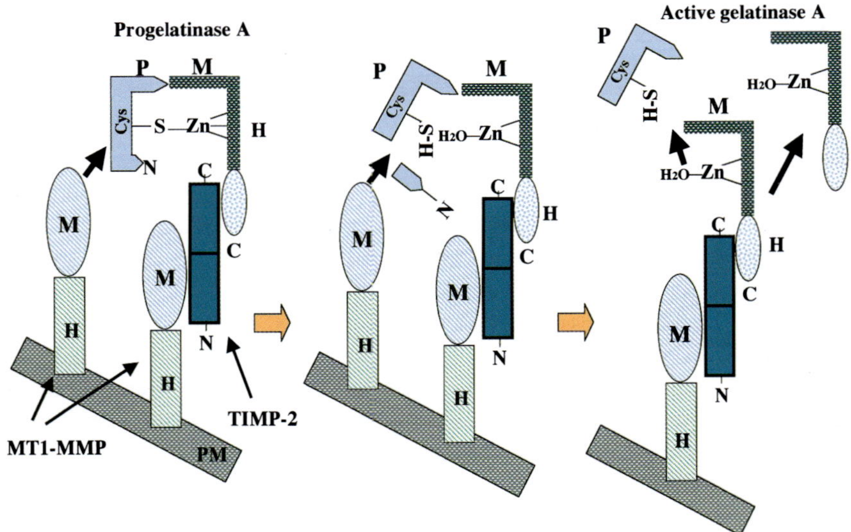

FIGURE 3 Schematic presentation of progelatinase A activation by MT1-MMP. Free MT1-MMP activates progelatinase A trapped in a trimolecular complex of MT1-MMP/TIMP-2/progelatinase A. P, prodomain; M, catalytic domain; H, hemopexin-like domain; PM, plasma membrane; C, C-terminal; N, N-terminal.

This activation process requires both active MT1-MMP and the TIMP-2-bound MT1-MMP. The TIMP-2 in the latter complex binds, through its C-terminal domain, to the hemopexin domain (GelA PEX) of progelatinase A, which is assumed to localize the zymogen close to the active MT1-MMP (Strongin *et al.*, 1995; Butler *et al.*, 1996; Kinoshita *et al.*, 1998; Nagase and Woessner, 1999). In other words, MT1-MMP/TIMP-2 can act as a mechanism to concentrate the substrate on the cell surface, where the activator enzyme is available (Kinoshita *et al.*, 1998). It is reported that MT1-MMP can also activate procollagenase 3, although it is not at all clear whether this process also requires the presence of TIMP-2 (Knäuper *et al.*, 1996). Moreover, because collagenase 3 activates progelatinase B, the cell surface activation cascade that is initiated by MT1-MMP may exist (Knäuper *et al.*, 1997).

In addition to activation of progelatinase A, MT1-MMP can directly degrade broad-spectrum ECM proteins, such as type I, II, and III collagens, fibronectin, laminin 5, aggrican, and others (Ohuchi *et al.*, 1997). This direct effect on ECM might be important, as MT1-MMP-deficient mice cause impaired endochondral ossification and osteopenia, despite the fact that gelatinase A-deficient mice show almost normal phenotypes (Itoh *et al.*, 1998; Holmbeck *et al.*, 1999; Zhou *et al.*, 2000). GPI-anchored MT4-MMP cleaves a peptide corresponding to the ectodomain cleavage site of protumor necrosis factor (TNF)-α and sheds pro-TNF-α when cotransfected in COS-7 cells (English *et al.*, 2000). MMP-1 and MMP-9 also cleave the human pro-TNF-α peptide. MMPs and MT-MMPs may be involved in pericellular signal networks via cleavage of non-ECM substrates. Furthermore, because GPI-anchored proteins are concentrated into the microdomains (lipid rafts) on the living cell surface in a GPI moiety-dependent manner, MT4-MMP may cooperate with other GPI-anchored proteins, such as uPAR (Itoh *et al.*, 1999).

III. ADAMs AND ADAMTSs

As mentioned earlier, ADAMs are transmembrane proteinases, whereas ADAMTSs are soluble extracellular matrix proteinases. There are fundamental differences between these two types of enzymes. The known substrates of ADAMs are other transmembrane proteins, and those of ADAMTs are other extracellular matrix proteins (Primakoff and Myles, 2000). Most ADAM and ADAMTS family members have a furin cleavage motif at the end of the propeptide (Tang, 2001). Thus, they are generally thought to be activated intracellularly by PCs and expressed on the cell surface or secreted as an active form. However, secretion and deposition of ADAMTS-1 in two forms, a larger precursor form with the prodomain and a smaller mature form without it, have been reported (Kuno and Matsushima, 1998). More interestingly, the prodomain in conjunction with the proteinase domain has ECM-binding capacity, as the precursor form of a deletion mutant with only these domains intact could still be found in the ECM.

A. ADAMs

The ADAM gene family has 29 members to date, although the function of most ADAM gene products is unknown (Primakoff and Myles, 2000; Tang, 2001; White *et al.*, 2001). Among them, 17 have a metalloproteinase catalytic site with the correct amino acid sequence (HESGHXXGXXHD) and thus are predicted to be active proteinases [ADAMs 1, 8–10, 12, 13 (33), 15–17, 19–21, 24–26, 28, and 30] (Table II). The other ADAMs are believed to lack proteinase activity, as their amino acid sequences reveal one or more residues in the active-site region that are incompatible with metalloprotease activity (Primakoff and Myles, 2000; White *et al.*, 2001). ADAM proteases are inhibited by EDTA and *o*-phenanthroline, which both chelate Zn^{2+} ions, but not by TIMP-1, an endogenous MMP inhibitor, as far as we know (Black and White, 1998). Three active ADAM proteinases, ADAM 17/TACE, ADAM 10/Kuzbanian, and ADAM 9/MDC9, have been identified to be sheddases (secretases), which cleave and thereby release the extracellular domain of plasma membrane-anchored cytokines, growth factors, receptors, adhesion molecules, and enzymes. The other predicted proteinases lack

TABLE II Features of ADAM and ADAMTS Proteins[a,b]

ADAMs	
Identification in human	ADAM 1–3, 7–12, 15, 17–23, 28–30, and 33 (human 13)
Active metalloproteinase	ADAM 1, 8–10, 12, 13 (33), 15–17, 19–21, 24–26, 28, and 30
Sheddase	ADAM 9 (MDC9), 10 (Kuzbanian), 17 (TACE)
Cell fusion related	ADAM 1, 2 (fertilin α, β), 12 (meltrin α)
Binding to integrin	ADAM 2—integrin α6β3
	ADAM 15 (metargidin)—integrin αvβ3
Testis-specific expression	ADAM 2, 3, (5), (6), 16, 18, 20, 21, 24–26, 29, and 30
ADAMTSs	
Known physiological substrates	ADAMTS2 (procollagen N-proteinase); ADAMTS4 and 5/11 (aggrecanase-1 and -2)
Tissue development	ADAMTS1
Tumorigenesis	ADAMTS1 (colon carcinoma); ADAMTS4 (glioma)
Antiangiogenesis	ADAMTS1 (METH-1) and 8 (METH-2)

[a]Resources to track current and new ADAM and ADAMTS family members include http://www.people.virginia.edu/~jag6n/whitelab.html; http://www.uta.fi/~loika; and http://www.gene.ucl.ac.uk/agi-bin/nomenclature/searchgenes.pl?

[b]Terms in parentheses are aliases.

an identified endogenous substrate. ADAM 17 proteolytically cleaves a 26-kDa membrane-anchored TNF-α protein and releases the soluble and active 17-kDa form, acting as a TNF-α converting enzyme (TACE) (Moss *et al.*, 1997). TACE knockout mice also cannot release embryonic transforming growth factor α (TGF-α), resulting in embryonic lethality. Moreover, functional TACE has been shown to be responsible for shedding of the TNF receptor, the adhesion molecule L-selectin, and amyloid protein precursor (APP), suggesting that TACE is a sheddase with multiple substrates (Peschon *et al.*, 1998; Primakoff and Myles, 2000). ADAM 10, also called Kuzbanian, releases a soluble form of Delta, a Notch ligand (Qi *et al.*, 1999). ADAM 10 can also cleave Notch, but this cleavage does not result in release of the Notch extracellular domain (Artavanis-Tsakonas *et al.*, 1999). Proper formation of the central nervous system (CNS) requires that only a limited number of cells with potential for neural differentiation actually become mature neurons: surrounding cells are prevented from adopting this fate by the process of lateral inhibition (Black and White, 1998). The Notch-Delta system and thus also ADAM 10 are involved in this lateral inhibition and consequent normal development of the CNS. ADAM 9/MDC9 (metalloprotease disintegrin cysteine-rich 9) sheds the heparin-binding EGF-like growth factor (Izumi *et al.*, 1998). For these shedding processes, it is possible that the ADAM might initially adhere to its substrate using its disintegrin domain and subsequently cleave the substrate proteolytically. To date, however, this has not been shown yet (Primakoff and Myles, 2000). Based on the domain structure of ADAMs, their participation in cell adhesion is also expected. ADAM 12/meltrin α promotes myoblast fusion into myotubes, a cell–cell fusion process that occurs after myogenesis begins, through its disintegrin domain (Yagami-Hiromasa *et al.*, 1995). Another example is the sperm surface protein fertilin, which is a heterodimer that contains two subunits, ADAM 1 and 2 (also called fertilin α and β, respectively). It has been hypothesized that a fertilin disintegrin domain, especially that of ADAM 2, could act to bind sperm to the egg (Primakoff *et al.*, 1987; Primakoff and Myles, 2000). In this process, sperm ADAM 2 and egg integrin α6β1 are suggested to be adhesion partners (Chen and Sampson, 1999). Additionally, the disintegrin domain of ADAM 15/Metargidin is also shown to bind integrin αvβ3 (Zhang *et al.*, 1998). ADAM 15 is the only ADAM that has an RGD sequence in its predicted disintegrin domain active site. Because integrins have signaling domains on both sides, involvement of ADAMs in bidirectional signaling is possible.

B. ADAMTSs

Eleven members of the ADAMTS family are currently known according to the human genome organization (HUGO) gene nomenclature committee (HGNC) family resources (http://www.gene.ucl.ac.uk/nomenclature/). All the ADAMTS described to date have the catalytic site concensus of a Zn-binding peptidase HEXXH, and therefore are presumed to be catalytically active (Tang, 2001). However, physiological substrates have been identified only for ADAMTS2, 4, and 5/11 (Table II). ADAMTS2 is known as a procollagen N-proteinase, which proteolytically removes amino peptides in the processing of type I and type II procollagens to collagens (Colige *et al.*, 1997). Because collagen is synthesized as precursors in the form of procollagen triple helices flanked at the N terminus and C terminus by propeptides, the removal of these propeptides by specific proteinases is necessary for the subsequent assembly of collagen fibrils in the ECM (Tang, 2001). In accordance with this, deficiency of ADAMTS2 is responsible for the inherited connective tissue disorder, type VIIC Ehlers–Danros syndrome (Colige *et al.*, 1997). Although ADAMTS3 is very homologous to ADAMTS2, it is unknown at present if it has procollagen N or C proteinase activity (Tang, 2001). ADAMTS4 and ADAMTS11 (the alias for ADAMTS5) have been identified as aggrecanase-1 and -2 (Tortorella *et al.*, 1999; Abbaszade *et al.*, 1999). Aggrecan is a proteoglycan that maintains the mechanical properties of cartilage, and its depletion due to increased proteolytic cleavage is responsible for arthritis. Aggrecan contains two N-terminal globular domains, G_1 and G_2, separated by a proteolytically sensitive interglobular domain, followed by a glycosaminoglycan attachment region and a C-terminal globular domain, G_3. Cleavage of aggrecan has been shown to occur at Asn^{341}-Phe^{342} and Glu^{373}-Ala^{374} within the interglobular domain, and cleavage at the latter site is suggested to be responsible for the increased aggrecan degradation observed in inflammatory joint disease (Abbaszade *et al.*, 1999). While cleavage at Asn^{341}-Phe^{342} has been shown by several MMPs (MMP-1, 2, 3, 7, 8, 9, and 13) (Sandy *et al.*, 1991), only aggrecanase-1 and -2 have been responsible for the cleavage at Glu^{373}-Ala^{374} so far (Tortorella *et al.*, 1999; Abbaszade *et al.*, 1999), indicating their pivotal role in arthritis. The thrombospondin motif of aggrecanase-1 is shown to be critical for substrate recognition and cleavage: the thrombospondin motif binds to the glycosaminoglycans (GAG) of aggrecan, and aggrecanase-1 is not effective in cleaving GAG-free aggrecan. Furthermore, aggrecanase-1 with the thrombospondin motif deleted was not functional (Tortorella *et al.*, 2000). To date, no physiological substrates have been identified for ADAMTS1, although its biological role is well shown through the study on the phenotype of the ADAMTS1 knockout mouse (Shindo *et al.*, 2000). The knockout exhibited significant growth retardation, with ureteropelvic junction obstruction due to fibrosis there and consequent hydronephrosis. The severity of this phenotype of the ADAMTS1 knockout mouse contrasts those of the soluble MMP knockouts, which are rather mild (Shapiro, 1998; Tang, 2001). This may indicate nonredundant and thus essential roles of ADAMTS in tissue development. In this light, identification of specific physiological substrates is awaited. Other suggested functions are shown in Table II.

IV. MMP INDUCTION BY CELL–CELL INTERACTION: A ROLE FOR EMMPRIN

In addition to extracellular and intracellular regulation by proteolytic activation of the proenzyme and by the inhibitory effects of TIMPs, MMPs are also regulated transcriptionally as a result of the action of various cytokines, growth factors, and hormones (Nabeshima *et al.*, 1999b). However, studies suggest that not only these soluble factors but also cell–cell or cell–matrix interactions play a key role. These direct interactions are mediated by integrins, nonintegrin cell adhesion molecules, and other transmembrane proteins. For example, the $\alpha 5\beta 1$ integrin–fibronectin interaction upregulates gelatinase B expression in macrophages (Xie *et al.*, 1998), and cellular interaction with reconstituted type I collagen through $\alpha 2\beta 1$ integrin mediates MT1-MMP expression in endothelial cells (Haas *et al.*, 1998) and fibroblasts (Seltzer *et al.*, 1994). Regulation by non-integrin type cell adhesion molecules is observed during interactions between lymphoid and endothelial cells: (i) induction of gelatinase B in T lymphoma cells by adhesion to endothelial cells via interaction between leukocyte function-associated antigen-1 (LFA-1) and intercellular adhesion molecule-1 (ICAM-1) (Aoudjit *et al.*, 1998) and (ii) induction of gelatinase A in T cells through very late antigen 4 (VLA-4)-vascular cell adhesion molecule-1 (VCAM-1)-mediated adhesion to endothelial cells (Romanic and Madri, 1994). Noncell adhesion molecule type transmembrane proteins include emmprin (for extracellular matrix metalloproteinase inducer), which was discovered by the Biswas laboratory via a functional approach and shown to be a surface molecule on tumor cells that stimulates nearby fibroblasts to produce MMP-1, 2, and 3 (Biswas, 1982; Biswas *et al.*, 1995; Nabeshima *et al.*, 1999b). It has been suggested that emmprin also stimulates the production of pro-MMP-2 activators, MT1- and MT2-MMP, by fibroblasts (Fig. 4) (Sameshima *et al.*, 2000a). Thus, at the invasion front where tumor cells contact fibroblasts, emmprin stimulates pericellular proteolysis of ECM through the activation of progelatinase A on the cell surface.

A. Emmprin in Tumorigenesis

Upregulation of emmprin has been demonstrated in urinary bladder (Muraoka *et al.*, 1993), breast (Polette *et al.*, 1997; Caudroy *et al.*, 1999; Dalberg *et al.*, 2000), lung (Polette *et al.*, 1997; Caudroy *et al.*, 1999), and oral squamous cell carcinomas Bordador *et al.*, 2000), malignant melanoma in the early invasive phase van den Oord *et al.*, 1997), gliomas (Sameshima *et al.*, 2000b), and malignant lymphomas (K. Nabeshima and J. Suzumiya, unpublished results) compared with their normal counterparts. In gliomas, its expression levels correlate with astrocytoma progression (Sameshima *et al.*, 2000b). Furthermore, a role for emmprin

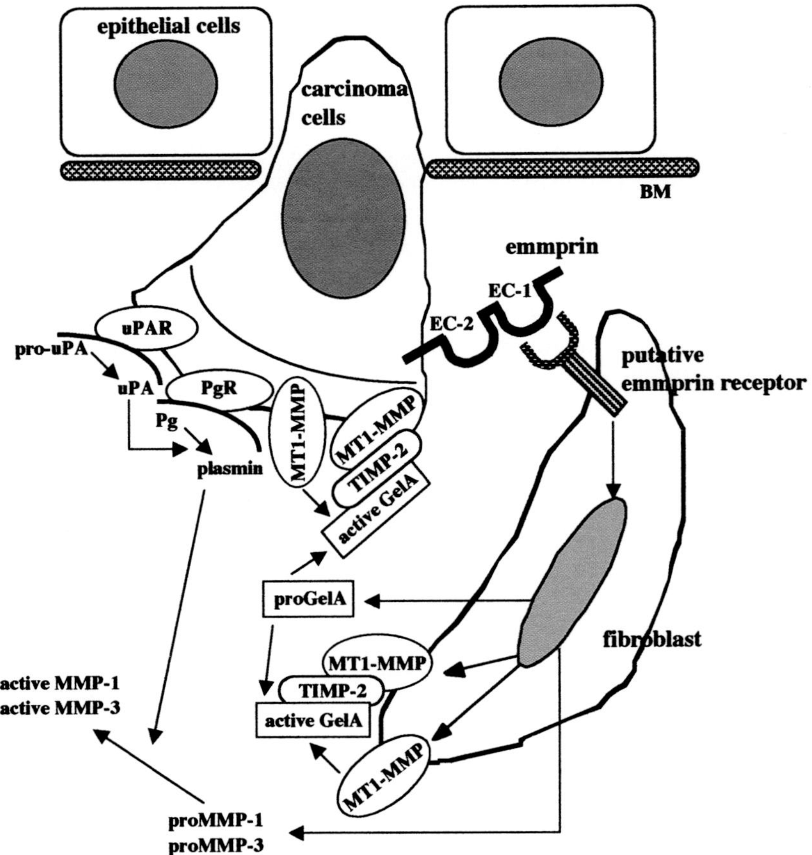

FIGURE 4 Induction and activation of MMPs via emmprin-dependent cancer cell–fibroblast interaction. uPA, urokinase-type plasminogen activator; uPAR, uPA receptor; Pg, plasminogen; PgR, plasminogen receptor; GelA, gelatinase A; BM, basement membrane.

in cancer progression has been shown by implantation of breast cancer cells that are transfected with emmprin cDNA into mouse mammary tissue: the transfected cancer cell clones were considerably more tumorigenic and invasive than plasmid-transfected cancer cells (Zucker *et al.*, 2001). Upregulation of emmprin is not restricted to neoplasms, but it also occurs in an inflammatory process such as rheumatoid arthritis (Konttinen *et al.*, 2000).

The cDNA for human emmprin encodes a 269 amino acid residue polypeptide that includes a signal peptide of 21 amino acid residues and a 185 amino acid extracellular domain consisting of two regions characteristic of the immunoglobulin (Ig) superfamily (EC-I and II) (Fig. 4), followed by a 24 amino acid residue transmembrane domain and a 39 amino acid cytoplasmic domain (Biswas *et al.*, 1995). The functional site for the metalloproteinase stimulatory activity of emmprin is supposed to be localized in the Ig domain I (EC-I) region, as the mutated emmprin protein lacking EC-I lost reactivity with E11F4, which is an activity-blocking monoclonal antibody (Biswas *et al.*, 1995). Additionally, posttranslational glycosylation of the protein is critical for emmprin activity because recombinant emmprin (r-emmprin) produced by bacteria, which has a molecular mass of ~29 kDa and is not glycosylated, is functionally inactive, whereas r-emmprin isolated from Chinese hamster ovary cells transfected with emmprin cDNA, which is highly glycosylated and has a molecular mass of ~58 kDa, successfully stimulates the production of MMP-1-3 (Biswas *et al.*, 1995; Guo *et al.*, 1997). We have found that synthetic peptides carrying sequences of this active domain of emmprin (EC-I) effectively inhibit the emmprin-dependent stimulation of MMP production in cocultured glioma cells and fibroblasts. Especially, the peptide carrying the sequence that contains a putative N-glycosylation site (Asn44) was the most effective inhibitor. On the contrary, the whole EC-I, substituted with sugars [a chitobiose unit; $(GlcNAc)_2$], stimulated the production and activation of progelatinase A by fibroblasts (T. Sameshima, K. Nabeshima, T. Inoue, H. Hojo, Y. Nakahara, B.P. Toole, Y. Okada, manuscript in preparation). Specific inhibition of emmprin activity may give us a new approach to suppress MMPs that are upregulated under pathological conditions, including tumors and inflammation.

V. ROLE OF METALLOPROTEINASES IN CELL MIGRATION

Cell migration plays a key role in a plethora of biological events, including morphogenesis, wound healing, and tumor metastasis (Murphy and Gavrilovic, 1999). Based on studies in many systems, degradation of ECM by MMPs is assumed to be a prerequisite for the cells to migrate into native or provisional tissue matrix (Stetler-Stevenson, 1993): cell migration can be enhanced by the overexpression of MMPs (Deryugina *et al.*, 1997), whereas overexpression of TIMPs or use of MMP inhibitors results in reduced migration (George *et al.*, 1998; Hiraoka *et al.*, 1998). Moreover, analysis in knockout mice for MMPs has shown alterations in some migration-related processes, such as pathologic inflammatory reactions, reduced angiogenesis, or delayed tumor progression in injection or implantation models, supporting a function of individual MMPs in cell motility (Shapiro, 1998; Itoh *et al.*, 1998). ADAMs are also important candidate molecules for the integration of cell adhesion and migration, and related matrix remodelling (Friedl and Bröcker, 2000). In the nematode *Caenorhabditis elegans*, gonad shape and size are determined by the migration of a leader cell, which is at the tip of the growing gonad arm. A metalloprotease (GON-1, similar to murine ADAMTS1) secreted by the leader cell has been found essential for this process, preparing the way for migration of the cell via modification of ECM (Moerman, 1999). In this migration of distal tip cells (DTC), another murine ADAMTS1-like metalloproteinase, MIG-17, is necessary for their correct migration. The MIG-17 protein is secreted from muscle cells of the body wall and localizes in the basement membranes of the gonad along the migration path. This localization is dependent on the disintegrin-like domain of MIG-17 and its catalytic activity. These results indicate that MIG-17 is not required for DTC migration per se, but rather influences the route of migration (Nishiwaki *et al.*, 2000). Regulation of MMPs, including their expression and localization, is different according to the type of cell migration (Nabeshima, 2001). We first describe the types of cell migration and then mechanisms for cell surface localization of MMPs that lead to their activation on the cell surface and pericellular reorganization of ECM.

A. Types of Cell Migration

There are two types of cell migration: single-cell locomotion (SCL) and cohort migration (Nabeshima *et al.*, 1999a). The latter is cell movement *en mass*, keeping cell–cell contact with one another. The mechanisms by which cells move have been investigated predominantly using *in vitro* models in which cells move as single cells (Stoker and Gherardi, 1991; Nabeshima *et al.*, 1999a). However, in human surgical specimens, carcinoma cells, especially those of well to moderately differentiated types, frequently invade the stroma as coherent cell nests rather than as single cells (Sträuli and Weiss, 1977; Nabeshima *et al.*, 1999a), suggesting that there is a way by which carcinoma cells move together as coherent cell clusters. We have called this type of movement "cohort migration" and demonstrated that carcinoma cells can actually move *en mass*, keeping cell–cell contact with each other *in vitro* (Nabeshima *et al.*, 1995a,b). After our proposal, cell movement in well-differentiated squamous cell carcinoma of the esophagus (Sanders *et al.*, 1998), malignant melanoma (Li *et al.*, 2001), and uterine cervical carcinoma cells (Wong *et al.*, 2001) has been referred to cohort migration. Being compared with SCL, one characteristic feature of cohort migration is compartmentalized release from cell–cell adhesion: cells extend leading lamellae to move via this release while keeping cell–cell contact in other portions (Nabeshima *et al.*, 1995b, 1997). However, the most important feature that clearly differentiates cohort migration from SCL is the fact that migrating cells regulate expression and localization of MMPs via cell–cell contact within migrating cell sheets. In our model, MT1-MMP and gelatinase A are expressed and localized specifically at the front pathfinder cells of the migrating cell sheets, and reorganization of gelatin matrix by these MMPs is essential for this type of migration (Nabeshima *et al.*, 2000).

B. Coordinated Expression of MMPs and Its Regulation via Cell–Cell Contact among Migrating Cells During Cohort Migration

In two-dimensional cohort migration assays, when migration of colon carcinoma cells was induced with hepatocyte growth factor/scatter factor (HGF/SF), MT1-MMP and gelatinase A were expressed only at the front cells of migrating cell sheets, with the following migrating cells being negative (Nabeshima *et al.*, 2000). In contrast, when cell scattering was induced by stimulating cell migration in the presence of the anti-E-cadherin antibody, the front cell-specific pattern of expression observed during cohort migration was lost: individual scattering cells expressed both MT1-MMP and gelatinase A in their leading edges and cytoplasm (Fig. 5). This pattern, expression in front cells during cohort migration and in individual cells during scattering, was also the case in MT1-MMP mRNA expression that was demonstrated by *in situ* hybridization (Nabeshima, 2001), suggesting regulation at the gene expression level. When cohort migration was induced on gelatin-coated substratum, these MMPs expressed in the front cells degraded the gelatin matrix in a very organized manner, leaving the radially arrayed gelatin matrix at the sites of leading edges. Inhibition of this organized lysis with BB94, a synthetic inhibitor specific to MMPs, TIMP-1 and -2, and GelA PEX suppressed

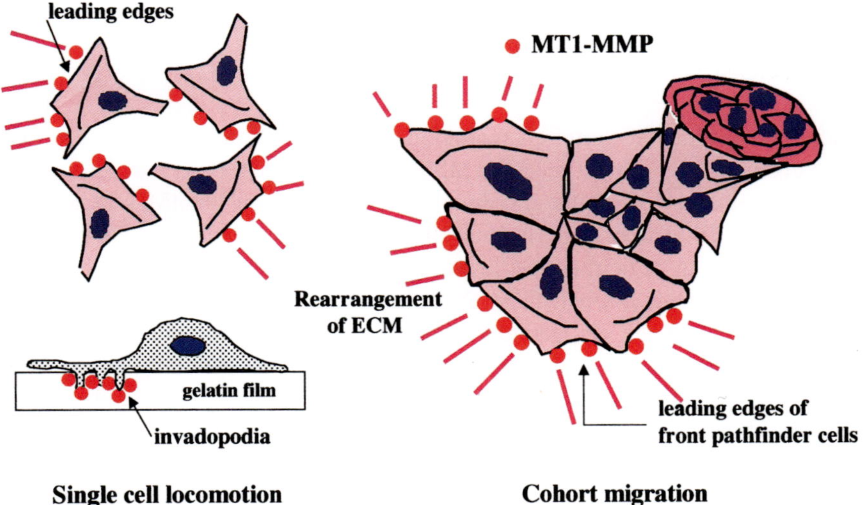

FIGURE 5 Cell surface expression of MT1-MMP in single cell locomotion (SCL) and cohort-type migration. Each migrating cell expresses MT1-MMP preferentially at the leading edges in SCL, whereas the protease is expressed at the leading edges of the front pathfinder cells of migrating cell sheets in cohort migration.

migration on gelatin matrix, indicating that the organized lysis by front cell MMPs is essential for cohort migration (Nabeshima *et al.*, 2000).

Although the precise mechanisms involved in the front cell-specific localization of these MMPs are currently unknown, gene expression of MT1-MMP in the following cells of migrating cell sheets appears to be downregulated via cell–cell contact among migrating cell sheets when considering its different expression patterns in cohort migration and scattering. Similarly, gene expression of MT1-MMP is reported to decrease in confluent cultures of mouse mammary gland epithelial cells compared with their sparse cultures (Tanaka *et al.*, 1997). The cellular binding of gelatinase A also reduces in confluent cultures compared with that in sparse ultures of breast carcinoma cells (Menashi *et al.*, 1998). However, expression of MT1-MMP and gelatinase A in the front pathfinder cells may be related to abundant interaction with ECM there. Fibronectin (FN) is one of the candidate ECM components, as FN is preferentially produced and deposited extracellularly by migrating cells during cohort migration (Nabeshima *et al.*, 1998; Shimao *et al.*, 1999; Inoue *et al.*, 2001) and that culturing of colon carcinoma cells on FN substratum stimulates production and activation of gelatinase A (K. Nabeshima, unpublished results). Taken together, the presence of abundant cell–cell contact within migrating cell sheets as in confluent cultures seems to suppress MT1-MMP expression and gelatinase A binding, whereas specialized cell–ECM contact may facilitate expression of these enzymes in the front pathfinder cells.

C. Cell Surface Localization of MMPs and Their Concentration to Leading Edges

Studies on several different cell migration systems indicate that there are two mechanisms at the protein level for MMP localization: (i) a mechanism for cell surface association of MMPs and (ii) that for localization of MMPs to leading edges (Table III). The first mechanism that focuses MMP activity at the cell surface would appear to be the most logical mechanism to efficiently affect and regulate cell movement through ECM (Murphy and Gavrilovic, 1999). A relatively major function for the cell-bound enzyme in comparison to soluble MMP has been shown (Deryugina *et al.*, 1998). In human glioma and fibrosarcoma cells transfected with cDNA encoding MT1-MMP, their ability to contract collagen lattices was shown to be dependent on the MT1-MMP-mediated activation of progelatinase A and cell surface association of activated gelatinase A. Soluble gelatinase A failed to affect gel contraction. In cohort migration, this cell surface association mechanism also depends on MT1-MMP. In our study, GelA PEX effectively inhibited cohort migration on the gelatin matrix. GelA PEX, the carboxyl-end domain of gelatinase A, is involved in the trimolecular complex formation of MT1-MMP, TIMP-2, and gelatinase A on the cell surfaces as mentioned earlier and

TABLE III Mechanisms for Cell Surface Localization of MMPs

MMP	Binding cell surface molecules
Binding to cell surface	
MMP-1	α2 integrin subunit Emmprin (to α3β1 integrin?)
Gelatinase A	MT1-3, 5-MMP (MT1-MMP/TIMP-2 complex) αvβ3 integrin
Gelatinase B	CD44
Localization to leading edges or invadopodia	
MT1-MMP	CD44 (to actin cytoskeleton)
Gelatinase A	Via MT1-MMP
Gelatinase B	CD44 transported through actin-dependent pathway

competitively inhibits gelatinase A activation by MT1-MMP (Strongin *et al.*, 1995; Deryugina *et al.*, 1998). Thus, gelatinase A that is bound to and activated by MT1-MMP on the cell surfaces seems to play a major role during cohort migration. Effective inhibition by both TIMP-1 and TIMP-2 also supports this hypothesis because gelatinase A is known to be inhibited by these TIMPs, whereas MT1-MMP is inhibited by TIMP-2 but not by TIMP-1 (d'Ortho *et al.*, 1998). Other cell surface localization mechanisms include integrins. Integrin αvβ3 mediates cell surface binding of gelatinase A, and the α2 integrin subunit mediates that of MMP-1 (Brooks *et al.*, 1996; Murphy and Gavrilovic, 1999; Dumin *et al.*, 1999). Gelatinase A can be localized in a proteolytically active form on the surface of invasive melanoma cells, based on its ability to bind directly integrin αvβ3 via Gel A PEX (Brooks *et al.*, 1996). In tissue sections, colocalization of gelatinase A and integrin αvβ3 was shown on angiogenic blood vessels and invading melanoma cells. MMP-1 was shown to bind to the A domain of α2 integrin in a cation-dependent manner (Murphy and Gavrilovic, 1999; Dumin *et al.*, 1999). MMP-1 and type I collagen may bind at two independent sites on the α2 integrin, as the mutated A domain, which no longer binds type I collagen, can still bind to MMP-1. The transmembrane protein other than integrins can also bind MMPs. For example, emmprin forms a complex with MMP-1 at the tumor cell surface (Guo *et al.*, 2000). Because emmprin is reported to form a complex with α3β1 integrin (Berditchevski *et al.*, 1997), the emmprin–MMP-1 complex may bind α3β1 integrin if the binding sites do not compete each other. Through these bindings of MMPs to integrins, an extracellular pool of latent MMPs may be recruited into focal contacts and released by competing ECM ligands, thereby increasing the pericellular proteolytic potential close to relevant ECM interactions (Olson *et al.*, 1998).

On the cell surface, more specific localization of MMPs to the leading portions of invading tumor cells has been reported to be essential for migration. In human melanoma cells, the presence of active gelatinase A just on the cell surfaces was not enough, but its more specialized localization, together with MT1-MMP, to the invasion front of cells (invadopodia, specialized membrane extensions into the FN-coated gelatin matrix) was needed for their invasion (Nakahara *et al.*, 1997). Active gelatinase B was also found in association with the CD44v3,8-10 splice variant on the invadopodia of a breast cancer cell line (Bourguignon *et al.*, 1998). Similarly, localization of MMP to leading edges of migrating cells are reported to be necessary for cell migration in two-dimensional systems (Murphy and Gavrilovic, 1999). Cytoplasmic docking systems supporting these localizations are not yet fully understood, but binding of MT1-MMP to CD44 by its extracellular domain has been reported (Kajita *et al.*, 2001). CD44 and MT1-MMP colocalize at the leading edge of the motile cells, and MT1-MMP acts as a processing enzyme for CD44 there, shedding CD44H from the cell surface. This processing event stimulated cell motility. Because CD44 are connected to actin cytoskeleton of cells, CD44 may link the MT1-MMP to the cytoskeleton. Because accumulation of gelatinase B through an actin-dependent pathway to the leading edges is observed in migrating airway epithelial cells (Legrand *et al.*, 1999), a similar transportation system may also be present for MT1-MMP.

D. Reorganization of ECM by Migrating Cells

In general, ECM is thought to be a kind of barrier for tumor cell invasion, and more or less its cleavage and removal are necessary for tumor cells to migrate (Stetler-Stevenson, 1993). At the same time, however, ECM components provide cells with good substrate to move on. Studies, including ours, suggest that the proteolytic degradation of ECM is not just a path-clearing mechanism, but a way of reorganizing the matrix to facilitate cellular interactions. Additionally, this reorganization of ECM is done by each migrating cells in SCL, whereas it is done preferentially by front pathfinder cells in cohort migration.

Studies have demonstrated an essential role of enzymatic modification of the basal lamina of the endoneurium for peripheral nerve regeneration following injury. The neurite-promoting activity of endoneurial laminin is inhibited by a Schwann cell-derived chondroitin sulfate proteoglycans (CSPGs). Treatment of peripheral nerve sections with gelatinase A resulted in the removal of CSPGs and exposure of epitopes of laminin permissive for neurite extension (Zuo *et al.*, 1998). In cohort migration, removal of the gelatin matrix at the leading edges of the front cells of migrating cell sheets was not random or complete. Instead, it was performed in a very coordinate and organized manner, leaving the radially arrayed gelatin matrix at the frontmost part. This limited and organized clearing of gelatin matrix was essential for cell migration because MMP inhibitors efficiently inhibited migration. Thus, an important role of MMP is not just to remove ECM, but to rearrange it to suit cell migration. This is the same for SCL. In three-dimensional matrix-based models, fibroblastic-type tumor cells are reported to cause initial fiber traction at the leading edge followed by radial fiber alignment toward the cell, which then favors persistent migration in the direction of maximal traction (Friedl and Bröcker, 2000). Moreover, it is suggested that binding of integrins to this kind of prestressed ECM fibers would strengthen the linkage between those receptors and the force-generating cytoskeleton at that side of cells, thereby causing the cell to migrate along the direction of the rigid substrate (Sheetz *et al.*, 1998).

VI. CONCLUSIONS

The reorganization of ECM plays an important role in many aspects of cellular functions, including cell movement, through the change of cell–ECM interactions. Thus, inhibition of cell surface or pericellular metalloproteinase activities will lead to control of pathological conditions related to such activities. However, it is known that MMP inhibitors cause some side effects at the same time. In this light, regulation mechanisms for metalloproteinase localization at the cell surface, specifically to leading edges of pathfinder cells in the case of cell migration, will be another target. In addition, investigation of new roles of cell surface proteinases other than cleavage of ECM may give us more opportunities. Furthermore, elucidation of interplays or coordinated actions among MMP and ADAM family members will contribute to the understanding of and a new therapeutic approach to some pathological disorders.

ACKNOWLEDGMENTS

The author expresses sincere gratitude to Dr. Bryan P. Toole, Department of Anatomy and Cell Biology, Tufts University Medical School, and Dr. Y. Nawa, Department of Parasitology, Miyazaki Medical College, for their invaluable suggestions and discussion.

REFERENCES

Abbaszade, I., Liu, R. Q., Yang, F., Rosenfeld, S. A., Ross, O. H., Link, J. R., Ellis, D. M., Tortorella, M. D., Pratta, M. A., Hollis, J. M., Wynn, R., Duke, J. L., George, H. J., Hillman, M. C., Jr., Murphy, K., Wiswall, B. H., Copeland, R. A., Decicco, C. P., Bruckner, R., Nagase, H., Itoh, Y., Newton, R. C., Magolda, R. L., Trzaskos, J. M., and Burn, T. C. (1999). Cloning and characterization of ADAMTS11, an aggrecanase from the ADAMTS family. *J. Biol. Chem.* **274**, 23443–23450.

Aoudjit, F., Potworowski, E. F., and St-Pierre, Y. (1998). Bi-directional induction of matrix metalloproteinase-9 and tissue inhibitor of matrix metalloproteinase-1 during T lymphoma/endothelial cell contact: Implication of ICAM-1. *J. Immunol.* **160**, 2967–2973.

Artavanis-Tsakonas, S., Rand, M. D., and Lake, R. J. (1999). Notch signaling: Cell fate control and signal integration in development. *Science* **284**, 770–776.

Berditchevski, F., Chang, S., Bodorova, J., and Hemler, M. E. (1997). Generation of monoclonal antibodies to integrin-associated proteins: Evidence that alpha3beta1 complexes with EMMPRIN/basigin/OX47/M6. *J. Biol. Chem.* **272**, 29174–29180.

Biswas, C. (1982). Tumor cell stimulation of collagenase production by fibroblasts. *Biochem. Biophys. Res. Commun.* **109**, 1026–1034.

Biswas, C., Zhang, Y., DeCastro, R., Guo, H., Nakamura, T., Kataoka, H., and Nabeshima, K. (1995). The human tumor cell-derived collagenase stimulatory factor (renamed EMMPRIN) is a member of the immunoglobulin superfamily. *Cancer Res.* **55**, 434–439.

Black, R. A., and White, J. M. (1998). ADAMs: Focus on the protease domain. *Curr. Opin. Cell Biol.* **10**, 654–659.

Bordador, L. C., Li, X., Toole, B., Chen, B., Regezi, J., Zardi, L., Hu, Y., and Ramos, D. M. (2000). Expression of emmprin by oral squamous cell carcinoma. *Int. J. Cancer* **85**, 347–352.

Bourguignon, L. Y., Gunja-Smith, Z., Lida, N., Zhu, H. B., Young, L. J., Muller, W. J., and Cardiff, R. D. (1998). CD44v3,8-10 is involved in cytoskeleton-mediated tumor cell migration and matrix metalloproteinase (MMP-9) association in metastatic breast cancer cells. *J. Cell Physiol.* **176**, 206–215.

Brooks, P. C., Stromblad, S., Sanders, L. C., von Schalscha, T. L., Aimes, R. T., Stetler-Stevenson, W. G., Quigley, J. P., and Cheresh, D. A. (1996). Localization of matrix metalloproteinase MMP-2 to the surface of invasive cells by interaction with integrin $\alpha v\beta 3$. *Cell* **85**, 683–693.

Butler, G. S., Butler, M., Atkinson, S. J., Will, H., Tamura, T., Schade van Westrum, S., Crabbe, T., Clements, J., d'Ortho, M.-P., and Murphy, G. (1997). The TIMP-2-membrane type I metalloproteinase 'receptor' regulates the concentration and efficient activation of progelatinase A: A kinetic study. *J. Biol. Chem.* **273**, 871–880.

Carmeliet, P., Moons, L., Lijnen, R., Baes, M., Lemaitre, V., Tipping, P., Drew, A., Eeckhout, Y., Shapiro, S., Lupu, F., and Collen, D. (1997). Urokinase-generated plasmin activates matrix metalloproteinases during aneurysm formation. *Nature Genet.* **17**, 439–444.

Caudroy, S., Polette, M., Tournier, J. M., Burlet, H., Toole, B., Zucker, S., and Birembaut, P. (1999). Expression of the extracellular matrix metalloproteinase inducer (EMMPRIN) and the matrix metalloproteinase-2 in bronchopulmonary and breast lesions. *J. Histochem. Cytochem.* **47**, 1575–1580.

Chambers, A. F., and Matrisian, L. M. (1997). Changing views of the role of matrix metalloproteinases in metastasis. *J. Natl. Cancer Inst.* **89**, 1260–1270.

Chen, H., and Sampson, N. S. (1999). Mediation of sperm-egg fusion: Evidence that mouse egg alpha6 beta1 integrin is the receptor for sperm fertilin beta. *Chem. Biol.* **6**, 1–10.

Colige, A., Li, S. W., Sieron, A. L., Nusgens, B. V., Prockop, D. J., and Lapiere, C. M. (1997). cDNA cloning and expression of bovine procollagen I N-proteinase: A new member of the superfamily of zinc-metalloproteinases with binding sites for cells and other matrix components. *Proc. Natl. Acad. Sci. USA* **94**, 2374–2379.

Dalberg, K., Eriksson, E., Enberg, U., Kjellman, M., and Backdahl, M. (2000). Gelatinase A, membrane type 1 matrix metalloproteinase, and extracellular matrix metalloproteinase inducer mRNA expression: Correlation with invasive growth of breast cancer. *World J. Surg.* **24**, 334–340.

Deryugina, E. I., Bourdon, M. A., Luo, G.-X., Reisfeld, R. A., and Strongin, A. (1997). Matrix metalloproteinase-2 activation modulates glioma cell migration. *J. Cell Sci.* **110**, 2473–2482.

Deryugina, E. I., Bourdon, M. A., Reisfeld, R. A., and Strongin, A. (1998). Remodeling of collagen matrix by human tumor cells requires activation and cell surface association of matrix metalloproteinase-2. *Cancer Res.* **58**, 3743–3750.

d'Ortho, M.-P., Stanton, H., Butler, M., Atkinson, S. J., Murphy, G., and Hembry, R. M. (1998). MT1-MMP on the cell surface causes focal degradation of gelatin films. *FEBS Lett.* **421**, 159–164.

Dumin, J. A., Dickeson, K., Goldberg, G., Santoro, S. A., and Parks, W. C. (1999). Interaction among collagenase-1, collagen, and the integrin $\alpha 2\beta 1$ during reepithelialization. *Invest. Dermatol.* **112**, 536.

English, W. R., Puentes, X. S., Freije, J. M. P., Knäuper, V., Amour, A., Merryweather, A., Lopez-Otin, C., and Murphy, G. (2000). Membrane type 4 matrix metalloproteinase (MMP17) has tumor necrosis factor-α convertase activity but does not activate pro-MMP2. *J. Biol. Chem.* **275**, 14046–14055.

Friedl, P., and Bröcker, E. B. (2000) The biology of cell locomotion within three-dimensional extracellular matrix. *Cell Mol. Life Sci.* **57**, 41–64.

George, S. J., Johnson, J. L., Angelini, G. D., Newby, A. C., and Baker, A. H. (1998). Adenovirus-mediated gene transfer of the human TIMP-1 gene inhibits smooth muscle cell migration and neointimal formation in human saphenous vein. *Hum. Gene Ther.* **9**, 867–877.

Grant, G. A., Eisen, A. Z., Marmer, B. L. Roswit, W. T., and Goldberg, G. I. (1987). The activation of human skin fibroblast procollagenase: Sequence identification of the major conversion products. *J. Biol. Chem.* **262**, 5886–5889.

Guo, H., Li, R., Zucker, S., and Toole, B. P. (2000). EMMPRIN (CD147), an inducer of matrix metalloproteinase synthesis, also binds interstitial collagenase to the tumor cell surface. *Cancer Res.* **60**, 888–891.

Guo, H., Zucker, S., Gordon, M. K., Toole, B. P., and Biswas, C. (1997). Stimulation of matrix metalloproteinase production by recombinant extracellular matrix metalloproteinase inducer from transfected chinese hamster ovary cells. *J. Biol. Chem.* **272**, 24–27.

Haas, T. L., Davis, S. J., and Madri, J. A. (1998). Three-dimensional type I collagen lattices induce coordinate expression of matrix metalloproteinases MT1-MMP and MMP-2 in microvascular endothelial cells. *J. Biol. Chem.* **273**, 3604–3610.

Hiraoka, N., Allen, E., Apel, I. J., Gyetko, M. R., and Weiss, S. J. (1998). Matrix metalloproteinases regulate neovascularization by acting as pericellular fibrinolysins. *Cell* **95**, 365–377.

Holmbeck, K., Bianco, P., Caterina, J., Yamada, S., Kromer, M., and Kuznetsov, S. A. (1999). MT1-MMP-deficient mice develop dwarfism, osteopenia, arthritis, and connective tissue disease due to inadequate collagen turnover. *Cell* **99**, 81–92.

Inoue, T., Nabeshima, K., Shimao, Y., Meng, J.-Y., and Koono, M. (2001). Regulation of fibronectin expression and splicing in migrating epithelial cells: Migrating MDCK cells produce less, but more active, fibronectin. *Biochem Biophys. Res. Commun.* **280**, 1262–1268.

Itoh, T., Tanioka, M., Yoshida, H., Yoshioka, T., Nishimoto, H., and Itohara, S. (1998). Reduced angiogenesis and tumor progression in gelatinase A-deficient mice. *Cancer Res.* **58**, 1048–1051.

Izumi, Y., Hirata, M., Hasuwa, H., Iwamoto, R., Umata, T., Miyado, K., Tamai, Y., Kurisaki, T., Sehara-Fujisawa, A., Ohno, S., and Mekada, E. (1998). A metalloproteinase-disintegrin, MDC9/meltrin-gamma/ADAM9 and PKCdelta are involved in TPA-induced ectodomain shedding of membrane-anchored heparin-binding EDF-like growth factor. *EMBO J.* **17**, 7260–7272.

Itoh, Y., Kajita, M., Kinoh, H., Mori, H., Okada, A., and Seiki, M. (1999). Membrane type 4 matrix metalloproteinase (MT4-MMP, MMP-17) is a glycosylphosphatidylinositol-anchored proteinase. *J. Biol. Chem.* **48**, 34260–34266.

Kajita, M., Itoh, Y., Chiba, T., Mori, H., Okada, A., Kinoh, H., and Seiki, M. (2001). Membrane-type 1 matrix metalloproteinase cleaves CD44 and promotes cell migration. *J. Cell Biol.* **153**, 893–904.

Kaushal, G. P., and Shah, S. V. (2000). The new kids on the block: ADAMTSs, potentially multifunctional metalloproteinases of the ADAM family. *J. Clin. Invest.* **105**, 1335–1337.

Kinoshita, T., Sato, H., Okada, A., Ohuchi, E., Imai, K., Okada, Y., and Seiki, M. (1998). TIMP-2 promotes activation of progelatinase A by membrane-type 1 matrix metalloproteinase immobilized on agarose beads. *J. Biol. Chem.* **273**, 16098–16103.

Knäuper, V., Smith, B., Lopez-Otin, C., and Murphy, G. (1997). Activation of progelatinase B (proMMP-9) by active collagenase-3 (MMP-13). *Eur. J. Biochem.* **248**, 369–373.

Knäuper, V., Will, H., Lopez-Otin, C., Smith, B., Atkinson, S. J., Stanton, H., Hembry, R. M., and Murphy, G. (1996). Cellular mechanisms for human procollagenase-3 (MMP-13) activation: Evidence that MT1-MMP (MMP-14) and gelatinase A (MMP-2) are able to generate active enzyme. *J. Biol. Chem.* **271**, 17124–17131.

Konttinen, Y. T., Li, T. F., Mandelin, J., Liljestrom, M., Sorsa, T., Santavirta, S., and Virtanen, I. (2000). Increased expression of extracellular matrix metalloproteinase inducer in rheumatoid synovium. *Arthritis Rheum.* **43**, 275–280.

Kuno, K., and Matsushima, K. (1998). ADAMTS-1 protein anchors at the extracellular matrix through the thrombospondin type I motifs and its spacing region. *J. Biol. Chem.* **273**, 13912–13917.

Legrand, C., Gilles, C., Zahm, J.-M., Polette, M., Buisson, A.-C., Kaplan, H., Birembaut, P., and Tournier, J.-M. (1999). Airway epithelial cell migration dynamics: MMP-9 role in cell-extracellular matrix remodeling. *J. Cell Biol.* **146**, 517–529.

Li, G., Satyamoorthy, K., and Herlyn, M. (2001). N-cadherin-mediated intercellular interactions promote survival and migration of melanoma cells. *Cancer Res.* **61**, 3819–3825.

Lohi, J., Wilson, C. L., Roby, J. D., and Parks, W. C. (2001). Epilysin, a novel human matrix metalloproteinase (MMP-28) expressed in testis and keratinocytes and in response to injury. *J. Biol. Chem.* **276**, 10134–10144.

Menashi, S., Dehem, M., Souliac, I., Legrand, Y., and Fridman, R. (1998). Density-dependent regulation of cell-surface association of matrix metalloproteinase-2 (MMP-2) in breast-carcinoma cells. *Int. J. Cancer* **75**, 259–265.

Moerman, D. G. (1999) A metalloprotease prepares the way. *Curr. Biol.* **9**, R701–R703.

Moss, M. L., Jin, S.-L. C., Milla, M. E., Burkhart, W., Carter, H. L., Chen, W.-J., Clay, W. C., and Didsbury, J. R. (1997) Cloning of a disintegrin metalloproteinase that processes precursor tumor-necrosis factor-α. *Nature* **385**, 733–736.

Muraoka, K., Nabeshima, K., Murayama, T., Biswas, C., and Koono, M. (1993). Enhanced expression of a tumor-cell-derived collagenase-stimulatory factor in urothelial carcinoma: Its usefulness as a tumor marker for bladder cancers. *Int. J. Cancer* **55**, 19–26.

Murphy, G., Atkinson, S., Ward, R., Gavrilovic, J., and Reynolds, J. J. (1992). The role of plasminogen activators in the regulation of connective tissue metalloproteinases. *Ann. N.Y. Acad. Sci.* **667**, 1–12.

Murphy, G., and Gavrilovic, J. (1999). Proteolysis and cell migration: Creating a path? *Curr. Opin. Cell Biol.* **11**, 614–621.

Nabeshima, K. (2001). Coordinated expression of MMPs in pathfinder cells and reorganization of extracellular matrix during cohort migration. *Connect. Tissue* **33**, 275–283.

Nabeshima, K., Asada, Y., Inoue, T., Kataoka, H., Shimao, Y., Sumiyoshi, A., and Koono, M. (1997). Modulation of E-cadherin expression in TPA-induced cell motility: Human well-differentiated adenocarcinoma cells move as coherent sheets associated with phosphorylation of E-cadherin-catenin complex. *Lab. Invest.* **76**, 139–151.

Nabeshima, K., Inoue, T., Shimao, Y., Kataoka, H., and Koono, M. (1998). TPA-induced cohort migration of well differentiated human rectal adenocarcinoma cells: Cells move in a RGD-dependent manner on fibronectin produced by cells, and phosphorylation of E-cadherin/catenin complex is induced independently of cell-extracellular matrix interactions. *Virch. Arch.* **433**, 243–253.

Nabeshima, K., Inoue, T., Shimao, Y., Kataoka, H., and Koono, M. (1999a). Cohort migration of carcinoma cells: Differentiated colorectal carcinoma cells move as coherent cell clusters or sheets. *Histol. Histopathol.* **14**, 1183–1197.

Nabeshima, K., Inoue, T., Shimao, Y., Okada, Y., Itoh, Y., Seiki, M., and Koono, M. (2000). Front-cell-specific expression of membrane-type 1 matrix metalloproteinase and gelatinase A during cohort migration of colon carcinoma cells induced by hepatocyte growth factor/scatter factor. *Cancer Res.* **60**, 3364–3369.

Nabeshima, K., Kataoka, H., Toole, B. P., and Koono, M. (1999b). Activation and induction of collagenases. *In "Collagenases"* (W. K. Hoeffler, ed.), pp. 91–113. R. G. Landes Company, Austin, TX.

Nabeshima, K., Komada, N., Inoue, T., and Koono, M. (1995a). A two-dimensional model of cell movement: Well-differentiated human rectal adenocarcinoma cells move as coherent sheets upon TPA stimulation. *Pathol. Res. Pract.* **191**, 76–83.

Nabeshima, K., Moriyama, T., Asada, Y., Komada, N., Inoue, T., Kataoka, H., Sumiyoshi, A., and Koono, M. (1995b). Ultrastructural study of TPA-induced cell motility: Human well-differentiated rectal adenocarcinoma cell move as coherent sheets via localized modulation of cell-cell adhesion. *Clin. Exp. Metast.* **13**, 499–508.

Nagase, H., Enghild, J. J., Suzuki, K., and Salvesen, G. (1990). Stepwise activation mechanisms of the precursor of matrix metalloproteinase 3 (stromelysin) by proteinases and (4-aminophenyl) mercuric acetate. *Biochemistry* **29**, 5783–5789.

Nagase, H., and Woessner, J. F. (1999). Matrix metalloproteinases. *J. Biol. Chem.* **274**, 21491–21494.

Nakahara, H., Howard, L., Thompson, E. W., Sato, H., Seiki, M., Yeh, Y., and Chen, W. T. (1997). Transmembrane/cytoplasmic domain-mediated membrane type 1-matrix metalloproteinase docking to invadopodia is required for cell invasion. *Proc. Natl. Acad. Sci. USA* **94**, 7959–7964.

Nishiwaki, K., Hisamoto, N., and Matsumoto, K. (2000). A metalloproteinase disintegrin that controls cell migration in *Caenorhabditis elegans*. *Nature* **288**, 2205–2208.

Ohuchi, E., Imai, K., Fujii, Y., Sato, H., Seiki, M., and Okada, Y. (1997). Membrane type 1 matrix metalloproteinase digests interstitial collagens and other extracellular matrix macromolecules. *J. Biol. Chem.* **272**, 2446–2451.

Okada, Y., Morodomi, T., Enghild, J. J., Suzuki, K., Yasui, A., Nakanishi, I., Salvesen, G., and Nagase, H. (1990). Matrix metalloproteinase 2 from human rheumatoid synovial fibroblasts: Purification and activation of the precursor and enzymic properties. *Eur. J. Biochem.* **197**, 721–730.

Olson, M. W., Toth, M., Gervasi, D. C., Sado, Y., Ninomiya, Y., and Fridman, R. (1998). High affinity binding of latent matrix metalloproteinase-9 to the α2(IV) chain of collagen IV. *J. Biol. Chem.* **273**, 10672–10681.

Pei, D., Kang, T., and Qi, H. (2000). Cysteine array matrix metalloproteinase (CA-MMP)/MMP-23 is a type II transmembrane matrix metalloproteinase regulated by a single cleavage for both secretion and activation. *J. Biol. Chem.* **275**, 33988–33997.

Pei, D., and Weiss, S. J. (1995). Furin-dependent intracellular activation of the human stromelysin-3 zymogen. *Nature* **375**, 244–247.

Peschon, J. J., Slack, J. L., Reddy, P., Stocking, K. L., Sunnarborg, S. W., Lee, D. C., Russell, W. E., Castner, B. J., Johnson, R. S., Fitzner, J. N., Boyce, R. W., Nelson, N., Kozlosky, C. J., Wolfson, M. F., Rauch, C. T., Cerretti, D. P., Paxton, R. J., March, C. J., and Black, R. A. (1998). An essential role for ectodomain shedding in mammalian development. *Science* **282**, 1281–1284.

Polette, M., Gilles, C., Marchand, V., Lorenzato, M., Toole, B. P., Tournier, J. M., Zucker, S., and Birembaut, P. (1997). Tumor collagenase stimulatory factor (TCSF) expression and localization in human lung and breast cancers. *J. Histochem. Cytochem.* **45**, 703–709.

Primakoff, P., Hyatt, H., and Tredick-Kline, J. (1987). Identification and purification of a sperm surface protein with a potential role in sperm-egg membrane fusion. *J. Cell Biol.* **104**, 141–149.

Primakoff, P., and Myles, D. G. (2000). The ADAM gene family: Surface proteins with adhesion and protease activity. *Trends Genet.* **16**, 83–87.

Qi, H., Rand, M. D., Wu, X., Sestan, N., Wang, W., Rakic, P., Xu, T., and Artavanis-Tsakonas, S. (1999). Processing of the notch ligand delta by the metalloproteinase Kuzbanian. *Science* **283**, 91–94.

Romanic, A. M., and Madri, J. A. (1994). The induction of 72-kD gelatinase in T cells upon adhesion to endothelial cells is VCAM-1 dependent. *J. Cell Biol.* **125**, 1165–1178.

Sameshima, T., Nabeshima, K., Toole, B. P., Yokogami, K., Okada, Y., Goya, T., Koono, M., and Wakisaka, S. (2000a). Glioma cell extracellular matrix metalloproteinase inducer (EMMPRIN) (CD147) stimulates production of membrane-type matrix metalloproteinases and activated gelatinase A in cocultures with brain-derived fibroblasts. *Cancer Lett.* **157**, 177–184.

Sameshima, T., Nabeshima, K., Toole, B. P., Yokogami, K., Okada, Y., Goya, T., Koono, M., and Wakisaka, S. (2000b). Expression of EMMPRIN (CD147), a cell surface inducer of matrix metalloproteinases, in normal human brain and gliomas. *Int. J. Cancer* **88**, 21–27.

Sanders, D. S. A., Bruton, R., Darnton, S. J., Casson, A. G., Hanson, I., Williams, H. K., and Jankowski, J. (1998). Sequential changes in cadherin-catenin expression associated with the progression and heterogeneity of primary oesophageal squamous carcinoma. *Int. J. Cancer* **79**, 573–579.

Sandy, J. D., Neame, P. J., Boynton, R. E., and Flannery, C. R. (1991). Catabolism of aggrecan in cartilage explants: Identification of a major cleavage site within the interglobular domain. *J. Biol. Chem.* **266**, 8683–8685.

Sato, H., Takino, H., Okada, Y., Cao, J., Yamamoto, E., and Seiki, M. (1994). A matrix metalloproteinase expressed on the surface of invasive tumor cells. *Nature (Lond.)* **370**, 61–65.

Seiki, M. (1999). Membrane-type matrix metalloproteinases. *APMIS* **107**, 137–143.

Seltzer, J. L., Lee, A. Y., Akers, K. T., Sudbeck, B., Southon, E. A., Wayner, E. A., and Eisen, A. Z. (1994). Activation of 72-kDa type IV collagenase/gelatinase by normal fibroblasts in collagen lattices is mediated by integrin receptors but is not related to lattice contraction. *Exp. Cell Res.* **213**, 365–374.

Shapiro, S. D. (1998). Matrix metalloproteinase degradation of extracellular matrix: biological consequences. *Curr. Opin. Cell Biol.* **10**, 602–608.

Sheetz, M. P., Felsenfeld, D. P., and Galbraith, C. G. (1998). Cell migration: Regulation of force on extracellular-matrix-integrin complexes: *Trends Cell Biol.* **8**, 51–54.

Shimao, Y., Nabeshima, K., Inoue, T., and Koono, M. (1999). Role of fibroblasts in HGF/SF-induced cohort migration of human colorectal carcinoma cells: Fibroblasts stimulate migration associated with increased fibronectin production via upregulated TGF-β1. *Int. J. Cancer* **82**, 449–458.

Shindo, T., Kurihara, H., Kuno, K., Yokoyama, H., Wada, T., Kurihara, Y., Imai, T., Wang, Y., Ogata, M., Nishimatsu, H., Moriyama, N., Oh-hashi, Y., Morita, H., Ishikawa, T., Nagai, R., Yazaki, Y., and Matsushima, K. (2000). ADAMTS1: A metalloproteinase-disintegrin essential for normal growth, fertility, and organ morphology and function. *J. Clin. Invest.* **105**, 1345–1352.

Stetler-Stevenson W. G., Aznavoorian S., and Liotta, L. A. (1993). Tumor cell interactions with the extracellular matrix during invasion ad metastasis. *Annu. Rev. Cell Biol.* **9**, 541–573.

Stocker, W., Grams, F., Baumann, U., Reinemer, P., Gomis-Ruth, F. X., McKay, D. B., and Bode, W. (1995). The metzincins: Topological and sequential relations between the astacins, adamalysins, serralysins, and matrixins (collagenases) define a superfamily of zinc-peptidases. *Protein Sci.* **4**, 823–840.

Stoker, M., and Gherardi, E. (1991). Regulation of cell movement: The motogenic cytokines. *Biochim. Biophys. Acta.* **1072**, 81–102.

Sträuli, P., and Weiss, L. (1977). Cell locomotion and tumor penetration: Report on a workshop of the EORTC cell surface project group. *Eur. J. Cancer* **13**, 1–12.

Streuli, C. (1999). Extracellular matrix remodelling and cellular differentiation. *Curr. Opin. Cell Biol.* **11**, 634–640.

Strongin, A. Y., Collier, I., Bannikov, G., Marmer, B. L., Grant, G. A., and Goldberg, G. I. (1995).

Mechanism of cell surface activation of 72-kDa type IV collagenase: Isolation of the activated form of the membrane metalloproteinase. *J. Biol. Chem.* **270**, 5331–5338.

Tanaka, S. S., Mariko Y., Mori, H., Ishijima, J., Tachi, S., Sato, H., Seiki, M., Yamanouchi, K., Tojo, H., and Tachi, C. (1997). Cell-cell contact downregulates expression of membrane type metalloproteinase-1 (MT1-MMP) in a mouse mammary gland epithelial cell line. *Zool. Sci.* **14**, 95–99.

Tang, B. L. (2001). ADAMTS: A novel family of extracellular matrix proteases. *Int. J. Biochem. Cell Biol.* **33**, 33–44.

Tortorella, M., Pratta, M., Rui-Qin, L., Abbaszade, I., Ross, H., Burn, T., and Arner, E. (2000). The thrombospondin motif of aggrecanase-1 (ADAMTS4) is critical for aggrecan substrate recognition and cleavage. *J. Biol. Chem.* **275**, 25791–25797.

Tortorella, M. D., Burn, T. C., Pratta, M. A., Abbaszade, I., Hollis, J. M., Liu, R., Rosenfeld, S. A., Copeland, R. A., Decicco, C. P., Wynn, R., Rockwell, A., Yang, F., Duke, J. L., Solomon, K.,

George, H., Bruckner, R., Nagase, H., Itoh, Y., Ellis, D. M., Ross, H., Wiswall, B. H., Murphy, K., Hillman, M. C., Jr., Hollis, G. F., and Arner, E. C. (1999). Purification and cloning of aggrecanase-1: A member of the ADAMTS family of proteins. *Science* **284**, 1664–1666.

van den Oord, J. J., Paemen, L., Opdenakker, G., and de Wolf-Peeters, C. (1997). Expression of gelatinase B and the extracellular matrix metalloproteinase inducer EMMPRIN in benign and malignant pigment cell lesions of the skin. *Am. J. Pathol.* **151**, 665–670.

Van Wart, H. E., and Birkedal-Hansen, H. (1990). The cysteine switch: A principle of regulation of metalloproteinase activity with potential applicability to the entire matrix metalloproteinase gene family. *Proc. Natl. Acad. Sci. USA* **87**, 5578–5582.

White, J. M., Wolfsberg, T. G., and Gruenke, J. A. (2001). Table of the ADAMs. http://www.people.virginia,edu/~jag6n/whitelab.html

Will, H., Atkinson, S. J., Butler, G. S., Smith, B., and Murphy, G. (1996). The soluble catalytic domain of membrane type 1 matrix metalloproteinase cleaves the propeptide of progelatinase A and initiates autoproteolytic activation: Regulation by TIMP-2 and TIMP-3. *J. Biol. Chem.* **271**, 17119–17123.

Wong, A. S. T., Pelech, S. L., Woo, M. M. M., Yim, G., Rosen, B., Ehlen, T., Leung, P. C. K., and Auersperg., N. (2001). Coexpression of hepatocyte growth factor-Met: An early step in ovarian carcinogenesis? *Oncogene* **20**, 1318–1328.

Xie, B., Laouar, A., and Huberman, E. (1998). Fibronectin-mediated cell adhesion is required for induction of 92-kDa type IV collagenase/gelatinase (MMP-9) gene expression during macrophage differentiation: The signaling role of protein kinase C-beta. *J. Biol. Chem.* **273**, 11576–11582.

Yagami-Hiromasa, T., Sato, T., Kurisaki, T., Kamijo, K., Nabeshima, Y., and Fujisawa-Sehara, A. (1995). A metalloproteinase-disintegrin participating in myoblast fusion. *Nature* **377**, 652–656.

Yana, I., and Weiss, S. J. (2000). Regulation of membrane type-1 matrix metalloproteinase activation by proprotein convertases. *Mol. Biol. Cell* **11**, 2387–2401.

Zhang, X.-P., Kamata, T., Yokoyama, K., Puzon-McLaughlin, W., and Takada, Y. (1998). Specific interaction of the recombinant disintegrin-like domain of MDC-15 (Metargidin, ADAM-15) with integrin $\alpha v\beta 3$. *J. Biol. Chem.* **273**, 7345–7350.

Zhou, Z., Apte, S. S., Soininen, R., Cao, R., Baaklini, G. Y., and Rauser, R. W. (2000). Impaired endochondral ossification and angiogenesis in mice deficient in membrane-type matrix metalloproteinase I. *Proc. Natl. Acad. Sci. USA* **97**, 4052–4057.

Zucker, S., Hymowitz, M., Rollo, E. E., Mann, R., Conner, C. E., Cao, J., Foda, H. D., Tompkins, D. C., and Toole, B. P. (2001). Tumorigenic potential of extracellular matrix metalloproteinase inducer. *Am. J. Pathol.* **158**, 1921–1928.

Zuo, J., Ferguson, T. A., Hernandez, Y. J., Stetler-Stevenson, W. G., and Muir, D. (1998). Neuronal matrix metalloproteinase-2 degrades and inactivates a neurite-inhibiting chondroitin sulfate proteoglycan. *J. Neurosci.* **18**, 5203–5211.

CHAPTER 15

Transcriptional Regulation of Matrix Metalloproteinases

ULRIKE BENBOW*,[1] AND CONSTANCE E. BRINCKERHOFF*,†

*Departments of *Medicine and †Biochemistry, Dartmouth Medical School, Hanover, New Hampshire 03755*

I. INTRODUCTION

Matrix metalloproteinases (MMPs) are a family of enzymes that, collectively, degrade all components of the extracellular matrix (Table I). There are at least 26 members of this family: all are active at neutral pH, require Ca^{2+} for activity, and contain a central zinc atom as part of their structure. Most MMPs are secreted into the extracellular space in a latent form, requiring proteolytic cleavage in order to be enzymatically active. However, one subgroup of MMPs is anchored in the membrane (membrane-type MMPs; MT-MMPs), and these enzymes are activated intracellularly by a furin-like mechanism. MMP-11 (stromelysin 3) is the only nonmembrane-bound MMP that is also activated intracellularly (Hotary *et al.*, 2000; Nagase and Woessner, 1999; Nelson *et al.*, 2000; Ohuchi *et al.*, 1997; Parsons *et al.*, 1997; Pei, 1999; Velasco *et al.*, 2000; Vincenti *et al.*, 1996).

Most cells in the body can express MMPs, although the expression of certain enzymes is often associated with a particular cell type. For example, MMP-2 and MMP-9, gelatinases A and B, respectively, preferentially degrade the type IV

[1]Present address: Bristol Heart Institute, University of Bristol, Bristol BS9 1 HD.

Copyright 2003, Elsevier Science (USA). All rights reserved.

TABLE I Human Matrix Metalloproteinases

Name	MMP No.	Matrix substrates
Collagenase		
Collagenase 1	MMP-1	Collagens I, II, III, VII, X, gelatins, aggrecan, entactin
Neutrophil collagenase	MMP-8	Collagens I, II, III, aggrecan, link protein
Collagenase 3	MMP-13	Collagens I, II, III
MT1-MMP	MT1-MMP	Collagens I, II, III fibronectin, pro-MMP-1
Stromelysins		
1	MMP-3	Collagens III, IV, IX, gelatins, aggrecan, fibronectin, laminin, pro-MMP-1
2	MMP-10	Collagen IV, aggrecan, fibronectin, elastin, laminin
3	MMP-11	Gelatin, collagens I, IV, V, VII, X, fibronectin, laminin, elastin, aggrecan, vitronectin
Gelatinases		
A	MMP-2	Gelatin, collagens IV, V, XIV, elastin, aggrecan, vitronectin
B	MMP-9	Collagen IV, gelatins, aggrecan, fibronectin, elastin, laminin, elastin, pro-MMP-1
Membrane-type MMPs	MT2-MMP	Pro-MMP-2
	MT3-MMP	Pro-MMP-2
	MT4-MMP	Not known
	MT5-MMP	Not known
	MT6-MMP	Not known
Others		
Matrilysin	MMP-7	Gelatins, fibronectin, pro-MMP-1
Macrophage metalloelastase	MMP-12	Elastin
	MMP-19	Not known
Enamelysin	MMP-20	Amelogenin
None	MMP-23	Not known

collagen in the basement membrane and are usually expressed by endothelial cells, even though other cells, e.g., stromal fibroblasts, macrophages, and tumor cells, also synthesize these MMPs (Hotary *et al.*, 2000; Nagase and Woessner, 1999; Nelson *et al.*, 2000; Ohuchi *et al.*, 1997; Parsons *et al.*, 1997; Pei, 1999; Velasco *et al.*, 2000; Vincenti *et al.*, 1996). In normal cells, expression of MMPs is constitutively low, and this low level of basal expression probably contributes to the normal physiology of most tissues and to connective tissue remodeling (Benbow and Brinckerhoff, 1997; Borden and Heller, 1997; Borden *et al.*, 1996; Vincenti *et al.*, 1996).

However, in pathologic conditions, the level of MMP expression increases considerably, often resulting in aberrant connective tissue destruction (Basset *et al.*, 1997; Brinckerhoff *et al.*, 2000; Chambers and Matrisian, 1997; Curran and Murray, 1999; Koblinski *et al.*, 2000; McCawley and Matrisian, 2000;

Westermark and Kahari, 1999). Inflammatory cytokines and growth factors and malignant transformation lead to a marked increase in MMP expression (Borden and Heller, 1997; Westermarck *et al.*, 1998, 2000). Excess MMP production is, therefore, associated with many diseases, including periodontitis, atherosclerosis, tumor invasion and metastasis, and arthritic disease, and several MMPs may be involved in the pathophysiology of any of these diseases (Basset *et al.*, 1997; Brinckerhoff *et al.*, 2000; Chambers and Matrisian, 1997; Curran and Murray, 1999; Koblinski *et al.*, 2000; McCawley and Matrisian, 2000; Westermark and Kahari, 1999). Increasingly, it is evident that the mechanisms regulating expression of these enzymes under normal conditions and in the various diseases are complex, permitting both the coordinate expression of some MMPs and the tissue-specific expression of others. Posttranscriptional mechanisms, i.e., changes in mRNA stability, are used for controlling MMP gene expression (Vincenti *et al.*, 1996). Indeed, the 3′ UTR of many MMPs contains the AUUUA motif, which is associated with modulating the half-life of mRNAs. In the absence of inducers, this motif enhances the turnover rate of several MMPs, but in response to inducers, such as interleukin (IL)-1 and phorbol esters, these mRNAs become stabilized (Vincenti *et al.*, 1996). However, regulation of gene expression at the level of transcription appears to play the major role in regulating these genes, and therefore this chapter focuses on the *cis*-acting sequences within the MMP promoters and the *trans*-acting factors that regulate transcription of the MMPs in different cell types and tissues.

II. REGULATION OF MMP TRANSCRIPTION

Many MMPs share similar sequences in their promoters (Fig. 1) and yet in the same cells, e.g., within the same transcriptional environment, some MMPs are expressed whereas others are not (Benbow and Brinckerhoff, 1997; Borden and Heller, 1997; Vincenti *et al.*, 1996). Thus, there must be intricate pathways and mechanisms that specifically control the expression of each MMP within a particular cell. These mechanisms may include both "*cis*" and "*trans*," as manifested by the presence (or absence) of activating and repressor elements within the promoter, allowing appropriate transcription factors to bind to these elements (Benbow and Brinckerhoff, 1997). Chromatin structure and methylation status of DNA also play a role in regulating gene transcription (Singh *et al.*, 2000).

Generally, stimuli such as inflammatory cytokines (interleukin–1 and tumor necrosis factor α) and growth factors (fibroblast growth factor, platelet-derived growth factor, epidermal growth factor) increase MMP transcription (Benbow and Brinckerhoff, 1997; Borden and Heller, 1997; Westermark and Kahari, 1999). This increase is initiated by the interaction of the ligand (growth factor or cytokine) with its cognate receptor. The interaction results in a conformational

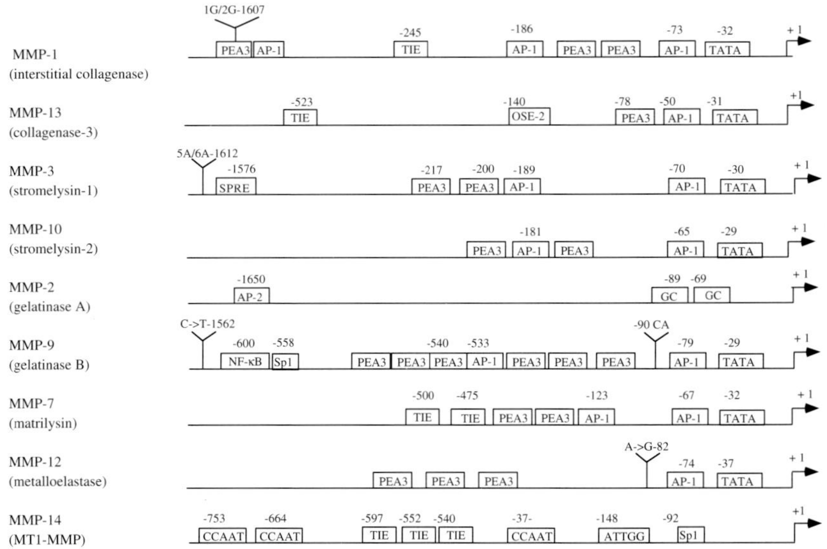

FIGURE 1 Model of MMP-1 promoter constructs. *cis*-acting elements known to participate in transcription are indicated. The Y indicates functional polymorphisms in MMP promoters. AP-1, activator protein-1 site; PEA3, polyomavirus enhancer A-binding protein 3 site; TIE, TGF-β inhibitory element; GC, Sp-1-binding site; OSE-2, osteoblast-specific element (also referred to as Runx-2 and Cbfa1).

change in the structure of the receptor on its cytoplasmic side, causing either autophosphorylation of the receptor or recruitment of secondary factors (adaptor molecules) to the site. There, they begin the signal/transduction cascade that culminates in the activation of transcription factors in the nucleus.

Mitogen-activated protein kinases (MAPKs) facilitate the signal/transduction pathways, and these kinases include c-Jun-N-terminal kinases (JNK), extracellular signal-regulating kinases (ERK1/2), and p38 kinases (Westermarck *et al.*, 1998, 2000; Westermark and Kahari, 1999). Interestingly, the relative contribution of each of these pathways to MMP gene expression varies considerably among cell types and inductive stimuli (Barchowsky *et al.*, 2000; Brauchle *et al.*, 2000; Westermarck *et al.*, 1998, 2000). For example, JNK has a central role in IL-1-induced MMP-1 gene expression in cultured synovial fibroblasts, where induction is completely blocked with a chemical inhibitor specific for JNK, suggesting that this pathway is critical (Barchowsky *et al.*, 2000). However, in animal studies with the same inhibitor, there is only partial inhibition of collagenase mRNA, indicating that other pathways, such as p38, are involved (Mengshol *et al.*, 2000, 2002). Indeed, MMP-1 expression in IL-1-stimulated fibroblasts and chondrocytic cells and in squamous cell carcinoma is mediated by both p38 and JNK (Brauchle *et al.*, 2000; Westermarck *et al.*, 1998, 2000). Thus, this cross talk among pathways is complex, and it is possible that repression of one pathway may result in the recruitment/activation of an alternative path in order to accomplish gene activation.

Furthermore, the MEK/ERK pathway was thought to only transduce growth factor-dependent proliferative signals, as these enzymes phosphorylate and activate Elk-1, a transcription factor that transactivates serum response elements (Westermark and Kahari, 1999). However, the ERK pathway can be blocked with the chemical inhibitor PD98059, which also blocks MMP-1 gene expression in A2058 melanoma cells (Tower *et al.*, 2002) and in IL-1-stimulated fibroblasts (Brauchle *et al.*, 2000; Mengshol *et al.*, 2002). In contrast, the p38 inhibitor SB203580 has no effect on MMP-1 expression in the melanoma cells, whereas it blocks expression in IL-1-stimulated fibroblasts (Barchowsky *et al.*, 2000), again emphasizing the complexity and diversity of these pathways in controlling MMP gene expression.

Once the signal has been transduced to the nucleus, it acts on the *cis*-acting sequences in the promoters of the MMP genes to alter transcription (Benbow and Brinckerhoff, 1997; Borden *et al.*, 1996; Vincenti *et al.*, 1996). These elements include the activator protein-1 site (AP-1; 5′-TGAG/CTCA-3′), located at about –70 bp in all the promoters, except MMP-2 (gelatinase A), MMP-11 (stromelysin-3), and MMP-14 (MT1-MMP). This site binds dimers of the Fos and Jun families and has a major role in both basal and induced transcription. Although earlier studies focused almost exclusively on the importance of the proximal AP-1 site, later studies emphasized that this site does not function alone. Rather, it must cooperate with a variety of *cis*-acting sequences found in the upstream regions of the MMP promoters. For instance, MMP-9 (gelatinase B) contains a series of upstream-*cis*-acting elements (AP-1, ETS, Sp1, and NF-κB), all located within 670 bp of promoter DNA, and depending on the stimulus, transcriptional induction requires cooperation between the proximal AP-1 site and the Sp-1 site or the NF-κB site (Sato *et al.*, 1993; Sato and Seiki, 1993). In addition, induction of MMP-1 by IL-1 in rabbit fibroblasts requires interaction between the AP-1 site at –77 bp and an NF-κB-like element located upstream at –3300 bp (Vincenti *et al.*, 1998). Perhaps it is the particular partnerships between the AP-1 site and the other elements upstream that help confer the specificity of MMP expression.

The MMP-13 (collagenase-3) promoter also contains a proximal AP-1 site. In contrast to many MMPs, this gene exhibits a pattern of expression that is considerably restricted, normally being limited to developing cartilage and bone. Mechanisms controlling this restricted expression are still unclear, but may involve the transcription factor Cbfa1/Runx-2/OSF2, which appears to be expressed almost exclusively in these developing tissues (Jimenez *et al.*, 1999; Mengshol *et al.*, 2001).

Among the MMPs, a Runx-2-binding site is unique to the MMP-13 promoter, and thus, the presence of Runx-2 protein in cartilage and bone may contribute to the tissue-specific expression of this gene, as induction of MMP-13 expression in chondrosarcoma cells requires AP-1 and Runx-2 (Porte *et al.*, 1999).

Additional AP-1 sites are scattered throughout MMP promoters and may participate in MMP gene expression. A second AP-1 site is found in the promoters of MMP-1 (collagenase), MMP-3 (stromelysin-1), MMP-7 (matrilysin), MMP-11 (stromelysin-2), and MMP-9 (gelatinase B) (Benbow and Brinckerhoff, 1997). Perhaps the best characterized is the AP-1 site (5′-TTAATCA-3′) at −186 bp in rabbit and human MMP-1 promoters. In contrast to the AP-1 site at ~ −70 bp, which has a major role in basal transcription, this site has only a modest effect on basal transcriptional activity, but it does increase transcriptional induction by phorbol esters substantially. These findings suggest distinct roles for these two AP-1 elements, which may be related to the differential binding of AP-1 family members to each site (Benbow and Brinckerhoff, 1997; Chamberlain *et al.*, 1993; White and Brinckerhoff, 1995). Further, sequence analysis of 4400 bp of human MMP-1 promoter DNA revealed several AP-1 sites, located between −2300 and −1600 bp, and at least one of these has been implicated in the transcriptional regulation of this gene (see later).

Multiple PEA3/ETS sites, which bind a wide assortment of transcription factors belonging to the Ets family (Wasylyk *et al.*, 1991), are also present in all the promoters, except MMP-2 (gelatinase A). However, the number of these sites and their spatial arrangement differ among the MMP family members, and these differences probably influence how the MMP genes are regulated. For example, there are two ETS sites at −217 and −200 bp in the MMP-3 promoter, and mutations in either of these reduce transcription, indicating a critical role for these elements (Buttice and Kurkinen, 1993). Initially, (Wasylyk *et al.*, 1990) demonstrated that AP-1 and adjacent ETS sites act cooperatively and suggested that this cooperativity is essential for transcription. Later studies have only corroborated this early finding (see later). In addition, the sequences immediately adjacent to the core ETS sequence, 5′-GGA-3′, probably influence which Ets family members bind (Benbow and Brinckerhoff, 1997). This differential binding may, in turn, contribute to transcriptional regulation due to differences in binding affinities of the proteins for DNA and variations in the partnering of different transcription factors.

The promoters of MMP-2 (gelatinase) and MMP-14 (MT1-MMP) differ from other MMPs in that they lack proximal AP-1 and ETS sites, as well as the traditional TATA box that is usually located at ~−30 bp from the transcription start site (Benbow and Brinckerhoff, 1997; Lohi *et al.*, 2000). The TATA box is a core transcriptional unit, which binds a group of general transcription factors that help initiate transcription by RNA polymerase. Its absence may contribute to the constitutive expression of these genes. Indeed, the fact that these enzymes are constitutively expressed and not usually subject to regulation has led to the suggestion that MMP-2 is a “housekeeping” gene (Benbow and

Brinckerhoff, 1997). Nonetheless, these MMPs have crucial roles in connective tissue metabolism (See later).

III. SINGLE NUCLEOTIDE POLYMORPHISMS IN MMP PROMOTERS

DNA polymorphisms are naturally occurring variations in the DNA sequence of a gene, resulting in more than one allele. Polymorphisms have a frequency in the normal human population of >1% and are estimated to be found in every 1000 bp throughout the genome (Sherry *et al.*, 1999). The vast majority of these polymorphisms are single nucleotide polymorphisms (SNPs) because of single base substitutions or deletions. Some SNPs either create or delete a restriction site, making them convenient markers for restriction fragment length polymorphisms (RFLPs), which may be helpful diagnostic markers for certain diseases. However, SNPs need not be associated with a RFLP, and these are detected by DNA sequence analysis. While most SNPs are functionally inert and do not contribute to gene expression, a few directly affect either the regulation of gene expression or the function of the encoded protein (Sherry *et al.*, 1999).

A number of SNPs have been described in four MMPs: MMP-1 (Rutter *et al.*, 1998) , MMP-3 (de Maat *et al.*, 1999; Humphries *et al.*, 1998; Terashima *et al.*, 1999; Ye *et al.*, 1996; Yoon *et al.*, 1999), MMP-9 (Peters *et al.*, 1999; Shimajiri *et al.*, 1999; Yoon *et al.*, 1999; Zhang *et al.*, 1999), and MMP-12 (Jormsjo *et al.*, 2000). Interestingly, these SNPs are found in the promoters of the genes, rather than in the coding region, perhaps because of the importance of strictly maintaining the amino acid sequence of the functioning enzymes. The SNP in the MMP-9 gene may be either a C→T substitution at –1562 bp (Peters *et al.*, 1999) or the presence of a microsatellite repeat of $(CA)_n$ at –90 bp (Peters *et al.*, 1999; Shimajiri *et al.*, 1999; Zhang *et al.*, 1999). The former appears to be associated with coronary atherosclerosis (Peters *et al.*, 1999), whereas the latter has been linked to intracranial or abdominal aortic aneurysms (Shimajiri *et al.*, 1999; Zhang *et al.*, 1999). The microsatellite polymorphism is multiallelic, with a varying number of repeats. The (CA)14 repeat has only 50 to 60% of the transcriptional activity in esophageal carcinoma cells and human fibroblasts, as do the (CA)21 and (CA)23 repeats, respectively (Shimajiri *et al.*, 1999; Yoon *et al.*, 1999). The SNP substitution at –1562 bp results in a decrease in the binding of nuclear proteins and an increase in transcriptional activity in DNA containing the T allele (Peters *et al.*, 1999). Because MMP-9 is involved in the degradation of type IV collagen in the basement membrane, an increase in MMP-9 gene expression may facilitate the ability of vascular smooth muscle cells to migrate and proliferate during atherogenesis (Galis, 1994; Yoon *et al.*, 1999).

The insertion/deletion polymorphism (5A/6A polymorphism) at –1612 bp in the MMP-3 promoter has also been connected to atherosclerosis. The track of 5 adenosines (5A) has greater promoter activity than the 6A track in both fibroblasts and vascular smooth muscle cells (Ye *et al.*, 1998). Gel mobility shift assays and DNase protection studies suggest the binding of two transciption factors to this site, one of which binds with higher affinity to the 6A allele (Ye *et al.*, 1998). These data have led to the suggestion that one protein may be a repressor that binds preferentially to the 6A allele. Although the identity of these transcription factors is unknown, it is possible that they contribute to the regulation of MMP-3 expression and to the pathophysiology of cardiovascular disease. For example, MMP-3 is present abundantly in atherosclerotic plaques where rupture is detected, and the A5 allele may heighten MMP-3 expression, thereby facilitating plaque rupture (Galis, 1994; Henney *et al.*, 1991; Ye *et al.*, 1998). Indeed a case-controlled study of acute myocardial infarctions in a population of more than 600 Japanese revealed that the frequency of the 5A allele was significantly higher ($P < 0.0001$) in patients vs controls, with 48.8% of patients harboring the 5A allele compared to 32.7% of controls (Terashima *et al.*, 1999). In contrast, the 6A allele has been associated with the accelerated growth of coronary atheromas, suggesting that decreased MMP-3 expression may contribute to plaque growth (Galis, 1994; Henney *et al.*, 1991; Ye *et al.*, 1998).

In the MMP-12 gene, the SNP at –82 bp results from an A→G substitution (Jormsjo *et al.*, 2000). This SNP is adjacent to the proximal AP-1 site, and gel shift analyses have demonstrated that this sequence change affects the binding affinity of AP-1 proteins. The A allele binds nuclear proteins with greater affinity and also has higher transcriptional activity in several macrophage cell lines in response to phorbol esters and insulin, and the A allele may be associated with increased narrowing of the coronary artery in diabetic patients who also have heart disease (Jormsjo *et al.*, 2000).

The SNP in the MMP-1 promoter is located at –1607 bp and is represented by the presence or absence of an extra guanine (G) at this site (Rutter *et al.*, 1998). Initially, several papers reported an ETS consensus site (5′-GGAA-3′) at –1607 bp in the human MMP-1 promoter (Aho *et al.*, 1997; Rutter *et al.*, 1997, 1998). However, an additional clone isolated from a human leukocyte genomic library failed to contain this site, having the sequence 5′GAA-3′ instead. Thus, the presence/absence of the extra guanine (G) at –1607 bp suggested the presence of a SNP at this position, and subsequent analysis of DNA from normal individuals revealed a distribution of approximately 25% 1G homozygotes, 25% 2G homozygotes, and 50% 1G/2G heterozygotes. This 1G/2G variation is, therefore, a true SNP and not a rare mutation (Rutter *et al.*, 1998).

The 2G SNP is located just upstream from an AP-1 site at –1602 bp, and it was hypothesized that these sites might cooperate to enhance transcription. To test this hypothesis, 4372 bp of MMP-1 promoter DNA containing either

1G or 2Gs linked to the luciferase reporter was tested for transcriptional activity in normal fibroblasts and in several tumor cell lines, which constitutively expressed MMP-1 (Rutter *et al.*, 1998). The only difference between DNAs in these constructs was the presence or absence of an extra G at –1607 bp. The 2G allele was expressed consistently at a higher level than the 1G allele in both normal fibroblasts and in tumor cells (Rutter *et al.*, 1998; Tower *et al.*, 2002). Furthermore, DNA/protein interactions, measured in gel mobility shift assays, revealed that the 2G DNA bound more nuclear proteins than 1G DNA and that this binding was of higher affinity. Taken together, these data indicate that the 2G allele heightened transcription of the MMP-1 gene.

Quite possibly, then, the 2G allele could contribute to the pathophysiology of diseases where MMP-1 plays a role, and indeed, this suggestion has received substantial support. Initially, it was reported that patients with colorectal or esophageal cancers whose tumors expressed MMP-1 had a poorer prognosis than patients with tumors not expressing MMP-1 (Murray *et al.*, 1996, 1998). Thus, there was an inverse correlation between MMP-1 expression and patient survival. More recently, five reports have associated the 2G polymorphism with aggressive cancers: ovarian (Kanamori *et al.*, 1999) endometrial (Nishioka *et al.*, 2000), lung (Zhu *et al.*, 2001), and melanoma (Rutter *et al.*, 1998; Ye *et al.*, 2001). In ovarian cancer, the number of patients either homozygous or heterozygous for the 2G allele was significantly greater than that observed in 150 individuals without this cancer ($P = 0.028$). In addition, the level of MMP-1 protein expressed in cancer tissues from patients carrying the 2G allele was increased significantly ($P = 0.038$) compared to 1G homozygotes (Kanamori *et al.*, 1999). Similarly, a study of endometrial cancers revealed an increased frequency of the 2G allele in 100 Japanese cancer patients compared to 150 normal individuals: 91% vs 80% ($P = 0.019$). By stimulating degradation of the extracellular matrix, increased production of MMP-1 appears to increase the incidence of these two cancers, at least in Japanese women (Kanamori *et al.*, 1999; Nishioka *et al.*, 2000).

Another study links the 2G allele in the MMP-1 promoter to enhanced susceptibility to lung cancer (Zhu *et al.*, 2001). These investigators tested the hypotheses that individuals with the 2G/2G genotype would (a) be an increased risk for lung cancer and (b) the risk would be elevated in smoking individuals. Genotyping 494 patients with lung cancer (404 Caucasian, 43 Mexican-Americans, 47 African-Americans) and 402 frequency-matched controls revealed a significant difference ($P < 0.05$) in the 2G/2G genotype distribution by ethnicity in controls, with 36.1% of Caucasians, 52.9% of Mexican-Americans, and 22.22% of African-Americans having this genotype. Overall there was a significant risk for lung cancer between the 2G/2G genotype in African-Americans, if the individuals were smokers, but not in Causcasians or Mexican-Americans. Thus, these data demonstrate the association of the MMP-1 polymorphism with the development of lung cancer in the presence of carcinogenic exposure.

Also reported is the influence of 2G SNP on the invasive behavior of cutaneous malignant melanoma (Rutter *et al.*, 1998; Ye *et al.*, 2001). The MMP-1 genotype was determined in 139 Caucasian patients with this disease. It was found that the insertion allele (the 2G allele) was associated with invasive disease and, thus, with a poorer prognosis ($P = 0.0333$), suggesting that the invasive behavior of melanoma is influenced by the increased level of MMP-1 expression that accompanies the presence of the 2G allele.

Finally, malignant melanoma was also used to study the frequency of the 2G allele in metastatic disease (Noll *et al.*, 2001). Loss of heterozygosity (LOH) at the MMP-1 locus (11q22.23) is a common occurrence in malignant melanoma (Herbst *et al.*, 1999, 2000), and this study tested the hypothesis that the 2G allele would be more frequent in metastatic tumors (Noll *et al.*, 2001). It was hypothesized that although loss of either the 1G or the 2G allele from 1G/2G heterozygotes is random, if the transcriptionally more active 2G allele were retained, this would favor tumor invasion and metastasis. Consequently, a higher proportion of metastases would contain the 2G genotype, compared to the 1G genotype. In fact, in 31 individuals with metastatic melanoma who were 1G/2G heterozygotes, 12 showed LOH in their metastatic tumors and 10 of these (83%) had undergone LOH with retention of the 2G allele, compared to one individual who retained the 1G allele ($P = 0.04$). Taken together, all of these reports strongly implicate 2G SNP in the MMP-1 promoter as playing a substantial role in cancer biology.

IV. BIOLOGY OF MMP GENE EXPRESSION

As our knowledge of the molecular biology of MMP gene expression increases, we are also learning more about the role of these enzymes in both normal physiology and in disease pathology. Once thought to have only limited functions, it is now apparent that MMPs are expressed ubiquitously and that they contribute to numerous biologic processes, such as development, wound healing, angiogenesis, and reproduction (Nagase and Woessner, 1999; Nelson *et al.*, 2000; Ohuchi *et al.*, 1997; Parsons *et al.*, 1997; Pei, 1999; Velasco *et al.*, 2000; Vincenti *et al.*, 1996).

Targeted disruption of specific genes in mice ("knockout" mice) and transgenic mouse technologies have been used to understand the biologic significance of MMP gene expression (Engsig *et al.*, 2000; Itoh *et al.*, 1998; Masson *et al.*, 1998; Mudgett *et al.*, 1998; Vu *et al.*, 1998). Interestingly, most of these knockout mice display little, if any, phenotype. For example, even though there is increased expression of MMP-3 in rheumatoid and osteoarthritis, MMP-3-deficient mice have no detectable phenotype. These mice develop normally, and even when challenged with an experimental model of arthritis, the knockout mice have the same severity of arthritic disease and cartilage degradation as wild-type mice (Mudgett *et al.*, 1998).

The gelatinases (MMP-2, MMP-9) have also been knocked out in mouse models (Itoh *et al.*, 1998; Masson *et al.*, 1998; Vu *et al.*, 1998). The MMP-9-deficient mouse shows abnormal vascularization of the growth plate that is overcome in adult mice (Engsig *et al.*, 2000; Vu *et al.*, 1998). The MMP-9 deficiency is also associated with a delay in osteoclast recruitment into hypertrophic cartilage (Engsig *et al.*, 2000) and reduced angiogenesis and tumor progression (Itoh *et al.*, 1998). Mice without the MMP-2 also develop normally but display reduced angiogenesis and tumorigenesis (Masson *et al.*, 1998). MMP-11 (stromelysin 3) deletion showed no apparent phenotype but did show a decrease in tumor "takes." In a syngeneic tumor model of MMP-11 knockout mice, injection of cancer cells did not affect neoangiogenesis or cell proliferation, but there was an increase in malignant cell death (Boulay *et al.*, 2001). All in all, the mild phenotypes suggest that these MMPs are not critical during development and that redundancies in MMP expression may be an important compensatory mechanism, with other MMP family members rescuing the enzymatic deficiencies.

In contrast, however, the MT1-MMP knockout mouse shows a severe phenotype. These mice develop dwarfism, osteopenia, and arthritis with a prominent fibrotic synovitis (Holmbeck *et al.*, 1999). They also have aberrant bone growth and angiogenesis (Zhou *et al.*, 2000). It is not known whether this phenotype results from the lack of activation of MMP-2, MMP-9, and MMP-13, which is accomplished by MT1-MMP, or to the loss of direct collagenolytic activity due to the absence of this enzyme. Although the phenotype seen in these studies is dramatic, it is possible the major effects of MT1-MMP loss are restricted to development, and thus the deficiency may not be so devastating in adults. A mouse with a conditional knockout of MT1-MMP could be an important tool for increasing our understanding of this novel MMP family member.

Transgenic mice have demonstrated a role for MMP-13 (interstitial collagenase 3) in the pathophysiology of arthritis. Overexpressing a constitutively active form of this enzyme in the hyaline cartilage of mice resulted in joint erosions resembling those seen in osteoarthritis, where type II collagen degradation is confined to cartilage and bone (Neuhold *et al.*, 2001). It has also been found in some cancers, particularly squamous cell carcinomas of the head and neck, and in some breast cancers (Johansson *et al.*, 1999; Kahari *et al.*, 1998; Tsukifuji *et al.*, 1999; Uitto *et al.*, 1998). MMP-13 is the major interstitial collagenase in mice and rats, as the genome of these animals does not contain MMP-1. As such, it is possible that the regulation of MMP-13 in these animals differs from that seen in humans, where both interstitial collagenases are present.

In humans, MMP-1 (collagenase-1) is the most ubiquitously expressed of the interstitial collagenases (Vincenti *et al.*, 1996). Most cells produce low basal levels, which contribute to normal physiology, and expression is increased readily and substantially upon exposure to growth factors and inflammatory cytokines (Vincenti *et al.*, 1996). MMP-8 (collagenase-2; neutrophil collagenase) is primarily the product of polymorphonuclear leukocytes, where it is synthesized and packaged in granules that are released in response to inflammatory stimuli. However, more recently, its expression has also been seen in osteoarthritic chondrocytes, and future studies may reveal new biologic roles for this enzyme (Shlopov *et al.*, 1997).

With its ubiquitous expression, MMP-1 has a role in the pathophysiology of many diseases. These include periodontal disease, rheumatoid and osteoarthritis, atherosclerosis, and tumor invasion/metastasis (Benbow and Brinckerhoff, 1997; Brinckerhoff *et al.*, 2000; Vincenti *et al.*, 1996). The contribution of MMP-1 to joint destruction in both rheumatoid and ostseoarthritis has been recognized for some time, but, more recently, attention has turned to the role of MMP-1 in tumor biology (Benbow *et al.*, 1999a,b; Brinckerhoff *et al.*, 2000; Durko *et al.*, 1997; Inoue *et al.*, 1999; Ito *et al.*, 1999; Murray *et al.*, 1996, 1998; Schoenermark *et al.*, 1999). Even though the degradation of type IV collagen in the basement membrane is an essential component of tumor invasion, this is not the only barrier that must be broken down. The stromal collagens, types I and III, are the major proteins in our body and must also be degraded if tumor cells are to migrate to other sites. Furthermore, because it is usually not the primary tumor that is lethal, the invasive potential of tumor cells becomes an important component of cancer biology. Increasingly, it is apparent that heightened production of MMP-1 is associated with progressing and advancing tumors, whether the enzyme is produced by the tumor cells and/or by the surrounding stromal cells (Benbow *et al.*, 1999a, b; Brinckerhoff *et al.*, 2000).

The specific role of MMP-1 in the degradation of stromal collagen by tumor cells and by stromal fibroblasts has been investigated in malignant melanoma (Benbow *et al.*, 1999a) and in breast cancer (Benbow *et al.*, 1999b). These studies have utilized an *in vitro* invasion assay, which quantitates the ability of the tumor cells to traverse a layer of type I collagen. These investigations were carried out with A2058 melanoma cells and MDA231 breast cancer cells, both of which have the 2G genotype and which produce copious amounts of MMP-1. Furthermore, MMP-1 is the predominant MMP synthesized by these cells, again emphasizing the role of this MMP in aggressive tumors. Although both of these cancer cells readily invade a matrix of type I collagen, they do so by different mechanisms. Despite the presence of large amounts of MMP-1, A2058 melanoma cells failed to invade the collagen matrix, probably because these cells did not produce MMP-3, an enzyme needed for the maximal activation of pro-MMP-1. However, when the tumor cells were cocultured with normal stromal fibroblasts or with conditioned medium derived from these fibroblasts, invasion occurred. Subsequent experiments demonstrated that the stromal cells produced MMP-3, which then activated pro-MMP-1, allowing invasion to proceed (Benbow *et al.*, 1999a). In contrast, the MDA-231 breast cancer cells invaded the collagen

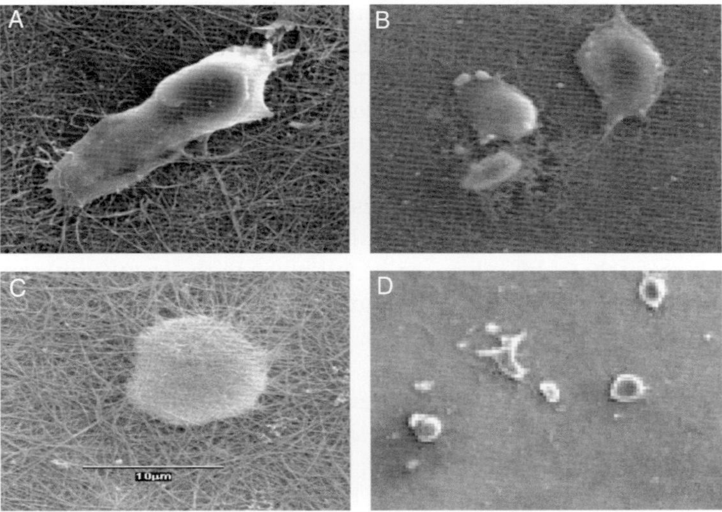

FIGURE 2 Invasion of type I collagen by MDA231 breast cancer cells. Cells were seeded on a collagen type I matrix. Scanning electron microscopy images were taken after (A) 8, (B) 24, and 48 in the absence (C) or presence of all-*trans* retinoic acid (D). In the absence of all-*trans* retinoic acid, invasion into the collagen type I matrix was observed after 24 h. After 48 h, cells completely invaded the matrix. In the presence of all-*trans* retinoic acid, invasion into the collagen layer was inhibited.

readily without the aid of stromal cells (Fig. 2). Experiments with Aprotinin, an inhibitor of serine proteinases, demonstrated that the cells secreted serine proteinase and activated the latent MMP-1. It is important to point out that even though the breast cancer cells invaded the collagen on their own, the degree of invasion was enhanced by the presence of stromal cell-conditioned medium (Benbow *et al.*, 1999b). This finding underscores the potential role of host cells in mediating the invasive behavior of tumor cells.

V. MMPs AS THERAPEUTIC TARGETS

Because overexpression of MMPs is directly linked to the pathology of numerous diseases, these enzymes represent attractive targets for therapies that block either their activity or their synthesis. Inhibiting enzyme activity with synthetic compounds has represented a substantial research effort. However, difficulties associated with inhibiting only a specific MMP, with delivery of the drug, with rates of clearance, and with achieving clinically effective concentrations have hampered progress (Greenwald *et al.*, 2000; Vincenti *et al.*, 1994).

Suppressing MMP synthesis by inhibiting the expression of MMP genes at the level of transcription is another approach (Borden and Heller, 1997; Schroen *et al.*, 1997; Vincenti *et al.*, 1996). The multifunctional growth factor, transforming growth factor β (TGF-β), glucocorticoid hormones, and vitamin A analogs all inhibit MMP synthesis (Benbow and Brinckerhoff, 1997; Heimbrook and Oliff, 1998; Kerr *et al.*, 1990; Mauviel *et al.*, 1996). Although these compounds can block MMP gene expression, difficulties associated with the goal of blocking expression of a particular MMP, rather than all of them, has been problematic. Perhaps some of these difficulties can be attributed to the fact that in contrast to transcriptional activation, which seems to depend on interaction of the proximal AP-1 site with numerous other elements, this AP-1 site plays a prominent role in repression by each of these agents.

TGF-β inhibits MMP-1 and MMP-3 gene expression by Fos- or Jun-dependent mechanisms, which operate at the AP-1 site and at a cooperating upstream motif (Kerr *et al.*, 1990; Mauviel *et al.*, 1996; White *et al.*, 2000). This motif is the TGF-β inhibitory element (TIE; 5′-GAGTTGGTGA-3′), located at −709 bp in the rat MMP-3 promoter and at −245 bp in both rabbit and human MMP-1 promoters. Treatment of cells with TGF-β alters the complex of proteins binding to the AP-1 site in the human MMP-1 promoter to include CREB (Kramer *et al.*, 1991). It also induces several nuclear proteins, including JunB (Kramer *et al.*, 1991), an AP-1 protein that antagonizes transactivation by c-Jun (Deng and Karin, 1993). However, despite the presence of a specific TIE, a sequence that can mediate repression of MMPs, the wide range of other effects of TGF-β on cells has precluded its successful use as an anti-invasive agent.

Glucocorticoid hormones also inhibit MMP gene expression, and these compounds are often used as therapeutic agents in rheumatoid arthritis, where several MMPs are overexpressed and can contribute to the pathophysiology of this disease (Heimbrook and Oliff, 1998). Because the promoters of MMPs do not contain a GRE, the suppressive action of these hormones is mediated by an indirect effect. The hormones bind to Fos and Jun proteins, which are complexed at the proximal AP-1 site, causing a conformational change in these proteins, with a subsequent reduction in transcription. However, glucocorticoids also have pleotropic effects on many cells, and although they may be effective therapeutically, there are serious concerns about side effects and toxicities associated with their use in long-term chronic conditions.

Similarly, the vitamin A analogs, retinoids (Schoenermark *et al.*, 1999; Schroen *et al.*, 1997; Sporn *et al.*, 1994; Vincenti *et al.*, 1996), have numerous effects on cells, including the inhibition of MMP synthesis, which results, for example, in the inhibition of tumor invasion (Fig. 2) (Schroen *et al.*, 1997; Spanjaard *et al.*, 1997; Sporn *et al.*, 1994). In contrast to untreated tumor cells, which migrate through the collagen, tumor cells treated with all-*trans* retinoic acid fail to invade and remain on top of the collagen (Benbow *et al.*, 1999b; Schoenermark *et al.*, 1999).

Retinoids bind to the nuclear hormone receptors, retinoic acid receptors (RARs) and retinoid X receptors (RXRs), which are members of the steroid hormone receptor superfamily (Schroen *et al.*, 1997; Sporn *et al.*, 1994). RAR/RXR heterodimers or RXR homodimers, in turn, bind to several retinoic acid response elements (RARE), with the motif AG[G/T]TCA. However, because most MMP promoters do not contain an RARE (Schroen *et al.*, 1997; Westermark and Kahari, 1999), retinoid-mediated repression occurs indirectly by a number of mechanisms. These include (a) induction of RAR mRNAs, (b) suppression of Fos and Jun mRNAs and protein, and (c) binding of RAR/RXR heterodimers to proteins complexed at the proximal AP-1 site in the promoter (Schroen *et al.*, 1997; Sporn *et al.*, 1994). Despite the broad range of target genes affected by RARs and RXRs, retinoids have a history of successful repressors of several kinds of malignancy, including squamous cell carcinoma of the head and neck and acute promyelocytic leukemia. Nonetheless, toxicity and teratogenicity have limited the enthusiasm for retinoids as therapeutic agents that reduce MMP gene expression.

Given the difficulties associated with TGF-β, glucocorticoid hormones, and retinoids, it is not surprising that novel therapeutic strategies are emerging. Among these are inhibitors of the signal/transduction pathways involved in MMP gene expression (Mengshol *et al.*, 2000a; Rowinsky *et al.*, 1999; Westermark and Kahari, 1999). Several of these therapies are in clinical trials where they may block MMP gene expression and halt cell proliferation as well. One important signal/transduction cascade involved in MMP gene expression is the Ras/mitogen-activated protein kinase (MAPK) pathway (Mengshol *et al.*, 2000, 2002; Rowinsky *et al.*, 1999; Westermark and Kahari, 1999). These novel inhibitors prevent farnesylation and geranylgeranylation of Ras that are required for its location in plasma membrane where it transduces signals through the pathways involved in the transcriptional regulation of oncogenes and MMPs (Buolamwini, 1999; Heimbrook and Oliff, 1998; Rowinsky *et al.*, 1999). In addition, inhibitors of MEK/ERK1/2 prevent the phosphorylation/activation of these signal/transducers, with subsequent inhibition of MMP gene expression (Mengshol *et al.*, 2000, 2002; Rowinsky *et al.*, 1999; Westermark and Kahari, 1999).

VI. CONCLUSION

Perhaps the best hope for successful therapies against MMPs lies in the use of combination therapies, where several drugs are used together. For example, an inhibitor of MMP activity has been used together with conventional chemotherapy to halt tumor progression (Zucker *et al.*, 2000). In addition, the development of compounds that inhibit MMP activity based on our knowledge of the crystal structure of the enzyme may permit the desired specificity. Similarly, the availability of ligands that are targeted to specific RARs or RXRs reports ("designer retinoids") may provide a means for successful therapies directed at inhibiting MMP synthesis (Schadendorf *et al.*, 1996; Schoenermark *et al.*, 1999; Spanjaard *et al.*, 1997). Our ever-increasing knowledge of mechanisms controlling MMP gene expression and of the roles of MMPs in physiology suggests that our ability to apply this information in beneficial ways will only increase. MMPs will become part of molecular medicine for the 21st century.

ACKNOWLEDGMENTS

Supported by grants from the NIH (AR 26599, CA 77267) and grants from the RGK Foundation (Austin, TX), The DOD, and The Susan G. Komen foundation.

REFERENCES

Aho, S., Rouda, S., Kennedy, S. H., Qin, H., and Tan, E. M. (1997). Regulation of human interstitial collagenase (matrix metalloproteinase-1) promoter activity by fibroblast growth factor. *Eur. J. Biochem.* **247**, 503–510.

Barchowsky, A., Frleta, D., and Vincenti, M. P. (2000). Integration of the NF-kappaB and mitogen-activated protein kinase/AP-1 pathways at the collagenase-1 promoter: Divergence of IL-1 and TNF-dependent signal transduction in rabbit primary synovial fibroblasts. *Cytokine* **12**, 1469–1479.

Basset, P., Okada, A., Chenard, M. P., Kannan, R., Stoll, I., Anglard, P., Bellocq, J. P., and Rio, M. C. (1997). Matrix metalloproteinases as stromal effectors of human carcinoma progression: Therapeutic implications. *Matrix Biol.* **15**, 535–541.

Benbow, U., and Brinckerhoff, C. E. (1997). The AP-1 site and MMP gene regulation: What is all the fuss about? *Matrix Biol.* **15**, 519–526.

Benbow, U., Schoenermark, M. P., Mitchell, T. I., Rutter, J. L., Shimokawa, K., Nagase, H. and Brinckerhoff, C. E. (1999a). A novel host/tumor cell interaction activates matrix metalloproteinase 1 and mediates invasion through type I collagen. *J. Biol. Chem.* **274**, 25371–25378.

Benbow, U., Schoenermark, M. P., Orndorff, K. A., Givan, A. L., and Brinckerhoff, C. E. (1999b). Human breast cancer cells activate procollagenase-1 and invade type I collagen: Invasion is inhibited by all-trans retinoic acid. *Clin. Exp. Metast.* **17**, 231–238.

Borden, P., Solymar, D., Sucharczuk, A., Lindman, B., Cannon, P., and Heller, R. A. (1996). Cytokine control of interstitial collagenase and collagenase-3 gene expression in human chondrocytes. *J. Biol. Chem.* **271**, 23577–23581.

Borden, P., and Heller, R. A. (1997). Transcriptional control of matrix metalloproteinases and the tissue inhibitors of matrix metalloproteinases. *Crit. Rev. Eukaryot. Gene Expr.* **7**, 159–178.

Boulay, A., Masson, R., Chenard, M. P., El Fahime, M., Cassard, L., Bellocq, J. P., Sautes-Fridman, C., Basset, P., and Rio, M. C. (2001). High cancer cell death in syngeneic tumors developed in host mice deficient for the stromelysin-3 matrix metalloproteinase. *Cancer Res.* **61**, 2189–2193.

Brauchle, M., Gluck, D., Di Padova, F., Han, J., and Gram, H. (2000). Independent role of p38 and ERK1/2 mitogen-activated kinases in the upregulation of matrix metalloproteinase-1. *Exp. Cell Res.* **258**, 135–144.

Brinckerhoff, C. E., Rutter, J. L., and Benbow, U. (2000). Interstitial collagenases as markers of tumor progression. *Clin. Cancer Res.* **6**, 4823–4830.

Buolamwini, K. K. (1999). Novel anticancer drug discovery *Curr. Opin. Chem. Biol.* **3**, 500–509.

Buttice, G., and Kurkinen, M. (1993). A polyomavirus enhancer A-binding protein-3 site and Ets-2 protein have a major role in the 12-O-tetradecanoylphorbol-13-acetate response of the human stromelysin gene. *J. Biol. Chem.* **268**, 7196–7204.

Chamberlain, S. H., Hemmer, R. M., and Brinckerhoff, C. E. (1993). Novel phorbol ester response region in the collagenase promoter binds Fos and Jun. *J. Cell Biochem.* **52**, 337–351.

Chambers, A. F., and Matrisian, L. M. (1997). Changing views of the role of matrix metalloproteinases in metastasis. *J. Natl. Cancer Inst.* **89**, 1260–1270.

Curran, S., and Murray, G. I. (1999). Matrix metalloproteinases in tumour invasion and metastasis. *J. Pathol.* **189**, 300–308.

de Maat, M. P., Jukema, J. W., Ye, S., Zwinderman, A. H., Moghaddam, P. H., Beekman, M., Kastelein, J. J., van Boven, A. J., Bruschke, A. V., Humphries, S. E., Kluft, C., and Henney, A. M. (1999). Effect of the stromelysin-1 promoter on efficacy of pravastatin in coronary atherosclerosis and restenosis. *Am. J. Cardiol.* **83**, 852–856.

Deng, T., and Karin, M. (1993). JunB differs from c-Jun in its DNA-binding and dimerization domains, and represses c-Jun by formation of inactive heterodimers. *Genes Dev.* **7**, 479–490.

Durko, M., Navab, R., Shibata, H. R., and Brodt, P. (1997). Suppression of basement membrane type IV collagen degradation and cell invasion in human melanoma cells expressing an antisense RNA for MMP-1. *Biochim. Biophys. Acta* **1356**, 271–280.

Engsig, M. T., Chen, Q. J., Vu, T. H., Pedersen, A. C., Therkidsen, B., Lund, L. R., Henriksen, K., Lenhard, T., Foged, N. T., Werb, Z., and Delaisse, J. M. (2000). Matrix metalloproteinase 9 and vascular endothelial growth factor are essential for osteoclast recruitment into developing long bones. *J. Cell Biol.* **151**, 879–890.

Galis, Z. S., Sukhova, G. K., Lark, M. W., and Libby, P. (1994). Increased expression of matrix metalloproteinases and matrix degrading activity in vulnerable regions of human atherosclerotic plaques. *J. Clin. Invest.* **94**, 2493–2503.

Greenwald, R. A., Zucker, S., and Golub, L. M. (2000). Inhibition of matrix metalloproteinases: Therapeutic application. *Ann. N. Y. Acad. Sci.* 878.

Heimbrook, D. C., and Oliff, A. (1998). Therapeutic intervention and signaling. *Curr. Opin. Cell Biol.* **10**, 284–288.

Henney, A. M., Wakeley, P. R., Davies, M. J., Foster, K., Hembry, R., Murphy, G., and Humphries, S. (1991). Localization of stromelysin gene expression in atherosclerotic plaques by in situ hybridization. *Proc. Natl. Acad. Sci. USA* **88**, 8154–8158.

Herbst, R. A., Gutzmer, R., Matiaske, F., Mommert, S., Casper, U., Kapp, A., and Weiss, J. (1999). Identification of two distinct deletion targets at 11q23 in cutaneous malignant melanoma. *Int. J. Cancer* **80**, 205–209.

Herbst, R. A., Mommert, S., Casper, U., Podewski, E. K., Kiehl, P., Kapp, A., and Weiss, J. (2000). 11q23 allelic loss is associated with regional lymph node metastasis in melanoma. *Clin. Cancer Res.* **6**, 3222–3227.

Holmbeck, K., Bianco, P., Caterina, J., Yamada, S., Kromer, M., Kuznetsov, S. A., Mankani, M., Robey, P. G., Poole, A. R., Pidoux, I., Ward, J. M., and Birkedal-Hansen, H. (1999). MT1-MMP-deficient mice develop dwarfism, osteopenia, arthritis, and connective tissue disease due to inadequate collagen turnover. *Cell* **99**, 81–92.

Hotary, K., Allen, E., Puntuieri, A., Yana, I., and Weiss, S. J. (2000). Regulation of cell invasion and morphogenesis in a three dimensional type I collagen matrix by membrane-type matrix metalloproteinases 1, 2, and 3. *J. Cell Biol.* **149**, 1309–1323.

Humphries, S. E., Luong, L. A., Talmud, P. J., Frick, M. H., Kesaniemi, Y. A., Pasternack, A., Taskinen, M. R., and Syvanne, M. (1998). The 5A/6A polymorphism in the promoter of the stromelysin-1 (MMP-3) gene predicts progression of angiographically determined coronary artery disease in men in the LOCAT gemfibrozil study: Lipid Coronary Angiography Trial. *Atherosclerosis* **139**, 49–56.

Inoue, T., Yashiro, M., Nishimura, S., Maeda, K., Sawada, T., Ogawa, Y., Sowa, M., and Chung, K. H. (1999). Matrix metalloproteinase-1 expression is a prognostic factor for patients with advanced gastric cancer. *Int. J. Mol. Med.* **4**, 73–77.

Ito, T., Ito, M., Shiozawa, J., Naito, S., Kanematsu, T., and Sekine, I. (1999). Expression of the MMP-1 in human pancreatic carcinoma: Relationship with prognostic factor. *Mod. Pathol.* **12**, 669–674.

Itoh, T., Tanioka, M., Yoshida, H., Yoshioka, T., Nishimoto, H., and Itohara, S. (1998). Reduced angiogenesis and tumor progression in gelatinase A-deficient mice. *Cancer Res.* **58**, 1048–1051.

Jimenez, M. J., Balbin, M., Lopez, J. M., Alvarez, J., Komori, T., and Lopez-Otin, C. (1999). Collagenase 3 is a target of Cbfa1, a transcription factor of the runt gene family involved in bone formation. *Mol. Cell Biol.* **19**, 4431–4442.

Johansson, N., Vaalamo, M., Grenman, S., Hietanen, S., Klemi, P., Saarialho-Kere, U., and Kahari, V. M. (1999). Collagenase-3 (MMP-13) is expressed by tumor cells in invasive vulvar squamous cell carcinomas. *Am. J. Pathol.* **154**, 469–480.

Jormsjo, S., Ye, S., Moritz, J., Walter, D. H., Dimmeler, S., Zeiher, A. M., Henney, A., Hamsten, A., and Eriksson, P. (2000). Allele-specific regulation of matrix metalloproteinase-12 gene activity is associated with coronary artery luminal dimensions in diabetic patients with manifest coronary artery disease. *Circ. Res.* **86**, 998–1003.

Kahari, V. M., Johansson, N., Grenman, R., Airola, K., and Saarialho-Kere, U. (1998). Expression of collagenase-3 (MMP-13) by tumor cells in squamous cell carcinomas of the head and neck. *Adv. Exp. Med. Biol.* **451**, 63–68.

Kanamori, Y., Matsushima, M., Minaguchi, T., Kobayashi, K., Sagae, S., Kudo, R., Terakawa, N., and Nakamura, Y. (1999). Correlation between expression of the matrix metalloproteinase-1 gene in ovarian cancers and an insertion/deletion polymorphism in its promoter region. *Cancer Res.* **59**, 4225–4227.

Kerr, L. D., Miller, D. B., and Matrisian, L. M. (1990). TGF-beta 1 inhibition of transin/stromelysin gene expression is mediated through a Fos binding sequence. *Cell* **61**, 267–278.

Koblinski, J. E., Ahram, M., and Sloane, B. F. (2000). Unraveling the role of proteases in cancer. *Clin. Chim. Acta* **291**, 113–135.

Kramer, I. M., Koornneef, I., de Laat, S. W., and van den Eijnden-van Raaij, A. J. (1991). TGF-beta 1 induces phosphorylation of the cyclic AMP responsive element binding protein in ML-CCl64 cells. *EMBO J.* **10**, 1083–1089.

Lohi, J., Lehti, K., Valtanen, H., Parks, W. C., and Keski-Oja, J. (2000). Structural analysis and promoter characterization of the human membrane-type matrix metalloproteinase-1 (MT1-MMP) gene. *Gene* **242**, 75–86.

Masson, R., Lefebvre, O., Noel, A., Fahime, M. E., and Chenard, M. P., Wendling, C., Kebers, F., LeMeur, M., Dierich, A., Foidart, J. M., Basset, P., and Rio, M. C. (1998). In vivo evidence that the stromelysin-3 metalloproteinase contributes in a paracrine manner to epithelial cell malignancy. *J. Cell Biol.* **140**, 1535–1541.

Mauviel, A., Chung, K. Y., Agarwal, A., Tamai, K., and Uitto, J. (1996). Cell-specific induction of distinct oncogenes of the Jun family is responsible for differential regulation of collagenase gene expression by transforming growth factor-beta in fibroblasts and keratinocytes. *J. Biol. Chem.* **271**, 10917–10923.

McCawley, L. J., and Matrisian, L. M. (2000). Matrix metalloproteinases: Multifunctional contributors to tumor progression. *Mol. Med. Today* **6**, 149–156.

Mengshol, J. A., Vincenti, M. P., and Brinckerhoff, C. E. (2001). Transcriptional regulation of collagenase-3 (MMP-13) by IL-1 chondrocytic cells requires stable chromatin integration and the transcription factors Runx-2 and AP-1. *Nucleic Acid Res.* **29**, 4361–4372.

Mengshol, J. A., Vincenti, M. P., Coon, C. I., Barchowsky, A., and Brinckerhoff, C. E. (2000a). Interleukin-1 induction of collagenase 3 (matrix metalloproteinase 13) gene expression in chondrocytes requires p38, c-Jun N-terminal kinase, and nuclear factor kappaB: Differential regulation of collagenase 1 and collagenase 3. *Arthritis Rheum.* **43**, 801–811.

Mengshol, J. M., Mix, K. S., and Brinckerhoff, C. E. (2002). Matrix metalloproteinases as therapeutic targets in arthritic diseases: Bull's eye or missing the mark? *Arthritis Rheum.* **46**, 13–20.

Mudgett, J. S., Hutchinson, N. I., Chartrain, N. A., Forsyth, A. J., McDonnell, J., Singer, I., Bayne, E. K., Flanagan, J., Kawka, D., Shen, C. F., Stevens, K., Chen, H., Trumbauer, M., and Visco, D. M. (1998). Susceptibility of stromelysin 1-deficient mice to collagen-induced arthritis and cartilage destruction. *Arthritis Rheum.* **41**, 110–121.

Murray, G. I., Duncan, M. E., O'Neil, P., Melvin, W. T., and Fothergill, J. E. (1996). Matrix metalloproteinase-1 is associated with poor prognosis in colorectal cancer. *Nature Med.* **2**, 461–462.

Murray, G. I., Duncan, M. E., O'Neil, P., McKay, J. A., Melvin, W. T., and Fothergill, J. E. (1998). Matrix metalloproteinase-1 is associated with poor prognosis in oesophageal cancer. *J. Pathol.* **185**, 256–261.

Nagase, H., and Woessner, J. F., Jr. (1999). Matrix metalloproteinases. *J. Biol. Chem.* **274**, 21491–21494.

Nelson, A. R., Fingleton, B., Rothenberg, M. L., and Matrisian, L. M. (2000). Matrix metalloproteinases: Biologic activity and clinical implications. *J. Clin. Oncol.* **18**, 1135–1149.

Neuhold, L. A., Killar, L., Zhao, W., Sung, M. L., Warner, L., Kulik, J., Turner, J., Wu, W., Billinghurst, C., Meijers, T., Poole, A. R., Babij, P., and DeGennaro, L. J. (2001). Postnatal expression in hyaline cartilage of constitutively active human collagenase-3 (MMP-13) induces osteoarthritis in mice. *J. Clin. Invest.* **107**, 35–44.

Nishioka, Y., Kobayashi, K., Sagae, S., Ishioka, S., Nishikawa, A., Matsushima, M., Kanamori, Y., Minaguchi, T., Nakamura, Y., Tokino, T., and Kudo, R. (2000). A single nucleotide polymorphism in the matrix metalloproteinase-1 promoter in endometrial carcinomas. *Jpn. J. Cancer Res.* **91**, 612–615.

Noll, W. W., Belloni, D. R., Rutter, J. L., Storm, C. A., Schned, A. R., Titus-Ernstoff, L., Ernstoff, M. S., and Brinckerhoff, C. E. (2001). Loss of heterozygosity on chromosome 11q22–23 in melanoma is associated with retention of the insertion polymorphism in the matrix metalloproteinase-1 promoter. *Am. J. Pathol.* **158**, 691–697.

Ohuchi, E., Imai, K., Fuji, Y., Sato, H., Seiki, M., and Okada, Y. (1997). Membrane type I matrix metalloproteinase digests interstitial collagenase and other extracellular matrix macromolecules. *J. Biol. Chem.* **272**, 2446–2451.

Parsons, S. L., Watson, S. A., Brown, P. D., Collins, H. M., and Steele, R. J. (1997). Matrix metalloproteinases. *Br. J. Surg.* **84**, 160–166.

Pei, D. (1999). Identification and characterization of the fifth membrane-type matrix metalloproteinase MT5-MMP. *J. Biol. Chem.* **274**, 8925–8933.

Peters, D. G., Kassam, A., St Jean, P. L., Yonas, H., and Ferrell, R. E. (1999). Functional polymorphism in the matrix metalloproteinase-9 promoter as a potential risk factor for intracranial aneurysm. *Stroke* **30**, 2612–2616.

Porte, D., Tuckermann, J., Becker, M., Baumann, B., Teurich, S., Higgins, T., Owen, M. J., Schorpp-Kistner, M., and Angel, P. (1999). Both AP-1 and Cbfa1-like factors are required for the induction of interstitial collagenase by parathyroid hormone. *Oncogene* **18**, 667–678.

Rowinsky, E. K., Windle, J. J., and Von Hoff, D. D. (1999). Ras protein farnesyltransferase: A strategic target for anticancer therapeutic development. *J. Clin. Oncol.* **17**, 3631–3652.

Rutter, J. L., Benbow, U., Coon, C. I., and Brinckerhoff, C. E. (1997). Cell-type specific regulation of human interstitial collagenase-1 gene expression by interleukin-1 beta (IL-1 beta) in human fibroblasts and BC-8701 breast cancer cells. *J. Cell Biochem.* **66**, 322–336.

Rutter, J. L., Mitchell, T. I., Buttice, G., Meyers, J., Gusella, J. F., Ozelius, L. J., and Brinckerhoff, C. E. (1998). A single nucleotide polymorphism in the matrix metalloproteinase-1 promoter creates an Ets binding site and augments transcription. *Cancer Res.* **58**, 5321–5325.

Sato, H., Kita, M., and Seiki, M. (1993). v-Src activates the expression of 92-kDa type IV collagenase gene through the AP-1 site and the GT box homologous to retinoblastoma control elements: A mechanism regulating gene expression independent of that by inflammatory cytokines. *J. Biol. Chem.* **268**, 23460–23468.

Sato, H., and Seiki, M. (1993). Regulatory mechanism of 92 kDa type IV collagenase gene expression which is associated with invasiveness of tumor cells. *Oncogene* **8**, 395–405.

Schadendorf, D., Kern, M. A., Artuc, M., Pahl, H. L., Rosenbach, T., Fichtner, I., Nurnberg, W., Stuting, S., von Stebut, E., Worm, M., Makki, A., Jurgovsky, K., Kolde, G., and Henz, B. M. (1996). Treatment of melanoma cells with the synthetic retinoid CD437 induces apoptosis via activation of AP-1 in vitro, and causes growth inhibition in xenografts in vivo. *J. Cell Biol.* **135**, 1889–1898.

Schoenermark, M. P., Mitchell, T. I., Rutter, J. L., Reczek, P. R., and Brinckerhoff, C. E. (1999). Retinoid-mediated suppression of tumor invasion and matrix metalloproteinase synthesis. *Ann. N. Y. Acad. Sci.* **878**, 466–486.

Schroen, D. J., Chen, J. D., Vincenti, M. P., and Brinckerhoff, C. E. (1997). The nuclear receptor corepressor SMRT inhibits interstitial collagenase (MMP-1) transcription through an HRE-independent mechanism. *Biochem. Biophys. Res. Commun.* **237**, 52–58.

Sherry, S. T., Ward, M., and Sirotkin, K. (1999). dbSNP-database for single nucleotide polymorphisms and other classes of minor genetic variation. *Genome Res.* **9**, 677–679.

Shimajiri, S., Arima, N., Tanimoto, A., Murata, Y., Hamada, T., Wang, K. Y., and Sasaguri, Y. (1999). Shortened microsatellite d(CA)21 sequence down-regulates promoter activity of matrix metalloproteinase 9 gene. *FEBS Lett.* **455**, 70–74.

Shlopov, B. V., Lie, W. R., Mainardi, C. L., Cole, A. A., Chubinskaya, S., and Hasty, K. A. (1997). Osteoarthritic lesions: Involvement of three different collagenases. *Arthritis Rheum.* **40**, 2065–2074.

Singh, H., Sekinger, E. A., and Gross, D. S. (2000). Chromatin and cancer: Causes and consequences. *J. Cell Biochem.* **Suppl.** 61–68.

Spanjaard, R. A., Ikeda, M., Lee, P. J., Charpentier, B., Chin, W. W., and Eberlein, T. J. (1997). Specific activation of retinoic acid receptors (RARs) and retinoid X receptors reveals a unique role for RARgamma in induction of differentiation and apoptosis of S91 melanoma cells. *J. Biol. Chem.* **272**, 18990–18999.

Sporn, M. B., Roberts, A. B., and Goodman, D. S. (1994). The Retinoids. *Raven Press*, New York.

Terashima, M., Akita, H., Kanazawa, K., Inoue, N., Yamada, S., Ito, K., Matsuda, Y., Takai, E., Iwai, C., Kurogane, H., Yoshida, Y., and Yokoyama, M. (1999). Stromelysin promoter 5A/6A polymorphism is associated with acute myocardial infarction. *Circulation* **99**, 2717–2719.

Tower, G. B., Benbow, U., and Brinckerhoff, C. E. (2002). ERK1/2 differentially regulates the expression from 1G/2G single nucleotide polymorphism in the MMP-1 promoter in melanoma cells. *Biochim. Biophys. Acta* **1586**, 265–274.

Tsukifuji, R., Tagawa, K., Hatamochi, A., and Shinkai, H. (1999). Expression of matrix metalloproteinase-1, -2 and -3 in squamous cell carcinoma and actinic keratosis. *Br. J. Cancer* **80**, 1087–1091.

Uitto, V. J., Airola, K., Vaalamo, M., Johansson, N., Putnins, E. E., Firth, J. D., Salonen, J., Lopez-Otin, C., Saarialho-Kere, U., and Kahari, V. M. (1998). Collagenase-3 (matrix metalloproteinase-13) expression is induced in oral mucosal epithelium during chronic inflammation. *Am. J. Pathol.* **152**, 1489–1499.

Velasco, G., Cal, S., Merlos-Suarez, A., Ferrando, A. A., Alvarez, S., Nakano, A., Arribas, J., and Lopez-Otin, C. (2000). Human MT6-matrix metalloproteinase: Identification, progelatinase A activation, and expression in brain tumors. *Cancer Res.* **60**, 877–882.

Vincenti, M. P., Clark, I. M., and Brinckerhoff, C. E. (1994). Using inhibitors of metalloproteinases to treat arthritis: Easier said than done. *Arthritis Rheum.* **37**, 1115–1126.

Vincenti, M. P., Coon, C. I., and Brinckerhoff, C. E. (1998). Nuclear factor kappaB/p50 activates an element in the distal matrix metalloproteinase 1 promoter in interleukin-1beta-stimulated synovial fibroblasts. *Arthritis Rheum.* **41**, 1987–1994.

Vincenti, M. P., White, L. A., Schroen, D. J., Benbow, U., and Brinckerhoff, C. E. (1996). Regulating expression of the gene for matrix metalloproteinase-1 (collagenase): Mechanisms that control enzyme activity, transcription, and mRNA stability. *Crit Rev Eukaryot. Gene Expr.* **6**, 391–411.

Vu, T. H., Shipley, J. M., Bergers, G., Berger, J. E., Helms, J. A., Hanahan, D., Shapiro, S. D., Senior, R. M., and Werb, Z. (1998). MMP-9/gelatinase B is a key regulator of growth plate angiogenesis and apoptosis of hypertrophic chondrocytes. *Cell* **93**, 411–422.

Wasylyk, B., Wasylyk, C., Flores, P., Begue, A., Leprince, D., and Stehelin, D. (1990). The c-ets proto-oncogenes encode transcription factors that cooperate with c-Fos and c-Jun for transcriptional activation. *Nature* **346**, 191–193.

Wasylyk, C., Gutman, A., Nicholson, R., and Wasylyk, B. (1991). The c-Ets oncoprotein activates the stromelysin promoter through the same elements as several non-nuclear oncoproteins. *EMBO J.* **10**, 1127–1134.

Westermarck, J., Holmstrom, T., Ahonen, M., Eriksson, J. E., and Kahari, V. M. (1998). *Matrix Biol.* **17**, 547–557.

Westermark, J., and Kahari, V. M. (1999). Regulation of matrix metalloproteinase expression in tumor invasion. *FASEB J.* **291**, 113–135.

Westermarck, J., Li, S., Jaakkola, P., Kallunki, T., Grenman, R., and Kahari, V. M. (2000). Activation of fibroblast collagenase-1 expression by tumor cells of squamous cell carcinomas is mediated by p38 mitogen-activated protein kinase and c-Jun NH2-terminal kinase-2. *Cancer Res.* **60**, 7156–7162.

White, L. A., and Brinckerhoff, C. E. (1995). Two activator protein-1 elements in the matrix metalloproteinase-1 promoter have different effects on transcription and bind Jun D, c-Fos, and Fra-2. *Matrix Biol.* **14**, 715–725.

White, L. A., Mitchell, T. I., and Brinckerhoff, C. E. (2000). Transforming growth factor beta inhibitory element in the rabbit matrix metalloproteinase-1 (collagenase-1) gene functions as a repressor of constitutive transcription. *Biochim. Biophys. Acta* **1490**, 259–268.

Ye, S., Dhillon, S., Turner, S. J., Bateman, A. C., Theaker, J. M., Pickering, R. M., Day, I., and Howell, W. M. (2001). Invasiveness of cutaneous malignant melanoma is influenced by matrix metalloproteinase 1 gene polymorphism. *Cancer Res.* **61**, 1296–1298.

Ye, S., Eriksson, P., Hamsten, A., Kurkinen, M., Humphries, S. E., and Henney, A. M. (1996). Progression of coronary atherosclerosis is associated with a common genetic variant of the human stromelysin-1 promoter which results in reduced gene expression. *J. Biol. Chem.* **271**, 13055–13060.

Ye, S., Humphries, S., and Henney, A. (1998). Matrix metalloproteinases: Implication in vascular matrix remodelling during atherogenesis. *Clin. Sci. (Colch)* **94**, 103–110.

Yoon, S., Tromp, G., Vongpunsawad, S., Ronkainen, A., Juvonen, T., and Kuivaniemi, H. (1999). Genetic analysis of MMP3, MMP9, and PAI-1 in Finnish patients with abdominal aortic or intracranial aneurysms. *Biochem. Biophys. Res. Commun.* **265**, 563–568.

Zhang, B., Ye, S., Herrmann, S. M., Eriksson, P., de Maat, M., Evans, A., Arveiler, D., Luc, G., Cambien, F., Hamsten, A., Watkins, H., and Henney, A. M. (1999). Functional polymorphism in the regulatory region of gelatinase B gene in relation to severity of coronary atherosclerosis. *Circulation* **99**, 1788–1794.

Zhou, Z., Apte, S. S., Soininen, R., Cao, R., Baaklini, G. Y., Rauser, R. W., Wang, J., Cao, Y., and Tryggvason, K. (2000). Impaired endochondral ossification and angiogenesis in mice deficient in membrane-type matrix metalloproteinase I. *Proc. Natl. Acad. Sci. USA* **97**, 4052–4057.

Zhu, Y., Spitz, M. R., and Wu, X. (2001). A single nucleotide polymorphism in the matrix metalloproteinase 1 (MMP-1) promoter enhances lung cancer susceptibility. *Cancer Res.* **61**, 7825–7829.

Zucker, S., Cao, J., and Chen, W. T. (2000). Critical appraisal of the use of matrix metalloproteinase inhibitors in cancer treatment. *Oncogene* **19**, 6642–6650.

CHAPTER 16

Knockout Mice of Matrix Metalloproteinase Genes

AKIKO OKADA
Sekiguchi Bio-matrix Signaling Project, Exploratory Research for Advanced Technology, Japan Science and Technology Corporation, c/o Aichi Medical University, Aichi 480-1195, Japan

MOTOHARU SEIKI
Department of Cancer Cell Research, Institute of Medical Science, University of Tokyo, Tokyo 113-8655, Japan

I. INTRODUCTION

Matrix metalloproteinases (MMPs) are believed to play pivotal roles during development, organogenesis, and tissue remodeling. Excessive or inappropriate expression of MMPs may contribute to the pathogenesis of many tissue destructive processes, such as arthritis, chronic pulmonary obstractive disease, and tumor progression.

Many experimental observations have been reported in which MMP activity is modulated in culture. A drawback of these studies is that they rely on an inhibitor-based approach, yet most of the inhibitors are not completely specific to MMPs. In addition, the function discerned for an MMP in the isolated context of *in vitro* studies may not be reflective of its true function in the complex environment of the whole animal. Analysis of transgenic mice that have a gain-and-loss of function of MMPs has given insights into the biological roles that these enzymes play in developmental and pathological processes.

Extracellular Matrix and the Liver
Copyright 2003, Elsevier Science (USA). All rights reserved.

II. TARGETED DISRUPTION OF MMP GENES IN MICE

Since the late 1990s, many of the MMPs have undergone gene-targeting experiments, and at least seven MMP genes have been disrupted individually (Table I).

MMPs, usually undetectable in cells under normal circumstances, are expressed prominently during a variety of biological processes, such as reproduction (Fata *et al.*, 2000). On the maternal side, MMP expression is associated with menstruation, ovulation, uterine implantation, parturition, and postpartum uterine and mammary gland involution (Hulboy *et al.*, 1997).

Development of the placenta starts with the invasion and migration of trophoblasts into the maternal tissue to establish connection with the maternal circulation. MMPs are believed to be required not only for trophoblast implantation, but also for embryonic growth and tissue morphogenesis. However, none of the individual MMP mutant mice generated to date has had an embryonic lethal phenotype. MMP-9 is highly expressed during embryonic development by trophoblast cells at implantation (Alexander *et al.*, 1996), but homozygous mice with a null mutation in the MMP-9 gene are viable (Vu *et al.*,

TABLE I Phenotypes of MMP-Deficient Mice

Gene	Phenotypes (results)	Reference
MMP-2 gelatinase A	Unaltered secretion of β-amyloid precursor protein	Itoh *et al.* (1997)
	Reduced angiogenesis and tumor progression	Itoh *et al.* (1998)
MMP-3 stromelysin-1	No effect on collagen-induced arthritis	Singer *et al.* (1997)
MMP-7 matrilysin	Decreased intestinal tumorigenesis	Wilson *et al.* (1997)
MMP-9 gelatinase B	Impaired primary angiogenesis in bone growth plates	Vu *et al.* (1998)
	Resistant to bullous pemphigoid	Liu *et al.* (2000)
	Reduced keratinocyte hyperproliferation and incidence of invasive tumors	Coussens *et al.* (2000)
MMP-11 stromelysin-3	Decreased chemical-induced mutagenesis	Masson *et al.* (1998)
MMP-12 macrophage elastase	Impaired macrophage recruitment and protection from emphysema	Hautamaki *et al.* (1997)
MMP-14 MT1-MMP	Craniofacial dysmorphism, dwarfism, osteopenia	Holmbeck *et al.* (1999)

1998). All MMP-deficient mice to date are capable of delivering and nurturing healthy pups.

III. POSTNATAL DEVELOPMENT OF KNOCKOUT MICE OF MMPs

MMP-9- or MMP-14-deficient mice demonstrate morphologic abnormalities during postnatal development, especially in bone development (Vu *et al.*, 1998; Holmbeck *et al.*, 1999). Bone is created by both endochondral ossification in which there is a chondroid phase and membranous ossification in which there is a direct transformation of fibrous tissue into bone. In the typical tubular bone, length is created by endochondral ossification, but the shape is formed circumferentially by periosteal (membranous) new bone formation. Bones that are formed predominantly by periosteal new bone formation or membranous bone formation are the bones of the skull, the scapula, and the ilium. There are subtle differences in these two types of mineralization. In the epiphyseal growth plate of long bone, the epiphyseal cartilage matrix consists mostly of proteoglycan and type II collagen. In the osteoid calcification that occurs in periosteal new bone formation, the osteoid consists principally of type I collagen with lesser amounts of noncollagenous proteins (Vigorita, 1999).

Homozygous mice with MMP-9 gene disruption exhibit an abnormal pattern of skeletal growth plate vascularization and ossification. Although hypertrophic chondrocytes develop normally, apoptosis, vascularization, and ossification in the angiogrowth plate are delayed, resulting in progressive lengthening of the growth plate. After 3 weeks postnatal, aberrant apoptosis, vascularization, and ossification compensate to remodel the enlarged growth plate and ultimately produce an axial skeleton of normal appearance. The major defect in MMP-9-deficient mice is delayed long bone growth and development. Growth plates from MMP-9 null mice in culture also show a delayed release of an angiogenic activator. MMP-9 is required to initiate primary angiogeneisis in the cartilage growth plate, probably through generation of an angiogenic signal (Vu *et al.*, 1998). However, the substrate for MMP-9 in modulating angiogenesis has not been identified. Interestingly, abnormal degradation of a structural or adhesive matrix protein was not obvious in MMP-9-deficient mice, observed in other MMP-deficient mice generated, except MMP-14-mutated mice.

MMP-14-deficient mice show the strongest phenotype among MMP mutant mice generated to date. MMP-14 is a membrane-bound matrix metalloproteinase (MT-MMP) capable of mediating pericellular proteolysis of extracellular matrix components. MMP-14 is therefore thought to be an important molecular tool for cellular remodeling of the surrounding matrix.

MMP-14 deficiency causes craniofacial dysmorphism, arthritis, osteopenia, dwarfism, and fibrosis of soft tissues due to ablation of a collagenolytic activity that is essential for the modeling of skeletal and extraskeletal connective tissues. MMP-14 plays pivotal roles in connective tissue metabolism (Holmbeck *et al.*, 1999).

During kidney organogenesis, many kinds of MMP are expressed, but no differences between MMP-9-deficient and control kidneys were detected and renal function was normal in MMP-9 mutants. In addition, embryonic kidneys of MMP-9 mutants developed normally in organ culture (Andrews *et al.*, 2000).

IV. MMP KNOCKOUT MICE AND DISEASE

A. Wound Healing

Healing of a skin wound requires several processes similar to development, such as cell migration, extracellular matrix (ECM) production and degradation, and tissue reorganization. Keratinocytes at the edge of the wound migrate to reepithelialize the wound surface. The dermis also contributes by new vascularization and contraction to facilitate wound closure. In animal studies, MMP activity is implicated in keratinocyte migration and dermal neovascularization and contraction (Okada *et al.*, 1997). Direct evidence for a role of MMPs in wound healing comes from studies of MMP-3 null mice.

Excisional wounds in MMP-3-deficient mice failed to contract and healed more slowly than those in wild-type mice. Cellular migration and epithelialization were unaffected in stromelysin-1-deficient animals. The functional defect in these mice is failure of contraction during the first phase of healing because of inadequate organization of actin-rich stromal fibroblasts (Bullard *et al.*, 1999). Wound healing in MMP-2-deficient mice was also slower than in wild-type mice, but in this case, cellular migration and epithelialization were affected and granulation tissue formation was very poor (Okada *et al.*, unpublished data).

B. Pulmonary Emphysema

A major component of obstructive pulmonary disease is destruction and enlargement of peripheral air spaces of the lung. Cigarette smoking is the most commonly identified correlate with chronic bronchites during life and extent of emphysema at postmortem. Prolonged cigarette smoking leads to inflammatory cell recruitment and activation with release of elastases, in excess of inhibitors.

ECM degradation coupled with abnormal repair results in lung destruction characteristic of emphysema. The contribution of neutrophil elastase is thought to be most important; however, it is possible that other neutrophil proteinases or enzymes form the more abundant macrophages also contribute to lung damage associated with prolonged cigarette smoking. In macrophage elastase (MMP-12)-deficient mice exposed to cigarette smoke, in contrast to wild-type mice, there are no increased numbers of lung macrophages and the animals do not develop emphysema in response to long-term exposure. Smoke-exposed MMP-12-deficient mice that received monthly intratracheal instillations of monocyte chemoattractant protein-1 accumulated alveolar macrophages but did not develop air space enlargement (Hautamaki *et al.*, 1997). Thus, macrophage elastase is probably sufficient for the development of emphysema that results from chronic inhalation of cigarette smoke. Additionally, MMP-12-deficient mice failed to recruit monocytes into their lungs in response to cigarette smoke. However, the concept that proteolytically generated elastin fragments mediate monocyte chemotaxis was first shown more than a decade ago. MMP-12 produced by macrophages cleaves elastin. This positive feedback loop could perpetuate macrophage accumulation and lung destruction.

Neutrophil influx into the alveolar space in metalloelastase-deficient animals was reduced to approximately 50% of that observed in parent strain mice following the induction of injury by immune complexes. In addition, lung permeability in metalloelastase-deficient mice was approximately 50% of that of injured parent strain animals with normal levels of metalloelastase. This was correlated with histological evidence of less lung injury in metalloelastase-deficient animals. Metalloelastase decreased neutrophil influx into the alveolar space (Warner *et al.*, 2001b). MMP-3 and MMP-9 are involved in the development of experimental acute lung injury, but the mechanisms by which these individual MMPs function appear to differ (Warner *et al.*, 2001a).

C. Rheumatoid Arthritis

It has long been proposed that MMP-3 is one of the major degradative matrix metalloproteinases responsible for the loss of cartilage in rheumatoid arthritis and osteoarthritis. However, knockout mice deficient in MMP-3 developed collagen-induced arthritis, as did the wild-type mice. Histologic analyses demonstrated no significant differences between wild-type and knockout mice in loss of articular cartilage and proteoglycan staining. Also, aggrecanase, which is distinct from MMPs, cleaves aggrecan at the MMP site (Singer *et al.*, 1997).

The disease gene for multicentric osteolysis with carpal and tarsal resorption, crippling arthritic changes, marked osteoporosis, and palmar and plantar subcutaneous nodules is localized to 16q12-21 where the gene locus encoding MMP-2 is found. No MMP-2 enzymatic activity in the serum or fibroblasts of affected family members was detected (Martignetti *et al.*, 2001). As mice deficient in MMP-2 do not show such a phenotype, other gene mutations may be necessary.

D. Cerebral Ischemia

MMPs are elevated after cerebral ischemia. After inducing cerebral ischemia in mice, ischemic lesion volumes were reduced significantly in MMP-9 knockout mice compared with wild-type littermates in male and female mice. In wild-type mice, the broad-spectrum MMP inhibitor BB-94 (batimastat) also reduced ischemic lesion size significantly. However, BB-94 had no detectable protective effect when administered to MMP-9 knockout mice subjected to focal cerebral ischemia (Asahi *et al.*, 2000).

E. Cancer

Tumor growth involves alterations in the stromal ECM, and malignant tumors often induce a fibroproliferative response in the adjacent stroma, characterized by an increased expression of type I and III procollagens. The formation of tumor stroma is often viewed as a nonspecific host attempt to wall off the tumor, and it is thought to have a negative influence on tumor progression. During the process of metastasis formation, malignant cells detach from the primary tumor, invade the stromal tissue, enter the circulation, arrest at the peripheral vascular bed, extravasate, invade the target organ interstitium and parenchyma, and form a metastatic colony (Stetler-Stevenson *et al.*, 1993). Tumor-induced angiogenesis is essential for growth of the primary tumor and metastases, and new blood vessels are also frequent sites for tumor cell entry into the circulation.

MMPs are believed to promote tumor progression by initiating carcinogenesis, enhancing tumor angiogenesis, disrupting local tissue architecture to allow tumor growth, and breaking down basement membrane barriers for cancer invasion and metastasis. Some MMPs, such as MMP-7 or MMP-13, are expressed by tumor cells.

MMP-7 is expressed in a high percentage of early stage human colorectal tumors and in mouse benign intestinal tumors of a mouse having a nonsense autosomal-dominant germline mutation in the adenomatous polyposis coli gene, the *Min* mouse. MMP-7 gene-targeted mice were generated and mated to *Min* mice. The absence of matrilysin resulted in a reduction in mean tumor multiplicity in mice of approximately 60% and a significant decrease in the average tumor diameter. These results may conclude that MMP-7 has a function in a capacity independent of matrix degradation (Wilson *et al.*, 1997).

MMPs are produced predominantly by surrounding host stromal and inflammatory cells in response to factors released by tumor cells. MMPs may then bind to tumor cells and angiogenic endothelial cells, advancing tumor progression. For example, MMP-2 binds through its carboxy-terminal domain to $\alpha v\beta 3$ integrin on melanoma cells and angiogenic blood vessels, enhancing tumor growth. Autolytic processing of MMP-2 with release of the carboxy-terminal domain competes with cell surface binding of the enzyme, inhibiting angiogenesis and tumor growth (Brooks *et al.*, 1998).

Mice devoid of MMP-2 generated by gene targeting develop normally, except for a subtle delay in their growth, thus providing a useful system to examine the role of MMP-2 in the cleavage and secretion of β-amyloid precursor protein (APP) *in vivo*. APP is cleaved within the β-amyloid region and secreted into the extracellular milieu of brain and cultured fibroblasts without MMP-2 activity. Therefore the cleavage and releasing process are not dependent upon MMP-2 (Itoh *et al.*, 1997).

MMP-2-deficient mice exhibit impaired primary tumor growth and decreased experimental metastases of B16-BL6 melanoma and Lewis lung carcinoma (LLC) cells (Itoh *et al.*, 1998).

MMP-11 is also a MMP expressed in mesenchymal cells located close to epithelial cells during physiological and pathological tissue remodeling processes. In human carcinomas, high MMP-11 levels are associated with a poor clinical outcome. MMP-11-deficient mice were fertile and did not exhibit obvious alterations in appearance and behavior, but the lack of MMP-11 altered malignant processes. Specifically, the absence of MMP-11 results in a decreased 7,12-dimethylbenzanthracene-induced tumorigenesis in MMP-11-deficient mice. Also, MMP-11-deficient fibroblasts have lost the capacity to promote implantation of MCF7 human malignant epithelial cells in nude mice (Masson *et al.*, 1998).

Mice lacking MMP-9 show reduced keratinocyte hyperproliferation at all neoplastic stages and a decreased incidence of invasive tumors. However, those carcinomas that do arise in the absence of MMP-9 exhibit a greater loss of keratinocyte differentiation, indicative of a more aggressive and higher grade tumor (Coussens *et al.*, 2000).

While MMPs commonly facilitate tumor progression, proteolytic cleavage products of MMPs may inhibit angiogenesis, limiting tumor progression. Angiostation, a plasminogen cleavage product containing kringle domain 1–4, inhibits endothelial cell proliferation and was isolated from the urine of mice with LLC cells. It inhibits endothelial cell proliferation and is believed to be responsible for maintaining LLC lung metastases in a dormant state (O'Reilly *et al.*, 1994). Generation of angiostatin in primary LLC tumors correlated with the presence of macrophages and macrophage elastase (Dong *et al.*, 1997). Several other MMPs are also capable of generating angiostatin (Cornelius *et al.*, 1998) and another antiangiogenic fragments of plasminogen. Thus, MMP may benefit the host or the tumor dependent on spatial expression, proteolytic capacity, and binding affinity for matrix and tumor cells.

V. CONCLUSION

Transgenic technology offers the possibility of determining whether MMPs contribute directly to developmental and pathological processes, and some phenotypes of MMP knockout mice have uncovered bonafide biological functions of MMPs. However, the transgenic technology sometimes has limitations. First, loss of a protein might alter many physical processes by compensation. Second, mutation of a gene may mask the real biological function of the gene product. Third, because of species differences, the results may be not generalized to human biology. Limitations of these techniques and powerful applications on the horizon are also presented as we embark on an era where controlled experiments can be performed in complex mammalian models. We must recognize these limitations, while utilizing transgenic technology as an effective tool.

REFERENCES

Alexander, C. M., Hansell, E. J., Behrendtsen, O., Flannery, M. L., Kishnani, N. S., Hawkes, S. P., and Werb, Z. (1996). Expression and function of matrix metalloproteinases and their inhibitors at the maternal-embryonic boundary during mouse embryo implantation. *Development* **122**, 1723–1736.

Andrews, K. L., Betsuyaku, T., Rogers, S., Shipley, J. M., Senior, R. M., and Miner, J. H. (2000). Gelatinase B (MMP-9) is not essential in the normal kidney and does not influence progression of renal disease in a mouse model of Alport syndrome. *Am. J. Pathol.* **157**, 303–311.

Asahi, M., Asahi, K., Jung, J. C., del Zoppo, G. J., Fini, M. E., and Lo, E. H. (2000). Role for matrix metalloproteinase 9 after focal cerebral ischemia: Effects of gene knockout and enzyme inhibition with BB-94. *J. Cereb. Blood Flow Metab.* **20**, 1681–1689.

Brooks, P. C., Silletti, S., von Schalscha, T. L., Friedlander, M., and Cheresh, D. A. (1998). Disruption of angiogenesis by PEX, a noncatalytic metalloproteinase fragment with integrin binding activity. *Cell* **92**, 391–400.

Bullard, K. M., Lund, L., Mudgett, J. S., Mellin, T. N., Hunt, T. K., Murphy, B., Ronan, J., Werb, Z., and Banda, M. J. (1999). Impaired wound contraction in stromelysin-1-deficient mice. *Ann. Surg.* **230**, 260–265.

Cornelius, L. A., Nehring, L. C., Harding, E., Bolanowski, M., Welgus, H. G., Kobayashi, D. K., Pierce, R. A., and Shapiro, S. D. (1998). Matrix metalloproteinases generate angiostatin: Effects on neovascularization. *J. Immunol.* **161**, 6845–6852.

Coussens, L. M., Tinkle, C. L., Hanahan, D., and Werb, Z. (2000). MMP-9 supplied by bone marrow-derived cells contributes to skin carcinogenesis. *Cell* **103**, 481–490.

Dong, Z., Kumar, R., Yang, X., and Fidler, I. J. (1997). Macrophage-derived metalloelastase is responsible for the generation of angiostatin in Lewis lung carcinoma. *Cell* **88**, 801–810.

Fata, J. E., Ho, A. T., Leco, K. J., Moorehead, R. A., and Khokha, R. (2000). Cellular turnover and extracellular matrix remodeling in female reproductive tissues: Functions of metalloproteinases and their inhibitors. *Cell Mol. Life Sci.* **57**, 77–95.

Hautamaki, R. D., Kobayashi, D. K., Senior, R. M., and Shapiro, S. D. (1997). Requirement for macrophage elastase for cigarette smoke-induced emphysema in mice. *Science* **277**, 2002–2004.

Holmbeck, K., Bianco, P., Caterina, J., Yamada, S., Kromer, M., Kuznetsov, S. A., Mankani, M., Robey, P. G., Poole, A. R., Pidoux, I., Ward, J. M., and Birkedal-Hansen, H. (1999). MT1-MMP-deficient mice develop dwarfism, osteopenia, arthritis, and connective tissue disease due to inadequate collagen turnover. *Cell* **99**, 81–92.

Hulboy, D. L., Rudolph, L. A., and Matrisian, L. M. (1997). Matrix metalloproteinases as mediators of reproductive function. *Mol. Hum. Reprod.* **3**, 27–45.

Itoh, T., Ikeda, T., Gomi, H., Nakao, S., Suzuki, T., and Itohara, S. (1997). Unaltered secretion of beta-amyloid precursor protein in gelatinase A (matrix metalloproteinase 2)-deficient mice. *J. Biol. Chem.* **272**, 22389–22392.

Itoh, T., Tanioka, M., Yoshida, H., Yoshioka, T., Nishimoto, H., and Itohara, S. (1998). Reduced angiogenesis and tumor progression in gelatinase A-deficient mice. *Cancer Res.* **58**, 1048–1051.

Martignetti, J. A., Aqeel, A. A., Sewairi, W. A., Boumah, C. E., Kambouris, M., Mayouf, S. A., Sheth, K. V., Eid, W. A., Dowling, O., Harris, J., Glucksman, M. J., Bahabri, S., Meyer, B. F., and Desnick, R. J. (2001). Mutation of the matrix metalloproteinase 2 gene (MMP2) causes a multicentric osteolysis and arthritis syndrome. *Nature Genet.* **28**, 261–265.

Masson, R., Lefebvre, O., Noël, A., Fahime, M. E., Chenard, M. P., Wendling, C., Kebers, F., LeMeur, M., Dierich, A., Foidart, J. M., Basset, P., and Rio, M. C. (1998). In vivo evidence that the stromelysin-3 metalloproteinase contributes in a paracrine manner to epithelial cell malignancy. *J. Cell Biol.* **140**, 1535–1541.

Okada, A., Tomasetto, C., Lutz, Y., Bellocq, J. P., Rio, M. C., and Basset, P. (1997). Expression of matrix metalloproteinases during rat skin wound healing: Evidence that membrane type-1 matrix metalloproteinase is a stromal activator of pro-gelatinase A. *J. Cell Biol.* **137**, 67–77.

O'Reilly, M. S., Holmgren, L., Shing, Y., Chen, C., Rosenthal, R. A., Moses, M., Lane, W. S., Cao, Y., Sage, E. H., and Folkman, J. (1994). Angiostatin: A novel angiogenesis inhibitor that mediates the suppression of metastases by a Lewis lung carcinoma. *Cell* **79**, 315–328.

Singer, I. I., Scott, S., Kawka, D. W., Bayne, E. K., Weidner, J. R., Williams, H. R., Mumford, R. A., Lark, M. W., McDonnell, J., Christen, A. J., Moore, V. L., Mudgett, J. S., and Visco, D. M. (1997). Aggrecanase and metalloproteinase-specific aggrecan neo-epitopes are induced in the articular cartilage of mice with collagen II-induced arthritis. *Osteoarthritis Cartilage* **5**, 407–418.

Stetler-Stevenson, W. G., Aznavoorian, S., and Liotta, L. A. (1993). Tumor cell interactions with the extracellular matrix during invasion and metastasis. *Annu. Rev. Cell Biol.* **9**, 541–573.

Vigorita, V. J. (1999). "Orthopaedic Pathology." Lippincott Williams & Wilkins, New York.

Vu, T. H., Shipley, J. M., Bergers, G., Berger, J. E., Helms, J. A., Hanahan, D., Shapiro, S. D., Senior, R. M., and Werb, Z. (1998). MMP-9/gelatinase B is a key regulator of growth plate angiogenesis and apoptosis of hypertrophic chondrocytes. *Cell* **93**, 411–422.

Warner, R. L., Beltran, L., Younkin, E. M., Lewis, C. S., Weiss, S. J., Varani, J., and Johnson, K. J. (2001a). Role of stromelysin 1 and gelatinase B in experimental acute lung injury. *Am. J. Respir. Cell Mol. Biol.* **24**, 537–544.

Warner, R. L., Lewis, C. S., Beltran, L., Younkin, E. M., Varani, J., and Johnson, K. J. (2001b). The role of metalloelastase in immune complex-induced acute lung injury. *Am. J. Pathol.* **158**, 2139–2144.

Wilson, C. L., Heppner, K. J., Labosky, P. A., Hogan, B. L., and Matrisian, L. M. (1997). Intestinal tumorigenesis is suppressed in mice lacking the metalloproteinase matrilysin. *Proc. Natl. Acad. Sci. USA* **94**, 1402–1407.

CHAPTER 17

Structures and Functions of Tissue Inhibitors of Metalloproteinases

TARO HAYAKAWA AND KYOKO YAMASHITA

Department of Biochemistry, Aichi-Gakuin University, School of Dentistry, Nagoya 464-8650, Japan

Tissue inhibitors of metalloproteinases (TIMPs) are secreted proteins that control the integrity of the extracellular matrix (ECM) through their inhibitory activities on matrix metalloproteinases (MMPs). The four members of the TIMP family differ in structure, biochemical properties, and expression, indicating that they have distinct physiological roles. The inhibitory spectra of the TIMP family, however, overlap each other, suggesting a double and triple security for their MMP inhibition. X-ray and nuclear magnetic resonance (NMR) studies have revealed two structural features: (1) the wedge-shaped inhibitor fits into the active site cleft of the MMP catalytic domain and (2) their N-terminal domains are folded into a β-barrel structure common to oligosaccharide/origonucleotide binding (OB)-fold proteins. In addition to their primary functions as intrinsic inhibitors of MMPs, TIMPs, especially TIMP-1 and TIMP-2, exert cell growth-modulating activity. Both TIMP-1 and TIMP-2 are identical to erythroid potentiating activity (EPA). TIMP-1 is also identical to embryogenesis-stimulating activity (ESA), and its complex with procathepsin L is identical to steroidogenesis-stimulating protein (STP). TIMP-1 and TIMP-2 protect cells from apoptosis, but TIMP-3 promotes apoptosis. It was found that both TIMP-1 and TIMP-2 directly stimulate the bone-resorbing activity of osteoclasts.

Extracellular Matrix and the Liver
Copyright 2003, Elsevier Science (USA). All rights reserved.

I. INTRODUCTION

Matrix metalloproteinases are a large family of Ca^{2+}-requiring zinc metallo-endopeptidases that collectively degrade most of the extracellular matrix components (Nagase, 1996; Woessner, 1998; Woessner and Nagase, 2000). MMPs play a pivotal role in both normal development and remodeling and in many pathological conditions, such as cancer invasion and metastasis, arthritis, glomerulonephritis, atherosclerosis, tissue ulceration, and periodontal diseases. The activity of MMPs is specifically controlled by their intrinsic inhibitors, called tissue inhibitors of metalloproteinases (Woessner and Nagase, 2000; Edwards *et al.*, 1996; Gomez *et al.*, 1997; Murate *et al.*, 1999). In addition to having potent inhibitory activity against MMPs, TIMPs have also been shown to have multiple functions represented by cell growth-modulating activity. The first half of this chapter covers the general aspects of structures and functions of TIMPs, especially putting weight on recent findings, and the second half focuses on unique functions of TIMPs, which are independent of their MMP-inhibitory activities.

II. GENERAL ASPECTS OF THE TIMP FAMILY

A. Structures of TIMPs

To date, four members of the human TIMP family (TIMP-1, TIMP-2, TIMP-3, and TIMP-4) are known (Table I). Some structural features are common to all four TIMPs. One of the important features is the special conservation of 12 cysteine residues that form intramolecular disulfide bonds folding the proteins into two domains, i.e., the large N-terminal domain (126 amino acids in the case of TIMP-1) and the small C-terminal domain with three loops in each domain, as shown in Fig. 1. It has been demonstrated that the N-terminal domain of TIMP-1 contains the inhibitory site for MMPs. The first 22 amino acids of the N-terminal domain of mature forms of all TIMPs are highly conserved (Fig. 2) and are known to be critical for MMP inhibitory activity.

X-ray studies of both MMP-3-TIMP-1 (Gomis-Rüth *et al.*, 1997) and MMP-14-TIMP-2 (Tuuttila *et al.*, 1998; Fernandez-Catalan *et al.*, 1998) complexes revealed two important structural features of those TIMPs: (1) the N-terminal segment (Cys1~Pro5) and a part of the C-connector loop (Ala65~Cys70 in TIMP-1 or Ala66~Cys72 in TIMP-2), which are linked by Cys1-Cys70 or Cys1-Cys72 disulfide bridge in TIMP-1 or TIMP-2, respectively, form a wedge-shaped edge and fit into the active site cleft of MMP catalytic domain; and (2) the N-terminal domain of either TIMP-1 or TIMP-2 is folded into a five-stranded β-pleated sheet rolled into a β barrel of conical shape common to the oligosaccharide/

TABLE I TIMP Family

	TIMP-1	TIMP-2	TIMP-3	TIMP-4
Chromosome	Xp 11.2–11.4	17q 23-25	22q 12.1-13.2	3 p 25
Gene structure	5 exons and 4 introns (4.3 kb)	?	5 exons and 4 introns (30 kb)	?
mRNA	0.9 kb	3.5, 1.0 kb	5.0 (2.4, 2.2) kb	1.4 kb
Protein size	25–36 kDa (20,685[a])	20–24 kDa (21,600[a])	27, 24 kDa (21,689[a])	(22.442[a])
Amino acids				
Signal peptide	23	26	23	29
Mature form	184	194	188	195
Sequence homology	TIMP-1/TIMP-2: 43%	TIMP-2/TIMP-3: 44%	TIMP-3/TIMP-4: 49%	
	TIMP-1/TIMP-3: 38%	TIMP-2/TIMP-4: 50%	TIMP-1/TIMP-4: 35%	
Glycosylation	+ (two sites)	—	+ (one site)	?
Localization mode	Diffusible	Diffusible	ECM bound	?
Complex with proMMP	pro-MMP-9	pro-MMP-1, pro-MMP-2	?	pro-MMP-2

[a]Calculated from their amino acid sequences.

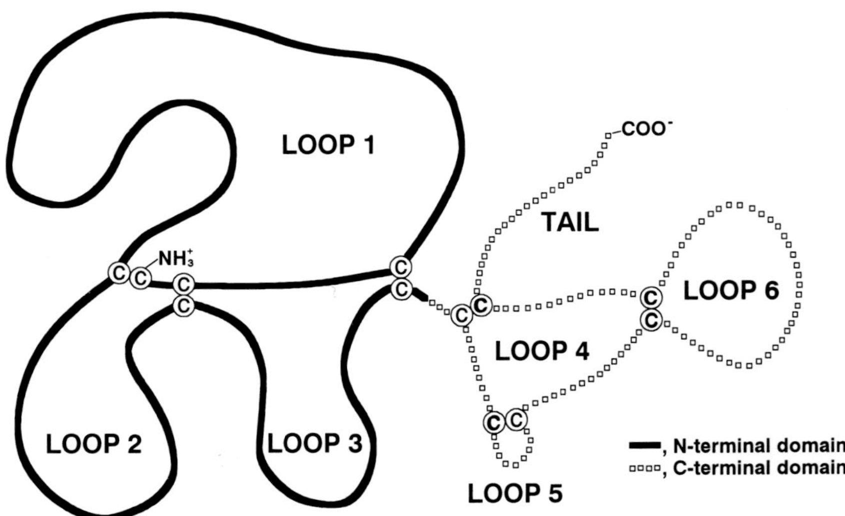

FIGURE 1 Schematic representation of the two-dimensional structure of TIMP. Six loop structures are described. C denotes a conserved cysteine residue. N terminus and C terminus are described as NH^{3+} and COO^{-}, respectively.

origonucleotide-binding fold (OB-fold) also found in certain DNA-binding proteins (Williamson *et al.*, 1994; Callebaut and Mornon, 1997).

NMR studies on both TIMP-1 and TIMP-2 in solution have revealed some features of the interaction between TIMP and MMP. An extended AB β-hairpin, which is a feature of TIMP-2, has a key role in the binding of TIMP-2 to MMPs (Muskett *et al.*, 1998; Williamson *et al.*, 1999). TIMP-1 exhibited a localized induced fit in the recognition of MMPs (Wu *et al.*, 2000) and underwent microsecond to millisecond motions at the site of MMP-induced fit (Gao *et al.*, 2000).

B. Functions of TIMPs

A principal physiological role of TIMPs is to regulate degradation of the basement membrane and extracellular matrix by MMPs during development and tissue remodeling. TIMP family members inhibit activated MMPs by binding in a 1:1 stoichiometry and in a noncovalent fashion. These complexes do not dissociate by gel filtration, indicating a fairly tight binding. For half-maximal inhibition, TIMP-1 was more effective than TIMP-2 at inhibiting MMP-1 (Howard *et al.*, 1991; Ward *et al.*, 1991) and MMP-3 (Nguyen *et al.*, 1994). TIMP-2, however, was more effective than TIMP-1 at inhibiting MMP-9, but conflicting results regarding these inhibitors with respect to MMP-2 have been reported (Ward *et al.*,

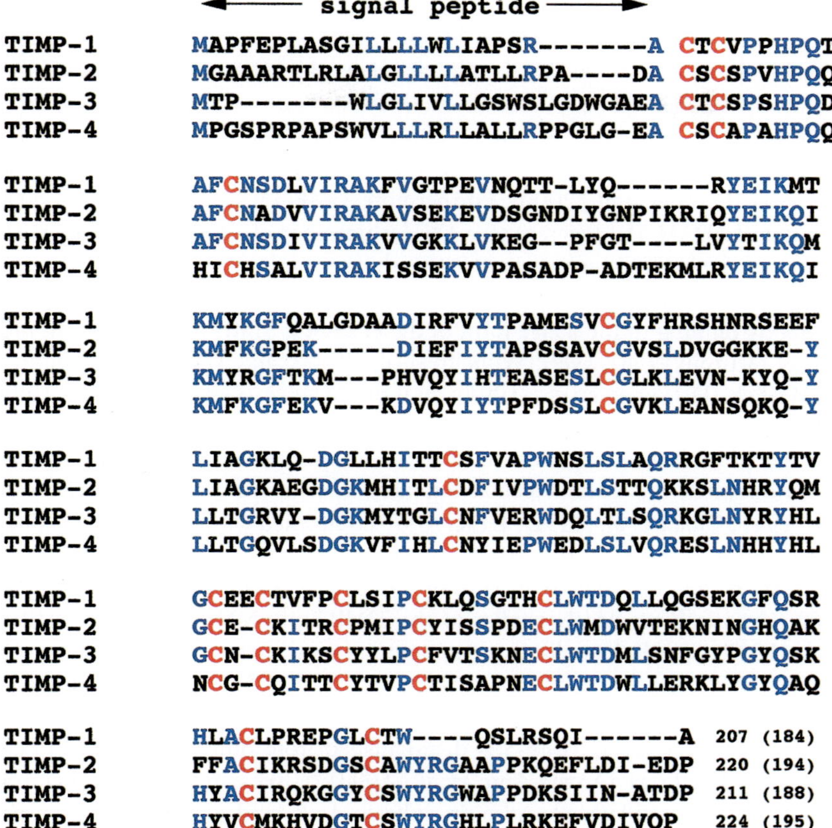

FIGURE 2 Primary structure of the human TIMP family. Amino acid sequences of human TIMPs are described. C (red) denotes 12 conserved residues. Letters in blue show conserved amino acids.

1991; Nguyen *et al.*, 1994). There was no significant difference between TIMP-1 and TIMP-3 at inhibiting MMPs-1, -2, -3, and -9 (Apte *et al.*, 1995). Both TIMP-1 and TIMP-3 were about five-fold more effective than TIMP-2 at inhibiting MMP-13 (Knäuper *et al.*, 1996). Interestingly, MT1-MMP was inhibited by TIMP-2 and TIMP-3, but was essentially not affected by TIMP-1 (Kinoshita *et al.*, 1996; Will *et al.*, 1996).

TIMP-2 (Lombard *et al.*, 1998) and TIMP-3 (Smith *et al.*, 1997) have been shown to inhibit the shedding of tumor necrosis factor (TNF)-α receptors in human colon adenocarcinoma cell lines, Colo 205 and DLD, respectively. So far, it has been accepted to be one of the criteria for a proteinase to be designated as a member of the MMP family that the proteinase activity should be inhibited by

TIMP (Nagase *et al.*, 1992). However, TIMPs are no longer known solely as intrinsic inhibitors of MMPs but also as having inhibitory activity against other zinc metalloendopeptidases, ADAM family, as shown by other studies; i.e., TIMP-3 inhibited TACE (ADAM-17) (Amour *et al.*, 1998) and both TIMP-1 and TIMP-3 inhibited ADAM-10 (Amour *et al.*, 2000). TIMP-3 also inhibited the secreted soluble form of human ADAM-12 (ADAM 12-S), but its inhibitory activity was very weak (Loechel *et al.*, 2000). Furthermore TIMP-3 was found to be a potent inhibitor of ADAM-TS4 (aggrecanase 1) (Hashimoto *et al.*, 2001; Kashiwagi *et al.*, 2001) and ADAM-TS5 (aggrecanase 2) (Kashiwagi *et al.*, 2001).

Another feature of TIMP-1 and TIMP-2 is that, in addition to inhibiting MMP activity by binding to the active form of the enzymes, TIMP-1 could bind to pro-MMP-9 (Murphy *et al.*, 1989; Wilhelm *et al.*, 1989), and TIMP-2 could bind to pro-MMP-1 (DeClerck *et al.*, 1991) and pro-MMP-2 (Goldberg *et al.*, 1989; Stetler-Stevenson *et al.*, 1989) in a 1:1 stoichiometry and in a noncovalent fashion. The physiological significance of these complexes has been revealed: both TIMPs inhibited the intramolecular autoactivation of the proenzymes (Howard *et al.*, 1991; Okada *et al.*, 1992). Either TIMP complexed with proenzyme still had the capability to inhibit active MMPs, suggesting formation of a tertiary complex. It has been demonstrated that TIMP-4 could also make a complex with pro-MMP-2 in a fashion closely similar to the TIMP-2-pro-MMP-2 interaction (Bigg *et al.*, 1997). TIMP-4 could not promote but did competitively reduce pro-MMP-2 activation by MT1-MMP (Bigg *et al.*, 2001; Hernandez-Barrantes *et al.*, 2001).

As shown in Table I, TIMP-3 is distinguished from other TIMPs by its tight binding to the ECM (Anand-Apte *et al.*, 1996). Glycosaminoglycans, such as heparin, heparan sulfate, chondroitin sulfates A, B, and C, and sulfated compounds, such as suramin and pentosan, have been demonstrated to make a stable complex with TIMP-3 through the β structure (A and B strand) of the N-terminal domain of the inhibitor. The colocalization of heparan sulfate and TIMP-3 has also been found in the endometrium subjacent to the lumen of the uterus (Yu *et al.*, 2000). These results are, however, inconsistent with the results of Langton *et al.* (1998), whose data suggested that TIMP-3 bound to ECM through its C-terminal domain.

Interestingly, mutations in the human TIMP-3 gene were found to be responsible for a dominantly inherited, adult-onset blindness called Sorsby's fundus dystrophy (SFD) (Anand-Apte *et al.*, 1996). In addition to five different mutations (S181C-, Y168C-, G167C-, G166C-, and S156C-TIMP-3) of this gene, a novel C-terminal truncated TIMP-3 mutant (F139X-TIMP-3) has been identified (Langton *et al.*, 2000). All of these mutant molecules were expressed and exhibited characteristics of the normal protein, including inhibition of MMPs and binding to ECM. However, unlike wild-type TIMP-3, they all formed dimers. These observations, together with the finding that the expression of TIMP-3 is increased, rather than decreased, in the eyes of SFD patients, provide compelling evidence that dimerized TIMP-3 plays an active role in the disease process by accumulating in the eye (Langton *et al.*, 2000). However, an SFD family without a mutation in the TIMP-3 gene has been found (Assink *et al.*, 2000).

Campbell *et al.* (1991) have defined serum and phorbol ester responsive elements lying just upstream of the major transcription start sites of the TIMP-1 gene that are binding sites for transcription factors AP-1 (Fos/Jun) and PEA3, a member of the *c-ets* protooncogene family (Edwards *et al.*, 1992; DeClerck *et al.*, 1994). Interestingly, the mouse TIMP promoter contains an AP-1 site and can be activated transcriptionally by many of the same agents that activate collagenase or stromelysin. This suggests that AP-1 is a critical factor that may serve to coordinate the responses of MMP and TIMP genes to diverse stimuli. Why would both a protease and its inhibitor be coordinately induced? The enzyme could digest the substrate to a very limited extent before being inactivated by the inhibitor, tightly controlling the extent of ECM degradation. Alternatively, there may be a slight difference in kinetics of the induction of enzymes and inhibitors that may also allow limited ECM degradation. A novel transcription factor, ssT, identified to be important in the basal transcription of murine TIMP-1 has been found (Phillips *et al.*, 1999). Also discovered were new elements named upstream TIMP-1 element-1 (UTE-1) in the human TIMP-1 promoter, which interacts with a 30-kDa nuclear protein (Trim *et al.*, 2000), and a repressive element within the TIMP-1 promoter and intron 1, which binds Sp1, Sp3, and an unidentified Ets-related factor to suppress the transcription of human TIMP-1 (Dean *et al.*, 2000).

The *in vivo* patterns of expression of the three TIMP genes in mice are distinct. In the mouse, the expression of TIMP-1 has been localized to areas of active ECM remodelling, such as sites of osteogenesis in the developing embryo and in the uterus and ovaries of adult females (Nomura *et al.*, 1989). TIMP-2 showed a distinctive accumulation in the placenta just prior to birth (Waterhouse *et al.*, 1993). However, TIMP-3 transcripts were found at high levels in certain adult tissues that lacked the expression of other TIMPs, e.g., the kidney and brain (Apte *et al.*, 1994). In the developing mouse embryo, TIMP-3 transcripts were abundant in the surface epithelia of organs such as the developing bronchial tree, kidney, colon, and esophagus (Apte *et al.*, 1994). These data support the notion that the individual TIMPs perform specific and noninterchangeable tasks.

The expressed protein of the *mig-5* gene found in WI-38 fibroblasts and HL-60 myeloid cells was demonstrated to be identical to TIMP-3, and its expression was not only induced in response to mitogenic stimulation, but was also expressed maximally at the mid-G1 phase of cell proliferation (Wick *et al.*, 1994).

In initial studies, no significant distinguishing differences were observed in TIMP-1 knockout mice with respect to embryonic development, viability, or fertility. Knockout mice at 4 months showed a significant ($p < 0.05$) increase in left ventricular (LV) mass, LV end-diastolic volume, and wall thickness compared with wild-type controls in comparative echocardiographic studies (Roton *et al.*, 2000). However, LV systolic pressure, ejection performance, and LV myocyte cross-sectional area were unchanged in the two groups of mice, but the myocardial fibrillar collagen content was reduced in the knockout mice. These findings suggest that constitutive TIMP-1 expression participates in the maintenance of normal LV myocardial structure. TIMP-1 knockout mice further showed a significant decrease in the length of the estrus period, significantly lower levels of uterine TIMP-3 mRNA expression, altered uterine morphology, significantly higher serum estradiol levels, and significantly lower serum progesterone levels compared with their wild-type counterparts (Nothnick, 2000).

III. MULTIPLE FUNCTIONS OF TIMPs

Although MMP inhibition is the primary function of TIMPs, accumulating studies have demonstrated a wide variety of other TIMP functions (Fig. 3), some of which have been known to be independent of their MMP inhibitory activities.

A. Erythroid Potentiating Activity (EPA)

The primary structure of human TIMP-1 deduced from its cDNA cloning (Docherty *et al.*, 1985) is known to be identical to that of EPA (Gasson *et al.*, 1985), which stimulated the growth of erythroid progenitor cells (BFU-E and CFU-E) (Westbrook *et al.*, 1984) and human erythroleukemia cell line K-562 cells

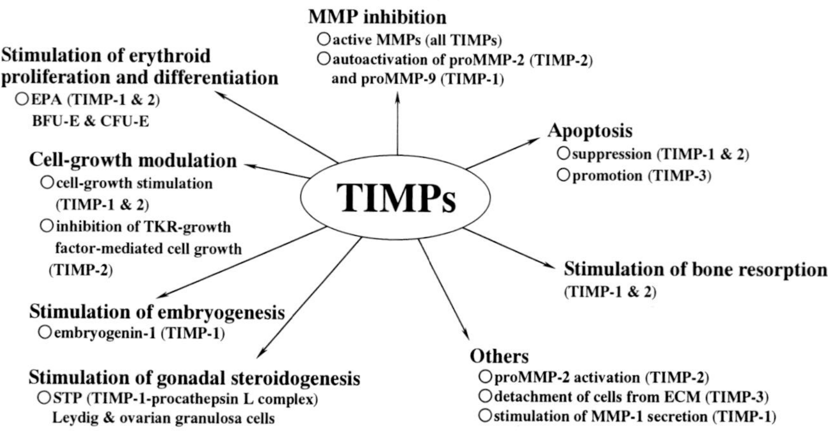

FIGURE 3 Multiple functions of the TIMP family.

(Avalos *et al.*, 1988). Hayakawa *et al.* (1990) demonstrated that TIMP-1 produced by human bone marrow stromal cell line KM 102 had both EPA activity and MMP inhibitory activity and proposed a bifunctional role for TIMP-1/EPA in the hematopoietic microenvironment, i.e., the maintenance of the integrity of the bone marrow matrix and the proliferation of erythroid progenitor cells in the matrix. Later, TIMP-2 was also reported to have EPA activity (Stetler-Stevenson *et al.*, 1992). N-terminal TIMP-1 point mutants (H7A-, Q9A- and H7A/Q9A-TIMP-1) lacking MMP inhibitory activity showed full EPA (BFU-E) activity equal to that of unmutated TIMP-1 (Chesler *et al.*, 1995) suggesting that both MMP inhibitory and EPA activities are distinct physically and functionally. C-terminal-truncated TIMP-1 (ΔC128-TIMP-1), however, showed fully both MMP inhibitory and EPA activities. The p38 MAP kinase pathway has been suggested to be involved in TIMP-1-induced erythroid differentiation (Petitfrère *et al.*, 2000). From results of a study on mouse erythroleukemia cell line ELM-I-1-3, we suggested the entity of EPA to be the combined function of the proliferating activity of TIMP-1 and the differentiating activity of erythropoietin on erythroid progenitor cells (Murate *et al.*, 1993).

B. Cell Growth-Modulating Activity

TIMP-1 has been demonstrated to be mitogenic for a wide range of human and bovine cells, including fibroblasts, keratinocytes, hematopoietic cells, and various carcinoma cell lines (Bertaux *et al.*, 1991; Hayakawa *et al.*, 1992, 1994; Yamashita *et al.*, 1996; Nemeth *et al.*, 1996; Kikuchi *et al.*, 1997; Murate *et al.*, 1997; Saika *et al.*, 1998; Luparello *et al.*, 1999), showing a maximal stimulation at approximately 100 ng/ml (3.6 n*M*) (Hayakawa *et al.*, 1994). Human TIMP-2 also had a potent cell growth-promoting activity for various types of human, bovine, and mouse cells, having an optimal concentration of 10 ng/ml (0.46 n*M*), i.e., about 10 times lower than that of TIMP-1 (Hayakawa *et al.*, 1994; Nemeth and Goolsby, 1993).

Neither TIMP-1 complexed with pro-MMP-9 nor TIMP-2 complexed with pro-MMP-2, both of which have fully inhibitory activity against active forms of MMPs (Kolkenbrock *et al.*, 1991; Ward *et al.*, 1991), showed any cell growth-promoting activity (Hayakawa *et al.*, 1994). On the contrary, both reductively alkylated TIMPs, having no MMP inhibitory activity, showed appreciable cell-proliferating activity. Even forming a stable (1:1) complex with MMP-2, a variant with an alanine appended to the amino terminus (Ala + TIMP-2) was inactive as an MMP inhibitor, but active as a growth factor to cause the proliferation of human skin fibroblasts (Hs 60) and also to inhibit the cell growth factor activity of basic fibroblast growth factor (bFGF) (Wingfield *et al.*, 1999), which is discussed later. These findings strongly indicate that the cell growth-modulating activity of TIMPs is independent of their MMP inhibitory activity.

The cell growth-promoting activity of both TIMPs is suggested to be a direct cellular effect mediated by a cell surface receptor. Avalos *et al.* (1988) demonstrated the presence of EPA/TIMP-1 receptors on K-562 cells, with a K_d of 62 (50–152) n*M* and a receptor number of 60,000 sites/cell. They also identified a 32-kDa receptor protein by using a cross-linker, disuccinimidyl suberate. Bertaux *et al.* (1991) reported that human keratinocytes had a TIMP-1 receptor with a K_d of 8.7 n*M* and a receptor number of 135,000 sites/cell. By using Raji cells, which express neither TIMP-1 nor TIMP-2, we demonstrated the presence of both high (K_d = 0.15 n*M*)- and low (35 n*M*)-affinity-binding sites for TIMP-2 with receptor numbers of 2.0×10^4 and 1.4×10^5 sites/cells, respectively (Hayakawa *et al.*, 1994).

Elevation of cAMP levels following activation of adenylate cyclase by a G-protein-coupled mechanism is suggested to be the signaling pathway involved in the mitogenic effects of TIMP-2 on human Hs68 fibroblasts and HT1080 fibrosarcoma cells (Corcoran and Stetler-Stevenson, 1995). However, we showed that H89, H7, bisindolylmaleimide, and K-252a, known to be potent inhibitors of protein kinase A, C, and myosin light chain kinase, had essentially no inhibitory effect on the mitogenic activity of TIMP-2 on human osteosarcoma cell line MG-63 cells stimulated with either TIMP, and we further demonstrated the crucial role of tyrosine kinase and the significant elevation of mitogen-activated protein (MAP) kinase activity in TIMP-dependent growth signaling (Yamashita *et al.*, 1996). One obvious difference between the aforementioned two studies was in the TIMP-2 concentration used; i.e., they used 0.5–1 μg/ml TIMP-2, which is 500–1,000 times higher than the level we used. Taking into account the presence of both high- and low-affinity-binding sites on the cell surface, as mentioned earlier, we were looking at signal transduction evoked through TIMP-2 binding to high- affinity binding sites, whereas Corcoran and Stetler-Stevenson (1995) studied another pathway involving transduction through low-affinity-binding sites. TIMP-3, which has been proposed to be a matrix-bound TIMP, as already mentioned, was reported to stimulate the proliferation of growth-retarded, non-transformed cells maintained under low serum conditions (Yang and Hawkes, 1992). These findings led us to conclude that cell growth-promoting activity is a common feature of members of the TIMP family.

Now we know that both TIMP-1 and TIMP-2 are constitutive components in human serum and also in fetal calf serum (FCS) (Hayakawa *et al.*, 1994; Kodama *et al.*, 1989; Fujimoto *et al.*, 1993). We demonstrated that cell proliferation was suppressed remarkably in TIMP-free FCS and was significantly recovered by the addition of TIMPs, suggesting that both TIMPs are newly recognized potent growth factors in serum (Hayakawa *et al.*, 1992, 1994).

Although several chemically defined synthetic media have been developed, the requirement for serum still remains for many vertebrate cell lines, hinting that one or more unknown serum factors may play key roles in regulating *in vivo* cell maintenance and proliferation. TIMPs may be such factors. The growth dependency on both TIMPs in FCS was, however, different from one cell line to another (Hayakawa *et al.*, 1994). Growth might depend on the amount of TIMPs that those cell lines produce by themselves, which would stimulate growth by an autocrine mechanism. It might also depend on the characteristics of the receptors of each cell line. TIMP-2 was found to inhibit the *in vitro* proliferation and migration of human microvascular endothelial cells stimulated with bFGF (Murphy *et al.*, 1993). As neither TIMP-1 nor a synthetic MMP inhibitor, BB-94, could take the place of TIMP-2, this ability of TIMP-2 seemed to be independent of its MMP inhibitory activity. In addition to its effect on bFGF, TIMP-2 was found to supress the mitogenic response to other tyrosine kinase-type growth factors, such as EGF and PDGF, inducing a decrease in phosphorylation of EGFR and a concomitant reduction of Grb-2 association (Hoegy *et al.*, 2001). However, taking into account that a fairly high concentration of TIMP-2 was required for the suppression, there is skepticism with respect to the *in vivo* significance of this inhibitor.

C. Embryogenesis-Stimulating Activity (ESA)

ESA, which is essential for the early *in vitro* development of bovine embryos, was found in serum-free conditioned media of bovine granulosa and oviductal epithelial cells as two molecular species. The primary structure of embryogenin-1, low molecular weight (30 K) ESA, determined by cDNA cloning, was found to be identical to that of bovine TIMP-1 (Satoh *et al.*, 1994). A conflicting result was reported showing that the elimination of TIMP-1 by immunoprecipitation from a bovine oviduct-conditioned medium did not affect the ESA activity of this medium (Vansteenbrugge *et al.*, 1997).

D. Steroidogenesis-Stimulating Protein (STP)

A 70-kDa protein complex designated as STP was isolated from the medium of rat Sertoli cell cultures. The complex, composed of 28- and 38-kDa proteins, was shown to be identical to TIMP-1 and the pro-form of cathepsin L, respectively, and to stimulate steroidogenesis by Leydig cells and ovarian granulosa cells in a dose-dependent and cAMP-independent manner (Boujrad *et al.*, 1995). TIMP-1 appeared to be responsible for the bioactivity, but procathepsin L was indispensable for maximal activity. Thus, a TIMP-1–procathepsin L complex is a potent activator of steroidogenesis and may regulate steroid concentrations and thus germ cell development in both males and females. TIMP-1-deficient mice were utilized to examine the STP activity of TIMP-1, but no evidence was found for the regulation of steroidogenesis *in vivo* (Nothnick *et al.*, 1997).

It has already been reported that cultured rat granulosa cells produced TIMP-1 and that this inhibitor acted in part to control the site and extent of follicular connective tissue remodeling associated with ovulation (Mann *et al.*, 1991). TIMP-1 together with TIMP-2 and TIMP-3 were demonstrated *in vivo* in pigs to be present in large and small follicles and to be produced by both granulosa and theca cells. Furthermore, an *in vitro* study suggests that TIMP-1 may regulate steroidogenesis in addition to controlling tissue remodeling (Shores and Hunter, 2000).

TIMP-1 is known to be expressed in the uterus of essentially all species. Mature female TIMP-1 null mice showed a significant decrease in the length of the estrus period, significantly high serum estradiol levels, and significantly lower serum progesterone levels compared with their wild-type counterparts. Also, during the estrus period, these null mice expressed significantly lower uterine TIMP-3 mRNA and showed altered uterine morphology (Nothnick, 2000). These results allow the conclusion that TIMP-1 has a mutifaceted role in regulating the murine reproductive cycle at both uterine and ovarian levels.

E. Apoptosis-Suppressing or Promoting Activity

It has been revealed that TIMP-1 suppressed apoptosis in Burkitt's lymphoma cell line inducing expression of Bcl-X, but not Bcl-2 as well as decreased NF-κB activity as comparted with controls (Guedez *et al.*, 1998a,b) and human breast epithelial cells possibly through constitutive activation of cell survival signaling path ways (Li *et al.*, 1999) and that TIMP-2 also suppressed that in B16F10 melanoma cells (Valente *et al.*, 1998). It was reported previously that transgenic mice overexpressing the TIMP-1 gene protected against the apoptosis of mammary epithelial cells differentiated during normal development of the mammary gland (Alexander *et al.*, 1996). Barasch *et al.* (1999) reported that TIMP-2 secreted by ureteric bud (UB) cells rescued metanephric mesenchymal cells from apoptosis as a metanephric mesenchymal growth factor. The growth factor activity of TIMP-2 seems to be independent of MMP inhibitory activity because ilomastat, a synthetic MMP inhibitor, had no effect on mesenchymal growth action. On the contrary, TIMP-3 was shown to induce cell death by stabilizing TNF-α receptors on the surface of human colon carcinoma (Smith *et al.*, 1997). It was further demonstrated that apoptosis in a human colon carcinoma cell line (Bian *et al.*, 1996), melanoma cells (Ahonen *et al.*, 1998), rat vascular smooth muscle cells through WAF 1 induction (Baker *et al.*, 1998), and several other cancer cell lines (Baker *et al.*, 1999) was induced by overexpression of TIMP-3. It was also demonstrated that the pro-death function of TIMP-3 was located within the N-terminal three loops and that the presence of functional MMP inhibitory activity was associated with the induction of apoptosis (Bond *et al.*, 2000). Formerly, one of the survival factors promoting the survival of a somatotropic cell line was purified from the conditioned medium of a mouse folliculo-stellate cell line culture and was identified to be TIMP-2 (Matsumoto *et al.*, 1993). The function of TIMP-2 as a survival factor might be ascribed to the antiapoptotic activity of TIMP-2.

F. TIMPs Stimulate Osteoclastic Bone Resorption

As both TIMP-1 and TIMP-2 have been reported to inhibit bone (Witty *et al.*, 1996; Shimizu *et al.*, 1990; Hill *et al.*, 1993), Shibutani *et al.* (1999) examined whether TIMP-1 and/or TIMP-2 in FCS, with which the culture medium was supplemented, would affect osteoclastic bone resorption *in vitro*. Contrary to their expectation, significant suppression of osteoclastic bone resorption was observed in a rabbit bone marrow culture system when TIMP-1 and TIMP-2 were removed from the FCS. Bone resorption was, however, almost fully restored by the addition of recombinant TIMPs. Using a mature osteoclast population with more than 95% purity, Sobue *et al.* (2001) revealed that both TIMPs directly stimulated the bone resorbing activity in a dose-dependent manner at concentrations less than 50 ng/ml, which is the same level as that for their cell growth-stimulating activity (Hayakawa *et al.*, 1992, 1994; Yamashita *et al.*, 1996; Nemeth *et al.*, 1996; Kikuchi *et al.*, 1997; Murate *et al.*, 1997; Saika *et al.*, 1998; Laparello *et al.*, 1999). In contrast, significantly higher concentrations (~μg/ml) were required to inhibit osteoclastic bone resorption (Witty *et al.*, 1996; Shimizu *et al.*, 1990; Hill *et al.*, 1993). The stimulating effects of TIMPs were abolished by the simultaneous addition of anti-TIMP antibodies. At higher concentrations, the stimulation of bone resorption decreased reversely to the control level. The magnitude of the stimulatory effect of TIMP-2 was more than that of TIMP-1. Metalloproteinase inhibitors, such as BE16627B and R94138, could not replace TIMPs with respect to the bone-resorbing activity, suggesting that the osteoclast-stimulating activity of TIMPs was independent of the inhibitory activity toward MMPs. The stimulatory signals from both TIMPs for osteoclastic bone resorption were demonstrated to be transduced through the tyrosine kinase/MEK/ERK pathway, which is also the pathway used for the cell growth-stimulating activity of both TIMPs (Yamashita *et al.*, 1996). Some TIMP-2 binding proteins on the plasma membrane of osteoclasts were detected by a cross-linking experiment. These findings demonstrate that TIMPs directly stimulate the bone-resorbing activity of isolated mature osteoclasts at their physiological concentrations and that the stimulatory action of TIMPs is likely to be independent of their activities as inhibitors of MMPs. As mentioned previously, TIMPs are so-called growth-modulating factors, and the osteoclast-stimulating activity of both TIMPs can also be included in this category as a novel function. Northern blot analysis demonstrated that osteoblasts/stromal cells expressed high levels of both TIMPs but that osteoclasts expressed only trace amounts of them. These facts suggest that osteoblasts/stromal cells stimulate mature osteoclasts in a paracrine fashion through the expression of both TIMPs, newly recognized mediators of osteoclast stimulation, besides via RANKL (Teitelbaum, 2000).

G. Other Functions

TIMP-3 appeared to accelerate the detachment of transforming cells from the ECM and their morphological changes (Yang and Hawkes, 1992). These peculiar functions of TIMP-3 might be related to the metastasis of cancer cells.

Clark *et al.* (1994) showed that TIMP-1 stimulated the secretion of MMP-1. The concentration of TIMP-1 needed to significantly stimulate MMP-1 release ranged from 1 to 8 μg/ml, which is sometimes achieved in some body fluids, such as amniotic fluid, at 16–18 weeks, gestation, sera after parturition, and synovial fluids from rheumatoid arthritis or osteoarthritis patients.

TIMP-2 was essential for the activation of pro-MMP-2 by MT1-MMP anchoring pro-MMP-2 on the cell surface (Nagase, 1998).

IV. CONCLUSIONS AND FUTURE PERSPECTIVES

TIMPs are really very amazing proteins, as they have many unique functions independent of their MMP inhibitory activities and those functions still continue to be discovered. As we already introduced, some binding studies have found evidence for specific, saturable binding of either TIMP-1 or TIMP-2 to high-affinity cell surface receptors (Avalos *et al.*, 1988; Bertaux *et al.*, 1991; Hayakawa *et al.*, 1994). However, trials to isolate and identify their receptor proteins have been unsuccessful so far. This is one of the big issues that should be addressed in the near future with relation to cell growth factor activity.

We have demonstrated that TIMP-1 accumulates in the nucleus of human gingival fibroblasts (Gin-1 cells) in accordance with the S phase of their cell cycle (Li *et al.*, 1995; Zhao *et al.*, 1998). This observation is of importance in suggesting an entirely new concept of TIMP-1 being possibly involved in cell growth processes. It was further reported later that TIMP-1 bound to the cell surface and was translocated to the nucleus of human MCF-7 breast carcinoma cells (Ritter *et al.*, 1999) and also that TIMP-1 was expressed constitutively in the nucleus of Sertoli cells, which expression was stimulated by 8-CPTcAMP (Grønning *et al.*, 2000). Although we have no evidence at this moment to explain what TIMP-1 is doing in these cell nuclei, one possibility is that TIMP-1 might be functioning in the nucleus as an MMP inhibitor, and another possibility is that TIMP-1 might bind to DNA, like a transcription factor, through its β-barrel structure as mentioned earlier (Gomis-Rüth *et al.*, 1997; Tuuttila *et al.*, 1998).

REFERENCES

Ahonen, M., Baker, A. H., and Kahali, V.-M. (1998). Adenovirus-mediated gene delivery of tissue inhibitor of metalloproteinases-3 inhibits invasion and induces apoptosis in melanoma cells. *Cancer Res.* **58**, 2310–2315.

Alexander, C. M., Howard, E. W., Bissell, M. J., and Werb, Z. (1996). Rescue of mammary epithelial cell apoptosis and entactin degradation by a tissue inhibitor of metalloproteinases-1 transgene. *J. Cell Biol.* **135**, 1669–1677.

Amour, A., Knight, C. G., Webster, A., Slocombe, P. M., Stephens, P. E., Knäuper, V., Docherty, A. J. P., and Murphy, G. (2000). The in vivo activity of ADAM-10 is inhibited by TIMP-1 and TIMP-3. *FEBS Lett.* **473**, 275–279.

Amour, A., Slocombe, P. M., Webster, A., Butler, M., Knight, C. G., Smith, B. J., Stephens, P. E., Shelley, C., Hutton, M., Knäuper, V., Docherty, A. J. P., and Murphy, G. (1998). TNF-α converting enzyme (TACE) is inhibited by TIMP-3. *FEBS Lett.* **435**, 39–44.

Anand-Apte, B., Bao, L., Smith, R., Iwata, K., Olsen, B. R., Zetter, B., and Apte, S. S. (1996). A review of tissue inhibitor of metalloproteinases-3 (TIMP-3) and experimental analysis of its effect on primary tumour growth. *Biochem. Cell Biol.* **74**, 853–862.

Apte, S. S., Hayashi, K., Seldin, M. F., Mattei, M.-G., Hayashi, M., and Olsen, B. R. (1994). Gene encoding a novel murine tissue inhibitor of metalloproteinases (TIMP), TIMP-3, is expressed in developing mouse epithelia, cartilage, and muscle, and is located on mouse chromosome 10. *Dev. Dyn.* **200**, 177–197.

Apte, S. S., Olsen, B. R., and Murphy, G. (1995). The gene structure of tissue inhibitor of metalloproteinases (TIMP)-3 and its inhibitory activities define the distinct TIMP gene family. *J. Biol. Chem.* **270**, 14313–14318.

Assink, J. J. M., de Backer, E., ten Brink, J. B., Kohono, T., de Jong, P. T. V. M., Bergen, A. A. B., and Meire, F. (2000). Sorsby fundus dystrophy without a mutation in the TIMP-3 gene. *Br. J. Ophthalmol.* **84**, 682–686.

Avalos, B. R., Kaufman, S. E., Tomonaga, M., Williams, R. E., Golde, D. W., and Gasson, J. C. (1988). K562 cells produce and respond to human erythroid-potentiating activity. *Blood* **71**, 1720–1725.

Baker, A. H., Zalsman, A. B., George, S. J., and Newby, A. C. (1998). Divergent effects of tissue inhibitor of metalloproteinase-1, -2, or -3 overexpression on rat vascular smooth muscle cell invasion, proliferation, and death in vitro. TIMP-3 promotes apoptosis. *J. Clin. Invest.* **101**, 1478–1487.

Baker, A. H., George, S. J., Zaltsman, A. B., Murphy, G., and Newby, A. C. (1999). Inhibition of invasion and induction of apoptotic cell death of cancer cell lines by overexpression of TIMP-3. *Br. J. Cancer* **79**, 1347–1355.

Barasch, J., Yang, J., Qiao, J., Tempst, P., Erdjument-Bromage, H., Leung, W., and Oliver, J. A. (1999). Tissue inhibitor of metalloproteinase-2 stimulates mesenchymal growth and regulates epithelial branching during morphogenesis of the rat metanephros. *J. Clin. Invest.* **103**, 1299–1307.

Bertaux, B., Hornebeck, W., Eisen, A. Z., and Dubertret, L. (1991). Growth stimulation of human keratinocytes by tissue inhibitor of metalloproteinases. *J. Invest. Dermatol.* **97**, 679–685.

Bian, J., Wang, Y., Smith, M. R., Kim, H., Jacobs, C., Jackman, J., Kung, H. F., Colburn, N. H., and Sun, Y. (1996). Suppression of *in vivo* tumor growth and induction of suspension cell death by tissue inhibitor of metalloproteinases (TIMP)-3. *Carcinogenesis* **17**, 1805–1811.

Bigg, H. F., Morrison, C. J., Butler, G. S., Bogoyevitch, M. A., Wang, Z., Soloway, P. D., and Overall, C. M. (2001). Tissue inhibitor of metalloproteinases-4 inhibits but does not support the activation of gelatinase A via efficient inhibition of membrane type 1-matrix metalloproteinase. *Cancer Res.* **61**, 3610–3618.

Bigg, H. F., Shi, Y. E., Liu, Y. E., Steffensen, B., and Overall, C. M. (1997). Specific, high affinity binding of tissue inhibitor of metalloproteinases-4 (TIMP-4) to the COOH-terminal hemopexin-like domain of human gelatinase A: TIMP-4 binds progelatinase A and the COOH-terminal domain in a similar manner to TIMP-2. *J. Biol. Chem.* **272**, 15496–15500.

Bond, M., Murphy, G., Bennett, M. R., Amour, A., Knäuper, V., Newby, A. C., and Baker, A. H. (2000). Localization of the death domain of tissue inhibitor of metalloproteinase-3 to the N terminus: Metalloproteinase inhibition is associated with proapoptotic activity. *J. Biol. Chem.* **275**, 41358–41363.

Boujrad, N., Ogwuegbu, S. O., Garnier, M., Lee, C. H., Martin, B. M., and Papadopoulos, V. (1995). Identification of a stimulator of steroid hormone synthesis isolated from testis. *Science* **268**, 1609–1612.

Callebaut, I., and Mornon, J. P. (1997). The human EBNA-2 coactivator p100: Multidomain organization and relationship to the staphylococcal nuclease fold and to the tudor protein involved in *Drosophila melanogaster* development. *Biochem. J.* **321**, 125–132.

Campbell, C. E., Flenniken, A. M., Skup, D., and Williams, B. R. (1991). Identification of a serum- and phorbol ester-responsive element in the murine tissue inhibitor of metalloproteinase gene. *J. Biol. Chem.* **266**, 7199–7206.

Chesler, L., Golde, D. W., Bersch, N., and Johnson, M. D. (1995). Metalloproteinase inhibition and erythroid potentiation are independent activities of tissue inhibitor of metalloproteinases-1. *Blood* **86**, 4506–4515.

Clark, I. M., Powell, L. K., and Cawston, T. E. (1994). Tissue inhibitor of metalloproteinases (TIMP-1) stimulates the secretion of collagenase from human skin fibroblasts. *Biochem. Biophys. Res. Commun.* **203**, 874–880.

Corcoran, M. L., and Stetler-Stevenson, W. G. (1995). Tissue inhibitor of metalloproteinase-2 stimulates fibroblast proliferation via a cAMP-dependent mechanism. *J. Biol. Chem.* **270**, 13453–13459.

Dean, G., Young, D. A., Edwards, D. R., and Clark, I. M. (2000). The human tissue inhibitor of metalloproteinases (TIMP)-1 gene contains repressive elements within the promoter and intron 1. *J. Biol. Chem.* **275**, 32664–32671.

De Clerck, Y. A., Darville, M. I., Eeckhout, Y., Rousseau, G. G. (1994). Characterization of the promoter of the gene encoding human tissue inhibitor of metalloproteinases-2 (TIMP-2). *Gene* **139**, 185–191.

De Clerck, Y. A., Yean, T.-D., Lu, H. S., Ting, J., and Langley, K. E. (1991). Inhibition of autoproteolytic activation of interstitial procollagenase by recombinant metalloproteinase inhibitor MI/TIMP-2. *J. Biol. Chem.* **266**, 3893–3899.

Docherty, A. J. P., Lyons, A., Smith, B. J., Wright, E. M., Stephens, P. E., Harris, T. J. R., Murphy, G., and Reynolds, J. J. (1985). Sequence of human tissue inhibitor of metalloproteinases and its identity to erythroid-potentiating activity. *Nature* **318**, 66–69.

Edwards, D. R., Beaudry, P. P., Laing, T. D., Kowal, V., Leco, K. J., Leco, P. A., and Lim, M. S. (1996). The roles of tissue inhibitors of metalloproteinases in tissue remodelling and cell growth. *Int. J. Obes.* **20** (Suppl. 3), S9–S15.

Edwards, D. R., Rocheleau, H., Sharma, R. R., Wills, A. J., Cowie, A., Hassell, J. A., and Heath, J. K. (1992). Involvement of AP1 and PEA3 binding sites in the regulation of murine tissue inhibitor of metalloproteinases-1 (TIMP-1) transcription. *Biochim. Biophys. Acta* **1171**, 41–55.

Fernandez-Catalan, C., Bode, W., Huber, R., Turk, D., Calvete, J. J., Lichte, A., Tschesche, H., and Maskos, K. (1998). Cryatal structure of the complex formed by the membrane type 1-matrix metalloproteinase with the tissue inhibitor of metalloproteinases-2, the soluble progelatinase A receptor. *EMBO J.* **17**, 5238–5248.

Fujimoto, N., Zhang, J., Iwata, K., Shinya, T., Okada, Y., and Hayakawa, T. (1993). A one-step sandwich enzyme immunoassay for tissue inhibitor of metalloproteinases-2 using monoclonal antibodies. *Clin. Chim. Acta* **220**, 31–45.

Gao, G., Semenchenko, V., Arumugam, S., and Van Doren, S. R. (2000). Tissue inhibitor of metalloproteinases-1 undergoes microsecond to millisecond motions at sites of matrix metalloproteinase-induced fit. *J. Mol. Biol.* **301**, 537–552.

Gasson, J. C., Golde, D. W., Kaufman, S. E., Westbrook, C. A., Hewich, R. M., Kaufman, R. J., Wong, G. G., Temple, P. A., Leary, A. C., Brown, E. L., Orr, E. C., and Clark, S. C. (1985). Molecular characterization and expression of the gene encoding human erythroid-potentiating activity. *Nature* **315**, 768–771.

Goldberg, G. I., Marmer, B. L., Grant, G. A., Eisen, A. Z., Wilhelm, S., and He, C. (1989). Human 72-kilodalton type IV collagenase forms a complex with a tissue inhibitor of metalloproteinases designated TIMP-2. *Proc. Natl. Acad. Sci. USA* **86**, 8207–8211.

Gomez, D. E., Alonso, D. F., Yoshiji, H., and Thorgeirsson, U. P. (1997). Tissue inhibitors of metalloproteinases: Structure, regulation and biological functions. *Eur. J. Cell Biol.* **7**, 111–122.

Gomis-Rüth, F.-X., Maskos, K., Betz, M., Bergner, A., Huber, R., Suzuki, K., Yoshida, N., Nagase, H., Brew, K., Bourenkov, G. P., Bartunik, H., and Bode, W. (1997). Mechanism of inhibition of the human matrix metalloproteinase stromelysin-1 by TIMP-1. *Nature* **389**, 77–81.

GrØnning, L. M., Wang, J. E., Ree, A. H., Haugen, T. B., Taskén, K., and Taskén, K. A. (2000). Regulation of tissue inhibitor of metalloproteinases-1 in rat Sertoli cells, Induction by germ cell residual bodies, interleukin-1α, and second messengers. *Biol. Reprod.* **62**, 1040–1046.

Guedez, L., Courtemanch, L., and Stetler-Stevenson, M. (1998). Tissue inhibitor of metalloproteinase (TIMP)-1 induces differentiation and an antiapoptotic phenotype in germinal center B cells. *Blood* **92**, 1342–1349.

Guedez, L., Stetler-Stevenson, W. G., Wolff, L., Wang, J., Fukushima, P., Mansoor, A., and Stetler-Stevenson, M. (1998). In vitro suppression of programmed cell death of B cells by tissue inhibitor of metalloproteinase-1. *J. Clin. Invest.* **102**, 2002–2010.

Hashimoto, G., Aoki, T., Nakamura, H., Tanzawa, K., and Okada, Y. (2001). Inhibition of ADAMTS4 (aggrecanase-1) by tissue inhibitors of metalloproteinases (TIMP-1, 2, 3 and 4). *FEBS Lett.* **494**, 192–195.

Hayakawa, T., Yamashita, K., Ohuchi, E., and Shinagawa, A. (1994). Cell growth-promoting activity of tissue inhibitor of metalloproteinases-2 (TIMP-2). *J. Cell Sci.* **107**, 2373–2379.

Hayakawa, T., Yamashita, K., Kishi, J., and Harigaya, K. (1990). Tissue inhibitor of metalloproteinases from human bone marrow stromal cell line KM 102 has erythroid-potentiating activity, suggesting its possibly bifunctional role in the hematopoietic microenvironment. *FEBS Lett.* **268**, 125–128.

Hayakawa, T., Yamashita, K., Tanzawa, K., Uchijima, E., Iwata, K. (1992). Growth-promoting activity of tissue inhibitor of metalloproteinases-1 (TIMP-1) for a wide range of cells. *FEBS Lett.* **298**, 29–32.

Hernandez-Barrantes, S., Shimura, Y., Soloway, P. D., Sang, Q. A., and Friedman, R. (2001). Differential roles of TIMP-4 and TIMP-2 in pro-MMP-2 activation by MT1-MMP. *Biochem. Biophys. Res. Commun.* **281**, 126–130.

Hill, P. A., Reynolds, J. J., and Meikle, M. C. (1993). Inhibition of stimulated bone resorption in vitro by TIMP-1 and TIMP-2. *Biochim. Biophys. Acta* **1177**, 71–74.

Hoegy, S. E., Oh, H.-R., Corcoran, M. L., and Stetler-Stevenson, W. G. (2001). Tissue inhibitor of metalloproteinases-2 (TIMP-2) suppresses TKP-growth factor signaling independent of metalloproteinase inhibition. *J. Biol. Chem.* **276**, 3203–3214.

Howard, E. W., Bullen, E. C., and Banda, M. J. (1991). Regulation of the autoactivation of human 72-kDa progelatinase by tissue inhibitor of metalloproteinase-2. *J. Biol. Chem.* **266**, 13064–13069.

Kashiwagi, M., Tortorella, M., Nagase, H., and Brew, K. (2001). TIMP-3 is a potent inhibitor of aggrecanase 1 (ADAM-TS4) and aggrecanase 2 (ADAM-TS5). *J. Biol. Chem.* **276**, 12501–12504.

Kikuchi, K., Kadono, T., Furue, M., and Tamaki, K. (1997). Tissue inhibitor of metalloproteinase 1 (TIMP-1) may be an autocrine growth factor in scleroderma fibroblasts. *J. Invest. Dermatol.* **108**, 281–284.

Kinoshita, T., Sato, H., Takino, T., Itoh, M., Akizawa, T., and Seiki, M. (1996). Processing of a precursor of 72-kilodalton type IV collagenase/gelatinase A by a recombinant membrane-type 1 matrix metalloproteinase. *Cancer Res.* **56**, 2535–2538.

Knäuper, V., Lopez-Otin, C., Smith, B., Knight, G., and Murphy, G. (1996). Biochemical characterization of human collagenase-3. *J. Biol. Chem.* **271**, 1544–1550.

Kodama, S., Yamashita, K., Kishi, J., Iwata, K., and Hayakawa, T. (1989). A sandwich enzyme immunoassay for collagenase inhibitor using monoclonal antibodies. *Matrix* **9**, 1–6.

Kolkenbrock, H., Orgel, D., Hecker-Kia, A., Noack, W., and Ulbrich, N. (1991). The complex between a tissue inhibitor of metalloproteinases (TIMP-2) and 72-kDa progelatinase is a metalloproteinase inhibitor. *Eur. J. Biochem.* **198**, 775–781.

Langton, K. P., Barker, M. D., and McKie, N. (1998). Localization of the functional domains of human tissue inhibitor of metalloproteinases-3 and the effects of a Sorsby's fundus dystrophy mutation. *J. Biol. Chem.* **273**, 16778–16781.

Langton, K. P., McKie, N., Curtis, A., Goodship, J. A., Bond, P. M., Barker, M. D., and Clarke, M. (2000). A novel tissue inhibitor of metalloproteinases-3 mutation reveals a common molecular phenotype on Sorsby's fundus dystrophy. *J. Biol. Chem.* **275**, 27027–27031.

Li, G., Fridman, R., and Kim, H.-R. C. (1999). Tissue inhibitor of metalloproteinase-1 inhibits apoptosis of human breast epithelial cells. *Cancer Res.* **59**, 6267–6275.

Li, H., Nishio, K., Yamashita, K., Hayakawa, T., and Hoshino, T. (1995). Cell cycle-dependent localization of tissue inhibitor of metalloproteinases-1 immunoreactivity in cultured human gingival fibroblasts. *Nagoya J. Med. Sci.* **58**, 133–142.

Loechel, F., Fox, J. W., Murphy, G., Albrechtsen, R., and Wewer, U. M. (2000). ADAM 12-S cleaves IGFBP-3 and IGFBP-5 and is inhibited by TIMP-3. *Biochem. Biophys. Res. Commun.* **278**, 511–515.

Lombard, M. A., Wallace, T. L., Kubicek, M. F., Petzold, G. L., Mitchell, M. A., Hendges, S. K., and Wilks, J. W. (1998). Synthetic matrix metalloproteinase inhibitors and tissue inhibitor of metalloproteinase (TIMP)-2, but not TIMP-1, inhibit shedding of tumor necrosis factor-α receptors in a human colon adenocarcinoma (Colo 205) cell line. *Cancer Res.* **58**, 4001–4007.

Luparello, C., Avanzato, G., Carella, C., and Pucci-Minafra, I. (1999). Tissue inhibitor of metalloproteinase (TIMP)-1 and proliferative behaviour of clonal breast cancer cells. *Breast Cancer Res. Treat.* **54**, 235–244.

Mann, J. S., Kindy, M. S., Edwards, D. R., and Curry, T. E., Jr. (1991). Hormonal regulation of matrix metalloproteinase inhibitors in rat granulosa cells and ovaries. *Endocrinology* **128**, 1825–1832.

Matsumoto, H., Ishibashi, Y., Ohtaki, T., Hasegawa, Y., Koyama, C., and Inoue, K. (1993). Newly established murine pituitary folliculo-stellate-like cell line (TtT/GF) secretes potent pituitary glandular cell survival factors, one of which corresponds to metalloproteinase inhibitor. *Biochem. Biophys. Res. Commun.* **194**, 909–915.

Murate, T., and Hayakawa, T. (1999). Multiple functions of tissue inhibitors of metalloproteinases (TIMPs): New aspects in hematopoiesis. *Platelets* **10**, 5–16.

Murate, T., Yamashita, K., Isogai, C., Suzuki, H., Ichihara, M., Hatano, S., Nakahara, Y., Kinoshita, T., Nagasaka, T., Yoshida, S., Komatsu, N., Miura, Y., Hotta, T., Fujimoto, N., Saito, H., and Hayakawa, T. (1997). The production of tissue inhibitors of metalloproteinases (TIMPs) in megakaryopoiesis: possible role of platelet- and megakaryocyte-derived TIMPs in bone marrow fibrosis. *Br. J. Haematol.* **99**, 181–189.

Murate, T., Yamashita, K., Ohashi, H., Kagami, Y., Tsushita, K., Kinoshita, T., Hotta, T., Saito, H., Yoshida, S., Mori, K. J., and Hayakawa, T. (1993). Erythroid potentiating activity of tissue inhibitor of metalloproteinases on the differentiation of erythropoietin-responsive mouse erythroleukemia cell line, ELM-I-1-3, is closely related to its cell growth potentiating activity. *Exp. Hematol.* **21**, 169–176.

Murphy, A. N., Unsworth, E. J., and Stetler-Stevenson, W. G. (1993). Tissue inhibitor of metalloproteinases-2 inhibits bFGF-induced human microvascular endothelial cell proliferation. *J. Cell Physiol.* **157**, 351–358.

Murphy, G., Ward, R., Hembry, R. M., Reynolds, J. J., Kuhn, K., and Tryggvason, K. (1989). Characterization of gelatinase from pig polymorphonuclear leukocytes: A metalloproteinase resembling tumour type IV collagenase. *Biochem. J.* **258**, 463–472.

Muskett, F. W., Frenkiel, T. A., Feeney, J., Freedman, R. B., Carr, M. D., and Williamson, R. A. (1998). High resolution structure of the N-terminal domain of tissue inhibitor of metalloproteinases-2 and characterization of its interaction site with matrix metalloproteinase-3. *J. Biol. Chem.* **273**, 21736–21743.

Nagase, H. (1996). Matrix metalloproteinases. *In*: "Zinc Metalloproteinases in Health and Disease". N. M. Hooper, (ed.), pp. 153–204. Taylor & Francis, London.

Nagase, H. (1998). Cell surface activation of progelatinase A (proMMP-2) and cell migration. *Cell Res.* **8**, 179–186.

Nagase, H., Barrett, A. J., and Woessner, J. F., Jr. (1992). Nomenclature and glossary of the matrix metalloproteinases. *Matrix* **Suppl**, 421–424.

Nemeth, J. A., and Goolsby, C. L. (1993). TIMP-2, a growth-stimulatory protein from SV-40 transformed human fibroblasts. *Exp. Cell Res.* **207**, 376–382.

Nemeth, J. A., Rafe, A., Steiner, M., and Goolsby, C. L. (1996). TIMP-2 growth-stimulatory activity: A concentration- and cell type-specific response in the presence of insulin. *Exp. Cell Res.* **224**, 110–115.

Nguyen, Q., Willenbrock, F., Cockett, M. I., O'Shea, M., Docherty, A. J. P., and Murphy, G. (1994). Different domain interactions are involved in the binding of tissue inhibitor of metalloproteinases to stromelysin-1 and gelatinase A. *Biochemistry* **33**, 2089–2095.

Nomura, S., Hogan, B. L. M., Wills, A. J., Heath, J. K., and Edwards, D. R. (1989). Developmental expression of tissue inhibitor of metalloproteinases (TIMP) RNA. *Development* **105**, 575–583.

Nothnick, W. B., Soloway, P., and Curry, T. E., Jr. (1997). Assessment of the role of tissue inhibitor of metalloproteinase-1 (TIMP-1) during the periovulatory period in female mice lacking a functional TIMP-1 gene. *Biol. Reprod.* **56**, 1181–1188.

Nothnick, W. B. (2000). Disruption of the tissue inhibitor of metalloproteinase-1 gene results in altered reproductive cyclicity and uterine morphology in reproductive-age female mice. *Biol. Reprod.* **63**, 905–912.

Okada, Y., Gonoji, Y., Naka, K., Tomita, K., Nakanishi, I., Iwata, K., Yamashita, K., and Hayakawa, T. (1992). Matrix metalloproteinase 9 (92-kDa gelatinase/type IV collagenase) from HT 1080 human fibrosarcoma cells. Purification and activation of the precursor and enzymic properties. *J. Biol. Chem.* **267**, 21712–21719.

Petitfrère, E., Kadri, Z., Boudot, C., Sowa, M.-L., Mayeux, P., Haye, B., and Billat, C. (2000). Involvement of the p38 mitogen-activated protein kinase pathway in tissue inhibitor of metalloproteinases-1-induced erythroid differentiation. *FEBS Lett.* **485**, 117–121.

Phillips, B. W., Sharma, R., Leco, P. A., and Edwards, D. R. (1999). A sequence-selective single-strand DNA-binding protein regulates basal transcription of the murine tissue inhibitor of metalloproteinases-1 (*Timp-1*) gene. *J. Biol. Chem.* **274**, 22197–22207.

Ritter, L. M., Garfield, S. H., and Thorgeirsson, U. P. (1999). Tissue inhibitor of metalloproteinases-1 (TIMP-1) binds to the cell surface and translocates to the nucleus of human MCF-7 breast carcinoma cells. *Biochem. Biophys. Res. Commun.* **257**, 494–499.

Roton, L., Nemoto, S., Simsic, J., Coker, M. L., Rao, V., Baicu, S., Defreyte, G., Soloway, P. J., Zile, M. R., and Spinale, F. G. (2000). Effects of gene deletion of the tissue inhibitor of the matrix metalloproteinase-type 1 (TIMP-1) on left ventricular geometry and function in mice. *J. Mol. Cell Cardiol.* **32**, 109–120.

Saika, S., Kawashima, Y., Okada, Y., Tanaka, S., Yamanaka, O., Ohnishi, Y., and Ooshiam, A. (1998). Recombinant TIMP-1 and -2 enhance the proliferation of rabbit corneal epithelial cells *in vitro* and the spreading of rabbit corneal epithelium *in situ*. *Curr. Eye Res.* **17**, 47–52.

Satoh, H., Kobayashi, K., Yamashita, S., Kikuchi, M., Sendai, Y., and Hoshi, H. (1994). Tissue inhibitor of metalloproteinases (TIMP-1) produced by granulosa and oviduct cells enhances *in vitro* development of bovine embryo. *Biol. Reprod.* **50**, 835–844.

Shibutani, T., Yamashita, K., Aoki, T., Iwayama, Y., Nishikawa, T., and Hayakawa, T. (1999). Tissue inhibitors of metalloproteinases (TIMP-1 and TIMP-2) stimulate osteoclastic bone resorption. *J. Bone Miner. Metab.* **17**, 245–251.

Shimizu, H., Sakamoto, M., and Sakamoto, S. (1990). Bone resorption by isolated osteoclasts in living versus devitalized bone: Differences in mode and extent and the effects of human recombinant tissue inhibitor of metalloproteinases. *J. Bone Miner. Res.* **5**, 411–418.

Shores, E. M., and Hunter, M. G. (2000). Production of tissue inhibitors of metalloproteinases (TIMPs) by pig ovarian cells *in vivo* and the effect of TIMP-1 on steroidgenesis *in vitro*. *J. Reprod. Fertil.* **120**, 73–81.

Smith, M. R., Kung, H.-F., Durum, S. K., Colburn, N. H., and Sun, Y. (1997). TIMP-3 induces cell death by stabilizing TNF-α receptors on the surface of human colon carcinoma cells. *Cytokine* **9**, 770–780.

Sobue, T., Hakeda, Y., Kobayashi, Y., Hayakawa, H., Yamashita, K., Aoki, T., Kumegawa, M., Noguchi, T., and Hayakawa, T. (2001). Tissue inhibitor of metalloproteinases-1 and directly stimulate the bone-resorbing activity of isolated mature osteoclasts. *J. Bone Miner. Res.* **16**, 2205–2214.

Stetler-Stevenson, W. G., Krutzsch, H. C., and Liotta, L. A. (1989). Tissue inhibitor of metalloproteinase (TIMP-2). A new member of the metalloproteinase inhibitor family. *J. Biol. Chem.* **264**, 17374–17378.

Stetler-Stevenson, W. G., Bersch, N., and Golde, D. W. (1992). Tissue inhibitor of metalloproteinase-2 (TIMP-2) has erythroid potentiating activity. *FEBS Lett.* **296**, 231–234.

Teitelbaum, S. L. (2000). Bone resorption by osteoclasts. *Science* **289**, 1504–1508.

Trim, J. E., Samra, S. K., Arthur, M. J. P., Wright, M. C., McAulay, M., Beri, R., and Mann, D. A. (2000). Upstream tissue inhibitor of metalloproteinases-1 (TIMP-1) element-1, a novel and essential regulatory DNA motif in the human TIMP-1 gene promoter, directly interacts with a 30-kDa nuclear protein. *J. Biol. Chem.* **275**, 6657–6663.

Tuuttila, A., Morgunova, E., Bergmann, U., Lindqvist, Y., Maskos, K., Fernandez-Catalan, C., Bode, W., Tryggvason, K., and Schneider, G. (1998). Three-dimensional structure of human tissue inhibitor of metalloproteinases-2 at 2.1 A resolution. *J. Mol. Biol.* **284**, 1133–1140.

Valente, P., Fassina, G., Melchiori, A., Masiello, L., Cilli, M., Vacca, A., Onisto, M., Santi, L., Stetler-Stevenson, W. G., and Albini, A. (1998). TIMP-2 over-expression reduces invasion and angiogenesis and protects B16F10 melanoma cells from apoptosis. *Int. J. Cancer* **75**, 246–253.

Vansteenbrugge, A., Van Langendonckt, A., Massip, A., and Dessy, F. (1997). Effect of estrus-associated glycoprotein and tissue inhibitor of metalloproteinase-1 secreted by oviduct cells on *in vitro* bovine embryo development. *Mol. Reprod. Dev.* **46**, 527–534.

Ward, R. V., Hembry, R. M., Reynolds, J. J., and Murphy, G. (1991). The purification of tissue inhibitor of metalloproteinase-2 from its 72 kDa progelatinase complex: Demonstration of the biochemical similarities of tissue inhibitors of metalloproteinases-2 and tissue inhibitor of metalloproteinases-1 *Biochem. J.* **278**, 179–187.

Waterhouse, P., Denhardt, D. T., and Khokha, R. (1993). Temporal expression of tissue inhibitor of metalloproteinases in mouse reproductive tissues during gestation. *Mol. Reprod. Dev.* **35**, 219–226.

Westbrook, C. A., Gasson, J. C., Gerber, S. E., Selsted, M. E., and Golde, D. W. (1984). Purification and characterization of human T-lymphocyte-derived erythroid-potentiating activity. *J. Biol. Chem.* **259**, 992–996.

Wick, M., Bürger, C., Brüsselbach, S., Lucibello, F. C., and Müller, R. (1994). A novel member of human tissue inhibitor of metalloproteinases (*TIMP*) gene family is regulated during G_1 progression, mitogenic stimulation, differentation, and senescence. *J. Biol. Chem.* **269**, 18953–18960.

Wilhelm, S. M., Collier, I. E., Marmer, B. L., Eisen, A. Z., Grant, G. A., and Goldberg, G. I. (1989). SV40-transformed human lung fibroblasts secrete a 92-kDa type IV collagenase which is identical to that secreted by normal human macrophages. *J. Biol. Chem.* **264**, 17213–17221.

Will, H., Atkinson, S. J., Butler, G. S., Smith, B., and Murphy, G. (1996). The soluble catalytic domain of membrane type 1 matrix metalloproteinase cleaves the propeptide of progelatinase A and initiates autoproteolytic activation. *J. Biol. Chem.* **271**, 17119–17123.

Williamson, R. A., Martorell, G., Carr, M. D., Murphy, G., Docherty, A. J. P., Freedman, R. B., and Feeney, J. (1994). Solution structure of the active domain of tissue inhibitor of metalloproteinases-2: A new member of the OB fold protein family. *Biochemistry* **33**, 11745–11759.

Williamson, R. A., Muskett, F. W., Howard, M. J., Freedman, R. B., and Carr, M. D. (1999). The effect of matrix metalloproteinase complex formation on the conformational mobility of tissue inhibitor of metalloproteinases-2 (TIMP-2). *J. Biol. Chem.* **274**, 37226–37232.

Wingfield, P. T., Sax, J. K., Stahl, S. J., Kaufman, J., Palmer, I., Chung, V., Corcoran, M. L., Kleiner, D. E., and Stetler-Stevenson, W. G. (1999). Biophysical and functional characterization of full-length, recombinant human tissue inhibitor of metalloproteinases-2 (TIMP-2) produced in *Escherichia coli*. *J. Biol. Chem.* **274**, 21362–21368.

Witty, J. P., Foster, S. A., Stricklin, G. P., Matrisian, L. M., and Stern, P. H. (1996). Parathyroid hormone-induced resorption in fetal rat limb bones is associated with production of the metalloproteinases collagenase and gelatinase B. *J. Bone Miner. Res.* **11**, 72–78.

Woessner, J. F., Jr. (1998). The matrix metalloproteinase family. *In*: "Matrix Metalloproteinases" (W. C. Parks and R. P. Mecham, eds.), pp. 1–14. Academic Press, San Diego.

Woessner, J. F., and Nagase, H. (2000). Matrix Metalloproteinases and TIMPs. Oxford Univ. Press, New York.

Wu, B., Arumugam, S., Gao, G., Lee, G. I., Semenchenko, V., Huang, W., Brew, K., and Van Doren, S. R. (2000). NMR structure of tissue inhibitor of metalloproteinases-1 implicates localized induced fit in inrecognition of matrix metalloproteinases. *J. Mol. Biol.* **295**, 257–268.

Yamashita, K., Suzuki, M., Iwata, H., Koike, T., Hamaguchi, M., Shinagawa, A., Noguchi, T., and Hayakawa, T. (1996). Tyrosine phosphorylation is crucial for growth signaling by tissue inhibitors of metalloproteinases (TIMP-1 and TIMP-2). *FEBS Lett.* **396**, 103–107.

Yang, T. T., and Hawkes, S. P. (1992). Role of the 21-kDa protein TIMP-3 in oncogenic transformation of cultured chicken embryo fibroblasts. *Proc. Natl. Acad. Sci. USA* **89**, 10676–10680.

Yu, W.-H., Yu, S. C., Meng, Q., Brew, K., and Woessner, J. F., Jr. (2000). TIMP-3 binds to sulfated glycosaminoglycans of the extracellular matrix. *J. Biol. Chem.* **275**, 31226–31232.

Zhao, W. Q., Li, H., Yamashita, K., Guo, X. K., Hoshino, T., Yoshida, S., Shinya, T., and Hayakawa, T. (1998). Cell cycle-associated accumulation of tissue inhibitor of metalloproteinases-1 (TIMP-1) in the nuclei of human gingival fibroblasts. *J. Cell Sci.* **111**, 1147–1153.

PART V

Matrix Metalloproteinases and Tissue Inhibitors of Metalloproteinases in Liver Fibrosis

CHAPTER 18

Gene Expression of Matrix Metalloproteinases in Acute and Chronic Liver Injuries

TERUMI TAKAHARA, YUTAKA YATA, LI PING ZHANG, AND AKIHARU WATANABE
Third Department of Internal Medicine, Toyama Medical and Pharmaceutical University, Toyama 930-0194, Japan

I. INTRODUCTION

During liver fibrosis, there is a pathologic accumulation of the extracellular matrix (ECM), which reflects alterations in the synthesis of matrix proteins, their degradation, or both (Arthur, 1995a,b, 2000). Matrix metalloproteinases (MMPs) specifically degrade extracellular matrix proteins and are involved in tissue remodeling during fibrotic and/or inflammatory processes. In addition, a number of studies have provided evidence for the involvement of MMPs and their inhibitors in various processes, including ovulation, embryogenic growth and differentiation, or tumor invasion and metastasis (Nagase,1999; Parks and Mecham,1998). Many different enzymes are involved in degrading extracellular matrix of the liver. Molecular biological progress has shown that several metalloproteinases in the liver share certain characteristics, such as some degree of sequence homology, a catalytic mechanism that depends on zinc, activation of the secreted zymogen form by proteolytic cleavage, and inhibition of the activated form by tissue inhibitors of metalloproteinases (TIMPs)(Brew *et al.*, 2000). The activity of MMPs

Copyright 2003, Elsevier Science (USA). All rights reserved.

may be regulated at the level of gene transcription, during proenzyme activation, or during binding of the proenzyme or active enzyme to specific inhibitors TIMPs (Birkedal-Hansen *et al.*, 1993). More than 23 MMPs (Nagase, 1999; Llano *et al.*, 1999; Pei, 1999; Park *et al.*, 2000) and 4 TIMPs (Brew *et al.*, 2000) have been identified thus far, and among them, MMP-1, -2, -3, -7, -8, -9, -10, -13, -14, -15, and -16 are expressed in the liver, especially in stellate cells, which are well known as main cell source of extracellular matrix, MMPs, and TIMPs (Friedman,1985; Arthur, 1995a,b, 2000; Knittel *et al.*, 1999, 2000). Moreover, the balance of MMPs and TIMPs is important, especially in the narrow spaces such as the pericellular or Disse's space. This chapter focuses on MMP expression in acute and chronic liver injury.

II. ACUTE LIVER INJURY

Wistar rats were given 50% CCl_4 in olive oil at a dose of 2 ml/kg of body weight via gastric tube and were sacrificed at 6 h, 12 h, 1 day, 2 days, 3 days, 5 days, and 7 days. The livers were examined routinely by light microscopy, and immunohistochemistry of type I collagen, type III collagen, type IV collage, and desmin was examined. The expression of collagens, MMPs, TIMPs, and transforming growth factor (TGF)-β1 genes was also studied by *in situ* hybridization and Northern blotting.

As reported previously (Takahara *et al.*, 1988, 1995b), sequential acute liver injury was detected, hepatocyte ballooning began 12 h after the intoxication around central veins, and hepatic necrosis occurred adjacent to the central veins. Zonal necrosis with infiltrating cells, such as macrophages and lymphocytes, was observed at 2 and 3 days around central veins. The necrotic change was diminished at 5 days and returned to normal at 7 days. Reticulin fibers were increased around the central areas at 3 days and decreased at 5 days. Many studies have been done to clarify the extracellular matrix-producing cells in acute liver injury and most implicate stellate cells, which are increased in number in the necrotic area. We have also detected the ECM in stellate cells by immunoelectron microscopy (Takahara *et al.*, 1988, 1992) and by *in situ* hybridization (Takahara *et al.*, 1995b; Nakatsukasa *et al.*, 1990).

We have also clarified the matrix degradation system because increased ECM all disappeared 7 days following acute liver injury. We extracted poly(A)$^+$ RNA from the liver, and Northern blotting was performed as described previously using ^{32}P-dCTP-labeled cDNA probes (Takahara *et al.*, 1995a,b). Figure 1a shows the time course of MMP and TIMP expression in acute liver injury induced by CCl_4. The relative mRNA expression is shown in Fig. 1b. At 6 h after the intoxication, the expression of MMP-13 (collagenase 3), which was seldom detected in normal liver, was clearly

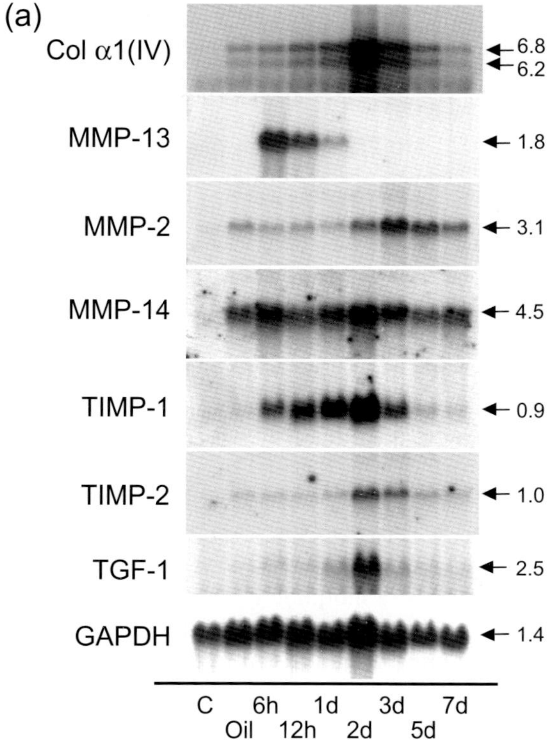

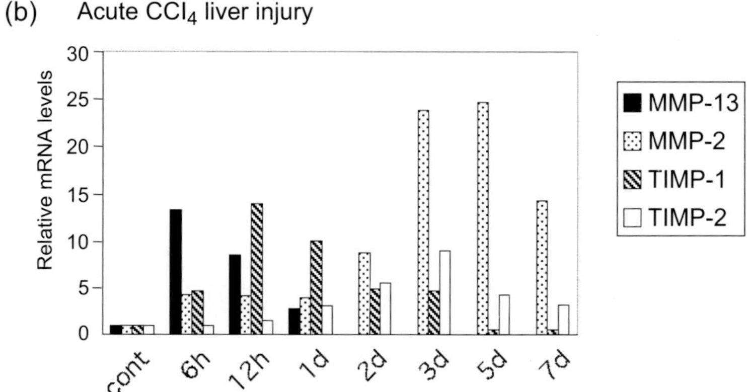

FIGURE 1 (a) Detection of MMPs, TIMPs, type IV collagen, and TGF-β mRNAs in liver specimens by Northern hybridization. poly(A)$^+$-enriched RNA was isolated from the liver of untreated control animals (C left), from animals treated with olive oil as controls (oil), or from animals subjected to acute CCl_4 injury and taken at the time points indicated. RNA (20 μg per lane) was electrophoresed, and the filters were hybridized with the indicated cDNAs. GAPDH was used as the internal control. (b) Relative abundance of each mRNA to control livers treated with olive oil only. Values are expressed as a fold increase compared to the control (cont).

increased, continued at 12 h, and decreased at 1 day. MMP-13 is known as a representative collagenase in rats because MMP-1 (collagenase 1) does not exist in this species. Then, MMP-2 (72 kDa type 4 collagenase, gelatinase A) is increased at 2 days, peaks at 3 days, and gradually decreases to the basal level at 7 days. MMP-14 (MT1-MMP) expression did not change during acute liver injury but was detected, constantly apart from a slight increase 6 h after intoxication. In addition, the expression of TIMP-1 and TIMP-2 was also increased in acute liver injury. Figure 1 shows the different time courses of expression between these two TIMPs. TIMP-1 was clearly detected 6 h after intoxication and peaked at 1 day and then decreased quickly at 3 days. In contrast, TIMP-2 was increased at 2 days with sustained expression until 5 days. Thus, in acute CCl_4 liver injury following a single dose, there was sequential induction of MMP-13, TIMP-1, MMP-2, and TIMP-2.

We also examined extracellular matrix components and TGF-β1, which is the principal fibrogenic cytokine in liver. Based on published data, ECM expression is increased at 2 and 3 days, together with TGF-β1. Our results also confirmed a parallel increase of type IV collagen expression at 2 days together with TGF-β1. The most severe inflammation and necrotic changes were observed 2 and 3 days following acute liver injury, and morphological changes of stellate cells were also detected at the same time. The increase of MMP-2, TIMP-1, TIMP-2, type IV collagen, and TGF-β1 during inflammation and recovery phases of acute liver injury resembles MMP/TIMP expression patterns and stellate cell morphology in chronic liver injury.

In contrast, early induction of tumor necrosis factor (TNF)-α was detected 6 and 12 h in acute CCl_4 liver injury (data not shown). We also detected early expression of TNF-α in acute galactosamine-induced liver injury, followed by the early induction of MMP-13 (Yata *et al.*, 1999). The induction of MMP-13 expression by TNF-α is also seen in cultured liver cells (Knittel *et al.*, 1999). In addition, the promotor of MMP-13 has a TPA responsive element, which is upregulated by TNF-α (Pendas *et al.*, 1997; Gutman *et al.*, 1999). The quick induction of MMP-13 after acute liver injury suggests that MMP-13 might be involved in liver regeneration and proliferation of hepatocytes. Iimuro *et al.* (1999) reported that adenoviral-mediated gene therapy of MMP-1 in rats increased hepatocyte proliferation, suggesting that collagenase may trigger hepatocyte proliferation. Thus, MMP-13 is expressed dramatically at an early stage, followed by MMP-2, which is reasonable because MMP-2 can degrade gelatin that is first generated by collagenase digestion. Thus, MMPs/TIMPs may be involved in the reconstitution of liver architecture in acute liver injury after destruction and absorption accompanying zonal necrosis and following regeneration associated with the degradation of excess ECM and hepatocyte proliferation. We have also examined MMP/TIMP expression in galactosamine-induced acute liver injury (Yata *et al.*, 1999), which resembles that of CCl_4.

III. REGENERATION AFTER PARTIAL HEPATECTOMY

To compare the changes of MMP/TIMP expression of acute liver injury with those of liver regeneration after partial hepatectomy, we examined the expressions of MMPs and TIMPs in liver regeneration. Wistar rats underwent 70% partial hepatectomy according to Higgins and Anderson (1931). poly(A)$^+$ RNA was extracted from the regenerated liver, Northern blotting was performed, and the expression was compared to sham-operated liver at 6 h, 12 h, 1 day, 2 days, 3 days, 5 days, 7 days, 10 days, and 14 days after hepatectomy. As already reported (Wolf *et al.*, 1992), hepatocyte proliferation was measured by immunostaining of PCNA, and the peak was at 36 h, with 25% of all cells positive. The liver weight was almost normal at 7 days, and liver histology was restored to normal at 14 days.

As shown in Fig. 2, MMP-13 expression was detected transiently at 6 h and disappeared afterward. When the same amount of mRNA of the liver in acute CCl_4 liver injury (6 h) and hepatectomy (6 h) was examined by Northern blotting, the relative expression of MMP-13 was almost 4-fold greater in acute liver injury (data not shown). MMP-2 expression was constant after hepatectomy, but it was increased 5-fold at 5 days and was sustained afterward. However, MMP-14 was increased dramatically during regeneration, peaking at 12 h and then decreasing gradually. The expression pattern of MMP-14 in hepatectomized liver was different from that in acute liver injury because MMP-14 did not change in acute liver injury. TIMP-1 expression was also increased transiently and then decreased, and the peak was at 2 days. However, the relative expression of TIMP-1 in hepatectomy to acute CCl_4 liver injury is also quite low. TIMP-2 expression was constant after hepatectomy, and the expression was increased to only 1.5-fold after 7 days. In this respect, the sequential expression of MMPs/TIMPs during the regenerative process lacking necrosis and inflammation was almost similar with that in acute liver injury, but expression was relatively lower in hepatectomized liver, in agreement with the report of Knittel *et al.* (2000). The expression of MMP-14 and MMP-13, although MMP-13 was rather low, was increased transiently at an early stage of regeneration, which might be important for proliferative signaling in hepatic cells. Reports indicate that MMP-14, apart from its role in MMP-2 activation, can directly degrade a number of matrix macromolecules at high efficiency. MMP-14 present on the cell surface focal degrades gelatin films (Dortho *et al.*, 1998). In addition, MMP-14 can activate MMP-13 (Knauper *et al.*, 1996).

The urokinase-type plasminogen activator (uPA) is an important factor for regeneration at early stages because it is expressed very rapidly after hepatectomy, and uPA knockout mice have impaired hepatic regeneration (Roselli *et al.*, 1998; Li *et al.*, 2001). uPA also activates latent prohepatocyte growth factor (pro-HGF)

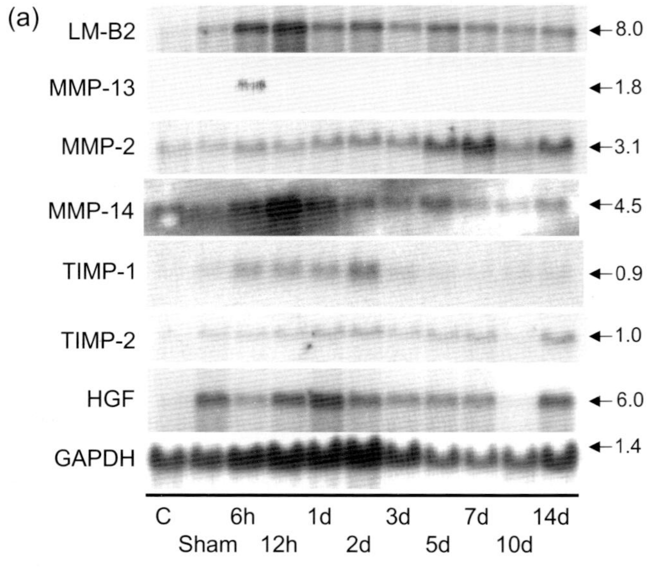

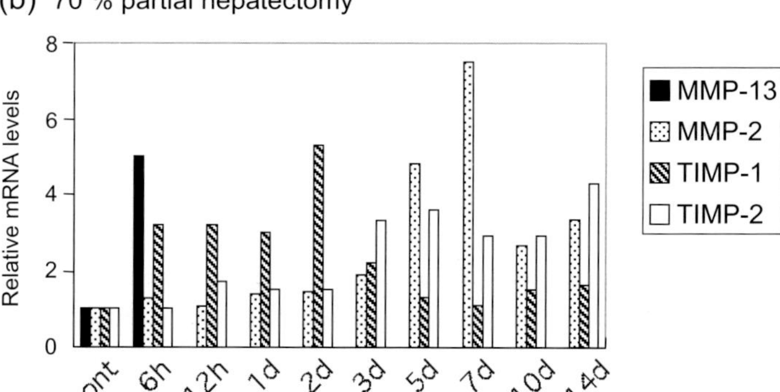

FIGURE 2 (a) Detection of MMPs, TIMPs, laminin B2, and HGF mRNAs in liver specimens by Northern hybridization. poly(A)$^+$-enriched RNA was isolated from the liver of untreated control animals (C left), sham-operated animals (sham), or from animals subjected to 70% partial hepatectomy and taken at the time points indicated. RNA (20 μg per lane) was electrophoresed, and the filters were hybridized with the indicated cDNAs. GAPDH was used as the internal control. (b) Relative abundance of each mRNAs to sham operation. Values are expressed as a fold increase compared to the control (cont).

to the active factor (Shimizu *et al.*, 2001). In addition, uPA can activate pro-MMPs (Nagase, 1997).

We examined the expression of extracellular matrix molecules during liver regeneration, and Fig. 2 shows laminin B2 as an example. The expression of laminin B2 was increased 6 and 12 h after hepatectomy. Compared to other extracellular matrix molecule, such as type I collagen and type III collagen, increased laminin expression was observed at an early stage of regeneration, possibly because laminin is an early signal for hepatocyte proliferation (Martinez-Hernandez and Amenta, 1995; Kim *et al.*, 1997). We also examined the expression of HGF, which was increased at 1 day.

IV. CHRONIC CCl_4 LIVER INJURY

Male Wistar rats, weighing 200 g, were subjected to chronic liver injury by subcutaneous injections of CCl_4/olive oil (1:1) twice weekly for 7, 9, or 12–14 weeks at a dose of 2 ml/kg of body weight and were sacrificed 7 days after the last injection. Liver histology and MMP/TIMP expression were examined using the same method as in acute CCl_4 liver injury. To evaluate gelatinase activity, liver samples were homogenized and the supernatants were examined by gelatin zymography as described previously (Takahara *et al.*, 1995a).

The animals showed fibrotic change at 7 weeks, and nodule formation was established at 12 weeks as reported previously (Takahara *et al.*, 1988, 1995).

Hybridizations with MMP-2, -13, -14 (MT-1-MMP), -15 (MT-2-MMP), and -16 (MT-3-MMP) cDNAs are shown in Figs. 3 and 4. During liver fibrosis, the expressions of MMPs except MMP-13 were increased and higher in intermediate stages than at early and late stages. The relative expression of MMP-2 was 7- to 12-fold higher than the control. Relative mRNA levels of MMP-14, -15, and -16 were also increased, as shown in Fig. 4. In contrast, the expression of MMP-13 is different from other MMPs. MMP-13 expression was increased at 7 and 9 weeks; however, it was suppressed at 12 weeks. Biochemical analyses of collagenase activity have been performed, and many studies described decreased collagenase activity during liver fibrosis (Maruyama *et al.*, 1982; Perez-Tamayo *et al.*, 1987), which coincides with MMP-13 gene expression. In comparison with MMP-13, TIMP-1 and TIMP-2 expression was increased during liver injury. So, the relative ratio of MMPs to TIMPs becomes lower during the process of liver fibrosis, and fibrosis is progressive in part through the net reduction in ECM degradation (Benyon *et al.*, 1996; Iredale, *et al.*, 1996).

In addition, gelatinase expression was examined by zymography (Takahara *et al.*, 1995a). Gelatinolytic activities at 92, 65, or 62 kDa were detected, which correspond to latent MMP-9, latent MMP-2, and active MMP-2, respectively. Total gelatinase expression rates were elevated from 13 to 28 times during the process of fibrosis. However, the amount of active 62-kDa gelatinase, which was not detected in control livers, increased during fibrosis and was maximal at 9 weeks. The ratio of 62 kDa/65 kDa+62 kDa was maximal at 9 weeks (~30%) and was decreased at 14 weeks (~16%). These results indicate that degradation due to the

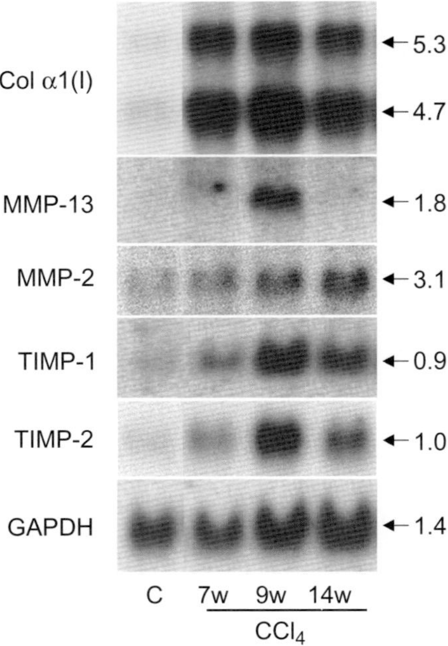

FIGURE 3 Detection of MMPs, TIMPs, and type I collagen in liver specimens by Northern hybridization. poly(A)$^+$-enriched RNA was isolated from the liver of control animals treated with olive oil (C left) or from animals subjected to chronic CCl_4 injury and taken at the time points indicated. RNA (20 μg per lane) was electrophoresed, and the filters were hybridized with the indicated cDNAs. GAPDH was used as the internal control.

active MMP-2 was maximal at the intermediate stage of fibrosis and was decreased in the cirrhotic stage.

We have shown that the expressions of three different membrane type MMPs, MMP-14, MMP-15, and MMP-16, are all increased during liver fibrosis. Figure 5 shows the expression of MMP-14, MMP-15, and MMP-16 in primary hepatic cells derived from normal rat liver. MMP-14 was expressed in stellate cells and Kupffer cells, MMP-15 in hepatocytes, and MMP-16 in stellate cells. However, it is not known which cell types are responsible for the expression of these genes during liver fibrosis.

V. LOCALIZATION OF MMP-2 AND MMP-14 IN HUMAN LIVER

Human liver samples of chronic hepatitis and liver cirrhosis were examined by *in situ* hybridization or immunohistochemistry as described previously (Takahara

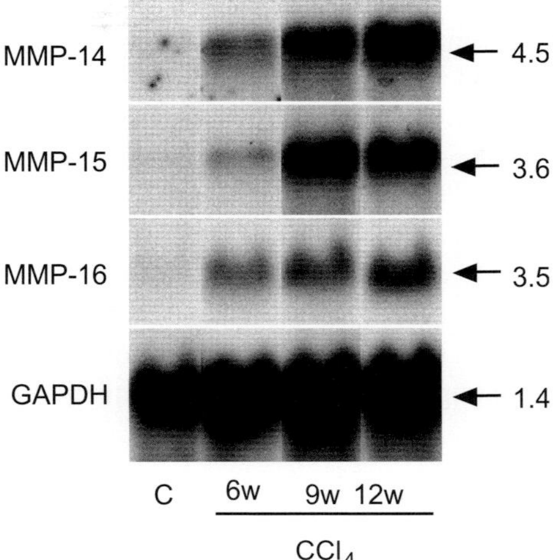

FIGURE 4 Detection of MMP-14, MMP-15, and MMP-16 in liver specimens by Northern hybridization as shown in Fig. 3. RNA (20 μg per lane) was electrophoresed, and the filters were hybridized with the indicated cDNAs.

et al., 1997). *In situ* hybridization was performed using cRNA probes labeled with ^{35}S-UTP. Antisense and sense probes were developed by *in vitro* transcription of MMP-2 and MMP-14 using T_3 and T_7 polymerase-binding sites of the cloning vector and an *in vitro* transcription kit. When *in situ* hybridization was combined with immunohistochemistry, sections were also incubated with α smooth muscle

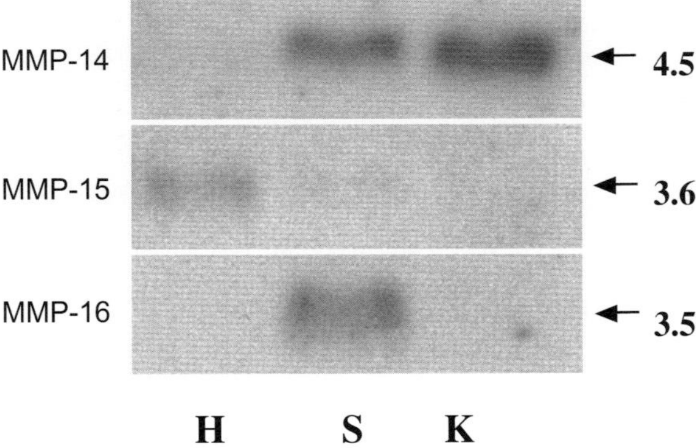

FIGURE 5 Detection of MMP-14, MMP-15, and MMP-16 in primary cultured cells from normal liver of Wistar rats. Cells were obtained by pronase/collagenase digestion and separated by density cushions using Nicodenz (Zhang *et al.*, 1999). H, hepatocyte; S, stellate cell; K, Kupffer cell.

actin (α-SMA) and were processed routinely using the ABC method. For immunoelectron microscopy, immunostained frozen sections were postfixed further in glutaraldehyde and OsO_4, dehydraded, and mounted in Epon. These samples were examined by transmission electron microscopy without staining.

In control liver sections, MMP-2 transcripts were not detected by *in situ* hybridization. However, silver granules of MMP-2 transcripts were observed in CH and LC in the elongated cells seen in the periportal and portal fibrous areas (Figs. 6a and 6b). Signals were clearly seen in fibrous septa of LC in elongated cells comprising fibers and around proliferating bile ducts and on capillary endothelial cells (Fig. 6a). In the lobules and periportal areas, positive granules were also detected in elongated cells (Fig. 6b). Combined *in situ* hybridization

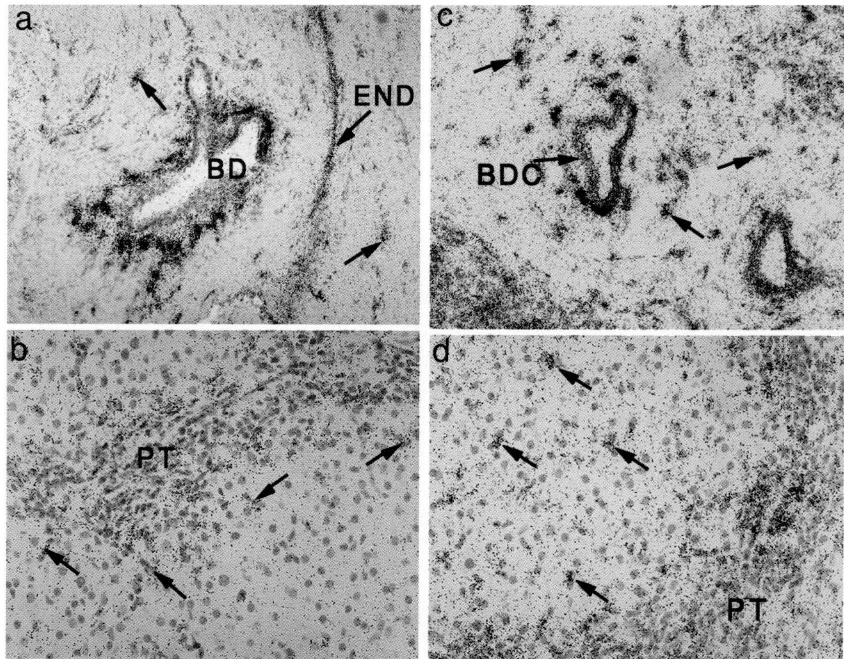

FIGURE 6 *In situ* hybridization of MMP-2 and MT1-MMP in liver sections using a ^{35}S cRNA probe. Silver granules are considered positive signals for transcripts. (a,b) Antisense probe for MMP-2, (c,d) antisense probe for MT1-MMP, (a,c) fibrous septa in LC, (b,d) periportal area of fibrotic liver. (a) MMP-2 gene transcripts are distributed on elongated cells around bile ducts (BD) and in the fibers (arrowheads). Signals are also detected on capillary endothelial cells (END). (b) Positive MMP-2 signals are found in many elongated cells (arrows) seen in the lobules and periportal areas. (c) MT1-MMP gene transcripts are clearly distributed on elongated cells (arrowheads), around bile ducts, and within fibers. Bile duct cells (BDC) are also positive for MT1-MMP. (d) Positive MT1-MMP signals are found in many elongated cells (arrows) and proliferating bile ducts (arrowhead) in fibrotic septa. Elongated cells in the lobules and periportal areas are also positive. PT, portal tract.

followed by immunohistochemistry simultaneously demonstrated MMP-2 mRNA and α-SMA in cells of the lobules, periportal areas, and fibrous septa, suggesting that these cells may represent stellate cells of the lobules and periportal areas.

Examination of normal liver by *in situ* hybridization for MMP-14 demonstrated only few specific granules in bile duct cells and elongated cells of portal tracts. However, MMP-14 transcripts were clearly found in CH and LC specimens. Silver granules were typically observed on the elongated cells found in and around portal fibers and periportal areas, on bile duct cells, and around bile duct cells (Figs. 6c and 6d). In addition, substantial staining was found on the elongated cells of the lobules and periportal areas (Fig. 6c). Combination *in situ* hybridization and immunohistochemistry also demonstrated that the MMP-14-positive elongated cells of the lobules and periportal area simultaneously expressed α-SMA, indicating that these cells may be stellate cells. Immunoelectron microscopy localized both MMP-2 and MMP-14 widely in sinusoidal cells, fibroblasts, and activated stellate cells. In addition, MMP-14 was clearly detected on the surface of extending processes of cytoplasm, suggesting MMP-2 activation on the surface of these cells (Nakahara *et al.*, 1997).

VI. REGULATION OF MMPs

Molecular biological studies have characterized promoter sequences of MMPs, which point to potential regulatory mechanisms genes by transcriptional factors downstream of cytokine or growth factor signaling. Many MMP promoters (except MMP-2 or MMP-14) have a TATA box and TPA responsive element (TRE). Activating protein (AP)-1 families such as c-Jun and c-Fos bind to the DNA-binding site of TRE, and its binding activities can be upregulated by inflammatory cytokines such as interleukin (IL)-1, TNF-α, and TGF-β1. In addition, these promoters have binding sites of Ets families (Ets-1, Ets-2, Egr, Elk-1, SP-1, E1AF), which regulate gene expression of MMPs together with AP-1 factors.

MMP-1 and MMP-13, which both degrade interstitial collagen, have similar promoter sequences (Angel *et al.*, 1987; Gutman *et al.*, 1990; Pendas *et al.*, 1997). However, MMP-1 contains a TGF-β inhibitory element (TIE), which suppresses the expression of MMP-1, whereas MMP-13 is upregulated by TGF-β because its promoter lacks this element, and AP-1 families are activated by TGF-β (Uria *et al.*, 1998).

As noted, the promotors of MMP-2 and MMP-14 do not have TATA box/TRE, as is observed in housekeeping genes, which are expressed universally in normal organs and many cell types. MMP-2 expression is not upregulated by TPA, but is expressed universally in many mesenchymal cells because expression is suppressed by epithelial cell-specific silencer elements (Frisch *et al.*, 1990). The MMP-14 promoter also does not have TATA box/TRE but a Egr-1-binding site; however, its role has not yet been clarified (Haas *et al.*, 1999).

VII. CONCLUSIONS

We have characterized the expression of MMPs/TIMPs in acute and chronic liver injury. During liver fibrosis, the relative ratio of MMPs to TIMPs is decreased. Thus fibrosis may be progressive because of a decrease in matrix degradation. Because chronic liver injury is caused by repeated tissue injury events, it is possible that the TGF-β-driven induction of MMP-2 and TIMPs dominates the tissue repair reaction, which lacks an appropriate TNF-mediated induction of MMP, as observed in the early phase of the single CCl_4 injury model. Activated stellate cells in fibrotic liver show a low expression of MMPs (except MMP-2) and a high expression of TIMPs and respond to TGF-β1 with further upregulation of TIMP, reflecting the MMPs/TIMPs expression pattern detected in fibrotic livers. However, expression of MMP-13 or MMP-14 during liver fibrosis has not been completely clarified. Further studies will elucidate the mechanisms of matrix degradation in liver fibrosis.

REFERENCES

Angel, P., Baumann, I., Stein, B., Delius, H., Rahmsdorf, H. J., and Herrlich, P. (1987). 12-O-tetradecanoyl-phorbol-13-acetate induction of the human collagenase gene is mediated by an inducible enhancer element located in the 5′-flanking region. *Mol. Cell Biol.* **7**, 2256–2266.

Arthur, M. J. (1995a). Collagenases and liver fibrosis. *J. Hepatol.* **22**, S43–S48.

Arthur, M. J. (1995b). Role of Ito cells in the degradation of matrix in liver. *J. Gastroenterol. Hepatol.* **10**, S57–S62.

Arthur, M. J. (2000). Fibrogenesis. II. Metalloproteinases and their inhibitors in liver fibrosis. *Am. J. Physiol. Gastrointest. Liver Physiol.* **279**, G245–G249.

Benyon, R. C., Iredale, J. P., Goddard, S., Winwood, P. J., and Arthur, J. P. (1996). Expression of tissue inhibitor of metalloproteinases 1 and 2 is increased in fibrotic human liver. *Gastroenterology* **110**, 821–831.

Birkedal-Hansen, H., Moore, W. G. I., Bodden, M. K., Windsor, L. J., Birkedal-Hansen, B., DeCarlo, A., and Engler, J. A. (1993). Matrix metalloproteinases: A review. *Crit. Rev. Oral Biol. Med.* **4**, 197–250.

Brew, K., Dinakarpandian, D., and Nagase, H. (2000). Tissue inhibitors of metalloproteinases: Evolution, structure and function. *Biochim. Biophys. Acta* **1477**, 267–283.

Dortho, M. P., Stanton, H., Butler, M., Atkinson, S. J., Murphy, G., and Hembry, R. M. (1998). MT1-MMP on the cell surface causes focal degradation of gelatine films. *FEBS Lett.* **421**, 159–164.

Friedman, S. L., Roll, F. J., Boyles, J., and Bissel, D. M. (1985). Hepatic lipocytes: The principal collagen producing cells of normal rat liver. *Proc. Natl. Acad. Sci. USA* **82**, 8681–8685.

Frisch, S. M., and Morisaki, J. H. (1990). Positive and negative transcriptional elements of human type IV collagenase gene. *Mol. Cell Biol.* **10**, 6524–6532.

Gutman, A., and Wasylyk, B. (1990). The collagenase gene promotor contains a TPA and oncogene-responsive unit encompassing the PEA3 and AP-1 binding sites. *EMBO J.* **9**, 2241–2246.

Higgins, G. M., and Anderson, R. M. (1931). Experimental pathology of the liver: Restoration of the liver of the white rat following partial surgical removal. *Arch. Pathol.* **12**, 186–202.

Iredale, J. P., Benyon, R. C., Arthur, M. J., Ferris, W. F., Alcolado, R., Winwood, P. J., Clark, N., and Murphy, G. (1996). Tissue inhibitor of metalloproteinase-1 messenger RNA expression is enhanced relative to interstitial collagenase messenger RNA in experimental liver injury and fibrisus. *Hepatology* **24**, 176–184.

Kim, T. H., Mars, W. M., Stolz, D. B., Petersen, B. E., and Michalopoulos, G. K. (1997). Extracellular matrix remodeling at the early stages of liver regeneration in the rat. *Hepatology* **26**, 470–476.

Knauper, V., Will, H., Lopez-Otin, C., Smith, B., Atkinson, S. J., Stanton, H., Hembry, R. M., and Murphy, G. (1996). Cellular mechanisms for human procollagenase-3 (MMP-13) activation: Evidence that MT1-MMP (MMP-14) and gelatinase a (MMP-2) are able to generate active enzyme. *J. Biol. Chem.* **271**, 17124–17131.

Knittel, T., Mehde, M., Grundmann, A., Saile, B., Scharf, J. G., and Ramadori, G. (2000). Expression of matrix metalloproteinases and their inhibitors during hepatic tissue repair in the rat. *Histochem. Cell Biol.* **113**, 443–453.

Knittel, T., Mehde, M., Kobold, D., Saile, B., Dinter, C., and Ramadori, G. (1999). Expression patterns of matrix metalloproteinases and their inhibitors in parenchymal and nonparenchymal cells of rat liver: Regulation by TNF-α and TGF-β1. *J. Hepatol.* **30**, 48–60.

Haas, T. L., Stitelman, D., Davis, S. J., Apte, S. S., and Madri, J. A. (1999). Egr-1 mediates extracellular matrix-driven transcription of membrane-type 1 matrix metalloproteinase in endothelium. *J. Biol. Chem.* **274**, 22679–22685.

Iimuro, Y., Fujii, H., Oe, S., Hirose, T., Morimoto, T., Brenner, D. A., and Yamaoka, Y. (1999). Overexpression of matrix metalloproteinase-1 attenuates liver fibrosis caused by thioacetamidein the rat. *Hepatology* **30**, 299A.

Li, W., Liang, X., Leu, J. I., Kovalovich, K., Ciliberto, G., and Taub, R. (2001). Global changes in interleukin-6-dependent gene expression patterns in mouse livers after partial hepatectomy. *Hepatology* **33**, 1377–1386.

Llano, E., Pendas, A. M., Freije, J. P., Nakano, A., Knauper, V., Murphy, G., and Lopez-Otin, C. (1999). Identification and characterization of human MT5-MMP, a new membrane-bound activator of progelatinase overexpressed in brain tumors. *Cancer Res.* **59**, 2570–2576.

Martinez-Hernandez, A., and Amenta, P. S. (1995). Liver regeneration: The extracellular matrix in hepatic regeneration. *FASEB J.* **9**, 1401–1410.

Maruyama, K., Feinman, L., Fainsilber, Z., Nakano, M., Okazaki, I., and Lieber, C. S. (1982). Mammalian collagenase increases in early alcoholic liver disease and decreases with cirrhosis. *Life Sci.* **30**, 1379–1384.

Nagase, H. (1997). Activation mechanisms of matrix metalloproteinases. *J. Biol. Chem.* **378**, 151–160.

Nagase, H., and Woessner, J. F., Jr. (1999). Matrix metalloproteinases. *J. Biol. Chem.* **274**, 21491–21494.

Nakahara, H., Howard, L., Thompson, E. W., Sato, H., Seiki, M., Yeh, Y., and Chen, W. T. (1997). Transmembrane/cytoplasmic domain-mediated membrane type 1-matrix metalloproteinase docking to invadopodia is required for cell invasion. *Proc. Natl. Acad. Sci. USA* **94**, 7959–7964.

Nakatsukasa, H., Nagy, P., Evarts, R. P., Hsia, C., Marsden, E., and Thorgeirsson, S. (1990). Cellular distribution of transforming growth factor-β1 and procollagen types I, III and IV transcripts in carbon tetrachloride-induced rat liver fibrosis. *J. Clin. Invest.* **85**, 1833–1842.

Park, H. I., Ni, J., Gerkema, F. E., Liu, D., Belozerov, V. E., and Sang, Q. X. (2000). Identification and characterization of human endometase (matrix metalloproteinase-26) from endometrial tumor. *J. Biol. Chem.* **275**, 20540–20544.

Parks, W. C., and Mecham, R. P. (1998). "Matrix Metalloproteinases." Academic Press, San Diego.

Pei, D. (1999). Leukolysin/MMP25/MT6-MMP: A novel matrix metalloproteinase specifically expressed in the leukocyte lineage. *Cell Res.* **9**, 291–303.

Pendas, A. M., Balbin, M., Llano, E., Jimenez, M. G., and Lopez-Otin, C. (1997). Structural analysis and promotor characterization of the human collagenase-3 gene (MMP-13). *Genomics* **40**, 222–233.

Perez-Tamayo, R., Montfort, I., and Gonzalez, E. (1987). Collagenolytic activativity in experimental cirrhosis of the liver. *Exp. Mol. Pathol.* **47**, 300–308.

Roselli, H. T., Su, M., Washington, K., Kerins, D. M., Vaughan, D. E., and Russell, W. E. (1998). Liver regeneration is transiently impaired in urokinase-deficient mice. *Am. J. Physiol.* **275**, G1472–G1479.

Shimizu, M., Hara, A., Okuno, M., Matsuno, H., Okada, K., Ueshima, S., Matsuo, O., Niwa, M., Akita, K., Yamada, Y., Yoshimi, N., Uematsu, T., Kojima, S., Friedman, S. L., Moriwaki, H., and Mori, H. (2001). Mechanism of retarded liver regeneration in plasminogen activator-deficient mice: Impaired activation of hepatocyte growth factor after Fas-mediated massive hepatic apoptosis. *Hepatology* **33**, 569–576.

Takahara, T., Kojima, T., Miyabayashi, C., Inoue, K., Sasaki, H., Muragaki, Y., and Ooshima, A. (1988). Collagen production in fat-storing cells after carbon tetrachloride intoxication in the rat: Immunoelectron microscopic observation of type I, type III collagens, and prolyl hydroxylase. *Lab. Invest.* **59**, 509–521.

Takahara, T., Nakayama, Y., Itoh, H., Miyabayashi, C., Watanabe, A., Sasaki, H., Inoue, K., Muragaki, Y., and Ooshima, A. (1992). Extracellular matrix formation in piecemeal necrosis: Immunoelectron microscopic study. *Liver* **12**, 368–380.

Takahara, T., Furui, K., Funaki, J., Nakayama, Y., Itoh, H., Miyabayashi, C., Sato, H., Seiki, M., Ooshima, A., and Watanabe, A. (1995a). Increased expression of matrix metalloproteinase-II in experimental liver fibrosis in rats. *Hepatology* **21**, 787–795.

Takahara, T., Furui, K., Yata, Y., Jin, B., Zhang, L. P., Nambu, S., Sato, H., Seiki, M., and Watanabe, A. (1997). Dual expression of matrix metalloproteinase-2 and membrane-type matrix metalloproteinase in fibrotic human livers. *Hepatology* **26**, 1521–1529.

Takahara, T., Sollberg, S., Muona, P., and Uitto, J. (1995b). Type VI collagen gene expression in experimental liver fibrosis: Quantitation and spatial distribution of mRNAs, and immunodetection of the protein. *Liver* **15**, 78–86.

Uria, A., Jimenez, M. G., Balbin, M., Freije, J. M., and Lopez-Otin, C. (1998). Differential effects of transforming growth factor-b on the expression of collagenase-1 and collagenase-3 in human fibroblasts. *J. Biol. Chem.* **273**, 9769–9777.

Wolf, H. K., and Michalopoulos, G. K. (1992). Hepatocyte regeneration in acute fulminant and non-fulminant hepatitis: A study of proliferating cell nuclear antigen expression. *Hepatology* **15**(4), 707–713.

Yata, Y., Takahara, T., Furui, K., Zhang, L. P., and Watanabe, A. (1999). Expression of matrix metalloproteinase-13 and tissue inhibitor of metalloproteinase-1 in acute liver injury. *J. Hepatol.* **30**, 419–425.

Zhang, L. P., Takahara, T., Yata, Y., Furui, K., Jin, B., Kawada, N., and Watanabe, A. (1999). Increased expression of plasminogen activator and plasminogen activator inhibitor during liver fibrogenesis of rats: Role of stellate cells. *J. Hepatol.* **31**, 703–711.

CHAPTER 19

Role of Tissue Inhibitors of Metalloproteinases (TIMPs) in the Progression and Regression of Liver Fibrosis: Regulation of TIMP-1 Gene Expression in Hepatic Stellate Cells

MICHAEL J. P. ARTHUR

Liver Research Group, Division of Infection, Inflammation, and Repair, University of Southampton, Hampshire SO16 6YD, United Kingdom

Liver fibrosis is characterized by the activation of hepatic stellate cells (HSC) with an increased synthesis of matrix proteins and decreased extracellular matrix degradation. Matrix metalloproteinases (MMPs) are the principal family of enzymes involved in the cleavage and degradation of matrix proteins. Because of their ability to destroy normal matrices and cause tissue injury, the extracellular activity of MMPs is carefully regulated. They are secreted initially as proenzymes, which are converted by a number of mechanisms to active forms following cleavage of a C-terminal propeptide. Active MMPs can be inhibited by the action of tissue inhibitors of metalloproteinases (TIMPs), which bind to their catalytic site. Studies of culture-activated HSC have demonstrated that these cells are a major source of TIMP-1 and TIMP-2 and that their secretion leads to net inhibition of MMP activity. Studies in human liver disease and in animal models of liver fibrosis indicate that TIMPs play a major role in regulating matrix degradation in the liver. Alteration of the balance between TIMPs and MMPs contributes significantly to the progression of liver fibrosis. In animal models of regression of liver fibrosis, improvement occurs because of the combination of apoptosis of activated hepatic stellate cells and by the increased degradation of the fibrillar liver matrix.

Copyright 2003, Elsevier Science (USA). All rights reserved.

Current evidence suggests that matrix degradation occurs as a consequence of a decreased expression of TIMP-1, which permits MMPs with activity against fibrillar collagens (collagenases) to exert their degradative action. Because of the importance of HSC-derived TIMP-1 in the pathogenesis of liver fibrosis, we have performed detailed studies of TIMP-1 gene regulation in these cells. We have found that the persistent upregulation of this gene observed in activated HSC is mediated via a combination of a Jun-D homodimer binding to an AP-1 site and also to an as yet uncharacterized transcription factor binding to a nearby novel regulatory element in the promoter region that we have named upstream TIMP element-1 (UTE-1). Taken together, these observations suggest that manipulating pathways of matrix degradation in liver offers a novel potential therapeutic strategy for the treatment of liver fibrosis.

I. TISSUE INHIBITORS OF METALLOPROTEINASES

The tissue inhibitors of metalloproteinases are a family of low molecular weight glycoproteins that play a critical role in regulating the extracellular activity of the MMP family of matrix-degrading enzymes. There are now four TIMPs described (TIMP-1 to TIMP-4), each encoded for by a separate gene. There is significant nucleotide sequence homology between the different members of the TIMP gene family, with each giving rise to a polypeptide of approximately 21 kDa with conservation of 12 cysteine residues, which via the formation of disulfide bridges lead to a characteristic six loop structure. The different TIMPs are then glycosylated to different degrees, giving rise to molecular masses ranging up to 28 kDa for TIMP-1. The sequence variation and the glycosylation profile confer the differing properties observed for each of the TIMPs.

The most important property of TIMPs is their ability to bind with 1:1 stoichiometry to the active catalytic site of MMPs and inhibit their ability to degrade the matrix in extracellular space. With some exceptions, all TIMPs are able to bind noncovalently to and inhibit all metalloproteinases with differing inhibitory constants. This binding is reversible under physiological conditions. TIMP-1 is particularly important in the inhibition of MMP-1 and MMP-9, whereas TIMP-2 is important in the inhibition of MMP-2. One exception to this rule is the inability of TIMP-1 to inhibit some of the membrane-type MMPs, including MT1-MMP, MT2-MMP, and MT5-MMP (Will *et al.*, 1996). This may have some importance for matrix degradation in fibrotic liver, as MT1-MMP is able to function as an interstitial collagenase (see later).

In addition to regulating the activity of MMPs, TIMPs have a diverse range of other biological functions and are involved in many different aspects of cell and matrix biology (for a detailed review, see Brew *et al.*, 2000). Their other key functions and properties include an ability to bind to matrix proteins (e.g., TIMP-3), involvement in the activation of some prometalloproteinases (e.g., TIMP-2 and pro-MMP2) or prevention of pro-MMP activation (e.g., TIMP-1 inhibition of pro-MMP1 activation by stromelysin/MMP3), regulation of cell proliferation (TIMP-1, TIMP-2), inhibition or promotion of apoptosis (TIMP-1 and TIMP-3, respectively), and inhibition of angiogenesis.

With this wide range of properties, TIMPs therefore play a key role in important physiological events, such as tissue remodeling, revascularization, and repair, and they have been widely implicated in the pathogenesis of a wide variety of inflammatory, fibrotic, and neoplastic disease processes. This chapter provides an update on the current state of knowledge of the role of TIMPs in the progression and regression of liver fibrosis and describes progress in our understanding of the regulation of TIMP-1 gene expression in hepatic stellate cells. Evidence for the involvement of TIMPs in liver fibrosis has been gathered from studies in cell culture, in animal models of liver disease, and in the study of human liver disease.

II. TIMP EXPRESSION IN CULTURED HEPATIC STELLATE CELLS

In primary culture, both rat and human HSC undergo a major phenotypic change (over the first 7–14 days) and develop the characteristic features of myofibroblasts. The culture-activated cells closely resemble those described in fibrotic liver *in vivo*, in animal models of liver fibrosis, and in human liver disease. These cells express α smooth muscle actin and demonstrate major increases in their expression of procollagen α1. From previous studies, it is also clear that these cells can (under certain conditions) express a number of metalloproteinases, including MMP-2, MMP-3, MMP-9, MMP-1/MMP-13, and MT1-MMP (Arthur *et al.*, 1989; Benyon *et al.*, 1999; Iredale *et al.*, 1995, 1996; Theret *et al.*, 1999; Vyas *et al.*, 1995), as well as their activation systems (Benyon *et al.*, 1999; Leyland *et al.*, 1996). HSC, particularly in their activated myofibroblastic phenotype, therefore have the potential ability to degrade extracellular matrix in the liver.

This cell culture model of hepatic stellate cell activation has been used to study their ability to regulate extracellular matrix degradation through the expression of TIMPs. In freshly isolated and early cultured HSC, TIMP-1 and TIMP-2 mRNA is initially undetectable, but expression of both TIMP-1 and TIMP-2 increases progressively as the cells become culture activated (Benyon *et al.*, 1996; Iredale *et al.*, 1992). In contrast, interstitial collagenase mRNA (MMP-13 in rat, MMP-1 in human) was detectable in freshly isolated and early cultured cells, but decreased significantly with HSC activation. Both TIMP-1 and TIMP-2 proteins could be detected by reverse zymography in supernatants of culture-activated HSC, and this combination of events led to a profound overall net inhibition of metalloproteinase activity. If TIMPs were separated by affinity chromatography from their respective MMPs, metalloproteinase activity was demonstrated to increase 25-fold in HSC culture media. These data give a clear indication of the overall importance of TIMP secretion in regulating matrix degradation by these cells. These results have led to the hypothesis that TIMPs released by activated HSC contribute significantly to the progression of liver fibrosis (see Fig. 1).

In culture-activated HSC, TIMP-1 gene expression is regulated in response to a number of exogenous cytokines and growth factors. The most important is probably transforming growth factor-β (TGF-β_1), which increases TIMP-1 but decreases TIMP-2 mRNA in cultured HSC (Herbst *et al.*, 1997). In other systems, TGF-β_1 has been shown to simultaneously increase the expression of TIMP-1 and collagen I, while decreasing the expression of MMP-1 and MMP-3, thus favoring the net accumulation of fibrillar interstitial matrix in tissues (Edwards *et al.*, 1987; Mackay *et al.*, 1992; Overall *et al.*, 1989). The growth regulatory cytokine oncostatin-M has also been described to increase the TIMP-1 mRNA content of cultured human HSC (Levy *et al.*, 2000). In chronically injured liver, the influence of these cytokines and growth factors in regulating HSC TIMP and collagen gene expression relative to expression of MMPs has a key influence on the progression of liver fibrosis.

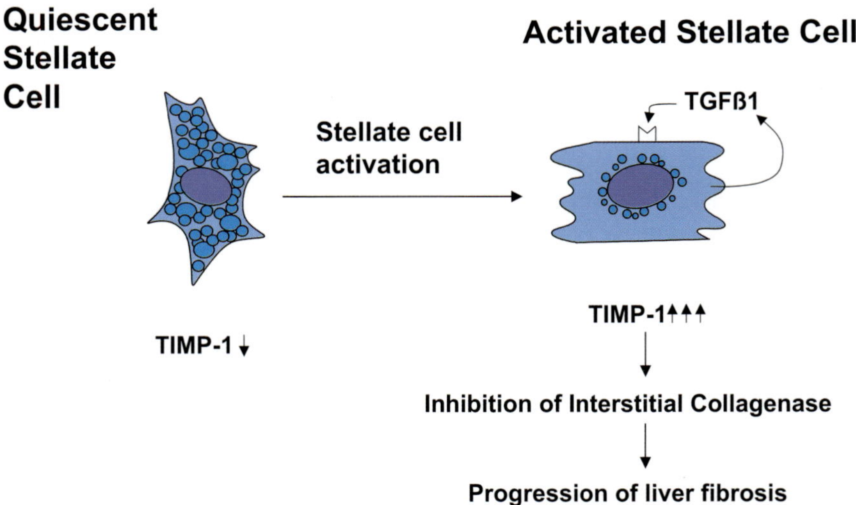

FIGURE 1 Hypothesis: Following activation to a myofibroblastic phenotype, hepatic stellate cells synthesize and secrete TIMP-1, which inhibits matrix degradation and contributes to the progression of liver fibrosis.

III. TIMP EXPRESSION BY HEPATOCYTES AND KUPFFER CELLS

Hepatocytes have also been reported to synthesize and release TIMPs, particularly in response to acute phase stimuli (Kordula *et al.*, 1992; Roeb *et al.*, 1993, 1994). TIMP-1 synthesis has been reported for HepG2 cells (a hepatoblastoma-derived cell line) in response to TGF-β_1 and interleukin-6 with similar observations subsequently reported in cultured primary rat hepatocytes. TIMP-3 expression has also been reported for hepatocytes (Knittel *et al.*, 1999), but this study confirmed that HSC were the major source of TIMP-1, whereas TIMP-2 could be found in cultured Kupffer cells. These data suggest that the expression of TIMPs can occur in other liver cells, but the current concensus is that in chronic liver disease and in progressive liver fibrosis, activated HSC are the most important source of TIMPs, particularly TIMP-1. Regulation of MMP activity in the pericellular space of HSCs, i.e., the space of Disse, is likely to be of key importance in determining the fate of freshly synthesized and secreted collagens and other matrix proteins in injured liver.

IV. ROLE OF TIMPs IN THE PROGRESSION OF LIVER FIBROSIS

Studies of the role of TIMPs in liver fibrosis have examined their expression in both human liver disease and in animal models of liver injury. In human liver disease, we have described a 3- to 5-fold increase in mRNA for TIMP-1 and TIMP-2 in total hepatic RNA prepared from fibrotic compared to normal liver in primary biliary cirrhosis, sclerosing cholangitis, biliary atresia, and autoimmune chronic active hepatitis (AICAH) (Benyon *et al.*, 1996; Iredale *et al.*, 1995). Changes in TIMPs were not accompanied by any significant change in the levels of mRNA for MMP-1 (interstitial collagenase), with the exception of AICAH. TIMP-1 protein levels measured in whole liver homogenates by ELISA were also increased 3- to 5-fold in these liver diseases. Other groups have found similar results; in needle biopsy specimens of chronic active hepatitis, the hepatic TIMP-1 level reflected the stage of disease with reported increases of 2.2-fold in stage 2A disease, 2.9-fold in stage 2B, and 4.1-fold in cirrhosis (Murawaki *et al.*, 1997). Increased mRNA for TIMP-1 and TIMP-2 has also been detected in fibrotic human liver by *in situ* hybridization, with the majority of transcripts detected in sinusoidal liver cells, which were probably HSC (Herbst *et al.*, 1997). These results support the hypothesis that the increased expression of TIMPs by activated HSC leads to net inhibition of degradation of fibrillar matrix proteins (collagen types I and III) and that this is most evident in the more advanced stages of human liver disease.

Studies in animal models of liver fibrosis provide additional information about the temporal pattern of expression of TIMPs in the early stages of liver injury and as liver fibrosis progresses. We have examined the expression of TIMP-1, MMP-13 (rat interstitial collagenase), and procollagen I in total RNA prepared from rat models of liver injury (Iredale *et al.*, 1996). The models examined included both carbon tetrachloride liver injury (8 weeks of repeated CCl_4 exposure) and the bile duct ligation model of biliary injury and fibrosis. Similar results were found in both models, with an increase in hepatic TIMP-1 mRNA content detected within 6 h of injury, becoming markedly increased 72 h after injury, and remaining elevated throughout the time course under study. In the CCl_4 model, increased TIMP expression was still evident after 8 weeks of intermittent exposure (CCl_4 given every 72 h). Throughout the time course, there was no significant increase in the expression of MMP-13 in either model. In both models, the increase in expression of TIMP-1 preceded the observed increase in the expression of procollagen I, which occurred at day 3. Using the same models (CCl_4 and bile duct obstruction), others have studied TIMP-1 and TIMP-2 mRNA transcripts in liver by *in situ* hybridization (Herbst *et al.*, 1997). Results of these studies were similar, with increased detection of TIMP-1 and TIMP-2 as early as 1–3 h after CCl_4 exposure, with the majority of transcripts located in activated HSC. TIMP-1 and TIMP-2 expression were increased significantly at 72 h and remained detectable throughout the time course of the study.

The expression of TIMP-2 and TIMP-3 has also been studied in the rat bile duct ligation model of liver fibrosis using Northern analysis (Kossakowska *et al.*, 1998). In this study, mRNA for TIMP-2 and TIMP-3 became detectable 10 days after bile duct ligation, whereas mRNA for TIMP-1 was detectable by day 2 and increased progressively to day 10. In this study, TIMP activities could not be detected in whole liver homogenate by reverse zymography, but this is a difficult technique to perform in complex biological samples and this result is difficult to interpret.

Other studies have performed experimental manipulations to increase the expression of TIMPs to determine if they accelerate the progression of liver fibrosis. In the first of these, CCl_4 treatment was combined with repeated (weekly) episodes of induction of the acute-phase response (APR) (Greenwel and Rojkind, 1997). This resulted in the APR-mediated induction of both procollagen I and TIMP-1 gene expression and led to accelerated development of liver fibrosis in this model. Compelling evidence for a key role for TIMP-1 in the pathogenesis of progressive liver fibrosis is provided by studies of transgenic mice that overexpress TIMP-1 in liver (Yoshiji *et al.*, 2000). In these mice, TIMP-1 gene expression was specifically driven in hepatocytes by use of a construct that incorporated the albumin promoter. In the absence of liver injury, there was no significant phenotype in these mice compared to their wild-type counterparts. Following administration of CCl_4, however, liver fibrosis occurred more rapidly and was more extensive in the TIMP-1 transgenic mice. This demonstrates that increased matrix synthesis (following CCl_4) and increased inhibition of matrix degradation (mediated by the TIMP-1 transgene) combined in this model to promote matrix deposition following liver injury. Although this is accentuated in this transgenic model, the implication is that similar mechanisms are relevant to progressive liver fibrosis in general.

The combined evidence from all of the studies reported to date suggest that in the progressive phase of liver fibrosis, matrix proteins are synthesized and secreted into an environment where the balance between TIMPs and MMPs favors inhibition of matrix degradation. The majority of evidence points to a key role for TIMP-1, but this may just reflect the fact that this molecule has been studied more frequently than TIMP-2 or TIMP-3. In studies that have included these latter molecules, similar results have been found and the relative importance of the different TIMPs in the pathogenesis of liver fibrosis remains to be determined. The field now requires careful analysis of mice in which TIMP genes have been deleted. Our initial attempts to study TIMP-1 null mice failed to find any significant phenotype after CCl_4 liver injury (J. P. Iredale and M. J. P. Arthur, unpublished observations), but in these mice there was significant upregulation of TIMP-2 in the liver. We speculate that this may reflect a compensatory mechanism conferring a survival advantage following the TIMP-1 gene deletion. On current evidence, it appears that further progress will require the development of conditional knockout mice in which the TIMP gene deletion may be affected after birth and following normal embryonic development.

V. ROLE OF TIMPs IN THE REGRESSION OF LIVER FIBROSIS

In recent years there has been increasing recognition of the dynamic nature of liver fibrosis and a clearer understanding of the fact that liver fibrosis may regress when liver injury is halted. Studies in patients with viral hepatitis who have responded to antiviral therapy and in patients undergoing surgical correction of biliary obstruction have shown that this can be an important aspect of human liver disease (Hammel *et al.*, 2001; Kweon *et al.*, 2000). We and others have therefore studied the mechanisms involved in the regression of liver fibrosis in animal models of reversibility of liver fibrosis.

In rat CCl_4 liver injury (doses administered 3× per week), there is significant fibrosis by the 4th week of exposure and this progresses such that early cirrhosis is established by 6 to 8 weeks. If, however, the exposure to CCl_4 is halted, liver fibrosis reverses and liver histology returns to normal. We have examined this series of events in rat liver after exposure to repeated doses of CCl_4 for 4 weeks, followed by a 4-week period of experimental observation (Iredale *et al.*, 1998). In this model of reversible liver fibrosis, there is a rapid reduction in the hepatic content of hydroxyproline and significant improvement in the histological extent of liver fibrosis within 10 days. By day 28, liver histology and hydroxyproline content have returned to near normal.

The observed regression of liver fibrosis is mediated by two interrelated mechanisms. First, the activated hepatic stellate cells undergo apoptosis within a short period after cessation of liver injury. Within 3 days of stopping CCl_4 treatment, the number of activated HSC was reduced by 50%, and by 10 days, the number of activated HSC was nearly back to the control level. In these studies, apoptosis was evaluated by counting apoptotic bodies in the fibrous bundles and by demonstrating TUNEL staining in α smooth muscle actin-positive-activated HSC. Second, the regression of liver fibrosis occurred as a result of increased matrix degradation, which was mediated by an alteration in the balance between TIMPs and MMPs. Our studies demonstrated that mRNA for procollagen I and TIMP-1 was reduced significantly by day 3 of recovery and was barely detectable by day 10. In contrast, MMP-13 (rat interstitial collagenase) mRNA continued to be detectable throughout the time course of the regression of liver fibrosis. These changes in the balance between TIMPs and MMPs were accompanied by a five-fold increase in hepatic collagenase activity, measured in whole liver homogenate.

Watanable and colleagues (2000) have also studied mechanisms of matrix degradation during the regression of liver fibrosis. Using a combination of polymerase chain reaction (RT-PCR) and *in situ* hybridization, they also demonstrated a decrease in the expression of TIMP-1, but found an increase in the expression of MMP-13 for a limited time window (days 2–5) in association with cells at the interface between fibrotic septae and adjacent liver parenchyma. While some of these cells may have been HSC, their work (see Chapter 20) suggests that MMP-13 may also be derived from liver stem cells.

The mechanisms by which TIMP-1 expression decreases in this rat model of regression of liver fibrosis are of considerable interest. As TIMP-1 is expressed predominantly by activated HSC, the explanation may rest in the observation that these cells are rapidly disappearing as a consequence of apoptosis. The possibility that altered TIMP-1 expression and HSC apoptosis are interrelated is raised by studies in other systems, which have shown that TIMP-1 can inhibit apoptosis in B lymphocytes (Guedez *et al.*, 1998) and breast epithelial cells (Li *et al.*, 1999). This raises the possibility that the decreased HSC expression of TIMP-1 (in response to cessation of liver injury) may be part of the mechanism that promotes the subsequent apoptosis of HSC and contributes to the resolution of liver fibrosis via this dual effect. These possibilities are currently under investigation by Iredale and colleagues in Southampton, with preliminary data indicating that TIMP-1 can act as a survival factor for hepatic stellate cells.

VI. REGULATION OF TIMP-1 GENE EXPRESSION IN ACTIVATED HEPATIC STELLATE CELLS

Because the accumulated evidence indicates that TIMP-1 derived from HSC plays a key role in the inhibition of matrix degradation in liver, we have examined the regulation of TIMP-1 gene expression in these cells. We initially mapped a minimal promoter using a strategy of transient transfection of primary HSC cultures with a series of TIMP-1 promoter–CAT constructs (Bahr *et al.*, 1999). These studies described a minimal promoter to –120 bp that contained a critical AP-1-binding site between –80 and –120. We have subsequently concentrated on nuclear transcription factors derived from activated HSC that are not present in quiescent cells and that bind to this AP-1 site in the TIMP-1 promoter. Initial studies revealed that nuclear extracts prepared from activated HSC (days 5–14) contained AP-1-binding activities of high mobility when analyzed by the electromobility shift assay (EMSA) that were not due to either c-Fos or c-Jun. The observed binding was sequence specific for the AP-1 site, and a combination of supershift and Western analyses indicated that the bound complex contained a combination of Jun-D and Fra-2 transcription factors. In fact, the only Jun family member detected in activated HSC nuclei was Jun-D. Mutation of the AP-1 site also led to a marked inhibition of promoter activity in transiently transfected HSC, indicating that this site was of key importance for TIMP-1 gene transcription. In subsequent work (Smart *et al.*, 2001), we have further characterized the role of Jun-D in TIMP-1 gene transcription in activated HSC. We have shown that activated HSC cotransfected with expression vectors for Jun family proteins demonstrate increased TIMP-1 promoter activity with Jun-D, but not with either c-Jun or Jun-B. The addition of a dominant-negative Jun-D, which is engineered to lack the transactivation domain and which competes with the native Jun-D, led to significant inhibition of TIMP-1 promoter activity in activated HSC. We also demonstrated that cotransfection with an engineered Jun-D homodimer (two Jun-D molecules bound together via EB1 homerization domains) led to significant increases in TIMP-1 promoter activity. These studies point to a key role for the Jun-D transcription factor in the regulation of TIMP-1 gene transcription in activated HSC. This observation that TIMP-1 gene regulation differs in activated HSC compared to previous reports in fibroblasts (Logan *et al.*, 1996) led us to search for other novel regulatory mechanisms. Using DNase I footprinting, we have identified a novel regulatory element in the minimal TIMP promoter, which we have called upstream TIMP element -1 (Trim *et al.*, 2000). Our studies have shown that the UTE-1 sequence is essential for TIMP-1 gene regulation in activated HSC, that this effect is sequence specific, and that in activated HSC it is mediated by binding of a 30-kDa nuclear protein (detected by UV cross-linking and southwestern analysis), which we are currently in the process of characterizing further. Using the UTE-1 sequence in EMSA studies, we have shown that nuclei from several

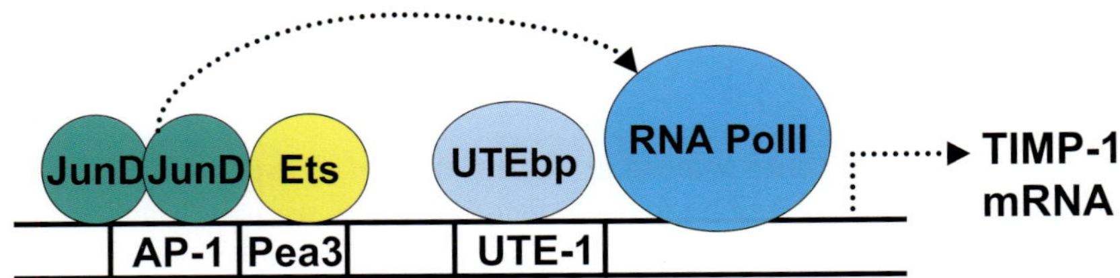

FIGURE 2 An overview of the key mechanisms of TIMP-1 gene regulation in activated hepatic stellate cells. The combination of the Jun-D homodimer binding to an AP-1 site and a binding protein (yet to be characterized) on the UTE-1 site is the key mechanism driving increased and persistent TIMP-1 gene expression in activated hepatic stellate cells.

different cell types (e.g., Jurkat T cells) are also able to form various protein–DNA complexes of several different mobilities, including some that match those found in activated HSC. These data demonstrate that UTE-1-binding is likely to be universally important in TIMP-1 gene regulation. We have also found a UTE-1-binding complex of lower mobility in the nuclei of quiescent HSC. This observation and the results obtained in other cells suggest that there are likely to be a number of different nuclear proteins that bind to UTE-1, which raises the possibility of a family of binding proteins that could act as either repressors or promoters of TIMP-1 gene regulation.

These studies of TIMP-1 gene regulation in HSC have therefore led to a number of novel observations that have shed light on TIMP-1 gene regulation in general. Our current overview of the transcriptional regulation of TIMP-1 gene expression in activated HSC is shown in Fig. 2. These studies open up the possibility of interfering with these TIMP-1 regulatory pathways as a potential therapeutic strategy for liver fibrosis.

VII. CONCLUSIONS

In chronic liver injury, liver fibrosis occurs as a consequence of both increased matrix synthesis and decreased matrix degradation. Current evidence indicates that decreased matrix degradation is of major pathological importance in progressive liver fibrosis and that this is mediated by an alteration in the balance between TIMPs and MMPs in the liver. TIMP-1, TIMP-2, and TIMP-3 have all been implicated. Although the majority of work points to a key role for TIMP-1, it is also possible that TIMP-2 and TIMP-3 also play a significant role. The major cellular source of TIMP-1 is activated HSC, but other liver cells (hepatocytes and Kupffer cells) may also participate. The overall importance of TIMPs is emphasized by the analysis of animal models of regression of liver fibrosis, in which matrix degradation occurs rapidly following a profound reduction in TIMP-1 gene expression. Finally, studies of TIMP-1 gene regulation in activated HSC have led to the finding that the persistent expression of TIMP-1 is mediated by a combination of transcription factors that bind to the AP-1 site and a novel binding element that we have identified and named UTE-1. These advances in our understanding of the role of altered matrix degradation in liver fibrosis lead to the possibility of novel therapeutic strategies for the treatment of chronic liver disease.

REFERENCES

Arthur, M. J. P., Friedman, S. L., Roll, F. J., and Bissell, D. M. (1989). Lipocytes from normal rat liver release a neutral metalloproteinase that degrades basement membrane (type IV) collagen. *J. Clin. Invest.* **84**, 1076–1085.

Bahr, M. J., Vincent, K., Arthur, M. J. P., Fowler, A. V., Smart, D. E., Wright, M. C., Clark, I. M., Benyon, R. C., Iredale, J. P., and Mann, D. A. (1999). Control of the tissue inhibitor of metalloproteinases-1 promoter in culture-activated rat hepatic stellate cells: Regulation by activator protein-1 DNA binding proteins. *Hepatology* **29**, 839–848.

Benyon, R. C., Hovell, C. J., Gaca, M. D. A., Jones, E. H., Iredale, J. P., and Arthur, M. J. P. (1999). Progelatinase A is produced and activated by rat hepatic stellate cells and promotes their proliferation. *Hepatology* **30**, 977–986.

Benyon, R. C., Iredale, J. P., Goddard, S., Winwood, P. J., and Arthur, M. J. P. (1996). Expression of tissue inhibitor of metalloproteinases-1 and -2 is increased in fibrotic human liver. *Gastroenterology* **110**, 821–831.

Brew, K., Dinakarpandian, D., and Nagase, H. (2000). Tissue inhibitor of metalloproteinases: Evolution, structure and function. *Biochim. Biophys. Acta* **1477**, 267–283.

Edwards, D. R., Murphy, G., Reynolds, J. J., Whitham, S. E., Docherty, A. J., Angel, P., and Heath, J. K. (1987). Transforming growth factor beta modulates the expression of collagenase and metalloproteinase inhibitor. *EMBO J.* **6**, 1899–1904.

Greenwel, P., and Rojkind, M. (1997). Accelerated development of liver fibrosis in CCl_4-treated rats by the weekly induction of acute phase response episodes: Upregulation of alpha1(I) procollagen and tissue inhibitor of metalloproteinase-1 mRNAs. *Biochim. Biophys. Acta* **1361**, 177–184.

Guedez, L., Stetler-Stevenson, W. G., Wolff, L., Wang, J., Fukushima, P., Mansoor, A., and Stetler-Stevenson, M. (1998). In vitro suppression of programmed cell death of B cells by tissue inhibitor of metalloproteinases-1. *J. Clin. Invest.* **102**, 2002–2010.

Hammel, P., Couvelard, A., O'Toole, D., Ratouis, A., Sauvanet, A., Flejou, J. F., Degott, C., Belghiti, J., Bernades, P., Valla, D., Ruszniewski, P., and Levy, P. (2001). Regression of liver fibrosis after biliary drainage in patients with chronic pancreatitis and stenosis of the common bile duct. *N. Engl. J. Med.* **344**, 418–423.

Herbst, H., Wege, T., Milani, S., Pellegrini, G., Orzechowski, H. D., Bechstein, W. O., and Neuhaus P, G. A., Schuppan, D. (1997). Tissue inhibitor of metalloproteinase-1 and -2 RNA expression in rat and human liver fibrosis. *Am J Pathol* **150**, 1647–1659.

Iredale, J. P., Benyon, R. C., Arthur, M. J., Ferris, W. F., Alcolado, R., Winwood, P. J., Clark, N., and Murphy, G. (1996). Tissue inhibitor of metalloproteinase-1 messenger RNA expression is enhanced relative to interstitial collagenase messenger RNA in experimental liver injury and fibrosis. *Hepatology* **24**, 176–184.

Iredale, J. P., Benyon, R. C., Pickering, J., McCullen, M., Northrop, M., Pawley, S., Hovell, C. and Arthur, M. J. P. (1998). Mechanisms of spontaneous resolution of rat liver fibrosis: Hepatic stellate cell apoptosis and reduced hepatic expression of metalloproteinase inhibitors. *J. Clin. Invest.* **102**, 538–549.

Iredale, J. P., Goddard, S., Murphy, G., Benyon, R. C., and Arthur, M. J. P. (1995). Tissue inhibitor of metalloproteinase-1 and interstitial collagenase expression in autoimmune chronic active hepatitis and activated human hepatic lipocytes. *Clin. Sci.* **89**, 75–81.

Iredale, J. P., Murphy, G., Hembry, R. M., Friedman, S. L., and Arthur, M. J. P. (1992). Human hepatic lipocytes synthesize tissue inhibitor of metalloproteinases-1 (TIMP-1): Implications for regulation of matrix degradation in liver. *J. Clin. Invest.* **90**, 282–287.

Knittel, T., Mehde, M., Kobold, D., Saile, B., Dinter, C., and Ramadori, G. (1999). Expression patterns of matrix metalloproteinases and their inhibitors in parenchymal and non-parenchymal cells of rat liver: Regulation by TNF-alpha and TGF-beta1. *J. Hepatol.* **30**, 48–60.

Kordula, T., Guttgemann, I., Rosejohn, S., Roeb, E., Osthues, A., Tschesche, H., Koj, A., Heinrich, P. C., and Graeve, L. (1992). Synthesis of tissue inhibitor of metalloproteinase-1 (TIMP- 1) in human hepatoma cells (HepG2): Upregulation by interleukin-6 and transforming growth factor-beta1. *FEBS Lett.* **313**, 143–147.

Kossakowska, A. E., Edwards, D. R., Lee, S. S., Urbanski, L. S., Stabbler, A. L., Zhang, C. L., Phillips, B. W., Zhang, Y., and Urbanski, S. J. (1998). Altered balance between matrix metalloproteinases and their inhibitors in experimental biliary fibrosis. *Am. J. Pathol.* **153**, 1895–1902.

Kweon, Y.O., Goodman, Z. D., Dienstag, J. L., Schiff, E. R., Brown, N. A., Burkhardt, E., Schoonhoven, R., Brenner, D. A., and Fried, M. W. (2001). Decreasing fibrogenesis: An immunohistochemical study of paired liver biopsies following lamivudine therapy for chronic hepatitis B. *J. Hepatol.* **35**(6), 749–755.

Levy, M. T., Trojanowska, M., Reuben, A., and Oncostatin, M. (2000). A cytokine upregulated in human cirrhosis, increases collagen production by human hepatic stellate cells. *J. Hepatol.* **32**, 218–226.

Leyland, H., Gentry, J., Arthur, M. J. P., and Benyon, R. C. (1996). The plasminogen-activating system in hepatic stellate cells. *Hepatology* **24**, 1172–1178.

Li, G. Y., Fridman, R., and Kim, H. R. C. (1999). Tissue inhibitor of metalloproteinase-1 inhibits apoptosis of human breast epithelial cells. *Cancer Res.* **59**, 6267–6275.

Logan, S. K., Garabedian, M. J., Campbell, C. E., and Werb, Z. (1996). Synergistic transcriptional activation of the tissue inhibitor of metalloproteinases-1 promoter via functional interaction of AP-1 and Ets-1 transcription factors. *J. Biol. Chem.* **271**, 774–782.

Mackay, A. R., Ballin, M., Pelina, M. D., Farina, A. R., Nason, A. M., Hartzler, J. L., and Thorgeirsson, U. P. (1992). Effect of phorbol ester and cytokines on matrix metalloproteinase and tissue inhibitor of metalloproteinase expression in tumor and normal cell lines. *Invasion Metast.* **12**, 168–184.

Murawaki, Y., Ikuta, Y., Idobe, Y., Kitamura, Y., and Kawasaki, H. (1997). Tissue inhibitor of metalloproteinase-1 in the liver of patients with chronic liver disease. *J. Hepatol.* **26**, 1213–1219.

Overall, C. M., Wrana, J. L., and Sudek, J. (1989). Independent regulation of collagenase, 72kD progelatinase, and metalloendoproteinase inhibitor expression in human fibroblasts by transforming growth factor-beta. *J. Biol. Chem.* **264**, 1860–1869.

Roeb, E., Graeve, L., Hoffmann, R., Decker, K., Edwards, D. R., and Heinrich, P. C. (1993). Regulation of tissue inhibitor of metalloproteinases-1 gene expression by cytokines and dexamethasone in rat hepatocyte primary cultures. *Hepatology* **18**, 1437–1442.

Roeb, E., Graeve, L., Mullberg, J., Matern, S., and Rosejohn, S. (1994). TIMP-1 protein expression is stimulated by IL-1 beta and IL-6 in primary rat hepatocytes. *FEBS Lett.* **349**, 45–49.

Smart, D. E., Vincent, K. J., Arthur, M. J. P., Eickelberg, O., Castellazzi, M., Mann, J., and Mann, D. A. (2001). JunD regulates transcription of the tissue inhibitors of metalloproteinases-1 and interleukin-6 genes in activated stellate cells. *J. Biol. Chem.* **276**, 24414–24421.

Theret, N., Lehti, K., Musso, O., and Clement, B. (1999). MMP2 activation by collagen I and concanavalin A in cultured human hepatic stellate cells. *Hepatology* **30**, 462–468.

Trim, J. E., Samra, S. K., Arthur, M. J., Wright, M. C., McAulay, M., Beri, R., and Mann, D. A. (2000). Upstream tissue inhibitor of metalloproteinases-1 (TIMP-1) element-1, a novel and essential regulatory DNA motif in the human TIMP-1 gene promoter, directly interacts with a 30-kDa nuclear protein. *J. Biol. Chem.* **275**, 6657–6663.

Vyas, S. K., Leyland, H., Gentry, J., and Arthur, M. J. P. (1995). Rat hepatic lipocytes synthesize and secrete transin (stromelysin) is expressed in early primary culture. *Gastroenterology* **109**, 889–898.

Watanabe, T., Niioka, M., Hozawa, S., Kameyama, K., Hayashi, T., Arai, M., Ishikawa, A., Maruyama, K., and Okazaki, I. (2000). Gene expression of interstitial collagenase in both progressive and recovery phase of rat liver fibrosis induced by carbon tetrachloride. *J. Hepatol.* **33**, 224–235.

Will, H., Atkinson, S. J., Butler, G. S., Smith, B., and Murphy, G. (1996). The soluble catalytic domain of membrane type 1 matrix metalloproteinase cleaves the propeptide of progelatinase A and initiates autoproteolytic activation: regulation by TIMP-2 and TIMP-3. *J. Biol. Chem.* **271**, 17119–17123.

Yoshiji, H., Kuriyama, S., Miyamoto, Y., Thorgeirsson, U. P., Gomez, D. E., Kawata, M., Yoshii, J., Ikenaka, Y., and Noguchi, R., Tsujinoue, H., Nakatani, T., Thorgeirsson, S. S., and Fukui, H. (2000). Tissue inhibitor of metalloproteinases-1 promotes liver fibrosis development in a transgenic mouse model. *Hepatology* **32**, 1248–1254.

CHAPTER 20

Stem Cells Expressing Matrix Metalloproteinase-13 mRNA Appear during Regression Reversal of Hepatic Cirrhosis

TETSU WATANABE,* MAKI NIIOKA,† SHIGENARI HOZAWA,† YOSHIHIKO SUGIOKA,* MASAO ARAI,† KATSUYA MARUYAMA,‡ HIDEYUKI OKANO,§ AND ISAO OKAZAKI†

*Departments of *Environmental Health and †Community Health, Tokai University School of Medicine, Kanagawa 259-1193, Japan; ‡Department of Internal Medicine, Kurihama National Hospital, Kanagawa 239-0841, Japan and §Department of Physiology, Keio University School of Medicine, Tokyo 160-8582, Japan*

The reversibility of hepatic fibrosis has been doubted. However, Perez-Tamayo (1965) described an interesting case in his review as follows: "Regression of human cirrhosis is not adequately documented although it can occur. Through the courtesy of Dr. Bennett I have examined a surgical biopsy specimen from a patient with hemochromatosis and advanced cirrhosis who was treated with repeated bleeding and P^{32}. A new surgical biopsy 10 years later revealed completely normal hepatic structure. The patient died later, however, with primary carcinoma of the liver, and at autopsy there was no cirrhosis." His description stimulated us to study the mechanism of the degradation of extracellular matrix in the fibrotic liver. Even today it is an interesting question whether liver cirrhosis is reversible or not, because liver cirrhosis is defined as irreversible fibrosis with distorted structure of the liver (Okuda, 2001).

This chapter describes a mechanism that underlies the degradation of newly formed fibrous bands in CCl_4-induced liver fibrosis and cirrhosis and demonstrates that neural progenitor cells derived from hematopoietic stem cells express interstitial collagenase in the recovery phase of experimental cirrhosis. The availability of hematopoietic stem cells for the treatment of human liver fibrosis or cirrhosis is also discussed.

Copyright 2003, Elsevier Science (USA). All rights reserved.

I. REVERSIBILITY OF HEPATIC FIBROSIS

The reversibility of liver fibrosis has been observed experimentally (Hutterer *et al.*, 1964; Jacques and McAdams, 1957; Morrione and Levine, 1967; Okazaki *et al.*, 1974; Quinn and Higginson, 1965; Takada *et al.*, 1967) since Cameron and Karunaratne (1936) reported this phenomenon in the carbon tetrachloride-induced cirrhosis model after removal of the toxic agent. Hepatic fibrosis induced by thioacetamide (Quinn and Higginson, 1965), α-naphthylisothiocyanate (Morrione and Levine, 1967), ethionine (Hutterer *et al.*, 1964), and a choline-deficient diet (Takada *et al.*, 1967), as well as by the ligation of bile duct (Jacques and McAdams, 1957; Zimmerman *et al.*, 1992), have been demonstrated to be reversible after removal of the causative agent.

The reversibility of liver fibrosis has also been observed after effective therapy among patients with alcoholic liver disease (Rubin and Hutterer, 1967), hemochromatosis (Perez-Tamayo, 1965; Rojkind and Dunn, 1979), primary biliary cirrhosis (Kaplan *et al.*, 1997; Poupon *et al.*, 1991), secondary biliary cirrhosis (Bunton and Cameron, 1963; Hammel *et al.*, 2001), hepatitis B virus (HBV)-related chronic hepatitis and cirrhosis (Maruyama *et al.*, 1981; Schiff *et al.*, 2000), hepatitis C virus (HCV)-related hepatitis and cirrhosis (Arai *et al.*, 1996; Bonis *et al.*, 1997; Dufour *et al.*, 1998; Lau *et al.*, 1998; Poynard *et al.*, 2000; Shiffman *et al.*, 1999; Shiratori *et al.*, 2000; Sobesky *et al.*, 1999), hepatitis D virus (HDV)-related hepatitis (Lau *et al.*, 1999), autoimmune hepatitis (Dufour *et al.*, 1997), and other liver diseases (Rojkind and Dunn, 1979).

The reversibility of liver fibrosis in patients with alcoholic cirrhosis and alcoholic fatty liver with fibrosis has been observed after alcohol abstinence (Rubin and Hutterer, 1967). Maruyama *et al.* (1992) reported that the histologically documented reversibility of liver fibrosis in alcoholic liver diseases was closely correlated with a decrease in serum fibrosis markers of type IV collagen. This work investigated patients who were hospitalized in an intensive care setting in the National Alcohol Treatment Center in Japan. Among out-clinic patients, however, regression of fibrosis due to alcohol consumption has not been clearly demonstrated (Lieber, 1999).

Rojkind and Dunn (1979) described 18 cases that recovered from liver cirrhosis. These included 15 patients with hemochromatosis treated with bleeding and P^{32}, 1 patient with Wilson's disease with cirrhosis, 1 patient with galactosemia with cirrhosis, and 1 patient with cirrhosis after intestinal bypass. In these cases, diagnosis and improvement of cirrhosis were confirmed by needle biopsy or surgical wedge biopsy or autopsy. Recovery of histologically documented cirrhosis to absent or minimal fibrosis with normal architecture still has several questions, including sampling error and claims on some cases. Rojkind and Dunn concluded that at least three cases showed complete cure of cirrhosis even taking these caveats into account. Because the period of observation was less than 5 years, the degradation of fibrosis seemed to occur in a surprisingly short time if the causes were removed.

In cases of liver fibrosis occurring in association with chronic pancreatitis and stenosis of the common bile duct, the regression of liver fibrosis was reported after biliary drainage (Hammel *et al.*, 2001). Recovery was also reported in an infant with cirrhosis in congenital biliary atresia after surgery (Bunton and Cameron, 1963). Reversibility of hepatic fibrosis and cirrhosis have been reported in some patients with primary biliary cirrhosis who respond to treatment (Kaplan *et al.*, 1997; Poupon *et al.*, 1991).

The authors demonstrated the disappearance of liver fibrosis during recovery from HBV-positive subacute hepatitis with massive fibrosis (Maruyama *et al.*, 1981). The improvement of fibrosis in patients with chronic hepatitis B after successful long-term lamivudine administration has also been reported in a larger cohort (Schiff *et al.*, 2000).

Clinical experience with α-interferon therapy for HCV antibody-positive chronic hepatitis and cirrhosis has documented the regression of hepatic fibrosis in many cases (Bonis *et al.*, 1997; Dufour *et al.*, 1998; Lau *et al.*, 1998; Poynard *et al.*, 2000; Shiffman *et al.*, 1999; Shiratori *et al.*, 2000; Sobesky *et al.*, 1999). The close correlation between the histologically documented grade of liver fibrosis and the serum type IV collagen levels has been reported in patients who responded to interferon therapy (Arai *et al.*, 1996; Capra *et al.*, 1993; Ishibashi *et al.*, 1996; Teran *et al.*, 1994). The authors observed higher serum matrix metalloproteinase (MMP)-1 levels and the higher ratio of MMP-1 to tissue inhibitors for MMPs (TIMP)-1 or TIMP-2 in patients who responded well to interferon therapy than those of nonresponders (Arai *et al.*, 1996). The effect of interferon on the recovery of liver fibrosis is not only due to the eradication of HCV, but may also result from a direct and indirect action of this cytokine on fibrogenesis (Arai *et al.*, 1996). That is, interferon itself decreases the production of collagen (Czaja *et al.*, 1989; Jimenez *et al.*, 1984) and attenuates the fibrogenic activity induced by transforming growth factor (TGF)-β (Castilla *et al.*, 1991; Varga *et al.*, 1990). Interferon can also induce collagenase activity (Berman and Duncan, 1989; Duncan and Berman, 1989).

It is now generally accepted that reversibility of liver fibrosis or cirrhosis occurs under some conditions if the cause of liver damage is removed or adequately treated.

II. EXPERIMENTAL MODEL

The experimental model the authors used for elucidation of the mechanism of spontaneous recovery from liver fibrosis is based on the original method described by Cameron and Karunaratne (1936). Wistar male rats were injected intraperitoneally with 1 mg/kg of a 30% solution of CCl_4 in olive oil twice a week for 8 or 12 weeks. These rats were sacrificed at 48 h, 5 days, and 7 days after the last injection of CCl_4 treatment for 8 or 12 weeks. Control rats were injected with the same amount of olive oil intraperitoneally. Hematoxylin and eosin staining, Azan Mallory staining, and silver staining were applied for the histological examination.

At 48 h after the last injection of 8-week CCl_4 intoxication, the liver showed moderate fibrosis, linking neighboring tracts completely and separating the lobules into small sublobules. In some portions, the newly formed fibrous bands connecting the neighboring portal triads extended within lobules with multiple-branched bands, which connected to the perihepatocellular fibrosis. Regenerative nodules were not seen (Okazaki *et al.*, 1974, 1982; Okazaki and Maruyama, 1974, 1980; Watanabe *et al.*, 2000, 2001).

Five days after the last injection of the 8-week treatment, the newly formed fibrous bands became thin or almost disappeared, and the hepatocytes recovered from fatty metamorphosis. The resorption of fibrous bands seemed to start at the side of perihepatocellular fibrosis. Enlargement of the portal triad still remained, but the extended multiple branched bands within lobules changed into fine needle-like bands or disappeared around hepatocytes without fatty metamorphosis. Fibrous bands during the resolution process showed a characteristic feature, somewhat like a pencil with a long needle. The findings at 7 days after the last injection of the 8-week treatment were almost the same as those at 5 days described earlier, although it is remarkable that the fibrous tissue decreased with the recovery of hepatocytes (Okazaki *et al.*, 1974, 1982; Okazaki and Maruyama, 1980; Watanabe *et al.*, 2000, 2001).

At 48 h after the last injection of 12-week CCl_4 intoxication, advanced fibrosis with regenerative nodules surrounded by thick fibrous bands was seen (Fig. 1A) (Okazaki *et al.*, 1973, 1974, 1982; Okazaki and Maruyama, 1974, 1980; Watanabe *et al.*, 2000, 2001). At 5 and 7 days after the last injection of 12-week CCl_4 intoxication, the spontaneous resolution of newly formed fibrous bands was observed even if the distortion of the liver structure with various sizes of regenerative nodules was still present (Fig. 1B) (Okazaki *et al.*, 1974, 1982; Okazaki and Maruyama, 1974, 1980; Watanabe *et al.*, 2000, 2001).

To quantitate the degree of fibrosis, the content of hydroxyproline in the homogenate of rat liver was estimated (Okazaki *et al.*, 1982), and Azan-stained sections were used for morphometric analysis (Watanabe *et al.*, 2000, 2001). The hydroxyproline content and fibrotic rates in the rat liver on days, 5 and 7 in the recovery phase in both 8- and 12-week CCl_4 intoxication decreased significantly.

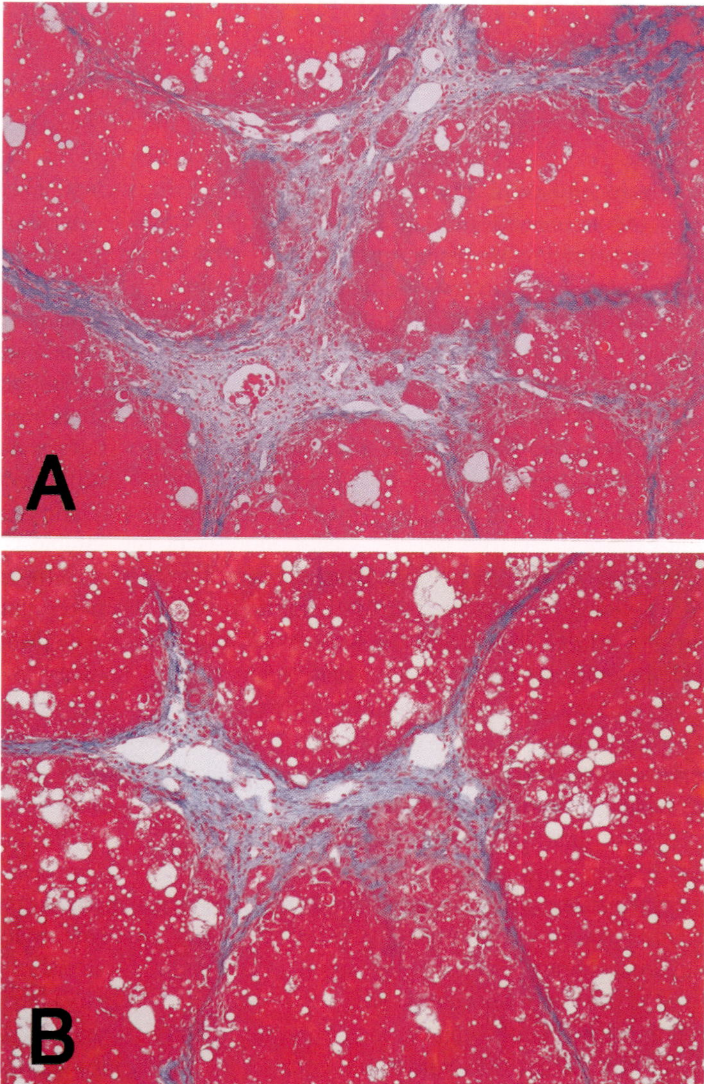

FIGURE 1 Azan staining of rat liver at day 2 (A) and at day 5 (B) after the last injection of CCl_4 for 12 weeks. Newly formed broad bands and regenerative nodules are seen (A), but remarkable resolution of fibrous bands is seen after the discontinuation of toxic reagents (B). Original magnification: 25×.

III. COLLAGENASE IN RECOVERY FROM LIVER CIRRHOSIS

Reversibility of liver fibrosis means the destruction of newly formed fibrous bands in the liver. Since Gross and Lapiere (1962) discovered tadpole collagenase, interstitial collagenase has been isolated from skin, uterus, granulocytes, macrophages, and other organs and cells (Harris and Krane, 1974; Okazaki and Maruyama, 1980). This enzyme is synthesized *de novo*, is excreted extracellularly, and attacks the collagen molecule at three-quarters distance from the amino-terminal end under neutral pH. This collagenase plays an important role in remodeling the extracellular matrix in growth, inflammation, tumor development, and other physiological and pathological conditions (Harris and Krane, 1974; Okazaki and Maruyama, 1980). The authors had postulated the presence of interstitial collagenase in the liver and demonstrated collagenase activity in experimental liver fibrosis for the first time (Okazaki and Maruyama, 1974). Specifically, collagenolysis was identified around the explant of a slice of rat fibrotic liver on a collagen gel film (Fig. 2), demonstrating the typical collagenase attack pattern against neutral salt-extracted collagen by disc electrophoresis of the sample collected from the reacted collagen gels (Okazaki and Maruyama, 1974). This disc gel revealed β^A (3/4-length of β chain), α^A (3/4-length of α chain), and α^B (1/4-length of α chain) products, which are the typical products of limited collagen degradation by mammalian collagenase on polyacrylamide gels.

Using this tissue culture technique and a semiquantitative assay, the authors measured collagenase activity in the recovery from liver fibrosis and cirrhosis. Contrary to our expectations, the activity decreased at days 5 and 7 after 4-, 8-, and 12-week CCl_4 intoxication (Maruyama *et al.*, 1976).

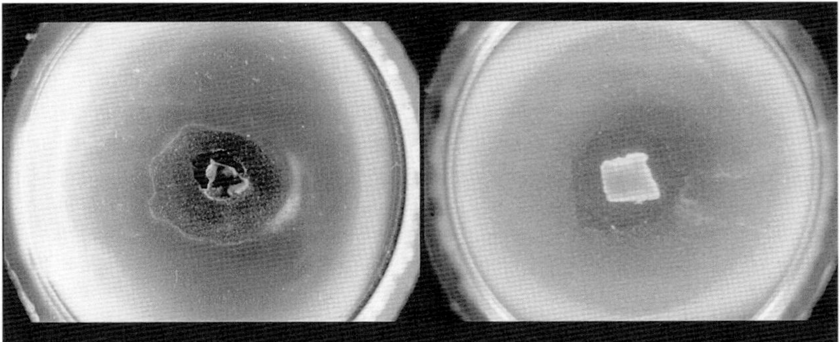

FIGURE 2 Collagenolysis by tadpole tail skin (left) and rat liver at day 2 after the last injection of CCl_4 for 3 weeks (right) on collagen film on which they were cultured for 3 days. Methods reported previously (Maruyama *et al.*, 1978; Okazaki and Maruyama, 1974).

In the following study, the authors used type I collagen that had been purified from acid-soluble collagens extracted from rabbit skin (Maruyama *et al.*, 1981, 1982). The enzyme source was the liver homogenate of baboons fed an alcohol-containing diet. The reaction mixture containing type I collagen as a substrate, the homogenate of baboon liver as an enzyme, and 3 m*M* *p*-chloromercuribenzoate to inhibit thiol proteinase activity and to convert procollagenase into the active form was incubated under neutral pH at 27°C, assayed by viscometer, and monitored by disc electrophoresis (Maruyama *et al.*, 1981). Levels of the reaction products of β^A and α^A and their ratio showed the predicted degradation pattern of interstitial collagenase. Using this assay method and the same substrate, the authors demonstrated increased activity of interstitial collagenase in the early stage of hepatic fibrosis in baboons that had been fed alcohol chronically (Maruyama *et al.*, 1981, 1982) and in patients with alcoholic hepatic fibrosis (Maruyama *et al.*, 1982).

The authors also measured the level of collagenase activity in the liver of rats fed alcohol chronically (Okazaki *et al.*, 1977, 1983) and in rats that had been treated chronically with carbon tetrachloride (Okazaki *et al.*, 1982). However, the level of collagenase activity in these rats was quite low and there was a slight, but not significant, increase in collagenase activity in the early stage of fibrosis. Other researchers subsequently demonstrated the presence of increased collagenase activity in the early stage of liver fibrosis and reduced collagenase activity in advanced fibrosis (Carter *et al.*, 1982; Lindblad and Fuller, 1983; Montfort *et al.*, 1990). The activity, however, in the recovery phase from liver fibrosis and cirrhosis was not accompanied by upregulation of interstitial collagenase even by this sensitive and quantitative assay (Okazaki *et al.*, 1982).

IV. GENE EXPRESSION OF MMPs AND TIMPs IN RECOVERY FROM LIVER CIRRHOSIS

The metabolism of extracellular matrix is regulated by matrix metalloproteinases in close association with tissue inhibitors for MMPs. In the case of rats, sequence homology analysis revealed that except for human MMP-13, there is no sequence in rats that shows more than 90% similarity with the sequence of MMP-1 in humans (Freije *et al.*, 1994; Knauper *et al.*, 1996). As the cDNA of rat MMP-1 has not yet been cloned, rat interstitial collagenase should be considered to be the rat homologue of human MMP-13. MMPs other than MMP-1, MMP-8, and MMP-13 cannot degrade type I collagen, which is very stable, and a net deposition of type I collagen has been observed in progressive hepatic fibrosis (Arthur, 1995, 2000; Friedman, 1993, 1997; Okazaki *et al.*, 1998, 1999, 2000, 2001; Rojkind, 1999).

Arthur *et al.* (1989, 1992) reported that hepatic stellate cells secrete a neutral metalloproteinase that can degrade type IV collagen (a component of the basement membrane), which appears to be MMP-2. Both MMP-2, a potent gelatinase, and membrane type 1 (MT1)-MMP, an activator of MMP-2, can cleave native type I collagen (Aimes and Quigley, 1995; Ohuchi *et al.*, 1997), but with less efficiency than MMP-1 (Aimes and Quigley, 1995). Interstitial collagenase is a key enzyme involved in the degradation of fibrosis, and its expression may be important in the recovery phase of liver fibrosis. The authors have examined whether the interstitial collagenase gene is upregulated in the recovery phase. The second question has been to identity cells expressing interstitial collagenase even if the participation of this enzyme in the recovery of liver fibrosis is less than that of MMP-2 and MT1-MMP in total activity. The authors succeeded in answering these questions by clarifying the cells expressing MMP-13 mRNA in the recovery from liver fibrosis and cirrhosis.

A. Gene Expression of MMP-13

The authors have demonstrated upregulation of MMP-13 mRNA at day 5 after the last injection of 8-week CCl_4 intoxication, i.e., in the early recovery phase of liver fibrosis, by reverse transcription–polymerase chain reaction (RT-PCR) (Watanabe *et al.*, 2000), whereas downregulation of MMP-13 is noted with the progress of liver fibrosis (Watanabe *et al.*, 2000) (Figs. 3A and 3B). Transcripts of MMP-13 were analyzed by RT-PCR, followed by Southern blot analysis. Although the downregulation of MMP-13 in the progress of liver fibrosis has been reported by Iredale *et al.* (1996, 1998), the authors' report is the first to demonstrate the upregulation of MMP-13 in the very early recovery phase, i.e., at day 5 after the last injection of CCl_4 for 8 weeks. Iredale *et al.* (1996, 1998) did not observe any increase in MMP-13 mRNA transcription but demonstrated an increase in TIMP mRNA transcripts in rat hepatic fibrosis induced by chronic CCl_4 intoxication. They hypothesized that the increase of TIMPs relative to MMP-13 may lead to the deposition of type I collagen. In the recovery phase, they did not observe increased MMP-13 gene expression. They postulated the importance of this balance of the downregulation of TIMPs without any increase in MMP-13. Differences in both the method and the time schedule used to detect MMP-13 mRNA between their experiment and ours may have affected the results. First, because the level of MMP-13 expression was quite low, we used RT-PCR analysis instead of conventional Northern blot hybridization or ribonuclease protection assay to detect the transcripts. Second, the authors examined the rat liver at day 5 after the last injection of 8-week CCl_4 intoxication, but they did not examine the gene expression at that time point. Intriguingly, the time course study of the gene expression of MMP-13 during the recovery phase is quite similar to our previous histochemical demonstration on lysosomal enzyme activities, showing a dramatic upregulation of enzyme activity at day 5 after discontinuing CCl_4 intoxication (Maruyama *et al.*, 1978; Okazaki *et al.*, 1974).

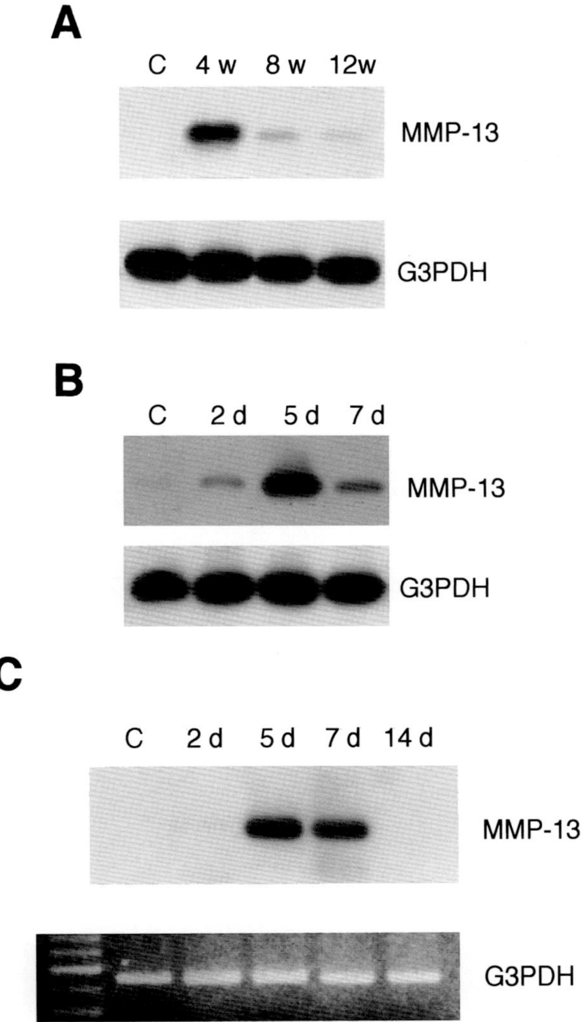

FIGURE 3 MMP-13 mRNA expression in the progressive and recovery phases of liver fibrosis. Gene expression was analyzed by RT-PCR followed by Southern blot hybridization. In the aggressive phase of fibrosis, MMP-13 mRNA was increased in the liver of rat treated with CCl_4 for 4 weeks (A). In the recovery phase of fibrosis, transcripts of MMP-13 increased markedly at 5 days after the last injection of 8-week (B) or 12-week (C) CCl_4 treatment.

The authors observed the intense gene expression at day 5 after the last injection of 12-week CCl_4 intoxication (Fig. 3C). It is very interesting why the intense gene expression of MMP-13 appears in the recovery from liver cirrhosis. As discussed later, this intense expression may be related to the appearance of

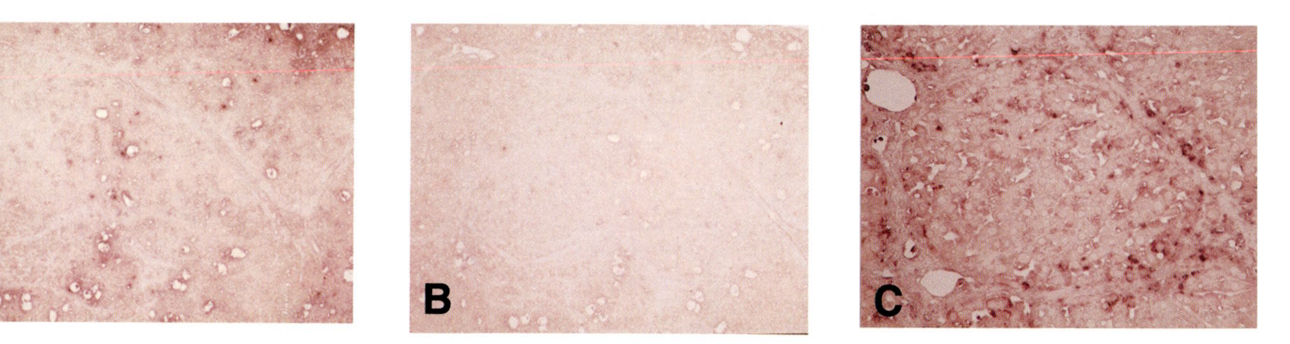

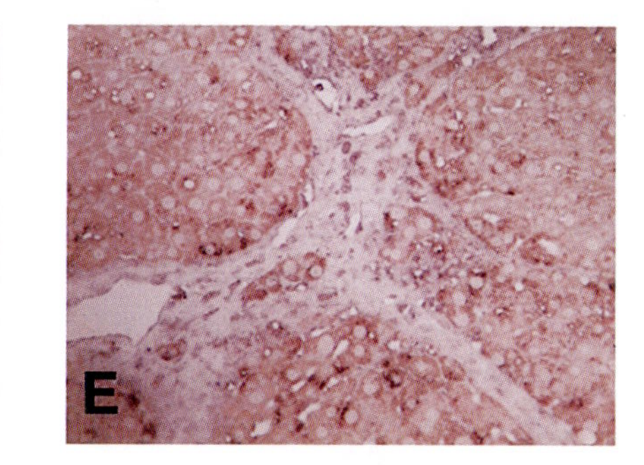

FIGURE 4 *In situ* hybridization of MMP-13 mRNA in CCl_4-treated rat liver. Liver samples were obtained at day 2 (A, B) and at day 5 (C) after the last injection of 8-week CCl_4 treatment or at day 2 (D) and at day 5 (E) after 12-week CCl_4 treatment. Few signals of MMP-13 mRNA were observed at day 2 with the antisense probe (A, D) but not with the sense probe (B). Increased gene expression of MMP-13 was observed at day 5 after the last injection of 8-week (C) or 12-week (E) CCl_4 treatment. MMP-13 mRNA-positive cells were located along the border between the resolving fibrous band and the parenchyma. In particular, the liver of rats obtained 5 days after the last injection of 12 weeks of intoxication intense signals was observed along or within the newly formed fibrous bands (E). Original magnification: 50×.

hepatic stem cells. These cells may appear in rat liver cirrhosis more frequently than in liver fibrosis.

In the normal rat liver, no signal for MMP-13 mRNA was observed by *in situ* hybridization. In rats treated with CCl_4 for 8 weeks, signals for MMP-13 mRNA were observed in a few cells lining the fibrous septa (Figs. 4A and 4B). Some of these cells were stained with the α smooth muscle actin (α-SMA) antibody (Watanabe *et al.*, 2000). However, in the cirrhotic liver of rats treated with CCl_4 for 12 weeks, very weak signals of MMP-13 mRNA in stellate cells were observed (Fig. 4D). No hepatocyte in the liver revealed MMP-13 mRNA transcripts regardless of the length of CCl_4 treatment. These data were consistent with the results of RT-PCR described earlier.

The authors demonstrated that MMP-13 expression was enhanced during the recovery phase of liver fibrosis, and cells positive for MMP-13 were observed mainly at the interface between the resolving fibrous septa and the parenchyma by *in situ* hybridization (Fig. 4C). The original paper (Watanabe *et al.*, 2000) was based on the findings at days 5 and 7 after the last injection of 8-week CCl_4 intoxication. The authors found very intense signals for gene expression of MMP-13 at day 5 after the last injection of 12-week CCl_4 intoxication. This was consistent with the results of RT-PCR described earlier. In particular, cells expressing MMP-13 at day 5 after 12-week CCl_4 treatment were seen in the marginal cells of the nodules interfaced with the resolving fibrous bands (Fig. 4E).

Overlapping the images of *in situ* hybridization and immunohistochemical staining revealed that some, but not all, of the MMP-13-positive cells were stellate cells that were stained with the α-SMA antibody. This is the first report providing direct evidence of definite MMP-13 gene expression during the recovery phase, which is in contrast with the downregulation of MMP-13 expression during the progression of fibrosis mentioned earlier.

Most MMP-13-positive cells in the recovery phase were albumin negative, α-fetoprotein negative, α-SMA negative, and ED2 negative. It is reasonable that some of the MMP-13-positive cells in the recovery phase could be stem cells. In the present study, it was confirmed that some stem cells express MMP-13 mRNA (see later).

B. Gene Expression of MMP-2 and MT1-MMP

The enzymatic characteristics of MMP-2 are quite different from those of MMP-1 and MMP-13. Not only are there differences in substrate specificity (Yu *et al.*, 1995), but also the expression of these MMPs is differentially regulated. That is, phorbol esters (TPA), interleukin-1β, tumor necrosis factor (TNF)-α, and basic fibroblast-growth factor stimulate MMP-1 expression (Yu *et al.*, 1995); TGF-β1, interleukin (IL)-8, and concanavalin A induce MMP-2 expression (Bar-Eli, 1999;

Gervasi *et al.*, 1996; Hanemaaijer *et al.*, 1993; Stearns *et al.*, 1999). p53 upregulates MMP-2 expression, but downregulates MMP-1 expression (Bian and Sun, 1997; Sun *et al.*, 1999).

Upregulation of MMP-2 (Takahara *et al.*, 1995, 1997; Theret *et al.*, 1999) and MT1-MMP (Preaux *et al.*, 1999; Takahara *et al.*, 1997) has been reported during the remodeling of the extracellular matrix in experimental animal and human liver fibrosis. Ikeda *et al.* (1999) reported that MMP-2 is necessary for the proliferation and infiltration of hepatic stellate cells in the process of liver fibrosis. Therefore, MMP-2 has been thought to be a fibrogenic MMP, which acts in the formation of liver fibrosis (Arthur, 2000; Li and Friedman, 1999; Okazaki *et al.*, 2000, 2001).

The authors observed the upregulation of MMP-2 mRNA and MT1-MMP mRNA in the progressive phase of liver fibrosis followed by their downregulation in the recovery phase of liver fibrosis by RT-PCR as shown in Fig. 5 (Watanabe *et al.*, 2001). The results are very different from that for MMP-13, which showed downregulation in the progressive phase and upregulation in the recovery phase as mentioned previously. The upregulation of both MMP-2 mRNA and MT1-MMP mRNA in the progressive phase is consistent with the reports mentioned earlier. In the recovery phase, however, Takahara *et al.* (1995) showed that after the discontinuation of chronic CCl_4 treatment, there was increased MMP-2 expression on days 3 and 7 and reduced expression on day 14. These findings are not consistent with ours and are probably due to the different experimental design and the different probes of MMPs.

The authors initially assumed that MMP-2 does not participate in the recovery from liver fibrosis because both MMP-2 and MT1-MMP gene expression decreased during the process of fibrolysis in the liver. Against our expectation, however, data indicate that both MMP-2 and MT1-MMP play some role in fibrolysis, although their gene expression decreases in the recovery phase (Watanabe *et al.*, 2001).

In the livers of rats treated with CCl_4 for 8 or 12 weeks, cells positive for MMP-2 mRNA were mainly observed around the fibrous bands, and some positive cells were scattered in the lobules at day 2 after the last injection of CCl_4 (Figs. 5A and 5B). On day 5, MMP-2 mRNA-positive cells were seen exclusively along or within the resolving fibrous septa (Fig. 5C). That is, fewer MMP-2 mRNA-positive cells were seen in the lobules at day 5 than those at day 2. Seven days after the last injection, the positive signals were still identifiable, scattered in the mesenchymal cells, including Kupffer cells, within the lobules (Fig. 5D). Surprisingly, these dynamic changes in the distribution of cells expressing MMP-2 mRNA occurred in the very early days of recovery. The same dynamic change in cells expressing mRNA was seen in MT1-MMP, TIMP-1, and TIMP-2, but not in MMP-13.

The findings of MT1-MMP mRNA in all of the 8- and 12-week-treated groups were essentially the same as the findings for MMP-2 mRNA, although the level of MT1-MMP mRNA was lower than that of MMP-2 mRNA (Watanabe *et al.*, 2001). Cells expressing MMP-2 mRNA were almost the same as cells expressing

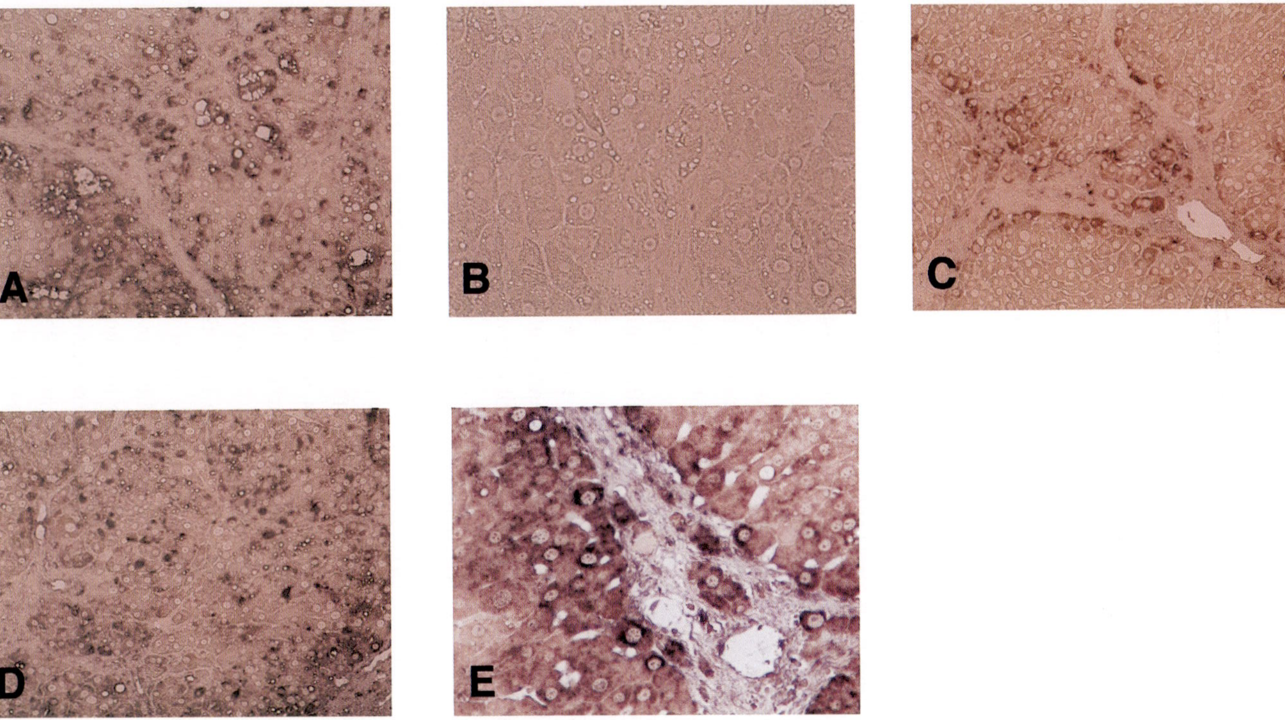

FIGURE 5 *In situ* hybridization of MMP-2 mRNA in CCl_4-treated rat liver. Intense signals were seen in rat liver obtained 2 days after the last injection of 8-week CCl_4 treatment with the antisense probe (A) but not with the sense probe (B). At day 5, MMP-2 mRNA-positive cells were seen along or within fibrous septa (C). At day 7, MMP-2-positive cells were seen in the lobules (D). Hepatocytes also expressed MMP-2 mRNA (E). *In situ* hybridization using antisense and sense probes was done in a serial section (A, B). Original magnification: A–D, 25×; E, 100×.

MT1-MMP mRNA at day 5. Both enzymes seem to degrade the newly formed fibrous bands in the recovery phase. Therefore, MMP-2 and MT1-MMP may play some role in the recovery from liver fibrosis.

MMP-2, a potent gelatinase, and MT1-MMP, an activator of MMP-2, can cleave native type 1 collagen (Aimes and Quigley, 1995; Ohuchi *et al.*, 1997), but with less efficiency than MMP-1 (Aimes and Quigley, 1995), as mentioned earlier. In addition to the colocalization of MMP-2-positive and MT1-MMP-positive cells around the fibrous bands at day 5, TIMP-2-positive cells also localized in the same area. Kinoshita *et al.* (1998) noted that TIMP-2, which is an inhibitor of MT1-MMP, paradoxically promotes activation of progelatinase A through MT1-MMP. From these results it is suggested that colocalized MT1-MMP and TIMP-2 can activate pro-MMP-2 around fibrous bands to resolve extracellular matrix.

At day 7, MMP-2- and/or MT1-MMP-positive cells were observed diffusely in the lobules, implying a role in the degradation of perihepatocellular fibrosis. Advanced liver fibrosis is associated with the appearance of perihepatocellular fibrosis, which contributes to the formation of sinusoidal capillarization (Okazaki *et al.*, 1973; Pooper and Udenfriend, 1970). Intralobular shunt vessels between portal vein tributaries and hepatic vein tributaries are formed, and it has been believed that there is no metabolic exchange of the blood with hepatocytes, leading to the irreversibility of liver fibrosis. The authors previously confirmed the reversibility of sinusoidal capitalization that appeared in the experimental liver fibrosis of rats induced by chronic CCl_4 intoxication (Maruyama *et al.*, 1976; Okazaki *et al.*, 1973). The authors also previously developed a method to assay type IV collagenase in the liver and demonstrated the reduced activity of type IV collagenase in human liver cirrhosis (Maruyama *et al.*, 1987, 1991; Okazaki *et al.*, 1998). Although the gene expression of both enzymes was downregulated, they may both participate in the destruction of perihepatocellular fibrosis, resulting in the recovery from liver fibrosis. Biological type IV collagenase activity includes net activities of MMP-2, MMP-3, and MMP-9 under the influence of their specific inhibitors, or TIMPs. Although the reports of Herbst *et al.* (1991), Vyas *et al.* (1995), and Winwood *et al.* (1995) showed that MMP-3 and MMP-9 are involved in perihepatocellular fibrosis, no study has reported gene expression of MT1-MMP and other MMPs in the context of fibrolysis in the liver.

Gelatinase activity of MMP-2 increased in the recovery phase of 8-week-treated rat liver by gelatin zymography, as shown in Fig. 6B (Watanabe *et al.*, 2001). An increase in gelatinase activity in the recovery phase is consistent with the observation of Takahara *et al.* (1995), except for the recovery phase of 12-week-treated rats, which they did not observe. A decrease in gelatinase activity in the recovery phase of 12-week-treated rats may reflect slow recovery from cirrhosis (Fig. 6B). The authors observed some discrepancy during mRNA expression and enzymatic activity in both progressive and recovery phases of liver fibrosis. Because gelatinase activity depends on a balance between MMP-2 and their inhibitors, the authors are now investigating this question.

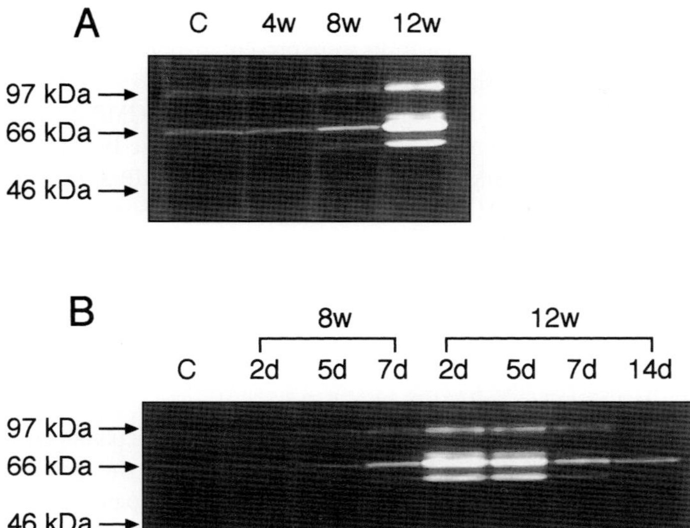

FIGURE 6 Detection of gelatinase activity by zymography. In the aggressive phase of liver fibrosis, liver samples obtained at 2 days after 8- and 12-week-CCl_4-treated rats showed elevated gelatinase activity (A). In the recovery phase of 8-week treatment, gelatinolytic activity increased at 5 and 7 days (B). In the recovery phase of 12-week treatment, however, gelatinolytic activity gradually decreased.

During the recovery phase of liver fibrosis, most of the cells positive for MMP-2 mRNA were negative for desmin and α-SMA (Watanabe *et al.*, 2001). Most of the cells positive for MT1-MMP mRNA were also negative for desmin and α-SMA (Watanabe *et al.*, 2001). Some MMP-2 mRNA-positive cells seemed to be hepatocytes, although they did not express albumin. Both mesenchymal cell (mainly hepatic stellate cells) and Kupffer cells were major sources of MMP-2 production during the recovery phase of liver fibrosis, although they did not express characteristic cell markers. A few hepatocytes within the fibrous bands also produced MMP-2 (Fig. 5E).

The size of cells expressing both enzymes seems to be larger than those positive for MMP-13, and the authors assume that both enzymes, in particular MMP-2 at least, may participate in the proliferation of hepatocytes, stellate cells, and Kupffer cells.

C. Gene Expression of TIMP-1 and TIMP-2

Iredale *et al.* (1996) and Roeb *et al.* (1997) demonstrated that TIMP-1 mRNA increased during the early phase of CCl_4 treatment and then decreased and remained at a low level during the recovery from liver fibrosis (Iredale *et al.*, 1998). The net activity of MMPs is determined by the balance between the activities of the MMPs and their inhibitors. Herbst *et al.* (1997) revealed that high levels of TIMP-1 and TIMP-2 transcripts were present in all fibrotic rat and human livers, predominantly in stellate cells.

The authors also observed upregulation of TIMP-1 and TIMP-2 mRNA in the progressive phase and downregulation in the recovery phase by RT-PCR. *In situ* hybridization showed strong signals for TIMP-1 and TIMP-2 mRNA in cells expressing MMP-2 and/or MT1-MMP at the interface between the newly formed fibrous bands and the parenchyma in the progressive phase.

In the recovery phase of liver fibrosis, gene expression of TIMPs gradually decreased. *In situ* hybridization revealed that the localization of TIMP-1- and TIMP-2-positive cells changed drastically over 5 and 7 days after discontinuance of chronic CCl_4 treatment. That is, at 5 days, mRNAs for both TIMPs were observed mainly in the parenchymal cells along the resolving fibrous bands, whereas weak signals of TIMPs were observed mainly in the hepatocytes at 7 days. Gene expression of TIMPs in hepatocytes may promote the proliferation of hepatocytes. These results indicate that TIMPs may participate in the recovery from liver fibrosis.

D. Comparison of MMPs and TIMPs Gene Expression in the Recovery Phase with Those in the Wound-Healing Process

In the recovery phase, i.e., at days 5 and 7 after the last injection of CCl_4 treatment for 8 or 12 weeks, gene expression of MMP-13 was upregulated dramatically (Watanabe *et al.*, 2000), whereas the gene expression of MMP-2/MT1-MMP decreased during these periods (Watanabe *et al.*, 2001). Although data are not shown, gene expression of TIMPs also decreased.

The reported gene expression of MMPs and TIMPs (Agren, 1999; Madlener *et al.*, 1998; Okada *et al.*, 1997; Soo *et al.*, 2000; Vaalamo *et al.*, 1999) was compared as shown in Fig. 7B. Although the results were not consistent, the same tendency is seen if both the progressive and the recovery phases of liver fibrosis are considered to be part of the wound-healing process.

V. STEM CELLS EXPRESSING MMP-13 IN RECOVERY FROM LIVER CIRRHOSIS

As described earlier, cells expressing MMP-13 mRNA by *in situ* hybridization appeared along the interface between the resolving fibrous bands that were newly formed and the parenchyma at day 5 after the last injection of CCl_4 for 8 or 12 weeks.

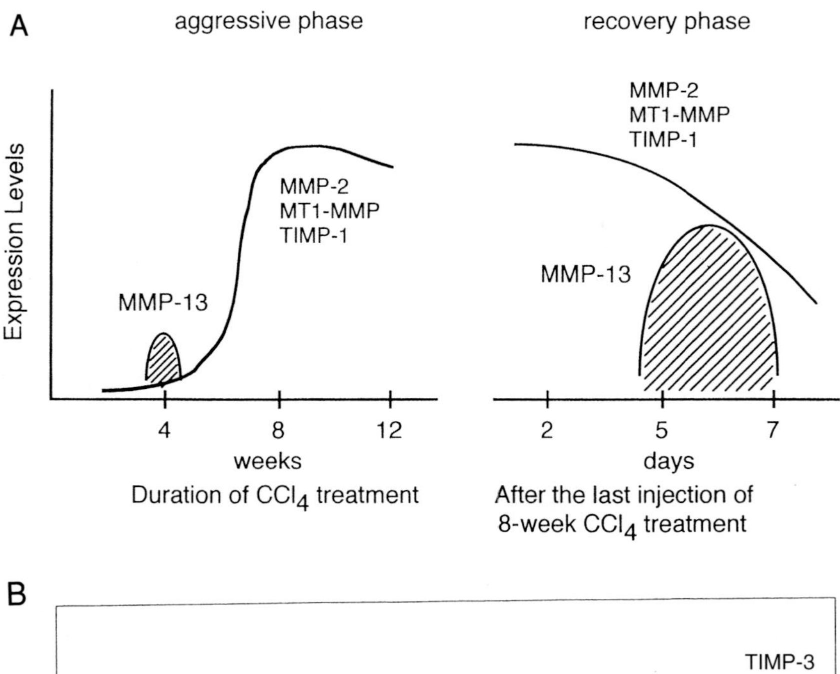

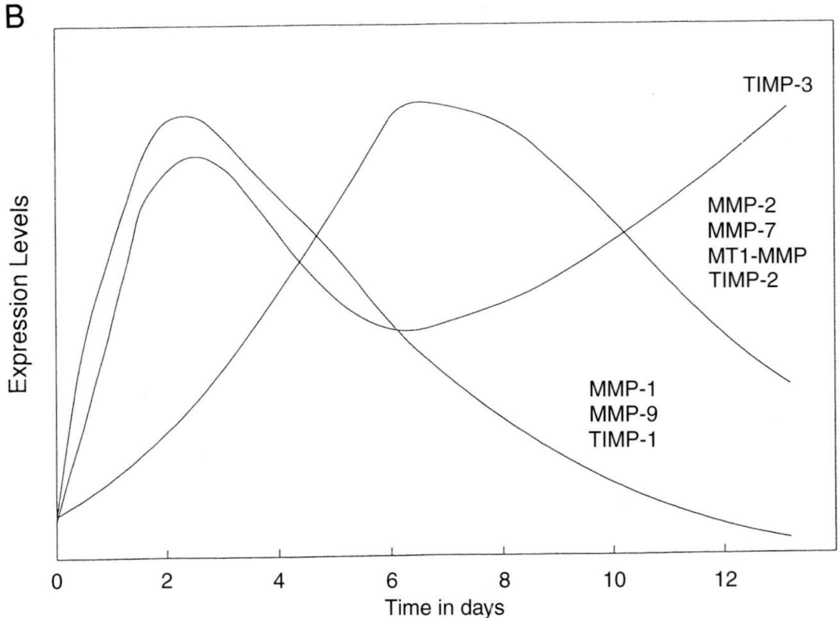

FIGURE 7 Comparison of MMP and TIMP gene expression in the recovery from liver fibrosis with those in wound healing. (A) Expression levels of MMP and TIMP mRNA in the aggressive (left) and recovery (right) phases of liver fibrosis. This diagram is presented by the permission of the Japanese Society of Hepatology [originally appeared in *Acta Hepatol. Jpn.* 41, 741–753 (2000)]. (B) Expression levels of MMP and TIMP mRNA in wound healing. The diagram is modified from Soo *et al.* (2000).

In particular, strong gene expression was observed at day 5 after the 12-week treatment. Some cells were confirmed to be hepatic stellate cells showing positive staining of α-SMA, but most cells did not show positive staining for α-SMA, albumin, α-fetoprotein, and ED2 immunohistochemically. From these results, the authors investigated whether cells expressing MMP-13 mRNA in the spontaneous resolution of liver fibrosis were stem cells, because recent evidence indicated that hematopoietic stem cells migrate in the liver and can differentiate into hepatic epithelium (Lagasse *et al.*, 2000).

The authors used antibodies specific for Musashi-1, nestin, and CD34. Musashi-1 protein, which is recognized as a neural RNA-binding protein, is involved in the development of neurons and glia by regulating gene expression at the posttranscriptional level (Sakakibara *et al.*, 1996). Nestin is an intermediate filament, which appears transiently in neural progenitor cells (Kawaguchi *et al.*, 2001). Neuronal-positive cells for Musashi-1 and nestin are considered to be progenitor cells in the central nervous system (Kawaguchi *et al.*, 2001; Roy *et al.*, 2000; Sakakibara *et al.*, 1996). These cells have not previously been observed in normal liver.

Putative stem cells showing expression of the aforementioned three markers were observed within the resolving fibrous bands as well as at the interface between the resolving fibrous bands and the parenchyma at day 5 after the last injection of 12-week CCl_4 intoxication, in the early phase of the recovery process of liver fibrosis (Watanabe *et al.*, submitted). The Musashi-1-positive cell was also positive for CD34, and the nestin-positive cell was positive for both CD34 and MMP-13 in serial sections (Figs. 8A–8D). By *in situ* hybridization, in mirror image sections paired with immunohistochemical studies, some MMP-13-positive cells expressed Musashi-1 (Figs. 8E–8G). Few putative stem cells were seen in both progressive and recovery phases of liver fibrosis after 8-week treatment. These cells may differentiate into hepatic stellate cells, although they did not express α-SMA at this time.

It has been reported that some hepatic stellate cells are stained with GFAP, nestin, and neural cell adhesion molecules (Knittel *et al.*, 1996, 1999; Neubauer *et al.*, 1996; Niki *et al.*, 1999). Therefore, putative stem cells expressing both MMP-13 and Musashi-1 may differentiate into hepatic stellate cells.

VI. HYPOTHESIS AND PROPOSAL OF A NEW STRATEGY FOR THE TREATMENT OF LIVER CIRRHOSIS

Neural progenitor cells that can be derived from hematopoietic stem cells express interstitial collagenase in the recovery phase of experimental liver cirrhosis.

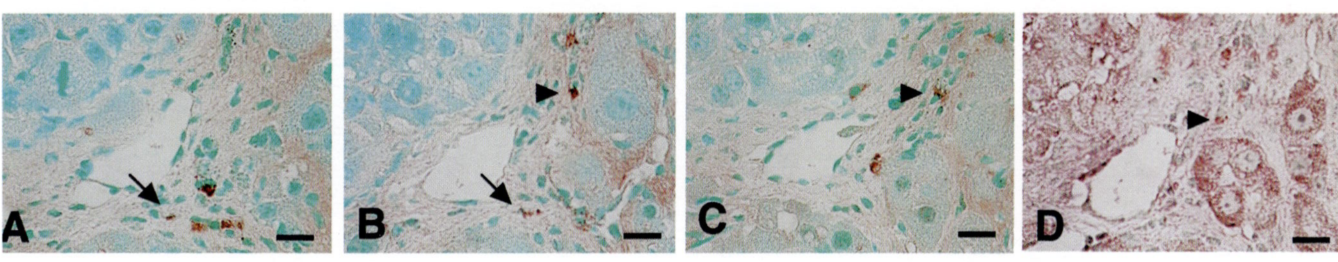

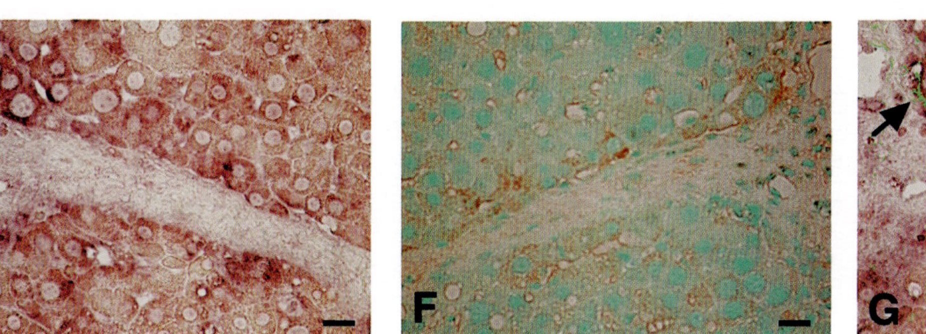

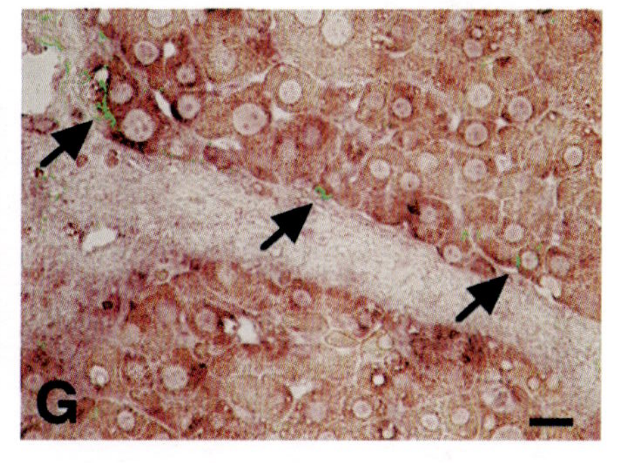

FIGURE 8 Photomicrographs showing the presence of stem cell markers detected by immunohistochemical staining and the gene expression of MMP-13 by *in situ* hybridization. Samples were obtained 5 days after the last injection of 12-week CCl_4 intoxication. Detection of Musashi-1 (A) CD34 (B), and nestin (C) by immunohistochemical staining and MMP-13 (D) by *in situ* hybridization in serial sections. Musashi-1-positive cells also express CD34 (arrow). Some nestin-positive cells express CD34 and MMP-13 (arrowhead). Immunohistochemical detection of Musashi-1 and detection of MMP-13 mRNA by *in situ* hybridization in mirror-image sections. *In situ* hybridization using the digoxigenin-labeled antisense RNA probe specific for MMP-13 (E) was performed in a mirror-image section paired with immunohistochemical detection of Musashi-1 (F), and the images were overlapped on the computer (G); positive signals for Musashi-1 were changed to green). Some of the cells expressing MMP-13 mRNA were positive for Musashi-1 (arrow). Scale bars: 50 μm.

We hypothesize that spontaneous resolution of experimental liver fibrosis and cirrhosis induced by chronic CCl_4 intoxication in rats occurs to be as follows.

Step 1. The preexisting liver cells (hepatocytes, stellate cells, endothelial cells, and Kupffer cells) recover, and stem cells derived from hematopoietic system, mainly bone marrow, appear. In particular, neural progenitor cells express MMP-13 mRNA.

Step 2. Stem cells with MMP-13 mRNA proliferate and differentiate into hepatocytes, stellate cells, and other cells.

Step 3. Some stem cells change phenotype and express MMP-2 and MT1-MMP, followed by weak expression of TIMPs. MMP-2, MT1-MMP, and an adequate amount of TIMP participate in the destruction of the extracellular matrix. MMP-2, MT1-MMP, and TIMPs proliferate liver cells, particularly hepatocytes and stellate cells.

Step 4. Hepatocyte growth factor (HGF) promotes the gene expression of MMP-13 as well as the proliferation of liver cells.

MMP-13 transcription in the recovery phase is transient, and stem cells expressing MMP-13 mRNA may differentiate into hepatocytes, stellate cells, and other cells. The authors did not observe cells expressing both MMP-13 mRNA and MMP-2 in the progressive and recovery phases of liver fibrosis. Most of the cells expressing MMP-2 and/or MT1-MMP mRNAs were not stained with albumin, α-fetoprotein, α-SMA, and ED2. Therefore, these cells could also be hepatic progenitor cells.

HGF is known to reduce experimental liver fibrosis (Ueki *et al.*, 1999; Yasuda *et al.*, 1996). Ueki *et al.* (1999) showed that in dimethylnitrosamine-induced liver fibrosis, HGF downregulated TGF-β expression, followed by the disappearance of predeposited fibrous tissue. Okazaki *et al.* (1999) proposed that HGF induces MMP-1 expression in LI 90 (cell line of human stellate cells) via Ets–1 in the promoter region of MMP-1. The recovery from liver fibrosis by external HGF treatment indicates that fibrolysis induced by increased collagenase activity could be related to regeneration of the liver.

The authors' results raise the possibility that hematopoietic stem cells could be used for the treatment of human liver fibrosis or cirrhosis. The authors are now investigating the possibility of gene therapy to expand stem cells in the liver and to express the MMP-1 gene in stem cells with or without transfusion of stem cells derived from bone marrow in patients with liver cirrhosis.

The most important problems to be resolved are whether stem cells with MMP-13 mRNA transdifferentiate into cancer cells and whether HGF promotes this process. The authors reported previously that the transient expression of interstitial collagenase was observed in differentiated, very early hepatocellular carcinomas (less than 2 cm), but not in moderately or undifferentiated hepatocellular carcinomas (Okazaki *et al.*, 1997).

The authors investigated previously the mechanism of collagenase production in liver cells using a monolayer culture of fibroblasts derived from the rabbit liver (Okazaki *et al.*, 1981, 1983, 1985). Different mechanisms of collagenase production were observed among fibroblasts derived from the synovium, gastric mucosa, and liver of the same rabbit. All fibroblasts used in this experiment were fourth passage cells in order to exclude macrophages and to obtain uniform cell lines. Synovial fibroblasts secreted a low level of collagenase without any treatment, and those treated with phorbol myristate acetate (PMA) produced a high level of collagenase. Gastric mucosal fibroblasts produced a high level of collagenase without any treatment. Upon treatment with PMA, MMP-1 production increased dramatically. Liver fibroblasts did not produce collagenase, even with PMA treatment (Okazaki *et al.*, 1985). The authors succeeded in inducing MMP-1 expression by coculturing fibroblasts and hepatocytes at a cell number ratio of 3:1. After a long latent period, a remarkably high level of collagenase synthesis was observed (Maruyama *et al.*, 1983; Okazaki *et al.*, 1985). The large quantity of collagenase produced by fibroblasts contributes to massive necrosis or tissue breakdown *in vivo*. As hepatic stellate cells can express high levels of MMP-1/MMP-13, MMP-1/MMP-13 production should be considered to be related to the activation of stellate cells.

The relationship between the activation process of hepatic stellate cells and the mechanism of production of MMP-1/MMP-13 has not been clarified. Activated stellate cells producing extracellular matrix seem not to express MMP-1/MMP-13 from the authors' observation (Watanabe *et al.*, 2000). Therefore, the mechanism of MMP-1/MMP-13 gene expression in stellate cell should be investigated. Moreover, the molecular mechanism of the transcriptional regulation of the MMP-1/MMP-13 gene has not been studied in hepatic stellate cells. Inflammatory cytokines such as tumor necrosis factor-α and interleukin-1β are known to be involved in the activation of hepatic stellate cells. Most such cytokines activate the retrovirus-associated DNA mitogen-activated protein kinase (Ras-MAPK) signaling pathway, including c-Jun NH2-terminal kinase (JNK) and extracellular signal-regulated kinase (ERK), which in turn activate the transcription of early genes such as c-*fos* and c-*jun*. The Fos and Jun proteins contribute to the induction of MMP-1 gene transcription by their binding to proximal AP-1 sites of the promoter in fibroblasts and immortalized cells (Grumbles *et al.*, 1997; Newberry *et al.*, 1997; Vincenti *et al.*, 1996). In addition to the role of AP-1, Vincenti *et al.* (1998) reported on the role of NF-κB in the induction of rabbit MMP-1 expression in IL1-treated synovial fibroblasts. TNF-α and IL-1β increase MAPK activity, including JNK and ERK in rat hepatic stellate cells, thereby stimulating AP-1 activity in hepatic stellate cells (Poulos *et al.*, 1997). Activation of hepatic stellate cells is also closely related to NF-κB activity (Elsharkawy *et al.*, 1999).

Activation of transcriptional factors such as AP-1 and NF-κB activity may also contribute to gene regulation of MMP-1 in hepatic stellate cells. The molecular mechanisms of the regulation of MMP-1 expression are still unknown. It should be determined how such transcriptional factors as NF-κB or AP-1 may be involved in MMP-1 gene expression in hepatic stellate cells. Hozawa *et al.* (2000) reported that Rac1 GTP may direct transcriptional induction of MMP-1 expression by TNF-α predominantly via NF-κB in human hepatic stellate cells. The authors hope that transfusion of stem cells derived from bone marrow should express MMP-1 using these mechanisms.

VII. CONCLUSION

When the authors started work on the relationship between lysosomal enzyme activity and the recovery from liver fibrosis, the histochemical finding of the lysosomal enzyme on the liver at day 5 after the last injection of 8- or 12-week CCl_4 treatment revealed a recovery of lysosomal enzyme in all hepatocytes (Okazaki *et al.*, 1974; Okazaki and Maruyama, 1980). Since then, three questions have been raised: (1) how the recovery occurs so early, (2) how hepatocytes participate in the recovery, particularly in the destruction of newly formed fibrous bands, and (3) whether the recovery from fibrosis is different from recovery from cirrhosis. The recovery may occur on any day in the progressive phase of liver fibrosis, but the process following the recovery does not continue. Hepatocytes expressing MMP-2 mRNA, MT1-MMP mRNA, and TIMPs mRNA were demonstrated by the authors (Watanabe *et al.*, 2001), and MMP-2 and MT1-MMP may participate in the degradation of the extracellular matrix. Cytoprotective treatment for hepatocytes has been used experimentally. This treatment may be effective to proliferate stem cells expressing MMP-13 mRNA and hepatocytes with MMP-2 and/or MT1-MMP mRNAs. Gene therapy, however, should be developed in order to reverse liver cirrhosis more effectively.

REFERENCES

Agren, M. S. (1999). Matrix metalloproteinases (MMPs) are required for re-epithelialization of cutaneous wounds. *Arch. Dermatol. Res.* **291**, 583–590.

Aimes, R. T., and Quigley, J. P. (1995). Matrix metalloproteinase-2 is an interstitial collagenase: Inhibitor-free enzyme catalyzes the cleavage of collagen fibrils and soluble native type I collagen generating the specific 3/4- and 1/4-length fragments. *J. Biol. Chem.* **270**, 5872–5876.

Arai, M., Niioka, M., Maruyama, K., Wada, N., Fujimoto, N., Nomiyama, T., Tanaka, S., and Okazaki, I. (1996). Changes in serum levels of metalloproteinases and their inhibitors by treatment of chronic hepatitis C with interferon. *Dig. Dis. Sci.* **41**, 995–1000.

Arthur, M., Friedman, S., Roll, F., and Bissell, D.(1989). Lipocytes from normal rat liver release a neutral metalloproteinase that degrades basement membrane (type IV) collagen. *J. Clin. Invest.* **84**, 1076–1085.

Arthur, M., Stanley, A., Iredale, J., Rafferty, J., Hembry, R., and Friedman, S. (1992). Secretion of 72kDa type IV collagenase/gelatinase by cultured human lipocytes. *Biochem. J.* **287**, 701–707.

Arthur, M. J. P. (1995). Role of Ito cells in the degradation of matrix in liver. *J. Gastroenterol. Hepatol.* **10**, 557–562.

Arthur, M. J. P. (2000). Fibrogenesis. II. Metalloproteinases and their inhibitors in liver fibrosis. *Am. J. Physiol. Gastrointest. Liver Physiol.* **279**, G245–G249.

Bar-Eli, M. (1999). Role of interleukin-8 in tumor growth and metastasis of human melanoma. *Pathobiology* **67**, 12–18.

Berman, B., and Duncan, M. R. (1989). Short-term keloid treatment in vivo with human interferon alpha-2b results in a selective and persistent normalization of keloidal fibroblast collagen, glycosaminoglycan, and collagenase production in vitro. *J. Am. Acad. Dermatol.* **21**, 694–702.

Bian, J., and Sun, Y. (1997). Transcriptional activation by p53 of the human type IV collagenase (gelatinase A or matrix metalloproteinase 2) promoter. *Mol. Cell. Biol.* **17**, 6330–6338.

Bonis, P. A., Ioannidis, J. P., Cappelleri, J. C., Kaplan, M. M., and Lau, J. (1997). Correlation of biochemical response to interferon alfa with histological improvement in hepatitis C: A meta-analysis of diagnostic test characteristics. *Hepatology* **26**, 1035–1044.

Bunton, G. L., and Cameron, R. (1963). Regeneration of liver after biliary cirrhosis. *Ann. N. Y. Acad. Sci.* **111**, 412–421.

Cameron, G.R., and Karunaratne, W.A.E. (1936). Carbon tetrachloride cirrhosis in relation to liver regeneration. *J. Path. Bact.* **42**, 1–21.

Capra, F., Casaril, M., Gabrielli, G. B., Tognella, P., Rizzi, A., Dolci, L., Colombari, R., Mezzelani, P., Corrocher, R., and De Sandre, G. (1993). alpha-Interferon in the treatment of chronic viral hepatitis: Effects on fibrogenesis serum markers. *J. Hepatol.* **18**, 112–118.

Carter, E., Mccarron, M., Alpert, E., and Isselbacher, K. (1982). Lysyl oxidase and collagenase in experimental acute and chronic liver injury. *Gastroenterology* **82**, 526–534.

Castilla, A., Prieto, J., and Fausto, N. (1991). Transforming growth factors beta1 and alpha in chronic liver disease: Effects of interferon alpha therapy. *N. Engl. J. Med.* **324**, 933–940.

Czaja, M. J., Weiner, F. R., Takahashi, S., Giambrone, M. A., van der Meide, P. H., Schellekens, H., Biempica, L., and Zern, M. A. (1989). Gamma-interferon treatment inhibits collagen deposition in murine schistosomiasis. *Hepatology* **10**, 795–800.

Dufour, J. F., DeLellis, R., and Kaplan, M. M. (1997). Reversibility of hepatic fibrosis in autoimmune hepatitis. *Ann. Int. Med.* **127**, 981–985.

Dufour, J. F., DeLellis, R., and Kaplan, M. M. (1998). Regression of hepatic fibrosis in hepatitis C with long-term interferon treatment. *Dig. Dis. Sci.* **43**, 2573–2576.

Duncan, M. D., and Berman, B. (1989). Differential regulation of glycosaminoglycan, fibronectin, and collagenase production in cultured human dermal fibroblasts by interferon-alpha, -beta, and gamma. *Arch. Dermatol. Res.* **281**, 11–18.

Elsharkawy, A. M., Wright, M. C., Hay, R. T., Arthur, M. J., Hughes, T., Bahr, M. J., Degitz, K., and Mann, D. A. (1999). Persistent activation of nuclear factor-kappaB in cultured rat hepatic stellate cells involves the induction of potentially novel Rel-like factors and prolonged changes in the expression of IkappaB family proteins. *Hepatology* **30**, 761–769.

Freije, J. M. P., Diez-Itza, I., Balbin, M., Sanchez, L. M., Blasco, R., and Tolivial, J. (1994). Molecular cloning and expression of collagenase-3, a novel human matrix metalloproteinase produced by breast carcinomas. *J. Biol. Chem.* **269**, 16766–16773.

Friedman, S. L. (1993). Seminars in medicine of the Beth Israel Hospital, Boston. The cellular basis of hepatic fibrosis. Mechanisms and treatment strategies. *N. Engl. J. Med.* **328**, 1828–1835.

Friedman, S. L. (1997). Molecular mechanisms of hepatic fibrosis and principles of therapy. *J. Gastroenterol. Hepatol.* **32**, 424–430.

Gervasi, D. C., Raz, A., Dehem, M., Yang, M., Kurkinen, M., and Friedman, R. (1996). Carbohydratemediated regulation of matrix metalloproteinase-2 activation in normal human fibroblasts and fibrosarcoma cells. *Biochem. Biophys. Res. Commun.* **228**, 530–538.

Gross, J., and Lapiere, C. M. (1962). Collagenolytic activity in amphibian tissues: A tissue culture assay. *Proc. Natl. Acad. Sci. USA* **48**, 1014–1022.

Grumbles, R. M., Shao, L., Jeffrey, J. J., and Howell, D. S. (1997). Regulation of the rat interstitial collagenase promoter by IL-1 beta, c- Jun, and Ras-dependent signaling in growth plate chondrocytes. *J. Cell Biochem.* **67**, 92–102.

Hammel, P., Couvelard, A., O'Toole, D., Ratouis, A., Sauvanet, A., Flejou, J. F., Degott, C., Belghiti, J., Bernades, P., Valla, D., Ruszniewski, P., and Levy, P. (2001). Regression of liver fibrosis after biliary drainage in patients with chronic pancreatitis and stenosis of the common bile duct. *N. Engl. J. Med.* **344**, 418–423.

Hanemaaijer, R., Koolwijk, P., le Clercq, L., de Vree, W. J., and van Hinsbergh, V. W. (1993). Regulation of matrix metalloproteinase expression in human vein and microvascular endothelial cells. Effect of tumor necrosis factor alpha, interleukin 1 and phorbol ester. *Biochem. J.* **296**, 803–809.

Harris, E. D., Jr., and Krane, S. M. (1974). Collagenases. *N. Engl. J. Med.* **291**, 557–563.

Herbst, H., Heinrichs, O., Schuppan, D., Milani, S., and Stein, H. (1991). Temporal and spatial patterns of transin/stromelysin RNA expression following toxic injury in rat liver. *Virch. Arch. B Cell. Pathol. Incl. Mol. Pathol.* **60**, 295–300.

Hozawa, S., Suzuki, H., Watanabe, T., Kagawa, S., Higuchi, A., Ishii, H., and Okazaki, I. (2000). Rho family GTPases regulate transcriptional induction of matrix metalloproteinase-1 by tumor necrosis factor α in human hepatic stellate cells. *Gastroenterology* **118** (Suppl. 2), A1013.

Hutterer, F., Rubin, E., and Popper, H. (1964). Mechanism of collagen resorption in reversible hepatic fibrosis. *Exp. Mol. Pathol.* **3**, 215–223.

Ikeda, K.,Wakahara, T.,Wang, Y.Q., Kadoya, H., Kawada, N., and Kaneda, N. (1999). In vitro migratory potential of rat quiescent hepatic stellate cells and its augumentation by cell activation. *Hepatology*, **29**, 1760–1767.

Iredale, J. P., Benyon, R. C., Arthur, M. J., Ferris, W. F., Alcolado, R., Winwood, P. J., Clark, N., and Murphy, G. (1996). Tissue inhibitor of metalloproteinase-1 messenger RNA expression is enhanced relative to interstitial collagenase messenger RNA in experimental liver injury and fibrosis. *Hepatology* **24**, 176–184.

Iredale, J. P., Benyon, R. C., Pickering, J., McCullen, M., Northrop, M., Pawley, S., Hovell, C., and Arthur, M. J. (1998). Mechanisms of spontaneous resolution of rat liver fibrosis: Hepatic stellate cell apoptosis and reduced hepatic expression of metalloproteinase inhibitors. *J. Clin. Invest.* **102**, 538–549.

Ishibashi, K., Kashiwagi, T., Ito, A., Tanaka, Y., Nagasawa, M., Toyama, T., Ozaki, S., Naito, M., and Azuma, M. (1996). Changes in serum fibrogenesis markers during interferon therapy for chronic hepatitis type C. *Hepatology* **24**, 27–31.

Jacques, W. E., and McAdams, A. I. (1957). Reversible biliary cirrhosis in rat after partial ligation of common bile duct. *Arch. Pathol.* **63**, 149–153.

Jimenez, S. A., Freundlich, B., and Rosenbloom, J. (1984). Selective inhibition of human diploid fibroblast collagen synthesis by interferons. *J. Clin. Invest.* **74**, 1112–1116.

Kaplan, M. M., DeLellis, R. A., and Wolfe, H. J. (1997). Sustained biochemical and histologic remission of primary biliary cirrhosis in response to medical treatment. *Ann. Intern. Med.* **126**, 682–688.

Kawaguchi, A., Miyata, T., Sawamoto, K., Takashita, N., Murayama, A., Akamatsu, W., Ogawa, M., Okabe, M., Tano, Y., Goldman, S. A., and Okano, H. (2001). Nestin-EGF transgenic mice: Visualization of the self-renewal and multipotency of CNS stem cells. *Mol. Cell. Neurosci.* **17**, 259–273.

Kinoshita, T., Sato, H., Okada, A., Ohuchi, E., Imai, K., Okada, Y., and Seiki, M. (1998). TIMP-2 promotes activation of progelatinase A by membrane-type 1 matrix metalloproteinase immobilized on agarose beads. *J. Biol. Chem.* **273**, 16098–16103.

Knauper, V., Lopez-Otin, C., Smith, B., Knight, G., and Murphy, G. (1996). Biochemical characterization of human collagenase-3. *J. Biol. Chem.* **271**, 1544–1550.

Knittel, T., Aurisch, S., Neubauer, K., Eichhorst, S., and Ramadori, G. (1996). Cell-type-specific expression of neural cell adhesion molecule (N-CAM) in Ito cells of rat liver: Up-regulation during in vitro activation and in hepatic tissue repair. *Am. J. Pathol.* **149**, 449–462.

Knittel, T., Kobold, D., Saile, B., Grundmann, A., Neubauer, K., Piscaglia, F., and Ramadori, G. (1999). Rat liver myofibroblasts and hepatic stellate cells: Different cell populations of the fibroblast lineage with fibrogenic potential. *Gastroenterology* **117**, 1205–1221.

Lagasse, E., Connors, H., Al-Dhalimy, M., Reitsma, M., Dohse, M., L., O., Wang, X., Finegold, M., Weissman, I. L., and Grompe, M. (2000). Purified hematopoietic stem cells can differentiate into hepatocytes in vivo. *Nature Med.* **6**, 1229–1234.

Lau, D. T., Kleiner, D. E., Ghany, M. G., Park, Y., Schmid, P., and Hoofnagle, J. H. (1998). 10-year follow-up after interferon-alpha therapy for chronic hepatitis C. *Hepatology* **28**, 1121–1127.

Lau, D. T., Kleiner, D. E., Park, Y., DiBisceglie, A. M., and Hoofnagle, J. H. (1999). Resolution of chronic delta hepatitis after 12 years of interferon alfa therapy. *Gastroenterology* **117**, 1229–1233.

Li, D., and Friedman, S. L. (1999). Liver fibrogenesis and the role of hepatic stellate cells: New insights and prospects for therapy. *J. Gastroenterol. Hepatol.* **14**, 618–633.

Lieber, C. S. (1999). Prevention and treatment of liver fibrosis based on pathogenesis. *Alcohol Clin. Exp. Res.* **23**, 944–949.

Lindblad, W., and Fuller, G. (1983). Hepatic collagenase activity during carbon tetrachloride-induced fibrosis. *Fundam. Toxicol* **3**, 34–40.

Madlener, M., Parks, W. C., and Werner, S. (1998). Matrix metalloproteinases (MMPs) and their physiological inhibitors (TIMPs) are differentially expressed during excisional skin wound repair. *Exp. Cell Res.* **242**, 201–210.

Maruyama, K., Feinman, L., Fainsilber, Z., Nakano, M., Okazaki, I., and Lieber, C. S. (1982). Mammalian collagenase increases in early alcoholic liver disease and decreases with cirrhosis. *Life Sci.* **30**, 1379–1384.

Maruyama, K., Feinman, L., Okazaki, I., and Lieber, C. S. (1981). Direct measurement of neutral collagenase activity in homogenates from baboon and human liver. *Biochim. Biophy. Acta* **658**, 124–131.

Maruyama, K., Okazaki, I., Kashiwazaki, K., Funatsu, K., Oda, M., Kamegaya, K., and Tsuchiya, M. (1978). Different appearance of collagenase and lysosomal enzymes in recovery of experimental hepatic fibrosis. *Biochem. Exp. Biol.* **14**, 191–201.

Maruyama, K., Okazaki, I., Kashiwazaki, K., Sonoda, I., Ishii, H., Tsuchiya, M., Shibata, T., and Inayama, S. (1987). Biosynthesis of type IV collagenase by hepatic sinusoidal cells in rats. *Acta Hep. Jap.* **28**, 973–974.

Maruyama, K., Okazaki, I., Kobayashi, T., Suzuki, H., Kashiwazaki, K., and Tsuchiya, M. (1983). Collagenase production by rabbit liver cells in monolayer culture. *J. Lab. Clin. Med.* **102**, 543–550.

Maruyama, K., Okazaki, I., Sato, S., Horie, Y., Miyaguchi, S., Okuyama, K., Takahashi, H., Takagi, S., Ishii, H., and Tsuchiya, M. (1992). Type IV collagen (IV-C) is a useful marker of hepatic fibrosis in patients with alcoholic liver diseases. *Alcohol Alcoholism* **27**, 176.

Maruyama, K., Okazaki, I., Takagi, T., and Ishii, H. (1991). Formation and degradation of basement membrane collagen. *Alcohol Alcoholism* **26**, 309–374.

Maruyama, K., Tsuchiya, M., Oda, M., Okazaki, I., and Funatsu, K. (1976). Role of collagenase in capillarization of hepatic sinusoid. *Microcirculation* **1**, 333–334.

Montfort, I., Perez-Tamayo, R., Alvizouri, A., and Tello, E. (1990). Collagenase of hepatocytes and sinusoidal liver cells in the reversibility of experimental cirrhosis of the liver. *Virch. Arch. B Cell Pathol.* **59**, 281–289.

Morrione, T. G., and Levine, J. (1967). Collagenolytic activity and collagen resorption in experimental cirrhosis. *Arch. Pathol.* **84**, 59–63.

Neubauer, K., Knittel, T., Aurisch, S., Fellmer, P., and Ramadori, G. (1996). Glial fibrillary acidic protein: A cell type specific marker for Ito cells in vivo and in vitro. *J. Hepatol.* **24**, 719–730.

Newberry, E. P., Willis, D., Latifi, T., Boudreaux, J. M., and Towler, D. A. (1997). Fibroblast growth factor receptor signaling activates the human interstitial collagenase promoter via the bipartite Ets-AP1 element. *Mol. Endocrinol.* **11**, 1129–1144.

Niki, T., Pekny, M., Hellemans, K., Bleser, P. D., Berg, K. V., Vaeyens, F., Quartier, E., Schuit, F., and Geerts, A. (1999). Class VI intermediate filament protein nestin is induced during activation of rat hepatic stellate cells. *Hepatology* **29**, 520–527.

Ohuchi, E., Imai, K., Fujii, Y., Sato, H., Okada, Y., and Seiki, M. (1997). Membrane type 1 matrix metalloproteinase digests interstitial collagenase and other extracellular macromolecules. *J. Biol. Chem.* **272**, 2446–2451.

Okada, A., Tomasetto, C., Lutz, Y., Bellocq, J. P., Rio, M. C., and Basset, P. (1997). Expression of matrix metalloproteinases during rat skin wound healing: Evidence that membrane type-1 matrix metalloproteinase is a stromal activator of pro-gelatinase A. *J. Cell Biol.* **137**, 67–77.

Okazaki, I., Brinckerhoff, C. E., Sinclair, J. R., Sinclair, P. R., Bonkowsky, H. L., and Harris, E. D., Jr. (1981). Iron increases collagenase production by rabbit synovial fibroblasts. *J. Lab. Clin. Med.* **97**, 396–402.

Okazaki, I., Feinman, L., Fainsilber, Z., Nakano, M., and Lieber, C. S. (1983). Development of an assay for hepatic mammalian collagenase and study of the effect of ethanol on enzyme activity. *In* "Biological Approach to Alcoholism" (C. S. Lieber, ed.), pp. 392–398. Public Health, Rockville, MD.

Okazaki, I., Feinman, L., and Lieber, C. S. (1977). Hepatic mammalian collagenase: Development of an assay and demonstration of increased activity after ethanol consumption. *Gastroenterology* **73**, 1236.

Okazaki, I., and Maruyama, K. (1974). Collagenase activity in experimental hepatic fibrosis. *Nature* **252**, 49–50.

Okazaki, I., and Maruyama, K. (1980). Mammalian collagenase in the process of hepatic fibrosis. *J. UOEH* **2**, 401–424.

Okazaki, I., Maruyama, K., Kashiwazaki, K., and Tsuchiya, M. (1985). Mechanism of collagenase production by liver cells. *In*: "Pathobiology of Hepatic Fibrosis" C. Hirayama and K. I. Kivirikko (eds.), pp. 141–149. Elsevier Science, Amsterdam.

Okazaki, I., Maruyama, K., Ono, A., Kobayashi, T., and Suzuki, H. (1982). Reversibility of hepatic fibrosis in toxic liver injury. *J. UOEH* **4**, 169–181.

Okazaki, I., Oda, M., Maruyama, K., Funatsu, K., Matsuzaki, S., Kamegaya, K., and Tsuchiya, M. (1974). Mechanism of collagen resorption in experimental hepatic fibrosis with special reference to the activity of lysosomal enzymes. *Biochem. Exp. Biol.* **11**, 15–28.

Okazaki, I., Tsuchiya, M., Kamegaya, K., Oda, M., Maruyama, K., and Oshio, C. (1973). Capillarization of hepatic sinusoids in carbon tetrachloride-induced hepatic fibrosis. *Bibl. Anat.* **12**, 476–483.

Okazaki, I., Wada, N., Nakano, M., Saito, A., Takasaki, K., Doi, M., Kameyama, K., Otani, Y., Kubochi, K., Niioka, M., Watanabe, T., and Maruyama, K. (1997). Difference in gene expression for matrix mettalloproteinase-1 between early and advanced hepatocellular carcinomas. *Hepatology* **25**, 580–584.

Okazaki, I., Watanabe, T., Hozawa, S., Arai, M., and Maruyama, K. (2000). Molecular mechanism of the reversibility of hepatic fibrosis: With special reference to the role of matrix metalloproteinases. *J. Gastroenterol. Hepatol.* **15**, D26–D32.

Okazaki, I., Watanabe, T., Hozawa, S., and Maruyama, K. (1998). Molecular pathology of liver fibrosis. *In* "Liver Cirrhosis" (M. Yamanaka and G. Toda, eds.), pp. 23–33. Excerpta Medica, Amsterdam.

Okazaki, I., Watanabe, T., Hozawa, S., Niioka, M., Arai, M., and Maruyama, K. (1999). Hepatic fibrolysis and hepatic sinusoidal cells. *Liver Dis. Hepatic Sinusoidal Cells* 242–251.

Okazaki, I., Watanabe, T., Hozawa, S., Niioka, M., Arai, M., and Maruyama, K. (2001). Reversibility of hepatic fibrosis: From the first report of collagenase in the liver to the possibility of gene therapy for recovery. *Keio J. Med.* **50**, 58–65.

Okuda, K. (2001) "Liver Cirrhosis." Blackwell, Oxford.

Perez-Tamayo, R. (1965). Some aspects of connective tissue of the liver. *Progr. Liver Dis.* **2**, 192–210.

Pooper, H., and Udenfriend, S. (1970). Hepatic fibrosis. Correlation of biochemical and morphological investigation. *Am. J. Med.* **49**, 707–721.

Poulos, J. E., Weber, J. D., Bellezzo, J. M., Di Bisceglie, A. M., Britton, R. S., Bacon, B. R., and Baldassare, J. J. (1997). AP-1 activity, and transin gene expression in rat hepatic stellate cells. *Am. J. Physiol.* **273**, G804–G811.

Poupon, R. E., Balkau, B., Eschwege, E., and Poupon, R. (1991). A multicenter, controlled trial of ursodiol for the treatment of primary biliary cirrhosis. *N. Engl. J. Med.* **324**, 1548–1554.

Poynard, T., McHutchinson, J., Davis, G. L., Esteban-Mur, R., Goodman, Z., Bedossa, P., and Albrecht, J. (2000). Impact of interferon alfa-2b and ribavirin on progression of liver fibrosis in patients with chronic hepatitis C. *Hepatology* **32**, 1131–1137.

Preaux, A. M., Mallat, A., Van Nhieu, J. T., D'Ortho, M. P., Hembry, R. M., and Mavier, P. (1999). Matrix metalloproteinase-2 activation in human hepatic fibrosis regulation by cell-matrix interactions. *Hepatology* **30**, 944–950.

Quinn, P. S., and Higginson, J. (1965). Reversible and irreversible changes in experimental cirrhosis. *Am. J. Pathol.* **47**, 353–369.

Rojkind, M. (1999). Role of metalloproteinases in liver fibrosis. Alcoholism. *Clin. Exp. Res.* **23**, 934–939.

Rojkind, M., and Dunn, M. A. (1979). Hepatic fibrosis. *Gastroenterology* **76**, 849–863.

Roy, N. S., Wang, S., Jiang, L., Benraiss, A., Harrison-Restelli, C., Fraser, R. A. R., Couldwell, W. T., Kawaguchi, A., Okano, H., Nedergaard, M., and Goldman, S. A. (2000). In vitro neurogenesis by progenitor cells isolated from the adult human hippocampus. *Nature Med.* **6**, 271–277.

Rubin, E., and Hutterer, F. (1967). Hepatic fibrosis: Studies in the formation and resorption. *Connect. Tissue Baltimore* 142–160.

Sakakibara, S., Imai, T., Hamaguchi, K., Okabe, M., Aruga, J., Nakajima, K., Yasutomi, D., Nagata, T., Kurihara, Y., Uesugi, S., Miyata, T., Ogawa, M., Mikoshiba, K., and Okano, H. (1996). Mouse-Musashi-1, a neural RNA-binding protein highly enriched in the mammalian CNS stem cell. *Dev. Biol.* **176**, 230–242.

Schiff, E. R., Heathcote, J., Dienstag, J. L., Hann, H.-W. L., Woessner, M., Brown, N., Dent, J. C., Gray, F., and Goldin, R. D. (2000). Improvements in liver histology and cirrhosis with extended lamivudine therapy. *Hepatology* **32**, 296A.

Shiffman, M. L., Hofmann, C. M., Contos, M. J., Luketic, V. A., Sanyal, A. J., Sterling, R. K., Ferreira-Gonzalez, A., Mills, A. S., and Garret, C. (1999). A randomized, controlled trial of maintenance interferon therapy for patients with chronic hepatitis C virus and persistent viremia. *Gastroenterology* **117**, 1164–1172.

Shiratori, Y., Imazeki, F., Moriyama, M., Yano, M., Arakawa, Y., Yokosuka, O., Kuroki, T., Nishiguchi, S., Sata, M., Yamada, G., Fujiyama, S., Yoshida, H., and Omata, M. (2000). Histologic improvement of fibrosis in patients with hepatitis C who have sustained response to interferon therapy. *Ann. Intern. Med.* **132**, 517–524.

Sobesky, R., Mathurin, P., Charlotte, F., Moussalli, J., Olivi, M., Vidaud, M., Ratziu, V., Opolon, P., and Poynard, T. (1999). Modeling the impact of interferon alfa treatment on liver fibrosis progression in chronic hepatitis C: A dynamic view. *Gastroenterology* **116**, 378–386.

Soo, C., Shaw, W. W., Zhang, X., Longaker, M. T., Howard, E. W., and Ting, K. (2000). Differential expression of matrix metalloproteinases and their tissue-derived inhibitors in cutaneous wound repair. *Plast. Reconstr. Surg.* **105**, 638–647.

Stearns, M. E., Garcia, F. U., Fudge, K., Rhim, J., and Wang, M. (1999). Role of interleukin 10 and transforming growth factor beta1 in the angiogenesis and metastasis of human prostate primary tumor line from orthotopic implants in severe combined immunodeficiency mice. *Clin. Cancer Res.* **5**, 711–720.

Sun, Y., Wenger, L., Rutter, J. L., Brinckerhoff, C. E., and Cheung, H. S. (1999). p53 down-regulates human matrix metalloproteinase-1 (collagenase-1) gene expression. *J. Biol. Chem.* 274, 11535–11540.

Takada, A., Porta, E. A., and Hartroft, W. S. (1967). The recovery of experimental dietary cirrhosis. II. Turnover changes of hepatic cells and collagen. *Am. J. Pathol.* 51, 959–976.

Takahara, T., Furui, K., Funaki, J., Nakayama, Y., Itoh, H., Miyabayashi, C., Sato, H., Seiki, M., Ooshima, A., and Watanabe, A. (1995). Increased expression of matrix metalloproteinase-II in experimental liver fibrosis in rats. *Hepatology* 21, 787–795.

Takahara, T., Furui, K., Yata, Y., Jin, B., Zhang, L. P., Nanbu, S., Sato, H., Seiki, M., and Watanabe, A. (1997). Dual expression of matrix metalloproteinase-2 and membrane-type 1-matrix metalloproteinase in fibrotic human livers. *Hepatology* 26, 1521–1529.

Teran, J. C., Mullen, K. D., Hoofnagle, J. H., and McCullough, A. J. (1994). Decrease in serum levels of markers of hepatic connective tissue turnover during and after treatment of chronic hepatitis B with interferon-alpha. *Hepatology* 19, 849–856.

Theret, N., Lehti, K., Musso, O., and Clement, B. (1999). MMP2 activation by collagen I and concanavalin A in cultured human hepatic stellate cells. *Hepatology* 30, 462–468.

Ueki, T., Kaneda, Y., Tsutsui, H., Nakanishi, K., Sawa, Y., Morishita, R., Matsumoto, K., Nakamura, T., Takahashi, H., Okamoto, E., and Fujimoto, J. (1999). Hepatocyte growth factor gene therapy of liver cirrhosis in rats. *Nature Med.* 5, 226–230.

Vaalamo, M., Leivo, T., and Saarialho-Kere, U. (1999). Differential expression of tissue inhibitors of metalloproteinases (TIMP-1, -2, -3, and -4) in normal and aberrant wound healing. *Hum. Pathol.* 30, 795–802.

Varga, J., Olsen, A., Herhal, J., Constantine, G., Rosenbloom, J., and Jimenez, S. A. (1990). Interferon-gamma reverses the stimulation of collagen but not fibronectin gene expression by transforming growth factor-beta in normal human fibroblasts. *Eur. J. Clin. Invest.* 20, 487–493.

Vincenti, M. P., White, L. A., Schroen, D. J., Benbow, U., and Brinckerhoff, C. E. (1996). Regulating expression of the gene for matrix metalloproteinase-1 (collagenase): Mechanisms that control enzyme activity, transcription, and mRNA stability. *Crit. Rev. Eukaryot. Gene Expr.* 6, 391–411.

Vyas, S. K., Leyland, H., Gentry, J., and Arthur, M. J. P. (1995). Rat hepatic lipocytes synthesize and secrete transin (stromelysin) in early primary culture. *Gastroenterology* 109, 889–898.

Watanabe, T., Niioka, M., Hozawa, S., Kameyama, K., Hayashi, T., Arai, M., Ishikawa, A., Maruyama, K., and Okazaki, I. (2000). Gene expression of interstitial collagenase in both progressive and recovery phase of rat liver fibrosis induced by carbon tetrachloride. *J. Hepatol.* 33, 224–235.

Watanabe, T., Niioka, M., Ishikawa, A., Hozawa, S., Arai, M., Maruyama, K., Okada, A., and Okazaki, I. (2001). Dynamic change of cells expressing MMP-2 mRNA and MT1-MMP mRNA in the recovery from liver fibrosis in the rat. *J. Hepatol.* 35, 465–473.

Winwood, P. J., Schuppan, D., Iredale, J. P., Kawser, C. A., Docherty, A. J. P., and Arthur, M. J. P. (1995). Kupffer cell-derived 95-kd typeIV collagenase/gelatinase B: Characterization and expression in cultured cells. *Hepatology* 22, 304–315.

Yasuda, H., Imai, E., Shiota, A., Fujise, N., Morinaga, T., and Higashio, K. (1996). Antifibrogenic effect of a deletion variant of hepatocyte growth factor on liver fibrosis in rats. *Hepatology* 24, 636–642.

Yu, A., Murphy, A., and Stetler-Stevenson, W. (1995). 72-kDa gelatinase (gelatinase A): Structure, activation, regulation, and substrate specificity. *Matrix Metalloprotein.* 85–113.

Zimmerman, H., Reichen, J., Zimmerman, A., Saegesser, H., Thenisch, B., and Hoeflin, F. (1992). Reversibility of secondary biliary fibrosis by biliodigestive anastomosis in the rat. *Gastroenterology* 103, 579–589.

PART VI

New Strategies for the Treatment of Liver Cirrhosis

CHAPTER 21

Retinoids in Liver Fibrosis: Induction of Proteolytic Activation of Transforming Growth Factor-β by Retinoic Acid and Its Therapeutic Control by Protease Inhibitors

Masataka Okuno,* Kuniharu Akita,* Soichi Kojima,† and Hisataka Moriwaki*

**First Department of Internal Medicine, Gifu University School of Medicine, Gifu 500-8705, Japan and †Laboratory of Molecular Cell Sciences, RIKEN, Wako, Japan*

Retinoic acid (RA, an active metabolite of vitamin A) has anti-inflammatory and scar formation effects in the process of wound healing. These multifunctional actions of RA may be linked to the controversy over its effects on liver fibrosis, depending on the models examined. RA exacerbates liver fibrosis induced by porcine serum that is not accompanied by hepatic necroinflammation. In this model, RA seems to act directly on hepatic stellate cells (HSCs); it enhances plasminogen activator/plasmin levels and thereby induces proteolytic activation of latent transforming growth factor-β (TGF-β) on the cell surface. The resultant TGF-β, in turn, autostimulates the production of collagen by the cells. This mechanism may provide a clue for the therapy of liver fibrosis. We have developed a protease inhibitor, camostat mesilate, to inhibit such TGF-β activation. The protease inhibitor suppresses TGF-β activation and thereby inhibits the transformation of HSCs in culture, leading to a reduction in matrix production by the cells. The compound is also effective in preventing hepatic fibrosis induced by porcine serum in rats when administered orally. Moreover, camostat mesilate reduces hepatic fibrosis even when it is given after the establishment of the fibrosis.

Copyright 2003, Elsevier Science (USA). All rights reserved.

Thus, proteolytic activation of TGF-β may be an important target for the therapy of hepatic fibrosis.

I. INTRODUCTION

Hepatic stellate cells (HSCs) play central roles in the storage of retinoids (vitamin A and its analogs) in the quiescent phase as well as in fibrogenesis of the liver in the activated phase (Friedman, 2000; Okuno *et al.*, 2000). During the progression of liver fibrosis, HSCs lose retinoid-containing lipid droplets from the cytoplasm, transform into myofibroblast-like cells, and start to produce a significant amount of extracellular matrices (ECMs). Thus, it is of great interest to understand the relationship between the loss of retinoids and the production of ECM in HSCs. However, there have been some controversies regarding the effects of retinoids on liver fibrosis. Some have shown an antifibrotic effect of retinoids, but others reported their fibrogenic action. For example, administration with retinyl palmitate (a storage form of vitamin A) suppressed experimental hepatic fibrosis in rats produced by carbon tetrachloride (CCl_4) and by porcine serum (Senoo and Wake, 1985). In support of this, vitamin A deficiency was shown to promote CCl_4-induced liver fibrosis (Seifert *et al.*, 1994). However, others have reported that vitamin A supplemented with ethanol stimulated the progression of liver fibrosis and cirrhosis in rats (Leo and Lieber, 1983). Furthermore, the incidence of human hepatic fibrosis was correlated with the amount of vitamin A intake by patients with vitamin A hepatotoxicity (Geubel *et al.*, 1991). In addition, advances in molecular biology have shown a suppression of some matrix-degrading enzyme gene promoters, including stromelysin and collagenase, by retinoic acid, an active metabolite of vitamin A (Nicholson *et al.*, 1991; Pan *et al.*, 1995), which may link to the fibrosis-enhancing effect. Thus, conflicting effects have been observed in the effect of retinoids on liver fibrosis depending on the experimental conditions. So far, it is not yet known if retinoid loss is required for HSC activation and which retinoid might accelerate or prevent hepatic fibrosis (Friedman, 2000). Moreover, it is also uncertain how retinyl esters (mostly retinyl palmitate) are lost and if retinyl esters might be converted to a certain retinoid metabolite(s) that may affect HSC activation.

This chapter discusses the ECM-producing effect of RA in HSCs. Generation of a certain isomer of RA may provoke HSC activation in culture and during liver fibrosis *in vivo*. This RA enhances proteolytic activation of TGF-β, a strong fibrogenic cytokine, by upregulating the plasminogen activator (PA)/plasmin system. Active TGF-β autostimulates its own synthesis by HSCs and enhances ECM production. This sequence of mechanisms of RA may suggest a clue to use protease inhibitors to sever an autoinduction of TGF-β. We have developed a drug that suppresses proteolytic activation of TGF-β and thereby inhibits hepatic fibrosis in rats.

II. RETINOIC ACID AND LIVER FIBROSIS

A. Stimulation of TGF-β Activation by RA

In the rat liver fibrosis model induced by porcine serum, pure fibrosis is generated without causing hepatic inflammation or parenchymal necrosis. A stable analog of RA exacerbated liver fibrosis in this model by enhancing the function of TGF-β (Okuno *et al.*, 1997). It has been reported that interstitial collagen and/or TGF-β production is increased, unchanged, or reduced by the exposure of rat HSCs to all-*trans*-RA or TGF-β (Davis *et al.*, 1987, 1990; Davis, 1988), implying that the difference(s) in the experimental conditions may account for the opposing effects of RA on liver fibrosis. We have demonstrated the ECM-producing effect of RA occurs via activating the most potent fibrogenic cytokine, TGF-β.

TGF-β is a major cytokine implicated in the pathogenesis of liver fibrosis and cirrhosis (Border and Noble, 1994). TGF-β stimulates HSCs to transform into myofibroblast-like cells, enhances their production of ECM proteins, and suppresses degradation of the ECM (Friedman, 2000; Okuno *et al.*, 2000). Three subtypes of TGF-β (TGF-β1, -β2, and -β3), whose biological properties are nearly identical, are found in mammals (Roberts and Sporn, 1990). TGF-β is synthesized and secreted in a biologically latent form (latent TGF-β), which needs to be activated to acquire binding capacity to their cognate receptors and to perform biological activities (Flaumenhaft *et al.*, 1993). Activation releases the 25-kDa TGF-β homodimetric molecule from the latent large complex. Plasmin-mediated activation occurs under physiological conditions, such as in the cocultures of vascular endothelial and smooth muscle cells (Sato *et al.*, 1990). In this system, activation occurs on the cell surface by plasmin generated from serum plasminogen by the action of PA (Kojima *et al.*, 1991).

The diverse activities of RA are mediated primarily by two families of nuclear receptors: RA receptors (RARs) and retinoid X receptors (RXRs) (Mangelsdorf *et al.*, 1994). They are ligand-dependent transcription factors that bind to *cis*-acting DNA sequences, called RA responsive elements or retinoid X responsive elements in the promoter region of the target genes. RARs bind to RA responsive elements in response to both all-*trans*-RA and 9-*cis*-RA, whereas RXRs bind to retinoid X responsive elements in response to 9-*cis*-RA only. Retinoids enhance the production of both PA in many cell types (Gudas *et al.*, 1994). In bovine vascular endothelial cells, the elevation of cell surface PA/plasmin levels by RA causes the formation of active TGF-β; this TGF-β subsequently mediates some of the effects of RA on the endothelial cells (Kojima *et al.*, 1993; Kojima and Rifkin, 1993).

We have demonstrated that RA also stimulates the formation of active TGF-β in rat HSC cultures via plasmin-mediated proteolytic cleavage of latent TGF-β on the cell surface (Fig. 1) (Okuno *et al.*, 1997). The resultant active TGF-β stimulates its own synthesis, resulting in the generation of a considerable amount of

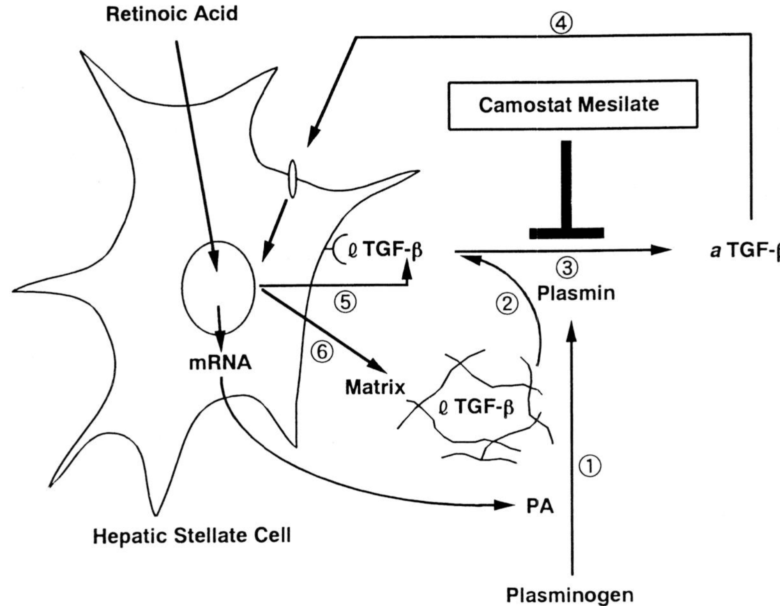

FIGURE 1 Schematic illustration of retinoic acid (RA)-induced promotion of liver fibrosis via TGF-β activation and its therapeutic control by a protease inhibitor, camostat mesilate. RA upregulates plasminogen activator (PA)/plasmin levels in hepatic stellate cells (HSCs) (1). This increase elaborates the release (2) and the activation (3) of latent TGF-β (lTGF-β) on the cell surface. The active TGF-β (aTGF-β) generated then stimulates the activation (or transformation) of HSCs (4), promotes its own synthesis (5) and increase in the amount of extracellular matrix (6), resulting in a cycle of TGF-β–extracellular matrix overexpressions and an exacerbation of liver fibrosis. Camostat mesilate inhibits this proteolytic activation of TGF-β and thereby prevents liver fibrosis. From Okuno *et al.* (2001), with permission.

TGF-β. In addition, TGF-β suppresses collagenase activities in HSC cultures. In keeping with the culture experiments, an *in vivo* liver fibrosis model induced by porcine serum also demonstrated an increased concentration of TGF-β in liver tissues as well as the exacerbation of hepatic fibrosis by the administration with RA. However, RA-treatment alone (without porcine serum) does not cause the increase of hepatic TGF-β nor liver fibrosis. Thus, the RA does not directly cause liver fibrosis by itself, but rather exacerbates the fibrosis induced by the other stimulus such as porcine serum. In other words, HSCs may need to be activated by some stimuli before they become sensitive to RA-treatment.

B. An Endogenous RA in Fibrotic Liver

Because the effect of RA on TGF-β activation has been examined by administering exogenous RA to the animals (Okuno *et al.*, 1997), the effect of endogenously produced RA remained to be elucidated. It is well known that retinylpalmitate contents, a storage form of retinoids in HSCs, decrease with the development of fibrosis (Blomhoff and Wake, 1991). This implies the possibility that the loss of retinylpalmitate in the fibrotic liver might be ascribed, in part, to the result of conversion of retinylpalmitate to some other metabolites, including RA. In support of this hypothesis, we have discovered an increase in 9,13-di-*cis*-RA generation, a major isomer of RA, in rat fibrotic livers induced by porcine serum (Okuno *et al.*, 1999b). Because 9,13-di-*cis*-RA is a major product arising from the *in vivo* isomerization of 9-*cis*-RA (Horst *et al.*, 1995; Kojima *et al.*, 1994) and has been suggested to be an indicator of preexisted 9-*cis*-RA (Arnhold *et al.*, 1996), the elevation of hepatic 9,13-di-*cis*-RA concentration suggests that 9-*cis*-RA might be generated during the development of fibrosis. In addition, we have found that 9,13-di-*cis*-RA itself can induce activation of latent TGF-β in a plasmin-dependent manner and enhance TGF-β biosynthesis (Fig. 2) (Imai *et al.*, 1997). This biological action of 9,13-di-*cis*-RA seems to be mediated by a nuclear RA receptor, RARα. However, a contradictory observation has also been reported, showing that RA contents as well as its signaling were diminished in the fibrotic liver induced by cholestasis (Ohata *et al.*, 1997). The reason for this discrepancy still remains

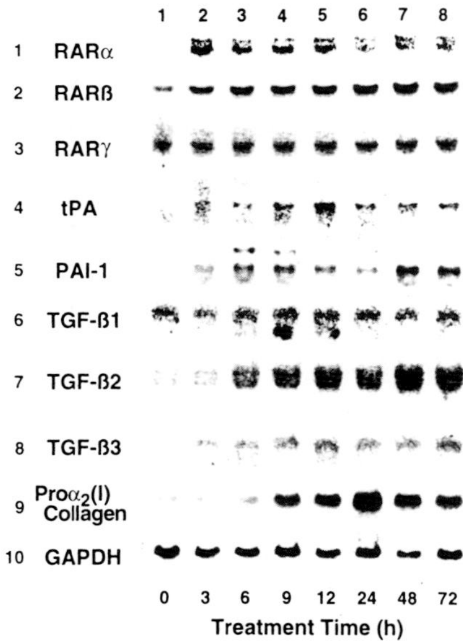

FIGURE 2 Changes in mRNA levels of retinoic acid receptor (RAR) α, β, γ, tissue plasminogen activator (tPA), TGF-β1, -β2, -β3, and type I procollagen following the exposure of human hepatic stellate cells to 9,13-di-*cis*-retinoic acid (9,13dcRA). Cell cultures were incubated either with vehicle (control; lane 1) or with 1 μ*M* 9,13dcRA (lanes 2–8) for up to 72 h. Total RNA was isolated from each cell sample, fractionated through 1% agarose–formaldehyde gels, transferred to nylon membranes, and hybridized with a ^{32}P-labeled probe for RAR α, β, γ, tPA, plasminogen activator inhibitor-1 (PAI-1), TGF-β1, -β2, -β3, pro-α2(I) collagen, or GAPDH. The radioactivity of each band was detected on an imaging analyzer.

to be elucidated; however, the difference in the animal models (immunologic stimulus by porcine serum vs cholestatic stimulus by bile duct ligation) might influence the retinoid absorption and metabolism in distinct manners, resulting in the difference in the experimental outcome.

III. PROTEASE INHIBITORS

A. Blockade of TGF-β Activation

Our studies demonstrate a fibrogenic aspect of RA in the liver via enhancing proteolytic activation of TGF-β. These studies may provide a clue for a novel therapy against liver fibrosis. Because the induction of TGF-β by RA in HSCs was initiated by the activation of latent TGF-β on the cell surface, this TGF-β activation can be a primary target for therapeutic strategy, and inhibitors of PA/plasmin could be potent antifibrogenic agents. In fact, the plasmin/TGF-β activation cascade has been shown to take place in human hepatic fibrosis (Inuzuka *et al.*, 1997). Interestingly, in previous studies, protease inhibitors and an inhibitor of surface plasmin have been used episodically as cytoprotective agents to prevent hepatic necrosis (Gressner and Bachem, 1995).

This idea led us to examine which protease inhibitors would be effective in suppressing TGF-β generation, employing cultured HSCs *in vitro* (Okuno *et al.*, 1998). Rat primary HSCs, isolated and cultured for 7 days, were incubated with plasmin in the absence and presence of each of 20 different protease inhibitors. Thereafter, concentrations of both active and total (active plus latent) TGF-β present in the culture medium were determined. To our surprise, all protease inhibitors tested significantly suppressed activation and generation of TGF-β. Similar results were obtained with HSCs stimulated by either RA or basic fibroblast growth factor (basic FGF), reagents that increase endogenous plasmin levels in the cells. Among 20 compounds examined, we selected one chemical, camostat mesilate (Fig. 3), and examined its effect on the activation

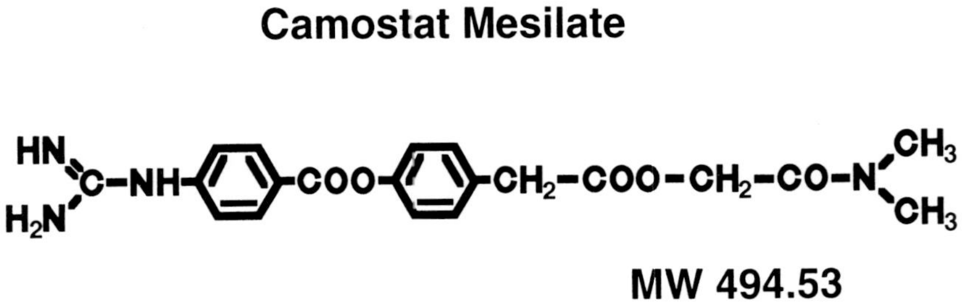

Camostat Mesilate

$(HN)(H_2N)C{-}NH{-}C_6H_4{-}COO{-}C_6H_4{-}CH_2{-}COO{-}CH_2{-}CO{-}N(CH_3)_2$

MW 494.53

FIGURE 3 Chemical structure of camostat mesilate. From Okuno *et al.* (2001), with permission.

(or transformation) of HSCs. Currently, camostat mesilate is used for the therapy of pancreatitis and reflux esophagitis, and its drug safety is widely confirmed from clinical experiences (Sugiyama *et al.*, 1997; Kobayashi *et al.*, 1998). Rat primary HSCs, cultured for 1 day after isolation, were incubated with camostat mesilate and cultured for an additional 7 days in the absence or presence of the compound. Camostat mesilate suppressed cell surface plasmin levels and thereby reduced the activation of TGF-β as well as its autoinduction (Fig. 4).

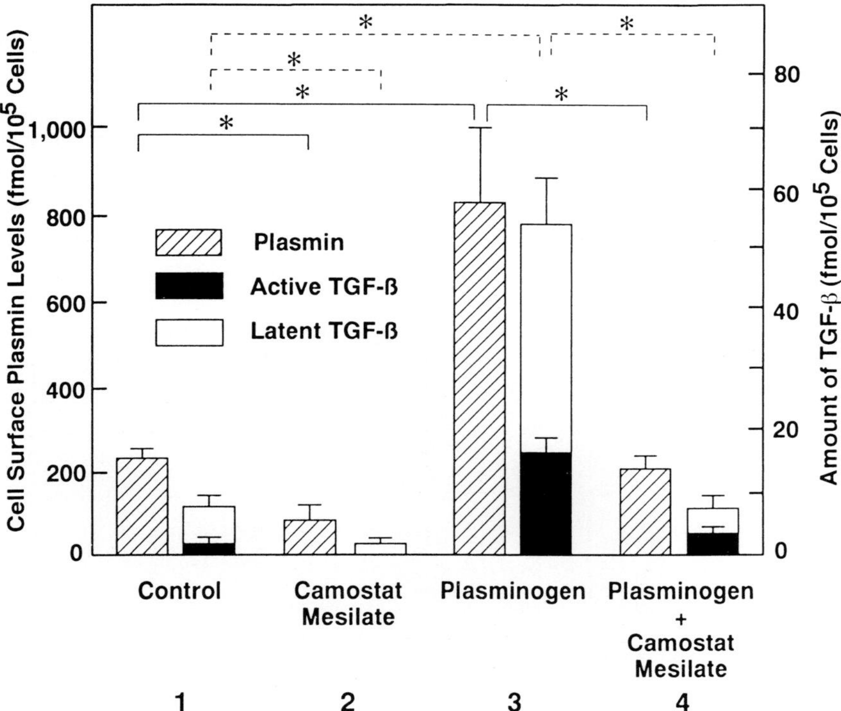

FIGURE 4 Suppression by camostat mesilate of both cell surface plasmin levels and TGF-β generation in primary rat HSC cultures. HSCs were isolated from a normal rat and cultured for 7 days in the absence or presence of 500 μ*M* camostat mesilate, 0.44 μ*M* plasminogen, or their combination. The plasmin activity in the cell surface extracts (slashed column) and the amounts of active (closed column) and total (closed plus open columns) TGF-β in the conditioned medium were determined. *Sample* 1, no reagent (control); *sample* 2, camostat mesilate; *sample* 3, plasminogen; and *sample* 4, plasminogen + camostat mesilate. Each value represents the mean ± SD (n = 5). An asterisk indicates a significant difference ($P < 0.01$) obtained by a comparison between the indicated pairs. Note that TGF-β activation as well as its autoinduction paralleled with cell surface plasmin levels that were enhanced by the inclusion with plasminogen in the medium. Camostat mesilate suppressed the plasmin levels and thereby reduced TGF-β activation in both control and plasminogen-included media. From Okuno *et al.* (2001), with permission.

Moreover, the compound inhibited morphological changes of HSCs (e.g., loss of lipid droplets from the cytoplasm), downregulated the expression of α smooth muscle actin, a marker of activated HSCs, and inhibited cellular proliferation (Okuno *et al.*, 2001). Thus, camostat mesilate inhibited not only the generation of active TGF-β, but also transformation of HSCs.

B. Prevention of Hepatic Fibrosis

These results suggest that the protease inhibitor may also be useful *in vivo*, suppressing the generation of TGF-β in the liver and subsequent development of hepatic fibrosis. Oral administration with camostat mesilate suppressed HSC activation and prevented porcine serum-induced fibrosis without causing any obvious adverse effects at the same dose range administered to humans (Fig. 5) (Okuno *et al.*, 2001). Plasmin has been believed to be antifibrogenic because it dissolves ECM directly or indirectly via activation of prometalloproteinases (Leyland *et al.*, 1996; Benyon *et al.*, 1999).

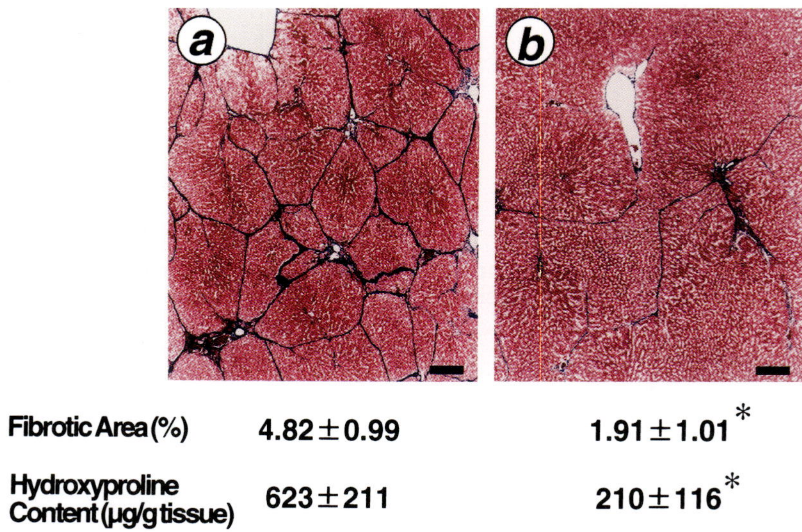

FIGURE 5 Prevention of porcine serum-induced hepatic fibrosis by camostat mesilate. Rats were injected with porcine serum intraperitoneally for 16 weeks. Half of the rats were fed a diet containing 2 mg/g camostat mesilate. Liver tissues were stained according to the method of Azan-Mallory. (a) Porcine serum and (b) porcine serum plus oral administration of camostat mesilate. Scale bar: 100 μm. The percentage area of fibrotic tissues was determined by an image analyzer. Tissue contents of hydroxyproline were also measured. An asterisk indicates a significant difference ($P < 0.05$) between the groups. From Okuno *et al.* (2001), with permission.

Expression of PA increases at the early stage of fibrosis; however, the expression of PA inhibitor-1 (PAI-1) is enhanced and overwhelms PA expression in the late phase, leading to a decrease in net fibrinolytic potential (Seki *et al.*, 1996; Zhang *et al.*, 1999). We reported that in the hepatic fibrosis model induced by porcine serum, plasmin exerts a fibrogenic effect in the early stage because it releases latent TGF-β from surrounding matrices and converts it to the active form (Okuno *et al.*, 1997, 2001). We have shown consistently that hepatic plasmin activity increases along with the development of hepatic fibrosis. Similar to *in vitro* data, camostat mesilate suppressed the increases in hepatic plasmin levels and TGF-β production probably by inhibiting TGF-β activation in HSCs (Okuno *et al.*, 2001). Based on these results, we suggest that camostat mesilate may inhibit both proteolytic release and activation of latent TGF-β also *in vivo*, leading to the reduction in the hepatic TGF-β content, thereby contributing to the maintenance of HSCs in a quiescent phenotype and to the prevention of hepatic fibrosis (Fig. 1).

Because camostat mesilate is a broad-spectrum protease inhibitor, the result might also suggest that this protease inhibitor blocked the activity not only of plasmin but other unknown proteases capable of generating TGF-β. To examine this possibility, we enhanced hepatic plasmin levels by *in vivo* transfection with the urokinase-type PA gene and examined if this treatment to enhance hepatic plasmin levels neutralizes the antifibrotic effect of camostat mesilate. In fact, upregulation of the plasmin levels abolished the suppressive effects of camostat mesilate on TGF-β contents and on fibrosis. We therefore concluded that the antifibrotic effect of camostat mesilate is dependent on its antiplasmin activity. Present data suggest that some other protease inhibitors that suppress TGF-β activation might also be beneficial for the therapy of liver fibrosis.

It may be possible that camostat mesilate might suppress matrix degradation via a direct and/or indirect inhibition of matrix-degrading enzymes, such as matrix metalloproteinase 2 (Leyland *et al.*, 1996; Benyon *et al.*, 1999), and therefore, the compound might be ineffective in the treatment of established hepatic fibrosis. However, we have confirmed the therapeutic (or fibrosis-reducing) effect of camostat mesilate by administering the compound to rats after hepatic fibrosis was established by porcine serum injection (Okuno *et al.*, 2001). Thus, the drug may have the potential not only to prevent but also to reduce hepatic fibrosis.

An advantage of camostat mesilate is that it can be administered orally. This advantage will be particularly favorable for long-term treatment for the patients with chronic liver diseases. It is of interest to test the synergistic effect of a combination of camostat mesilate with other antifibrotic agents. In addition, as TGF-β is known to be involved in fibrogenesis not only in the liver but also in other organs, such as the lung, kidney, and skin, we expect that camostat mesilate might also be effective in fibrogenesis in these organs.

IV. CONCLUSIONS

Based on our present and previous results, we hypothesize the following two possibilities regarding controversy about the effects of retinoid on liver fibrosis (Okuno *et al.*, 1999a). First, the overloading of retinylesters to HSCs may cause the excess accumulation of lipid droplets in the cytoplasm and would interfere with protein synthesis including ECM. RA administered would directly modulate certain gene expression through the interaction with RARs to stimulate ECM production. Second, the apparent discrepancy between hepatic fibrosis models might be ascribed to the presence or absence of inflammation in the liver. Because RA is known to suppress inflammation and accelerate scar formation in wounded tissues, it may inhibit fibrosis indirectly via modulating hepatic inflammation. These anti-inflammatory and subsequent antifibrotic effects were seen in the CCl_4-induced cirrhotic liver in rats, in which parenchymal necroinflammation is accompanied (Okuno *et al.*, 1990). RA may also possess a potential to stimulate hepatic fibrosis, particularly when it acts directly on HSCs. This effect may be observed in the pure fibrosis model, such as the one induced by porcine serum, in which hepatic inflammation is absent (Okuno *et al.*, 1997, 1999b). In the latter condition, RA seems to act directly on HSCs and induces the plasmin-dependent activation of TGF-β. Moreover, we propose that a protease inhibitor, camostat mesilate, can prevent as well as reduce hepatic fibrosis via suppressing TGF-β activation in an animal model. Our findings support the utility of orally administered, nontoxic protease inhibitors as a potential new therapy for patients with chronic liver disease.

In addition, another important function of TGF-β is to suppress hepatic parenchymal regeneration. Thus, the protease inhibitor might also be useful for the therapy of impaired liver regeneration often encountered after the surgical resection of cirrhotic livers. We have extended our idea and have obtained preliminary results supporting such beneficial effects of the compound. The compound restored hepatic regeneration after partial hepatectomy in rats, which was suppressed by the action of TGF-β secreted from HSCs by the stimulation of lipopolysaccharide (K. Akita *et al.*, 2002). Thus, activation of latent TGF-β is a key event for the pathogenesis of hepatic fibrosis, as well as impaired liver regeneration, and may be an important target for the therapy of such pathologic states.

ACKNOWLEDGMENTS

This study was supported partly by Grants-in-Aid from the Ministry of Education, Science, Sports and Culture (12670472 to M.O.; 13670579 to S.K. and 10557055 to H.M.).

REFERENCES

Akita, K., Okuno, M., Enya, M., Imai, S., Moriwaki, H., Kawada, N., Suzuki, Y., and Kojima, S. (2002). Impaired liver regeneration in mice by lipopolysaccharide via TNF-α/kallikrein-mediated activation of latent TGF-β. *Gastroenterology* **123**, 352–364.

Arnhold, T., Tzimas, G., Wittfoht, W., Plonait, S., and Nau, H. (1996). Identification of 9-*cis*-retinoic acid, 9,13-di-*cis*-retinoic acid, and 14-hydroxy-4,14-retro-retinol in human plasma after liver consumption. *Life Sci.* **59**, PL169–PL177.

Benyon, R. C., Hovell, C. J., Da Gaca, M., Jones, E. H., Iredale, J. P., and Arthur, M. J. (1999). Progelatinase A is produced and activated by rat hepatic stellate cells and promotes their proliferation. *Hepatology* **30**, 977–986.

Blomhoff, R., and Wake, K. (1991). Perisinusoidal stellate cells of the liver: Important roles in retinal metabolism and fibrosis. *FASEB J.* **5**, 271–277.

Border, W. A., and Noble, N. A. (1994). Transforming growth factor β in tissue fibrosis. *N. Engl. J. Med.* **331**, 1286–1292.

Davis, B. H. (1988). Transforming growth factor β responsiveness is modulated by the extracellular collagen matrix during hepatic Ito cell culture. *J. Cell Physiol.* **136**, 547–553.

Davis, B. H., Kramer, R. T., and Davidson, N. O. (1990). Retinoic acid modulates rat Ito cell proliferation, collagen and transforming growth factor β production. *J. Clin. Invest.* **86**, 2062–2070.

Davis, B. H., Pratt, B. M., and Madri, J. A. (1987). Retinol and extracellular collagen matrices modulate hepatic Ito cell collagen phenotype and cellular retinol binding protein levels. *J. Biol. Chem.* **262**, 10280–10286.

Flaumenhaft, R., Kojima, S., Abe, M., and Rifkin, D. B. (1993). Activation of latent transforming growth factor β. *In* "Advances in Pharmacology" (J. T. August, M. W. Anders, and F. Murad, eds.), Vol 24, pp. 51–76. Academic Press, San Diego.

Friedman, S. L. (2000). Molecular regulation of hepatic fibrosis, an integrated cellular response to tissue injury. *J. Biol. Chem.* **275**, 2247–2250.

Geubel, A. P., De Galocsy, C., Alves, N., Rahier, J., and Davis, C. (1991). Liver damage caused by therapeutic vitamin A administration: Estimate of dose-related toxicity in 41 cases. *Gastroenterology* **100**, 1701–1709.

Gressner, A. M., and Bachem, M. G. (1995). Molecular mechanisms of liver fibrogenesis: A homage to the role of activated fat-storing cells. *Digestion* **56**, 335–346.

Gudas, L. J., Sporn, M. B., and Roberts, A. B. (1994). Cellular biology and biochemistry of the retinoids. *In* "The Retinoids: Biology, Chemistry, and Medicine" (M. B. Sporn, A. B. Roberts, and D. S. Goodman, eds.), pp. 443–520. Raven, New York.

Horst, R. L., Reinhardt, T. A., Goff, J. P., Nonnecke, B. J., Gambhir, U. K., Fiorella, P. D., and Napoli, J. L. (1995). Identification of 9-*cis*, 13-*cis*-retinoic acid as a major circulating retinoid in plasma. *Biochemistry* **34**, 1203–1209.

Imai, S., Okuno, M., Moriwaki, H., Muto, Y., Murakami, K., Shudo, K., Suzuki, Y., and Kojima, S. (1997). 9,13-di-*cis*-Retinoic acid induces the production of tPA and activation of latent TGF-β via RARα in a human liver stellate cell line, LI90. *FEBS Lett.* **411**, 102–106.

Inuzuka, S., Ueno, T., Torimura, T., and Tanikawa, K. (1997). The significance of colocalization of plasminogen activator inhibitor-1 and vitronectin in hepatic fibrosis. *Scand. J. Gastroenterol.* **32**, 1052–1060.

Kobayashi, I., Ohwada, S., Ohya, T., Yokomori, T., Iesato, H., and Morishita, Y. (1998). Jejunal pouch with nerve preservation and interposition after total gastrectomy. *Hepato-Gastroenterol* **45**, 558–562.

Kojima, R., Fujimori, T., Kiyota, N., Toriya, Y., Fukuda, T., Ohashi, T., Sato, T., Yoshizawa, Y., Takeyama, K., Mano, H., Masushige, S., and Kato, S. (1994). *In vivo* isomerization of retinoic acids: Rapid isomer exchange and gene expression. *J. Biol. Chem.* **269**, 32700–32707.

Kojima, S., Harpel, P. C., and Rifkin, D. B. (1991). Lipoprotein(a) inhibits the generation of transforming growth factor β: An endogenous inhibitor of smooth muscle cell migration. *J. Cell Biol.* **113**, 1439–1445.

Kojima, S., Nara, K., and Rifkin, D. B. (1993). Requirement for transglutaminase in the activation of latent transforming growth factor-β in bovine endothelial cells. *J. Cell Biol.* **121**, 439–448.

Kojima, S., and Rifkin, D. B. (1993). Mechanism of retinoid-induced activation of latent transforming growth factor-β in bovine endothelial cells. *J. Cell Physiol.* **155**, 323–332.

Leo, M. A., and Lieber, C. S. (1983). Hepatic fibrosis after long-term administration of ethanol and moderate vitamin A supplementation in the rat. *Hepatology* **3**, 1–11.

Leyland, H., Gentry, J., Arthur, M. J., and Benyon, R. C. (1996). The plasminogen-activating system in hepatic stellate cells. *Hepatology* **24**, 1172–1178.

Mangelsdorf, D. J., Umesono, K., and Evans, R. M. (1994). The retinoid receptors. *In* "The Retinoids: Biology, Chemistry, and Medicine" (M. B. Sporn, A. B. Roberts, and D. S. Goodman, eds.), pp. 319–349. Raven, New York.

Nicholson, R. C., Mader, S., Nagpal, S., Leid, M., Rochette-Egly, C., and Chambon, P. (1991). Negative regulation of the rat stromelysin gene promoter by retinoic acid is mediated by an AP1 binding site. *EMBO J.* **9**, 4443–4454.

Ohata, M., Lin, M., Satre, M., and Tsukamoto, H. (1997). Diminished retinoic acid signaling in hepatic stellate cells in cholestatic liver fibrosis. *Am. J. Physiol.* **272**, G589–G596.

Okuno, M., Adachi, S., Akita, K., Moriwaki, H., Kojima, S., and Friedman, S. L. (2000). Liver fibrosis and hepatic stellate cells. *Connect. Tissue* **32**, 401–406.

Okuno, M., Akita, K., Moriwaki, H., Kawada, N., Ikeda, K., Kaneda, K., Suzuki, Y., and Kojima, S. (2001). Prevention of rat hepatic fibrosis by the protease inhibitor, camostat mesilate, via reduced generation of active TGF-β. *Gastroenterology* **120**, 1784–1800.

Okuno, M., Moriwaki, H., Imai, S., Muto, Y., Kawada, N., Suzuki, Y., and Kojima, S. (1997). Retinoids exacerbate rat liver fibrosis by inducing the activation of latent TGF-β in liver stellate cells. *Hepatology* **26**, 913–921.

Okuno, M., Muto, Y., Moriwaki, H., Kato, M., Ninomiya, M., Nishiwaki, S., Noma, A., Tagaya, O., Nozaki, Y., and Suzuki, Y. (1990). Inhibitory effect of acyclic retinoid (polyprenoic acid) in hepatic fibrosis in CCl_4-treated rats. *J. Gastroenterol.* **25**, 223–229.

Okuno, M., Moriwaki, H., Muto, Y., and Kojima, S. (1998). Protease inhibitors suppress TGF-β generation by hepatic stellate cells. *J. Hepatol.* **29**, 1031–1032.

Okuno, M., Nagase, S., Shiratori, Y., Moriwaki, H., Muto, Y., Kawada, N., and Kojima, S. (1999a). Retinoid and liver fibrosis. *In* "Liver Diseases and Hepatic Sinusoidal Cells." (K. Tanikawa and T. Ueno, eds.), pp. 232–241. Springer-Verlag, Tokyo.

Okuno, M., Sato, T., Kitamoto, T., Imai, S., Kawada, N., Suzuki, Y., Yoshimura, H., Moriwaki, H., Onuki, K., Masushige, S., Muto, Y., Friedman, S. L., Kato, S., and Kojima, S. (1999b). Increased 9,13-di-*cis*-retinoic acid in rat hepatic fibrosis: Implication for a potential link between retinoid loss and TGF-β mediated fibrogenesis *in vivo*. *J. Hepatol.* **30**, 1073–1080.

Pan, L., Eckhoff, C., and Brinckerhoff, C. E. (1995). Suppression of collagenase gene expression by all-*trans* and 9-*cis* retinoic acid is ligand dependent and requires both RARs and RXRs. *J. Cell Biochem.* **57**, 575–589.

Roberts, A. B., and Sporn, M. B. (1990) The transforming growth factor-βs. *In* "Handbook of Experimental Pharmacology" (M. B. Sporn and A. B. Roberts, eds.), Vol 95/I, pp. 419–472. Springer-Verlag, Berlin.

Sato, Y., Tsuboi, R., Lyons, R., Moses, H., and Rifkin, D. B. (1990). Characterization of the activation of latent TGF-β by co-cultures of endothelial cells and pericytes or smooth muscle cells: A self-regulating system. *J. Cell Biol.* **111**, 757–763.

Seifert, W. F., Bosma, A., Brouwer, A., Hendriks, H., Roholl, P. J. M., van Leeuwen, R. E. W., van Thiel-de Ruiter, G. C. F., Seifert-Bock, I., and Knook, D. L. (1994). Vitamin A deficiency promotes CCl_4 liver fibrosis. *Hepatology* **19**, 193–201.

Seki, T., Imai, H., Uno, S., Ariga, T., and Gelehrter, T. D. (1996). Production of tissue-type plasminogen activator (t-PA) and type-1 plasminogen activator inhibitor (PAI-1) in mildly cirrhotic rat liver. *Thromb. Haemost.* **75**, 801–807.

Senoo, H., and Wake, K. (1985). Suppression of experimental hepatic fibrosis by administration of vitamin A. *Lab. Invest.* **52**, 182–194.

Sugiyama, M., Atomi, Y., Wada, N., Kuroda, A., and Muto, T. (1997). Effect of oral protease inhibitor administration on gallbladder motility in patients with mild chronic pancreatitis. *J. Gastroenterol.* **32**, 374–379.

Zhang, L. P., Takahara, T., Yata, Y., Furui, K., Jin, B., Kawada, N., and Watanabe, A. (1999). Increased expression of plasminogen activator and plasminogen activator inhibitor during liver fibrogenesis of rats: Role of stellate cells. *J. Hepatol.* **31**, 703–711.

CHAPTER 22

Role of Growth Hormone Resistance and Impaired Insulin-like Growth Factor-I Bioavailability in the Pathogenesis of Cirrhosis and Its Complications

ANTHONY J. DONAGHY

A.W. Morrow Gastroenterology and Liver Center, Royal Prince Alfred Hospital, Camperdown, 2050 New South Wales, Australia

Growth hormone (GH) and insulin-like growth factor-I (IGF-I) are ubiquitous and potent growth factors that have a central role in cellular proliferation and differentiation, somatic growth, and the regulation of body composition in adult life. Stress, disease, and organ failure may lead to a state of acquired GH resistance that has deleterious effects on whole body homeostasis. The liver is the central organ of the endocrine GH:IGF:IGF-binding protein (BP) axis, and the development of cirrhosis has been associated with profound changes in this axis, which result in significant impairment of IGF-I bioavailability to its tissue receptors. This chapter reviews the reported changes in the GH:IGF:IGFBP axis in cirrhosis and their relationships to the pathogenesis of cirrhosis and the complications that influence the significant morbidity and mortality of this disease state.

Extracellular Matrix and the Liver
Copyright 2003, Elsevier Science (USA). All rights reserved.

I. LIVER AND THE GH:IGF:IGFBP AXIS

A. Introduction

The somatomedin hypothesis of Salmon and Daughaday (1957) established that the majority of growth-producing effects of pituitary-derived growth hormone were produced by the somatomedins, later termed insulin-like growth factors (IGFs). Under hypothalamic control, GH is secreted by the anterior pituitary gland, transported in the circulation by the high-affinity growth hormone-binding protein (GHBP), and acts through the hepatic GH receptor to regulate the production of the potent mitogenic growth factor IGF-I (Fig. 1). Although multiple tissues synthesize IGF-I, there is much evidence to support the liver as the primary source of circulating levels (D'Ercole *et al.*, 1984): biologically reactive IGF-I is secreted by perfused livers (Schalch *et al.*, 1979) (Schwander *et al.*, 1983), by cultured liver explants, and by normal hepatocytes in primary culture (Scott *et al.*, 1985a,b). Furthermore, estimated hepatic rates of IGF-I production are sufficient to account for the known turnover of IGF-I in the circulation

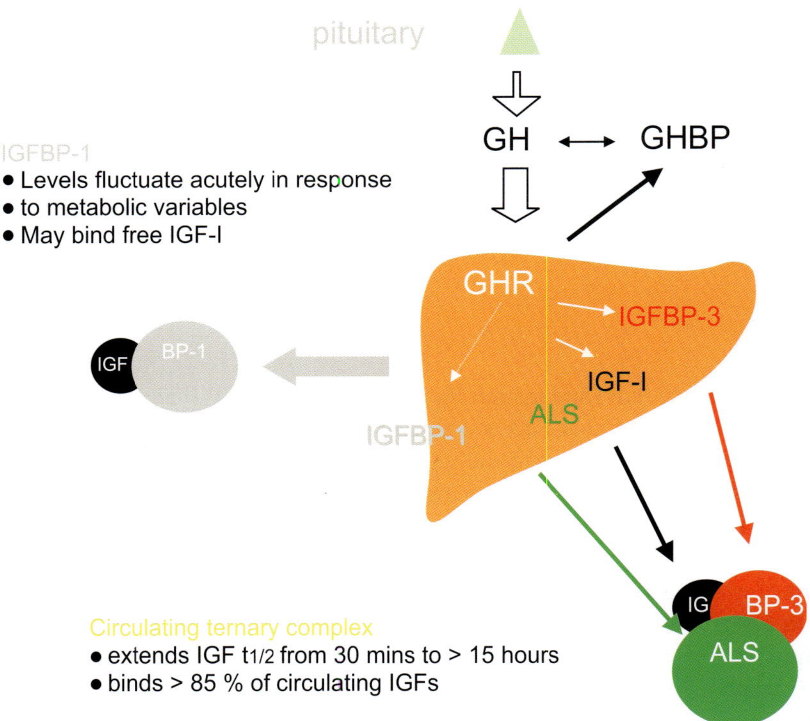

FIGURE 1 Schematic of the central role of the liver in the GH:IGF:IGFBP axis.

(Schwander *et al.*, 1983). Additionally, levels of IGF-I mRNA in liver are 50- to 100-fold greater than in most other tissues (Murphy *et al.* 1987).

IGF-I is an anabolic peptide that is vital to cell proliferation and differentiation in virtually every organ (Jones and Clemmons, 1995). IGF-I stimulates cellular differentiation and proliferation; is an essential cofactor for the progression through the cell cycle from phase G to S (Jones and Clemmons, 1995); is a dominant regulator of somatic growth before puberty (Guler *et al.*, 1988); and, after epiphyseal closure, plays an important role in the maintenance of body composition (Ho and Hoffman, 1993). The biologic role of IGF-II remains uncertain, although roles in fetal development and in tumor development have been defined. In addition to their functions as endocrine growth factors, IGFs are thought to exert growth-promoting effects at the local tissue level through paracrine and autocrine mechanisms.

B. IGF-Binding Proteins

IGFs are not secreted and stored in a specific organ, but circulate bound to insulin-like growth factor-binding proteins, of which six have been characterized and cloned (IGFBP-1 to -6) (Baxter, 1993). IGFBPs bind greater than 95% of circulating IGFs with less than 5% existing free in the circulation. IGFBPs have a major influence on the bioactivity of IGFs at the level of their tissue receptors by prolonging IGF half-life in the circulation, controlling the release of IGFs from the vascular to the extracellular space, and modulating the presentation of IGF to its receptor (Jones and Clemmons, 1995).

The largest proportion of serum IGF-I circulates in a 150-kDa heterotrimeric complex thought to comprise IGF-I, IGFBP-3, and an acid-labile subunit (ALS) (Baxter, 1990). The liver is thought to be the dominant source of the ALS, with hepatocytes the only cell type capable of producing ALS protein in primary cell cultures of rat (Scott and Baxter, 1991; Dai *et al.*, 1994; Scharf *et al.*, 1996) and human hepatocytes (Scharf *et al.*, 1995). This contrasted to the intrahepatic cellular distribution of IGFBP-3, which was documented in portal venous and sinusoidal endothelial cells and not in hepatocytes or other nonparenchymal cells, such as Kupffer or Ito cells (Chin *et al.*, 1994). Moser *et al.* (1992) had shown previously that IGFBP-3 is produced from endothelial cells in primary culture. These findings support the intrahepatic compartmentalization concept: different hepatic cellular sources of IGF-I and ALS (hepatocytes) and IGFBP-3 (nonparenchymal cell) allow only intravascular ternary complex formation and prevent intracellular formation of the circulating ternary complex, which, because of its large size (150 kDa), would not be able to leave the cellular cytoplasm and reach the circulation.

The ternary complex is thought to be restricted to the vascular space and extends the serum half-life of IGF-I from minutes to 12–18 h (Guler *et al.*, 1989). Approximately 20–30% of total serum IGF is bound to a smaller 40- to 50-kDa complex, which can leave the vascular compartment and has a much shorter halflife of 20–30 min (Guler *et al.*, 1989). All of the proteins of the circulating ternary complexes, i.e., IGF-I, IGFBP-3, IGFBP-5, and ALS, have been demonstrated to be GH dependent (Rajaram *et al.*, 1997).

Less than 1% of total serum IGF-I is thought to circulate in the free form, although this form may be the most biologically active and levels may vary in response to changes in metabolic variables through their impact on circulating IGFBP-1 levels (Lee *et al.*, 1997). IGFBP-1 was the first IGFBP to be identified and characterized (Lee *et al.*, 1993). Brinkman *et al.* (1988) identified a single 1.5-kb mRNA species in human fetal liver but not in any other fetal organ tissues. In adult humans, IGFBP-1 mRNA has been identified in only liver, placental membrane, and decidua (Brewer *et al.*, 1988). Insulin is the dominant regulatory influence on IGFBP-1 and has been shown to be a potent inhibitor of IGFBP-1 mRNA levels (Powell *et al.*, 1991). In most *in vitro* studies, IGFBP-1 has an inhibitory effect on IGF activity, and *in vivo* studies have shown an inverse relationship between IGF-I and growth. Many studies have suggested that a central role for IGFBP-1 might be limitation of the considerable hypoglycemic and growth-promoting potential of free IGF-I during periods of low substrate availability.

The IGFBP-1 gene is one of the most highly expressed immediate early genes in regenerating liver (Mohn *et al.*, 1991). Transcription is increased at 30 min, and IGFBP-1 mRNA is increased 250-fold 1 to 4 h after partial hepatectomy in rats pretreated with cycloheximide. *In situ* hybridization studies by Ross and co-workers (1996) showed the widespread presence of IGFBP-1 mRNA throughout the hepatic sinusoids in normal human liver but that expression was limited to hepatocytes in regenerative nodules of human cirrhotic liver.

C. IGFBPs and the Regulation of IGF Bioavailability

Studies have broadened the possible range of functions for these proteins (Rosenfeld *et al.*, 1999). Most IGFBPs bind IGF-I with greater affinity than the IGF receptor, thereby preventing the activation of intracellular signaling pathways. Posttranscriptional modifications such as proteolysis and phosphorylation have been shown to modify the binding affinities of the IGFs for their IGFBPs and may increase the quantity of free IGF-I available to the IGF-I receptor. In addition to modifying the access of IGFs to IGF receptors, several of the IGFBPs may be capable of increasing IGF action and some may be capable of IGF-independent regulation of cell growth. IGFBPs may associate with cell membranes or membrane receptors, and some IGFBPs have nuclear recognition sites and may be found within the cell nucleus (Jones and Clemmons, 1995; Rosenfeld *et al.*, 1999).

II. THE STATE OF ACQUIRED GROWTH HORMONE RESISTANCE

A. Definition of Acquired GH Resistance

GH resistance is classically defined as low circulating levels of IGF-I in the presence of increased secretion of GH from the anterior pituitary gland. It implies a failure of the target organ to respond fully to the stimulus of high circulating levels of GH. It also implies the absolute or relative failure of a target organ to produce "normal" quantities of IGF or other GH-dependent proteins in response to GH stimulation. The liver is the main source of circulating levels of IGF-I and as such is considered the main target organ determining the bioavailability of IGF-I as an endocrine growth factor. However, IGF-I is produced by most tissues, and while our knowledge of tissue production is far more limited, it is believed that IGF-I exerts powerful growth stimulatory influences locally via paracrine and autocrine mechanisms.

B. Evolutionary Advantage of GH Resistance

Acquired GH resistance is a commonly recognized feature of protein catabolic states, in which breakdown of lean body mass results in changes in body composition, loss of muscle strength, and impaired immune responses. It is considered to be a secondary adaptive response, contrasting with inherited conditions of "primary GH resistance," such as Laron dwarfism in which there is GH receptor deficiency or loss of function (Laron *et al.*, 1993). It is apparent in acquired GH resistance that there is a shift from the indirect growth-promoting actions of GH that are largely mediated by the IGFs to direct GH actions, such as lipolysis and insulin antagonism (Phillips and Unterman, 1984). In the setting of catabolism induced by injury or severe illness, these changes might provide substrates essential for short-term fuel needs but might be counterproductive in the mid- and longer term with respect to the maintenance of body composition.

III. THE GH:IGF:IGFBP AXIS IN THE PATHOGENESIS OF CIRRHOSIS

The pathogenesis of GH resistance in cirrhosis is multifactorial with contributions from hepatic factors such as hepatocyte loss, portosystemic shunting, and insulin resistance interplaying with extrahepatic factors such as nutritional intake and alcohol. Development of the cirrhotic lesion appears central to the pathogenesis of GH resistance, as patients with chronic, noncirrhotic liver disease do not exhibit such profound changes in the GH:IGF:IGFBP axis (Stewart *et al.*, 1983). Etiological relationships between changes in the GH:IGF:IGFBP axis and the development of cirrhosis, i.e., fibrogenesis, have been more difficult to define. The following section reviews the influence of both hepatic and extrahepatic factors on the GH-resistant state of cirrhosis, followed by a review of the evidence to date suggesting a direct link between low IGF bioavailability and liver injury, particularly the pathophysiological process of liver fibrosis.

A. Etiological Factors in the Pathogenesis of GH Resistance in Cirrhosis

1. Hepatocyte Loss

a. Hypothalamic and Pituitary Regulation of GH Secretion

Many authors have speculated that high serum GH levels are secondary to diminished feedback to the pituitary and perhaps also to the hypothalamus from low levels of hepatically produced IGF-I (Moller and Becker, 1992). Abnormalities at the hypothalamic–pituitary level of the GH:IGF axis in cirrhosis have been suggested by studies demonstrating abnormal GH responses to the growth hormone-releasing hormone (GHRH) (Salerno *et al.*, 1987), thyrotrophin-releasing hormone (TRH) (Zanobi *et al.*, 1983), and hyperglycemia (Muggeo *et al.*, 1979). Santolaria *et al.* (1995) demonstrated raised serum GH levels in 24 hospitalized alcoholic cirrhotics but noted that GHRH levels were not altered significantly. Stewart *et al.* (1983) showed that GH levels were increased in alcoholic cirrhosis but not in alcoholic hepatitis, suggesting that the development of cirrhosis was central to the development of GH resistance in these patients. Moller *et al.* (1993) described significantly elevated urinary GH levels in alcoholic cirrhotics reflecting the integrated serum GH concentration. Cuneo and colleagues (1995) documented increased frequency of GH secretory bursts, increased total daily GH secretion rates, and markedly impaired endogenous GH clearance in adult cirrhotic patients.

b. Serum GHBP

Earlier studies had demonstrated low serum levels of the high-affinity GHBP in cirrhosis, with decrements correlating with the degree of hepatocellular dysfunction (Baruch *et al.*, 1991; Hattori *et al.*, 1992). In contrast, the later study of Cuneo *et al.* (1995) demonstrated only a slight decrease in serum GHBP levels in their cirrhotic cohort. This discrepancy may be explained by the smaller number of subjects in this study group or by the different GHBP methodology used.

c. Hepatic GH Receptor Gene Expression and Labeled GH Binding

The high-affinity circulating GHBP is thought to be derived from the extramembranous domain of the GH receptor in humans (Barnard *et al.*, 1989) and it has been speculated that decreased GHBP levels might indicate influences of liver disease directly on the GH receptor. An early study by Chang *et al.* (1990) demonstrated decreased GH binding to cirrhotic liver by a radioreceptor assay. In a study of children with cirrhosis secondary to extrahepatic biliary atresia, Holt and co-workers (1997b) showed for the first time that hepatic GH receptor mRNA levels, measured by quantitative reverse transcriptase polymerase chain reaction, were reduced by 59% in cirrhotic liver.

d. Hepatic IGF-I Gene Expression

Gene expression studies in human cirrhotic liver have yielded somewhat conflicting results. Ross *et al.* (1996) identified multiple transcripts for IGF-I in cirrhotic liver by Northern analysis, with the 7.5-kb transcript dominant with smaller transcripts seen at 4 and 1.5 kb. This group found that IGF-I mRNA levels were highly variable in adult cirrhotic liver but not significantly different from that of normal liver. In contrast, the study of pediatric cirrhotic patients by Holt *et al.* (1997b) showed markedly reduced IGF-I gene expression. Possible explanations for this difference may have been that the latter group used an RNase protection assay and had a significantly larger number of control liver samples.

e. Serum IGF-I

The liver is thought to be the major source of circulating levels of IGF-I (Schwander *et al.*, 1983), and many historical studies have documented low levels of IGF-I in patients with liver disease of both alcoholic and nonalcoholic etiology (Russell, 1985). The lack of negative feedback of low IGF-I levels is thought to contribute to the high GH levels in this state of acquired GH resistance (Jenkins and Ross, 1996). The critical importance of the functional hepatocyte mass to the hepatic production of IGF-I has been demonstrated most dramatically in studies showing that low serum levels of IGF-I and IGFBP-3 return to control levels after successful orthotopic liver transplantation in children (Holt *et al.*, 1996) and adults (Schalch *et al.*, 1998).

Moller *et al.* (1993) showed markedly reduced serum IGF-I levels in alcoholic cirrhotics with IGF-I levels reflecting the degree of hepatic insufficiency. Our group has confirmed low serum levels of IGF-I in patients with a broad range of liver disease etiologies and severities, and we have demonstrated a correlation with the Childs Pugh score as a measure of disease severity (Donaghy *et al.*, 1995) (Fig. 2). Scharf and co-workers (1996b) confirmed low levels of both total and free IGF-I in adult cirrhotic patients. This group also demonstrated that the increase

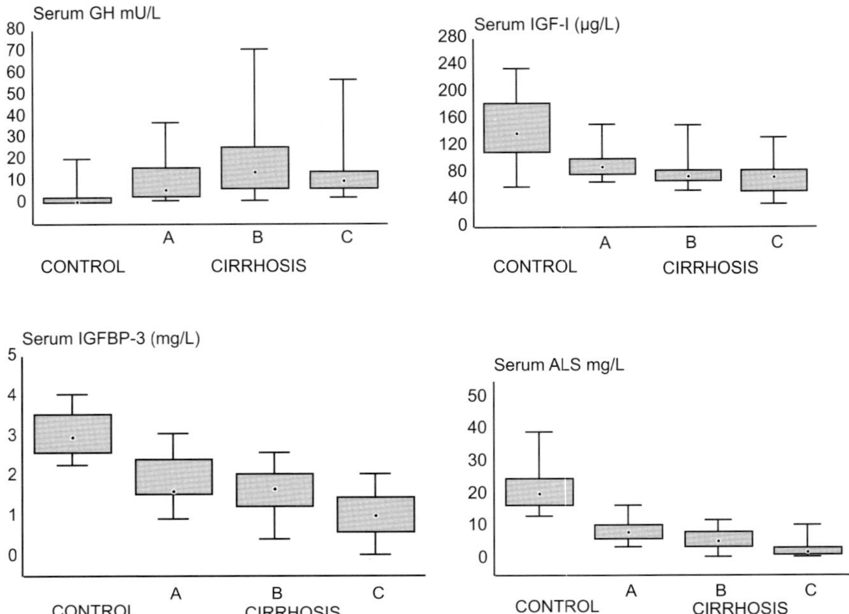

FIGURE 2 Serum levels of GH and the ternary complex proteins IGF-I and IGFBP-3 and ALS in cirrhosis. Box and whisker plots depict median, 25, and 75 percentiles and range. Data are depicted for controls and each of the Childs Pugh prognostic groups, with the Childs Pugh group C having the most severe disease.

in IGF-I levels in cirrhotic patients after GH injection was reduced markedly in patients with more severe disease (Childs groups B and C), reflecting a strong relationship between disease severity and this dynamic marker of GH resistance.

f. Hepatic IGFBP-3 Gene Expression

Ross *et al.* (1996) described the finding of a single 2.4-kb mRNA transcript for IGFBP-3 in both normal and cirrhotic human liver by Northern analysis. This group noted that IGFBP-3 mRNA levels tended to be higher in cirrhotic than normal liver but that this difference was not statistically different. In the study of pediatric cirrhosis, Holt *et al.* (1997b) did not detect any difference in IGFBP-3 mRNA levels between normal and cirrhotic liver by Northern analysis.

g. Serum IGFBP-3

In contrast to the findings of unchanged IGFBP-3 gene expression in cirrhotic liver, most authors have described significantly decreased serum levels of IGFBP-3.

The earliest report came from Zapf *et al.* (1980), who described significantly decreased IGF-binding activity, pointing to the liver as the site of production of the major IGF carrier protein later identified as IGFBP-3. Subsequent studies have confirmed low serum IGFBP-3 levels by immunoassay, and we and others have shown strong correlations with both clinical (ascites) and laboratory (albumin, INR) indicators of liver disease severity (Donaghy *et al.*, 1995; Scharf *et al.*, 1996; Shaarawy *et al.*, 1998).

The study of Scharf made further important observations about IGFBP-3 pathophysiology in cirrhosis. The significant increase in serum IGFBP-3 levels after a single dose of GH confirmed the growth hormone dependence of IGFBP-3. This group also noted profound alterations in IGF-binding protein complexes by gel filtration experiments. A reduction in the 150-kDa or ternary circulating complex was observed together with an increase in the 30- to 40-kDa or binary complex(es). GH injection did not alter ternary complex levels, but did increase those of the binary complex(es). These authors suggested that a decrease in serum ALS levels might be rate limiting for ternary complex production.

IGFBP-3 has been shown to undergo posttranslational modification, namely proteolysis, in states of human illness (Cwyfan *et al.*, 1992). However, two studies have failed to demonstrate IGFBP-3 proteolysis in cirrhosis (Moller *et al.*, 1995; Ross *et al.*, 1996). These studies, together with those showing undiminished IGFBP-3 gene expression in cirrhosis, have suggested that low IGFBP-3 serum levels may be due to increased clearance of IGFBP-3 from the circulation rather than proteolytic degradation.

h. Acid-Labile Subunit

ALS gene expression in cirrhotic human liver has not been described to date. Holt *et al.* (1998) showed that serum ALS levels were decreased significantly in children with end stage liver disease and, interestingly, remained low after liver transplantation. We have shown that serum levels of ALS are reduced significantly in cirrhosis and that ALS levels demonstrated the strongest correlation with disease severity of all of the components of the circulating ternary complex (unpublished data).

i. IGFBP-1 Gene Expression

The first report of IGFBP-1 gene expression in cirrhotic liver came from Ross *et al.* (1994), who described a single mRNA transcript at 1.5 kb in both cirrhotic and control liver samples. This group found IGFBP-1 mRNA levels to be highly variable in cirrhotic liver but not significantly different from controls. Serum insulin levels were also high in the cirrhotic group without the usual inverse relationship between insulin and IGFBP-1 mRNA levels. A later study from that group showed abundant IGFBP-1 gene expression in hepatocytes of regenerative nodules by *in situ* hybridization, providing possible further evidence, as has been suggested by *in vitro* studies, of increased IGFBP-1 mRNA levels in liver regeneration. In the study of pediatric cirrhotic liver, Holt *et al.* (1997b) also found no difference in IGFBP-1 mRNA levels between cirrhotic and control liver.

j. Serum IGFBP-1

The majority of reported studies have demonstrated increased serum levels of IGFBP-1 in cirrhosis (Ross *et al.*, 1994; Donaghy *et al.*, 1995; Holt *et al.*, 1996, 1997b). Ross *et al.* (1997) showed increased serum levels of IGFBP-1 in adult patients undergoing liver transplantation. These patients also had elevated serum insulin levels with a lack of the usual inverse relationship between insulin and IGFBP-1, suggesting hepatic insulin resistance (Donaghy *et al.*, 1995). Holt and coworkers (1997b) showed markedly increased serum IGFBP-1 levels in children with cirrhosis complicating extrahepatic biliary atresia. In a separate study, the same group showed that IGFBP-1 levels fell after liver transplantation and that the inverse correlation with insulin was restored only after transplantation (Holt *et al.*, 1996).

2. Portosystemic Shunting

The finding of low levels of serum IGF-I in hepatic schistosomiasis (a postinfective illness characterized by portal hypertension but normal hepatocyte function) (Assaad, 1990) suggested that portosystemic shunting might be a significant factor in the pathogenesis of the GH resistance of cirrhosis. In an elegant study, Moller *et al.* (1995) found strong correlations between the wedged hepatic venous pressure (a measure of portal hypertension) and high GH, low IGF-I, and low IGFBP-3 levels in a group of alcoholic cirrhotics. These findings suggested that the shunting of nutrient-rich blood from the portal vasculature away from the hepatic sinusoids had at least an additional effect to the hepatocyte loss of cirrhosis. A possible explanation for these findings may come from the early work of Baxter and Turtle (1978) demonstrating in a rat model that significant concentrations of insulin are required for optimal binding of GH to hepatocyte GH receptors.

3. Nutritional Intake

Jackson-Smith *et al.* (1995) showed that short-term caloric or protein restriction in normal humans resulted in decreased levels of IGF-I and IGFBP-3 in association with a deterioration in nitrogen balance. The pattern of dietary intake in cirrhotic patients, particularly those patients with alcoholic cirrhosis, is often altered substantially. With increasing alcohol intake, the percentage of energy derived from protein, fat, and carbohydrate decreases and the nutritional quality of the diet declines. Mendenhall *et al.* (1984) observed a direct correlation between the prevalence of malnutrition and daily total energy intakes in patients with alcoholic liver disease (ALD). Sarin *et al.* (1997) showed that intakes of protein and carbohydrate estimated by dietary recall were significantly lower in patients with ALD than controls, although they were not significantly lower than the intakes of nonalcoholic cirrhotics or alcoholics without liver disease. This study also demonstrated that patients with severe alcoholic hepatitis had the lowest intakes of protein, carbohydrate, and fat.

4. Alcohol

Alcohol has been shown to exert direct and deleterious effects on the GH:IGF:IGFBP axis separate and in addition to those seen in patients with the pathological lesion of cirrhosis. A number of animal studies have shown that both acute and chronic ethanol administration decreases circulating IGF-I levels. Sonntag and Boyd (1989) showed that chronic ethanol exposure decreased plasma IGF-I concentrations compared to either pair-fed or *ad libitum*-fed animals, but that this was not caused by a further reduction in overall GH secretory dynamics. In a later study, Xu *et al.* (1995) showed that ethanol inhibited GH-induced protein synthesis and IGF-I gene expression in rat liver slices, but did so without changing GH receptor number or the affinity of GH for its receptor. These findings importantly suggested that ethanol directly suppressed GH-induced signal transduction. Srivastava *et al.* (1995) showed that chronic ethanol-fed rats showed a significant decrease in serum IGF-I levels with a parallel decrease in the expression of hepatic IGF-I mRNA.

IGFBPs influence IGF bioavailability by regulating the delivery and binding of IGFs to tissue IGF receptors. Block *et al.* (1997) demonstrated that levels of a 39- to 42-kDa IGFBP, likely IGFBP-3, were decreased in a chronic ethanol rat model, suggesting the possible impact of ethanol on longer term IGF bioavailability through decreased levels of the circulating ternary complex and therefore a decreased IGF-I half-life.

Lang *et al.* (1998) showed that chronic alcohol feeding decreased serum IGF-I levels and IGF-I mRNA levels in liver and skeletal muscle of rats. In addition, this group demonstrated increased IGFBP-1 concentrations in plasma, liver, and skeletal muscle in the alcohol-consuming animals and that these changes were independent of changes in insulin, GH, or glucocorticoid levels. Flyvberg *et al.* (1997) showed that acute oral ethanol administration increased liver IGFBP-1 gene expression by 218% with a concomitant rise in circulating levels of IGFBP-I identified by immunoprecipitation and Western ligand blotting. Ethanol may increase levels of cyclic AMP, which is a potent transcriptional regulator of IGFBP-1 (Babajko, 1995). In a study of acute alcohol ingestion in nine healthy volunteers, Knip *et al.* (1995) demonstrated a dose-dependent increase in serum levels of

IGFBP-1, despite simultaneous hyperinsulinemia. Insulin is usually a potent inhibitor of IGFBP-1 gene transcription (Powell *et al.*, 1991), with these findings implying that ethanol has a direct stimulatory effect on hepatic IGFBP-1 synthesis. In a study of alcohol withdrawal in noncirrhotic humans Passilta *et al.* (1999) showed that, in addition to a rise in serum IGF-I levels, serum IGFBP-1 levels decreased by 59% and that this change correlated inversely with a rise in insulin levels. This study elegantly demonstrated the effects of alcohol on the regulation of IGF bioavailability and that inhibitory regulation of IGFBP-1 by insulin was maintained during the alcohol withdrawal period. High levels of IGFBP-1 may bind increasing quantities of free or bioavailable IGF-I, thereby inhibiting tissue growth and anabolism.

A number of studies have demonstrated that ethanol may inhibit the IGF-I receptor, through which IGF-I exerts its cellular mitogenic and proliferative properties. Most cells in primary culture require the interaction of IGF-I with the IGF-I receptor to enter the S phase of the cell cycle. Ethanol has been shown not to interfere with the binding of IGF-I to its receptor but to cause profound interference with ligand-mediated tyrosine autophosphorylation of the IGF-I receptor in a 3T3 fibroblast cell line (Resnicoff *et al.*, 1993) and in C6 rat glioblastoma cells (Resnicoff *et al.*, 1994). Tyrosine autophosphorylation of the IGF-I receptor is essential for its mitogenic function. The targeting of ethanol to the IGF-I receptor was further emphasized by the findings of inhibition of downstream targets for IGF-I receptor tyrosine kinase, such as insulin-related substrate-1 and phosphatidylinositol 3-kinase (Resnicoff *et al.*, 1994). In these studies, the ethanol-induced effects on IGF-I signal transduction affected profound decreases in IGF-I-mediated cell growth. These important findings imply an important interactive role for IGF-I in the deleterious effects of ethanol on important physiological processes, such as fetal development, hepatocellular regeneration, and the maintenance of lean body mass.

B. Direct Influences of GH:IGF-I on Liver Growth

1. Normal Liver Growth

The impaired growth of normal liver in GH-deficient transgenic mice is only partially corrected by overexpression of IGF-I in double transgenic animals, suggesting the importance of GH in the maintenance of normal liver growth, possibly mediated in an autocrine/paracrine manner by hepatic IGF-I that is expressed in response to circulating GH (Behringer *et al.*, 1988). Normal human liver has virtually no IGF-I-binding sites, but Caro *et al.* (1988) have demonstrated significant increases in IGF-I binding to hepatocytes from both fetal and regenerating rat liver, suggesting autocrine regulation of liver growth by IGF-I during physiological and pathological conditions of growth.

2. Liver Regeneration

GH and IGF-I influence normal liver growth with transgenic mouse studies showing the liver to be hypoplastic in GH-deficient mice (Behringer *et al.*, 1988). GH and IGF-I have also been shown to have significant effects on liver regeneration in experimental models. Unterman and Phillips (1986) showed that liver regeneration was retarded in hypophysectomized rats, and GH has been shown to accelerate liver regeneration, possibly by promoting the early initiation of hepatocyte growth factor gene expression (Ekberg *et al.*, 1992).

3. Fibrogenesis

A direct link between the GH:IGF-I axis and liver fibrogenesis has been more difficult to define; however, recent interesting studies have important implied direct relationships. In a study of IGF-I receptor numbers in stellate cells, Brenzel and Gressner (1996) showed that hepatocyte-generated IGF-I or IGFBPs might mediate stellate cell activation during the initial transformation to myofibroblasts. Svegliati-Baroni *et al.* (1999) showed that both IGF-I and insulin-stimulated hepatic stellate cell (HSC) proliferation in a dose-dependent fashion with IGF-I was five times more potent than insulin. IGF-I increased type I collagen gene expression and accumulation in HSC culture media through a P13-K- and ERK-dependent mechanism. These studies have suggested a direct role for IGF-I in stellate cell activation, proliferation, and collagen production.

In an experimental model of cirrhosis, Castilla *et al.* (1997a) showed that IGF-I had hepatoprotective effects by improving liver function and reducing oxidative liver damage and fibrosis in rats with compensated or advanced cirrhosis. This particular decrease in prooxidant liver injury and resulting fibrogenesis might be extremely important in conditions such as alcoholic cirrhosis where oxidative liver damage is a major pathogenic mechanism. A further recent study from that group showed that IGF-I reduced all of the studied parameters of fibrogenesis in cirrhotic rats: hydroxyproline content, prolyl hydroxlase activity, and collagen $\alpha 1(\mathrm{I})$ and $\alpha 1(\mathrm{III})$ mRNA expression (Muguerza *et al.*, 2001). IGF-I treatment was also associated with a decreased number of transformed myofibroblasts. These findings indicate that IGF-I appears to have opposite effects on fibrogenesis at either the cellular or the whole body level and raise the possibility that IGF-I therapy at the whole body level may induce other changes, such as in IGF-binding protein levels, that may alter IGF bioavailability significantly.

IV. GH:IGF:IGFBP PATHOPHYSIOLOGY AND CIRRHOSIS COMPLICATIONS

A number of studies have suggested that low levels of IGF-I, particularly in association with changes in the profiles of the IGFBPs, which further reduce IGF-I bioavailability, may play an important role in the pathogenesis of the complications of cirrhosis (Table I).

A. Starvation Metabolism

The impaired capacity of the cirrhotic liver to store glycogen contributes to a metabolic state in cirrhosis resembling starvation with early recruitment of alternative fuel stores during periods of fasting. Indeed the glycogen stores of the cirrhotic liver may be depleted in less than 6 h of fasting compared to 36–48 h in normals, necessitating the breakdown of tissue fat and protein stores to provide fuel substrates for gluconeogenesis (Owen *et al.*, 1983). Resistance to the actions of the anabolic growth factors GH and insulin is central to the pathogenesis of this starvation metabolism. An early study by Freemark *et al.* (1985) showed that somatomedin C, now known as IGF-I, was 20-fold more potent than insulin in stimulating hepatic glycogen formation. This interesting and infrequently quoted finding begs the speculation that low levels of bioavailable IGF-I may have a direct pathogenic role in the impairment of hepatic glycogen storage capacity, which is arguably the dominant metabolic lesion of cirrhosis.

B. Growth Retardation

Liver transplantation is now an established therapy for advanced liver disease in children, many of whom are significantly growth retarded. IGF-I is the dominant regulator of somatic growth before puberty, and reduced IGF bioavailability is believed to play a dominant pathogenic role. Quirk *et al.* (1994) showed that plasma levels of IGF-I and IGFBP capacity were reduced significantly in growth-retarded children with end stage liver disease compared to normally growing, age-matched children with cystic fibrosis. A number of studies have shown that serum levels of IGF-I, IGFBP-3, and IGFBP-1 can return to control values in the postorthotopic liver transplantation (OLT) period in children and adults (Holt *et al.*, 1996; Schalch *et al.*, 1998). OLT, however, did not normalize all components of the GH:IGF:IGFBP axis with, in particular, serum ALS levels not reaching control levels post-OLT (Holt *et al.*, 1997a).

Infante *et al.* (1998) showed an increase in height 4 years post-OLT in association with the increase in serum IGF-I levels and that catch-up growth was highly significant, particularly in the first year posttransplantation (Infante *et al.*, 1998).

TABLE I Summary of Pathogenic Factors Involving the GH:IGF:IGFBP Axis in Disease Pathogenesis and Cirrhosis Complications

Pathogenesis
Hepatocyte loss
Portosystemic shunting
Nutritional intake
Alcohol
Starvation metabolism
Growth retardation
Complications
Protein calorie malnutrition
Insulin resistance
Osteoporosis

Holt *et al.* (1997a) showed that serum levels of IGFBP-1 and IGFBP-2 correlated negatively with pretransplant nutritional status as measured by anthropometry. This study also noted that growth postorthotopic liver transplantation correlated positively with serum IGF-I and negatively with IGFBP-1. These findings overall strongly suggest that OLT may reverse the GH-resistant state of cirrhosis and that this improvement is integral to the catch-up growth observed in these patients.

C. Protein Calorie Malnutrition

The prevalence of protein calorie malnutrition has been reported to be as high as 100% in ALD (Mendenhall *et al.*, 1995), and the adverse impact of malnutrition on clinical outcome has been highlighted in many studies (Mendenhall *et al.*, 1986; Shaw *et al.*, 1985). In the Veterans Affairs Study of alcoholic hepatitis, low levels of IGF-I were noted but, interestingly, partial correlation analysis showed that malnutrition correlated with IGF-I independently of liver dysfunction and histopathological alterations (Mendenhall *et al.*, 1989). This finding has suggested that IGF-I levels might be regulated more dominantly by nutritional factors than liver function in ALD.

However, a later study of 64 hospitalized alcoholic patients (Caregaro *et al.*, 1997), while demonstrating a diminished 2-year survival in patients with a low IGF-I z score, did not note any significant correlation of IGF-I with anthropometric indices. These findings might be explained by the increased disease severity seen in hospitalized alcoholics, but probably highlight the multipathogenesis of the low IGF-I levels seen in advanced cirrhosis.

Inaba and colleagues (1999) have shown that disturbances in the GH:IGF axis can contribute to the disturbed protein metabolism observed in patients after hepatectomy. In this study, perioperative IGF-I levels were lower in those patients who developed postoperative complications, and the postoperative serum IGF-I level showed a positive correlation with the whole body protein turnover rate.

D. Insulin Resistance

The insulin resistance of alcoholic cirrhosis is of multifactorial pathogenesis and remains incompletely understood. While the current major deficit is thought to be of nonoxidative glucose disposal, i.e., impaired glycogen synthesis in skeletal muscle, there are many other suggested contributing factors (Petrides *et al.*, 1992), including high circulating levels of the counterregulatory hormones glucagon, cortisol, and GH. Studies have suggested a pathogenic role for the GH:IGF axis both directly and indirectly through relationships of malnutrition to insulin resistance.

Shankar *et al.* (1988) showed that suppression of GH with somatostatin in euglycemic clamp studies resulted in significantly improved insulin sensitivity. This study of alcoholic cirrhotic patients suggested a direct pathogenic role of high GH levels in the insulin resistance of these patients. However, in a later study, Shmueli *et al.* (1994) could not show any significant effect of GH suppression on whole body glucose uptake, forearm glucose uptake, or insulin sensitivity. The differences in these study findings may have been at least partially explained by the ongoing alcohol intake of the patients in the earlier study.

Shmueli *et al.* (1996) also showed that high serum IGFBP-1 levels seen in their alcoholic cirrhotic group showed a strong inverse correlation with insulin sensitivity. Serum levels of IGFBP-1 fluctuate in response to metabolic variables, predominantly insulin and substrate availability (Lee *et al.*, 1993), with high IGFBP-1 levels believed to decrease IGF-I bioavailability. This finding further strongly suggests the potential significance of low IGF-I bioavailability to the insulin resistance of cirrhosis.

Wahl *et al.* (1992) showed strong correlations of serum markers of protein energy malnutrition (PEM) to insulin sensitivity as measured by euglycemic insulin clamping. The finding that nutritional support improved insulin sensitivity in these alcoholic cirrhotic patients gives further evidence of a relationship of PEM to insulin resistance in this patient group. The close relationship of low IGF-I bioavailability to PEM suggests a possible additional indirect role for low IGF-I levels in the insulin resistance of cirrhosis.

E. Osteoporosis

Significant decreases in bone mineral density have long been described in patients with cholestatic liver disease, with malabsorption of vitamin D thought to be pathogenic. However, more recent studies have demonstrated osteopenia and osteoporosis in cirrhosis of noncholestatic etiology, suggesting that the cirrhotic lesion itself may be etiological. Gallego-Rojo *et al.* (1998) found low serum IGF-I levels in patients with cirrhosis complicating chronic viral hepatitis and noted that those patients with osteoporosis had the lowest serum IGF-I levels. These findings suggest that low serum IGF-I levels might play a role in the bone mass loss in these patients.

V. GROWTH HORMONE AND IGF-I AS THERAPY IN CIRRHOSIS

In severe catabolic illness, including cirrhosis, a number of studies have demonstrated that aggressive nutritional support alone does not prevent significant loss of lean body mass (Streat *et al.*, 1987). Recombinant human GH therapy has been shown to improve protein anabolism and to have an impact on short-term morbidity after abdominal surgery (Ward *et al.*, 1987), major trauma and severe sepsis (Douglas *et al.*, 1990), and severe burns (Herndon *et al.*, 1990). GH and IGF-I therapy have been reported in human and experimental cirrhosis (Table II).

A. GH Intervention in Human Cirrhosis

1. Effects on the GH:IGF:IGFBP Axis

In the first reported study of GH therapy in human liver disease, Moller *et al.* (1994) showed significant increases in serum levels of IGF-I in a group of male alcoholic cirrhotics treated with recombinant human GH for 6 weeks.

TABLE II Summary of Outcomes Assessed in GH:IGF-I Therapeutic Studies in Human and Experimental Cirrhosis

Human
GH:IGF:IGFBP axis
Body composition
Somatic growth
Nitrogen retention and food efficiency
Immunity and anabolism
Animal
Glucose metabolism
Immunity and anabolism
Osteopenia
Hypogonadism

Serum IGFBP-3 levels increased in both treatment and placebo groups, whereas no changes in serum IGFBP-1 levels were seen. In a study designed to assess the responses of GH:IGF:IGFBP parameters to a single dose of rhGH in adult cirrhotic patients, Scharf *et al.* (1996b) demonstrated that GH affected significant increases in serum levels of free and total IGF-I and IGFBP-3.

2. Nitrogen Balance and Body Composition

We have examined the therapeutic effect of high-dose exogenous GH therapy in a group of 20 patients of varying liver disease etiology and severity. We observed an attenuated but significant increase in serum levels of all of the ternary complex proteins (Fig. 3). We also observed a significant improvement in cumulative nitrogen balance in the GH-treated group, suggesting that the acquired GH resistance of cirrhosis could at least be partially overcome and that the improvement in IGF bioavailability could affect an improvement in nitrogen retention (Donaghy *et al.*, 1997). No significant side effects of GH therapy were observed in this study.

In a study of adult cirrhotics treated with rhGH for 1 month, Wallace and co-workers (1997) confirmed the GH-induced increases in IGF-I and IGFBP-3 shown in the previous adult cirrhosis study. This group also demonstrated that GH affected improvement in lean body mass as measured by total body potassium counting. Total body weight and body water were also increased in association with clinically significant fluid retention. This study also showed that hepatocyte function, as measured by the MEGX (Lignocaine extraction) test, was not altered by GH therapy.

3. Somatic Growth

In a placebo-controlled, double-blind, crossover study, Greer *et al.* (1998) saw no effect of high-dose rhGH therapy on the GH:IGF:IGFBP axis, anthropometry, or body composition measurements in 12 pediatric cirrhotic patients. In this study, it would appear that the degree of severity of the liver disease and therefore of the GH resistance in these children may have prevented a hepatically produced growth factor response in these patients. In contrast, Maghnie *et al.* (1998) showed a two- to threefold increase in growth rate in a group of two children with inherited liver disease treated for 1 year with high-dose rhGH therapy, but there was no control group in this study.

In a study of rhGH therapy for 1 year in eight growth-retarded children postliver transplantation, Sarna and colleagues (1996) reported significant increases in serum IGF-I and IGFBP-3 levels in association with accelerated growth rates. The change in the height standard deviation score correlated significantly with the serum IGF-I level at 6 months. While these findings stress the integral relationship between growth and the GH:IGF:IGFBP axis, there was no

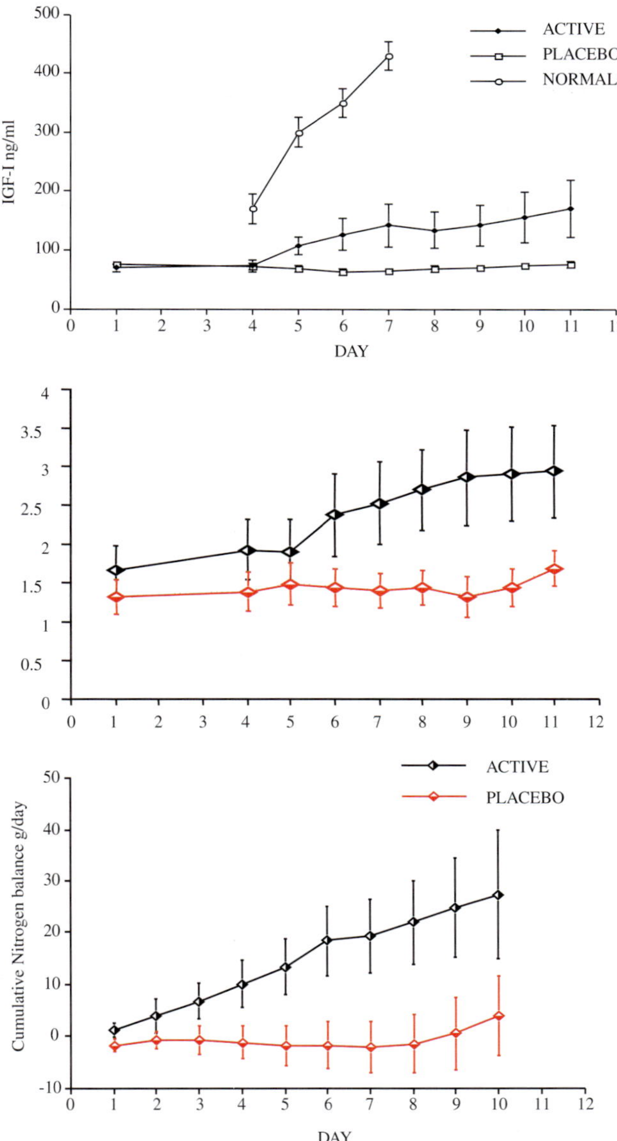

FIGURE 3 Impact of exogenous GH therapy on serum levels of IGF-I, IGFBP-3, and cumulative nitrogen balance in cirrhosis. Serum IGF-I (μg/liter) levels rose significantly in the active GH treatment group with no significant change seen in the placebo group. However, this rise was attenuated markedly compared with a historical group of normals. Serum IGFBP-3 (mg/liter) levels rose significantly in the active GH treatment group with no significant change seen in the placebo group. Total cumulative nitrogen balance increased significantly only in the GH-treated group with placebo arm patients remaining in net neutral nitrogen balance.

control group in this study to allow accurate comparison with unstimulated growth rates in these posttransplant children.

Studies of IGF-I therapy in human cirrhosis have not been reported, likely because of the risks of IGF-I-induced hypoglycemia in patients predisposed to this problem because of impaired glycogen storage.

B. GH and IGF-I Therapy in Animal Models

1. Nitrogen Retention and Food Efficiency

Picardi *et al.* (1997) demonstrated that low doses of IGF-I improved nitrogen retention and food efficiency in rats with early stage cirrhosis. Exogenous IGF-I therapy was associated with increased food intake and improved nutrient utilization, confirmed by the enhancement of the incorporation of labeled nitrogen into gastrocnemius muscle. In this study, possible mechanisms of IGF-I action included enhancement of appetite and improvements in substrate utilization for energy generation and muscle protein metabolism. This study also made the interesting observation that mild but statistically significant muscle wasting was already present in nonascitic cirrhotic rats, suggesting that protein calorie malnutrition might develop early in the evolution of cirrhosis independently of the cirrhotic complications of ascites, variceal hemorrhage, and sepsis, which undoubtedly contribute to its progression.

2. Immunity and Anabolism

Mendenhall and co-workers (1997) showed that both exogenous GH and IGF-I therapy partially reversed the weight loss seen in severely dietary-restricted, undernourished, alcohol-fed rats. This group also demonstrated that IGF-I but not GH produced significant improvements in immune parameters such as thymus and splenic T lymphocyte counts. An interesting further finding in this study was that neither GH nor IGF-I could restore the functional status of splenic T lymphocytes in the presence of ongoing caloric restriction and alcohol intake. The previously noted direct suppression by alcohol on IGF-I signal transduction might explain these findings.

3. Glucose Metabolism

In a well-designed intervention study, using the euglycemic clamp methodology in cirrhotic rats, Petersen *et al.* (1997) showed that IGF-I was able to improve the parameters of insulin resistance. In particular, during the IGF-I clamp, glucose turnover was increased threefold, rates of muscle glycogen synthesis were increased 7.4 times higher than during insulin stimulation, and endogenous hepatic glucose production was suppressed significantly in the cirrhotic rats. These findings strongly suggest an integral role for low IGF-I levels in the insulin resistance of cirrhosis.

4. Osteopenia

In a rat model of cirrhosis, Cemborain *et al.* (1998) noted significant reductions in bone weight, bone density, and markers of bone resorption. Twenty-one days of recombinant IGF-I therapy improved these parameters significantly and restored bone mass partially. These experimental findings are confirmatory of those described previously in human studies wherein the most severe osteoporosis was seen in those patients with the lowest IGF-I levels (Gallego-Rojo *et al.*, 1998).

5. Hypogonadism

A provocative study by Castilla *et al.* (1997a) reported that low doses of IGF-I were able to revert the testicular atrophy seen in rats with advanced cirrhosis. IGF-I therapy also partially restored gonadal hormone levels to control.

While the number of reported studies in humans remains small, the demonstration that GH-induced improvements in nitrogen balance in the short term can lead to improved nutritional status with prolonged therapy holds significant promise. Further studies of longer duration are required to assess the impact on clinical outcome in severely malnourished cirrhotics, many of whom succumb while awaiting liver transplantation. The improvements in indices of cirrhotic complications such as protein catabolism, insulin resistance, osteoporosis, and hypogonadism seen in IGF-I intervention studies in animal models of cirrhosis strongly support the role of reduced IGF-I bioavailability in their pathogenesis. Further studies are required to elucidate the role of changes in IGF bioavailability in the pathogenesis of fibrogenesis in cirrhosis.

REFERENCES

Babajko, S. (1995). Transcriptional regulation of insulin-like growth factor-binding protein-1 expression by insulin and cyclic AMP. *Growth Regul.* 5(2), 83–91.

Barnard, R., Quirk, P., *et al.* (1989). Characterization of the growth hormone-binding protein of human serum using a panel of monoclonal antibodies. *J. Endocrinol.* 123(2), 327–332.

Baruch, Y., Amit, T., *et al.* (1991). Decreased serum growth hormone-binding protein in patients with liver cirrhosis. *J. Clin. Endocrinol. Metab.* 73(4), 777–780.

Baxter, R. (1990). Circulating levels and molecular distribution of the acid labile subunit of the high molecular weight growth factor-binding protein complex. *J. Clin. Endocrinol. Metab.* 70, 1347–1353.

Baxter, R. (1993). Circulating binding proteins for the insulin-like growth factors. *Trends Endocrinol. Metab.* **4**, 91–96.

Baxter, R., and Turtle, J. (1978). Regulation of hepatic growth hormone receptors by insulin. *Biochem. Biophys. Res. Commun.* **84**(2), 350–357.

Behringer, R. R., Mathews, L. S., *et al.* (1988). Dwarf mice produced by genetic ablation of growth hormone-expressing cells. *Genes Dev.* **2**(4), 453–461.

Block, G., Styche, A., *et al.* (1997). Effects of ethanol on insulin-like growth factor binding proteins in rats. *Gastroenterology* **104**, A878.

Brenzel, A., and Gressner, A. M. (1996). Characterization of insulin-like growth factor (IGF)-I-receptor binding sites during in vitro transformation of rat hepatic stellate cells to myofibroblasts. *Eur. J. Clin. Chem. Clin. Biochem.* **34**(5), 401–409.

Brewer, M. T., Stetler, G. L., *et al.* (1988). Cloning, characterization, and expression of a human insulin-like growth factor binding protein [published erratum appears in *Biochem. Biophys. Res. Commun.* **155**(3), 1485 (1988)]. *Biochem. Biophys. Res. Commun.* **152**(3), 1289–1297.

Brinkman, A., Groffen, C., *et al.* (1988). Isolation and characterisation of a cDNA encoding the low molecular weight insulin-like growth factor binding protein (IBP-1). *EMBO J.* **7**, 2417–2423.

Caregaro, L., Alberino, F., *et al.* (1997). Nutritional and prognostic significance of insulin-like growth factor 1 in patients with liver cirrhosis. *Nutrition* **13**(3), 185–190.

Caro, J. F., Poulos, J., *et al.* (1988). Insulin-like growth factor I binding in hepatocytes from human liver, human hepatoma, and normal, regenerating, and fetal rat liver. *J. Clin. Invest.* **81**(4), 976–981.

Castilla, C. I., Garcia, M., *et al.* (1997a). Low Doses of Insulin-like Growth Factor (IGF-I) Revert the Testicular Atrophy Associated to Advanced Liver Cirrhosis. European Association for the Study of the Liver, London.

Castilla, C. I., Garcia, M., *et al.* (1997b). Hepatoprotective effects of insulin-like growth factor I in rats with carbon tetrachloride-induced cirrhosis. *Gastroenterology* **113**(5), 1682–1691.

Cemborain, A., Castilla-Cortazar, I., *et al.* (1998). Osteopenia in rats with liver cirrhosis: Beneficial effects of IGF-I treatment. *J. Hepatol.* **28**(1), 122–131.

Chang, T. C., Lin, J. J., *et al.* (1990). Absence of growth-hormone receptor in hepatocellular carcinoma and cirrhotic liver. *Hepatology* **11**(1), 123–126.

Chin, E., Zhou, J., *et al.* (1994). Cellular localisation and regulation of gene expression for components of the insulin-like growth factor ternary binding protein complex. *Endocrinology* **134**(6), 2498–2504.

Cuneo, R. C., Hickman, P. E., *et al.* (1995). Altered endogenous growth hormone secretory kinetics and diurnal GH-binding protein profiles in adults with chronic liver disease. *Clin. Endocrinol. Oxf.* **43**(3), 265–275.

Cwyfan, H. S., Cotterill, A. M., *et al.* (1992). The induction of specific proteases for insulin-like growth factor-binding proteins following major heart surgery. *J. Endocrinol.* **135**(1), 135–145.

Dai, J., Scott, C. D., *et al.* (1994). Regulation of the acid-labile subunit of the insulin-like growth factor complex in cultured rat hepatocytes. *Endocrinology* **135**(3), 1066–1072.

D'Ercole, A. J., Stiles, A. D., *et al.* (1984). Tissue concentrations of somatomedin C: Further evidence for multiple sites of synthesis and paracrine or autocrine mechanisms of action. *Proc. Natl. Acad. Sci. USA* **81**(3), 935–939.

Donaghy, A., Ross, R., *et al.* (1995). Growth hormone, insulinlike growth factor-1, and insulinlike growth factor binding proteins 1 and 3 in chronic liver disease. *Hepatology* **21**(3), 680–688.

Donaghy, A., Ross, R., *et al.* (1997). Growth hormone therapy in patients with cirrhosis: A pilot study of efficacy and safety. *Gastroenterology* **113**(5), 1617–1622.

Douglas, R., Humberstone, D., *et al.* (1990). Metabolic effects of recombinant human growth hormone: Studies in the postabsorptive state and during total parenteral nutrition. *Br. J. Surg.* **77**, 787–790.

Ekberg, S., Luther, M., *et al.* (1992). Growth hormone promotes early initiation of hepatocyte growth factor gene expression in the liver of hypophysectomised rats after partial hepatectomy. *J. Endocrinol.* **135**(1), 59–67.

Flyvberg, A., van Neck, J., *et al.* (1997). Acute, Oral Ethanol Administration Stimulates Hepatic and Renal IGFBP-1 mRNA and Circulating IGFBP-1 in Rats. Fourth International Symposium on the IGFs, Tokyo.

Freemark, M., D'Ercole, A. J., *et al.* (1985). Somatomedin-C stimulates glycogen synthesis in fetal rat hepatocytes. *Endocrinology* **116**(6), 2578–2582.

Gallego-Rojo, F. J., Gonzalez-Calvin, J. L., *et al.* (1998). Bone mineral density, serum insulin-like growth factor I, and bone turnover markers in viral cirrhosis. *Hepatology* **28**(3), 695–699.

Greer, R. M., Quirk, P., *et al.* (1998). Growth hormone resistance and somatomedins in children with end-stage liver disease awaiting transplantation. *J. Pediatr. Gastroenterol. Nutr.* **27**(2), 148–154.

Guler, H. P., Zapf, J., *et al.* (1988). Recombinant human insulin-like growth factor I stimulates growth and has distinct effects on organ size in hypophysectomized rats. *Proc. Natl. Acad. Sci. USA* **85**(13), 4889–4893.

Guler, H. P., Zapf, J., *et al.* (1989). Insulin-like growth factors I and II in healthy man: Estimations of half-lives and production rates. *Acta Endocrinol. Copenh.* **121**(6), 753–758.

Hattori, N., Kurahachi, H., *et al.* (1992). Serum growth hormone-binding protein, insulin-like growth factor-I, and growth hormone in patients with liver cirrhosis. *Metabolism* **41**(4), 377–381.

Herndon, D. N., Barrow, R. E., *et al.* (1990). Effects of recombinant human growth hormone on donor site healing in severely burned children. *Ann. Surg.* **212**(4), 424–431.

Ho, K. K., and Hoffman, D. M. (1993). Aging and growth hormone. *Horm. Res.* **40**(1–3), 80–86.

Holt, R. I., Baker, A. J., *et al.* (1998). The insulin-like growth factor and binding protein axis in children with end-stage liver disease before and after orthotopic liver transplantation. *Pediatr. Transplant.* **2**(1), 76–84.

Holt, R. I., Broide, E., *et al.* (1997a). Orthotopic liver transplantation reverses the adverse nutritional changes of end-stage liver disease in children. *Am. J. Clin. Nutr.* **65**(2), 534–542.

Holt, R. I., Crossey, P., *et al.* (1997b). Hepatic growth hormone receptor, insulin-like growth factor I and insulin-like growth factor binding protein messenger RNA expression in paediatric liver disease. *Hepatology* **26**, 1600–1606.

Holt, R. I., Jones, J. S., *et al.* (1996). Sequential changes in insulin-like growth factor I (IGF-I) and IGF-binding proteins in children with end-stage liver disease before and after successful orthotopic liver transplantation. *J. Clin. Endocrinol. Metab.* **81**(1), 160–168.

Inaba, T., Saito, H., *et al.* (1999). Growth hormone/insulin-like growth factor 1 axis alterations contribute to disturbed protein metabolism in cirrhosis patients after hepatectomy. *J. Hepatol.* **31**(2), 271–276.

Infante, D., Tormo, R., *et al.* (1998). Changes in growth, growth hormone, and insulin-like growth factor-I (IGF-I) after orthotopic liver transplantation. *Pediatr. Surg. Int.* **13**(5–6), 323–326.

Jackson Smith, W., Underwood, L., *et al.* (1995). Effects of caloric or protein restriction on insulin-like growth factor-I(IGF-I) and IGF-binding proteins in children and adults. *J. Clin. Endocrinol. Metab.* **53**, 1247–1250.

Jenkins, R., and Ross, R. (eds.) (1996). Acquired growth hormone resistance in catabolic states. *In* "Bailliere's Clinical Endocrinology and Metabolism." Bailliere Tindall, London.

Jones, I., and Clemmons, D. (1995). Insulin-like growth factors and their binding proteins: Biological actions. *Endocr. Rev.* **16**(11), 3–34.

Knip, M., Ekman, A. C., *et al.* (1995). Ethanol induces a paradoxical simultaneous increase in circulating concentrations of insulin-like growth factor binding protein-1 and insulin. *Metabolism* **44**(10), 1356–1359.

Lang, C. H., Fan, J., *et al.* (1998). Modulation of the insulin-like growth factor system by chronic alcohol feeding. *Alcohol. Clini. Exp. Res.* **22**(4), 823–829.

Laron, Z., Blum, W., *et al.* (1993). Classification of growth hormone insensitivity syndrome. *J. Pediatr.* **122**(2), 60–73.

Lee, P. D., Conover, C. A., *et al.* (1993). Regulation and function of insulin-like growth factor-binding protein-1. *Proc. Soc. Exp. Biol. Med.* **204**(1), 4–29.

Lee, P. D., Giudice, L. C., *et al.* (1997). Insulin-like growth factor binding protein-1: Recent findings and new directions. *Proc. Soc. Exp. Biol. Med.* **216**(3), 319–357.

Maghnie, M., Barreca, A., *et al.* (1998). Failure to increase insulin-like growth factor-I synthesis is involved in the mechanisms of growth retardation of children with inherited liver disorders. *Clin. Endocrinol.* **48**(6), 747–755.

Mendenhall, C. L., Anderson, S., *et al.* (1984). Protein calorie malnutrition associated with alcoholic hepatitis. *Am. J. Med.* **76**, 211–222.

Mendenhall, C. L., Chernhausek, S. D., *et al.* (1989). The interactions of insulin-like growth factor-1(IGF-1) with protein-calorie malnutrition in patients with alcoholic liver disease: V.A. Cooperative study on alcoholic hepatitis V1. *Alcohol Alcohol.* **24**(4), 319–329.

Mendenhall, C. L., Moritz, T. E., *et al.* (1995). Protein energy malnutrition in severe alcoholic hepatitis: Diagnosis and response to treatment: The VA Cooperative Study Group #275. *Jpn. J. Parenter. Enteral. Nutr.* **19**(4), 258–265.

Mendenhall, C. L., Roselle, G. A., *et al.* (1997). Effects of recombinant human insulin-like growth factor-1 and recombinant human growth hormone on anabolism and immunity in calorie-restricted alcoholic rats. *Alcohol Clin. Exp. Res.* **21**(1), 1–10.

Mendenhall, C. L., Tosch, T., *et al.* (1986). VA cooperative study on alcoholic hepatitis 11: Prognostic significance of protein calorie malnutrition. *Am. J. Clin. Nutr.* **43**, 213–218.

Mohn, K. L., Melby, A. E., *et al.* (1991). The gene encoding rat insulinlike growth factor-binding protein 1 is rapidly and highly induced in regenerating liver. *Mol. Cell. Biol.* **11**(3), 1393–1401.

Moller, S., and Becker, U. (1992). Insulin-like growth factor 1 and growth hormone in chronic liver disease. *Dig. Dis.* **10**(4), 239–248.

Moller, S., Becker, U., *et al.* (1994). Short-term effect of recombinant human growth hormone in patients with alcoholic cirrhosis. *J. Hepatol.* **21**(5), 710–717.

Moller, S., Gronbaek, M., *et al.* (1993). Urinary growth hormone (U-GH) excretion and serum insulin-like growth factor 1 (IGF-1) in patients with alcoholic cirrhosis. *J. Hepatol.* **17**(3), 315–320.

Moller, S., Juul, A., *et al.* (1995). Concentrations, release, and disposal of insulin-like growth factor (IGF)-binding proteins (IGFBP), IGF-I, and growth hormone in different vascular beds in patients with cirrhosis. *J. Clin. Endocrinol. Metab.* **80**(4), 1148–1157.

Moser, D. R., Lowe, W. J., *et al.* (1992). Endothelial cells express insulin-like growth factor-binding proteins 2 to 6. *Mol. Endocrinol.* **6**(11), 1805–1814.

Muggeo, M., Tiengo, A., *et al.* (1979). Altered control of growth hormone secretion in patients with cirrhosis of the liver. *Arch. Intern. Med.* **139**, 1157–1160.

Muguerza, B., Castilla-Cortazar, I., *et al.* (2001). Antifibrogenic effects in vivo of low doses of IGF-I in rats. *Biochim. Biophys. Acta* **1536**, 185–195.

Murphy, L. J., Bell, G. I., *et al.* (1987). Tissue distribution of insulin-like growth factor I and II messenger ribonucleic acid in the adult rat. *Endocrinology* **120**(4), 1279–1282.

Owen, O. E., Trapp, V. E., *et al.* (1983). Nature and quantity of fuels consumed in patients with alcoholic cirrhosis. *J. Clin. Invest.* **72**(5), 1821–1832.

Petersen, K. F., Jacob, R., *et al.* (1997). Effects of insulin-like growth factor-I on glucose metabolism in rats with liver cirrhosis. *Am. J. Physiol.* **273**, E1189–E1193.

Petrides, A. S., Strohmeyer, G., *et al.* (1992). Insulin resistance in liver disease and portal hypertension. *Prog. Liver Dis.* **10**(311), 311–328.

Phillips, L. S., and Unterman, T. G. (1984). Somatomedin activity in disorders of nutrition and metabolism. *Clin. Endocrinol. Metab.* **13**(1), 145–189.

Picardi, A., de Oliveira, A., *et al.* (1997). Low doses of insulin-like growth factor-I improve nitrogen retention and food efficiency in rats with early cirrhosis. *J. Hepatol.* **26**(1), 191–202.

Powell, D. R., Suwanichkul, A., *et al.* (1991). Insulin inhibits transcription of the human gene for insulin-like growth factor-binding protein-1. *J. Biol. Chem.* **266**(28), 18868–18876.

Quirk, P., Owens, P., *et al.* (1994). Insulin-like growth factors I and II are reduced in plasma from growth retarded children with chronic liver disease. *Growth Regul.* **4**(1), 35–38.

Rajaram, S., Baylink, D. J., *et al.* (1997). Insulin-like growth factor-binding proteins in serum and other biological fluids: Regulation and functions. *Endocr. Rev.* **18**(6), 801–831.

Resnicoff, M., Rubini, M., *et al.* (1994). Ethanol inhibits insulin-like growth factor-1-mediated signalling and proliferation of C6 rat glioblastoma cells. *Lab. Invest.* **71**(5), 657–662.

Resnicoff, M., Sell, C., *et al.* (1993). Ethanol inhibits the autophosphorylation of the insulin-like growth factor 1 (IGF-1) receptor and IGF-1-mediated proliferation of 3T3 cells. *J. Biol. Chem.* **268**(29), 21777–21782.

Rosenfeld, R. G., Hwa, V., *et al.* (1999). The insulin-like growth factor binding protein superfamily: New perspectives. *Pediatrics* **104**(4 Pt 2), 1018–1021.

Ross, R. J., Chew, S. L., *et al.* (1996). Expression of IGF-I and IGF-binding protein genes in cirrhotic liver. *J. Endocrinol.* **149**(2), 209–216.

Ross, R. J., Rodriguez, A. J., *et al.* (1994). Expression of IGFBP-1 in normal and cirrhotic human livers. *J. Endocrinol.* **141**(2), 377–382.

Russell, W. E. (1985). Growth hormone, somatomedins, and the liver. *Semin. Liver Dis.* **5**(1), 46–58.

Salerno, F., Locatelli, V., *et al.* (1987). Growth hormone hyperresponsiveness to growth hormone-releasing hormone in patients with severe liver cirrhosis. *Clin. Endocrinol. Oxf.* **27**(2), 183–190.

Salmon, W. D., and Daughaday, W. H. (1957). A hormonally controlled serum factor which stimulates sulphate incorporation in vitro. *J. Lab. Clin. Med.* **49**, 825–836.

Santolaria, F., Gonzalez, G. G., *et al.* (1995). Effects of alcohol and liver cirrhosis on the GH-IGF-I axis. *Alcohol Alcohol.* **30**(6), 703–708.

Sarin, S. K., Dhingra, N., *et al.* (1997). Dietary and nutritional abnormalities in alcoholic liver disease: A comparison with chronic alcoholics without liver disease. *Am. J. Gastroenterol.* **92**(5), 777–783.

Sarna, S., Sipila, I., *et al.* (1996). Recombinant human growth hormone improves growth in children receiving glucocorticoid treatment after liver transplantation. *J. Clin. Endocrinol. Metabol.* **81**(4), 1476–1482.

Schalch, D. S., Heinrich, U. E., *et al.* (1979). Role of the liver in regulating somatomedin activity: Hormonal effects on the synthesis and release of insulin-like growth factor and its carrier protein by the isolated perfused rat liver. *Endocrinology* **104**(4), 1143–1151.

Schalch, D. S., Kalayoglu, M., *et al.* (1998). Serum insulin-like growth factors and their binding proteins in patients with hepatic failure and after liver transplantation. *Metab. Clin. Exp.* **47**(2), 200–206.

Scharf, J., Ramadori, G., *et al.* (1996a). Synthesis of insulinlike growth factor binding proteins and of the acid-labile subunit in primary cultures of rat hepatocytes, of Kupffer cells, and in cocultures: Regulation by insulin, insulinlike growth factor, and growth hormone. *Hepatology* **23**(4), 818–827.

Scharf, J., Schmitz, F., *et al.* (1996b). Insulin-like growth factor-I serum concentrations and patterns of insulin-like growth factor binding proteins in patients with chronic liver disease. *J. Hepatol.* **25**, 689–699.

Scharf, J. G., Schmidt-Sandte, W., *et al.* (1995). Synthesis of insulin-like growth factor binding proteins and of the acid-labile subunit of the insulin-like growth factor ternary binding protein complex in primary cultures of human hepatocytes. *J. Hepatol.* **23**(4), 424–430.

Schwander, J. C., Hauri, C., *et al.* (1983). Synthesis and secretion of insulin-like growth factor and its binding protein by the perfused rat liver: Dependence on growth hormone status. *Endocrinology* **113**(1), 297–305.

Scott, C. D., and Baxter, R. C. (1991). Synthesis of the acid-labile subunit of the growth-hormone-dependent insulin-like-growth-factor-binding protein complex by rat hepatocytes in culture. *Biochem. J.*

Scott, C. D., Martin, J. L., *et al.* (1985a). Production of insulin-like growth factor 1 and its binding protein by adult rat hepatocytes in primary culture. *Endocrinology* **116**, 1094–1101.

Scott, C. D., Martin, J. L., *et al.* (1985b). Rat hepatocyte insulin-like growth factor I and binding protein: Effect of growth hormone in vitro and in vivo. *Endocrinology* **116**(3), 1102–1107.

Shaarawy, M., Fikry, M. A., *et al.* (1998). Insulin-like growth factor binding protein-3: A novel biomarker for the assessment of the synthetic capacity of hepatocytes in liver cirrhosis. *J. Clin. Endocrinol. Metab.* **83**(9), 3316–3319.

Shankar, T. P., Solomon, S. S., *et al.* (1988). Growth hormone and carbohydrate intolerance in cirrhosis. *Horm. Metab. Res.* **20**(9), 579–583.

Shaw, B. W., Wood, W. P., *et al.* (1985). Influence of selected patient variables and operative blood loss on six month survival following liver transplantation. *Semin. Liver Dis.* **5**, 385–393.

Shmueli, E., Miell, J. P., *et al.* (1996). High insulin-like growth factor binding protein 1 levels in cirrhosis: Link with insulin resistance. *Hepatology* **24**(1), 127–133.

Shmueli, E., Stewart, M., *et al.* (1994). Growth hormone, insulin-like growth factor-1 and insulin resistance in cirrhosis. *Hepatology* **19**(2), 322–328.

Sonntag, W. E., and Boyd, R. L. (1989). Diminished insulin-like growth factor-1 levels after chronic ethanol: Relationship to pulsatile growth hormone release. *Alcohol Clin. Exp. Res.* **13**(1), 3–7.

Srivastava, V., Hiney, J. K., *et al.* (1995). Effect of ethanol on the synthesis of insulin-like growth factor 1 (IGF-1) and the IGF-1 receptor in late prepubertal female rats: A correlation with serum IGF-1. *Alcohol Clin. Exp. Res.* **19**(6), 1467–1473.

Stewart, A., Johnston, D. G., *et al.* (1983). Hormone and metabolite profiles in alcoholic liver disease. *Eur. J. Clin. Invest.* **13**(5), 397–403.

Streat, S., Beddoe, A., *et al.* (1987). Aggressive nutritional support does not prevent protein loss despite fat gain in septic intensive care patients. *J. Trauma* **27**, 262–266.

Svetlani-Baroni, G., Ridolfi, F., *et al.* (1999). Insulin and insulin-like growth factor-I stimulate proliferation and type I collagen accumulation by human hepatic stellate cells: Differential effects on signal transduction pathways. *Hepatology* **29**, 1743–1751.

Unterman, T. G., and Phillips, L. S. (1986). Circulating somatomedin activity during hepatic regeneration. *Endocrinology* **119**(1), 185–192.

Wahl, D. G., Dollet, J. M., *et al.* (1992). Relationship of insulin resistance to protein-energy malnutrition in patients with alcoholic liver cirrhosis: Effect of short-term nutritional support. *Alcohol Clin. Exp. Res.* **16**(5), 971–978.

Wallace, J., Cuneo, R., *et al.* (1997). "Growth Hormone Treatment in Adults with Liver Cirrhosis: Effects on Body Composition and Metabolism." International Congress of Endocrinology, San Francisco.

Ward, H. C., Halliday, D., *et al.* (1987). Protein and energy metabolism with biosynthetic human growth hormone after gastrointestinal surgery. *Ann. Surg.* **206**(1), 56–61.

Xu, X., Ingram, R. L., *et al.* (1995). Ethanol suppresses growth hormone-mediated cellular responses in liver slices. *Alcohol Clin. Exp. Res.* **19**(5), 1246–1251.

Zanobi, A., Zecca, L., *et al.* (1983). Failure of inhibition by TRH of L-dopa stimulated GH secretion in patients with alcoholic cirrhosis of the liver. *Clin. Endocrinol.* **18**, 233–239.

Zapf, J., Morell, B., *et al.* (1980). Serum levels of insulin-like growth factor (IGF) and its carrier protein in various metabolic disorders. *Acta Endocrinol. Copenh.* **95**(4), 505–517.

CHAPTER 23

Resolution of Liver Fibrosis in Hepatitis C Patients by Interferon Therapy and Prevention of Hepatocellular Carcinoma

MASAO OMATA, YASUSHI SHIRATORI, AND HARUHIKO YOSHIDA
Department of Gastroenterology, University of Tokyo, Tokyo 113–8655, Japan

I. INTRODUCTION

The total number of deaths in Japan is currently about 1,000,000 per annum, with hepatocellular carcinoma (HCC) accounting for more than 30,000 of these deaths. Studies conducted since the early 1980s indicate that hepatitis type B and type C account for 18 and 59% of the total number of cases of HCC, respectively. A further study at the University of Tokyo conducted in the early 1990s indicated that hepatitis type B and type C accounted for 11 and 83% of the total number of cases, respectively, indicating an increase in the ratio of type C to type B. These data underscore the importance of the hepatitis viruses as etiological factors in the pathogenesis of HCC, with 94% of tumors involving viral infection (Shiratori *et al.*, 1995).

In Japan, the populations of patients with type B and C viral infections are each estimated to number 1,500,000, although there is an 8 to 1 ratio for type C to type B regarding the incidence of HCC. It is obvious that the incidence of HCC is caused mostly by the type C hepatitis virus at present.

Copyright 2003, Elsevier Science (USA). All rights reserved.

The natural history of chronic hepatitis C virus (HCV) infection suggests a sequential but slow progression from acute HCV infection to chronic infection and cirrhosis, leading to death from either liver failure or HCC (Di Bisceglie *et al.*, 1991; Fattovich *et al.*, 1997; Seeff *et al.*, 1992; Takano *et al.*, 1995; Tong *et al.*, 1995).

Thus, it should be clarified how the disease progresses to hepatocellular carcinoma in those with hepatitis B and C infections.

II. STEPWISE PROGRESSION OF HEPATIC FIBROSIS TOWARD HCC

In order to elucidate the course from chronic hepatitis to HCC, we undertook a clinical follow-up study in our outpatient clinic (Takano *et al.*, 1995). Typically, patients were healthy Japanese with an average age of 40 who felt subjectively well. However, they exhibited slightly abnormal liver function tests, such as an elevated transaminase (GPT) level of about 100. These patients underwent a liver biopsy with pathological examination revealing chronic hepatitis (not cirrhosis) and were subsequently followed up (Fig. 1). One hundred and twenty-seven patients had chronic hepatitis type B and 124 had chronic hepatitis type C. In about 6 years of follow-up, HCC was noted in 5 type B patients and 13 type C patients (Fig. 1).

The calculated annual rates of carcinogenesis for 100,000 people were 647 patients with type B and 1723 with type C. Ninety-six to 97% of Japanese citizens are thought not to be infected with the hepatitis viruses. In those persons without infection, the incidence rate could be assumed 1000-fold lower at about 1.7 cases per 100,000 people. The relative risk of pulmonary carcinoma for smokers is only 10 times that for nonsmokers. It is therefore apparent that HCC will be reduced drastically if the hepatitis virus could be eliminated effectively. Interestingly, a report of a vaccination program against type B hepatitis performed in Taiwan indicated that the incidence of HCC between the ages of 6 and 17 had fallen dramatically and became comparable to that of rare cerebral neoplasms (Chang *et al.*, 1997). As a single parameter, this figure thus represents the highest rate of carcinogenesis. The degree of fibrosis of the liver biopsy is now classified as slight (F1), moderate (F2), and severe (F3) (Desmet *et al.*, 1994). Thirty-six patients with type C were classified as F1 (slight) and only 1 patient developed carcinoma after 9.5 years (Fig. 1). The 45 patients in the F3 subgroup had a much higher rate of HCC, with 8 individuals developing the neoplasm after short periods of follow-up. Predictably, patients in the F2 (moderate fibrosis) subgroup exhibited a moderate incidence of HCC. This trend results from the proposed pathway of development of HCC from chronic hepatitis. This can be envisioned as a stepwise model with progression from mild fibrosis (F1) to

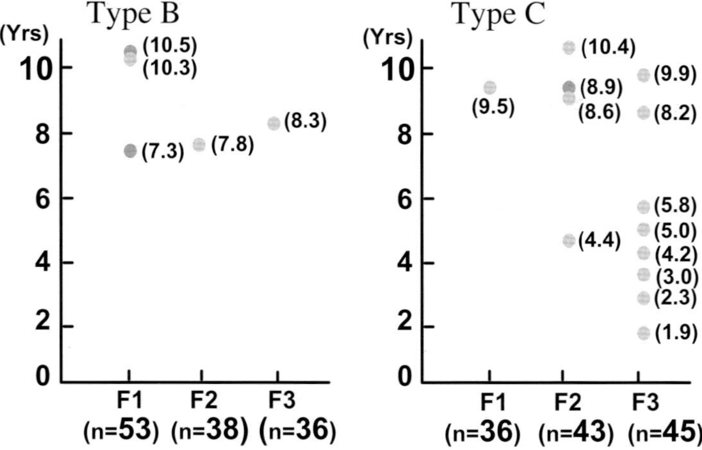

FIGURE 1 Follow-up study of 124 patients with C-viral and 127 patients with B-viral chronic hepatitis and occurrence of hepatocellular carcinoma.

hepatic cirrhosis (F4) and eventual HCC (F5). This model is applicable to type C but not type B. Indeed, 3 of the 53 type B patients in the chronic hepatitis with mild fibrosis (F1) subgroup developed HCC, whereas only 1 of 36 type B patients with severe fibrosis (F3) developed carcinoma. There are two potential mechanisms underlying the development of HCC with type B hepatitis. Tumors may develop following repeated bouts of necrosis, inflammation, and fibrosis, culminating in cirrhosis and cancer or, alternatively, tumors may occur independently of fibrosis. In type C, the onset of carcinoma is thought to occur after the development of fibrosis (cirrhosis) lasting over decades. Indeed, cirrhosis or hepatic fibrosis is typically present in liver specimens containing carcinomatous tissue. For example, of the 124 patients with type C, 13 developed carcinoma, 12 of whom exhibited cirrhosis (F4) (Fig. 1). Of the five patients with type B hepatitis, 2 patients (40%) did not exhibit cirrhosis when they developed tumors (Fig. 1). Therefore, the course of carcinogenesis for types B and C differs substantially.

The annual rates of carcinogenesis per 100,000 patients are about 1700 and 600 for all type C and type B with chronic hepatitis patients, respectively. The corresponding figures for the F1, F2, and F3 subgroups of type C are 457, 1450, and 3005. Thus, expressing the annual rate of carcinogenesis per 100,000 patients as a percentage yields respective numbers of 0.45, 1.5, and 3%. Because the average age of these patients is 45, one can calculate the estimated lifetime risk of developing hepatocarcinoma, in approximately 30 years. The annual rate of 3% in the "severe" chronic hepatitis subgroup (F3) gives an estimated lifetime rate of carcinogenesis of 90% (3×30). Conversely, the "slight" chronic hepatitis subgroup (F1) has a carcinogenesis rate of 13% (0.45×3.0), such that 87% of patients will not develop a tumor. Liver function tests such as alanine aminotransferase (ALT) cannot be used to identify individuals who are predisposed to the development of cancer in their lifetime. However, if the degree of hepatic fibrosis can be ascertained by analysis of a liver biopsy, then it may be possible to estimate the risk of developing carcinoma over a period of 30 years in patients with type C hepatitis.

In contrast, in type B patients, it remains difficult to predict who will develop carcinoma even following analysis and assessment of a liver biopsy.

Introduction of the F classification [for fibrosis classification: F1 (slight), F2 (medium), F3 (severe), and F4 (cirrhosis)] could be regarded as an appropriate classification for the multistep model toward cancer in patients with chronic hepatitis C virus infection that now predominate in clinical practice.

III. INTERFERON TREATMENT AND ITS RESPONSE

After the introduction of interferon (IFN) for patients with chronic hepatitis C in the mid-1980s with beneficial effects (Hoofnagle and Di Bisceglie, 1997), factors predictive of sustained response to IFN have been studied extensively (Hoofnagle and Di Bisceglie, 1997; Lau *et al.*, 1993; Poynard *et al.*, 1996; Shiratori *et al.*, 1997). Early studies used biochemical response to assess the efficacy of IFN therapy, but the biochemical response during and after IFN administration cannot predict sustained response, and virological evaluation was found to be superior to biochemical response in long-term follow-up studies (Lindsay, 1997).

Long-term follow-up studies revealed that patients who tested negative for HCV RNA 6 months after treatment remained in remission with normal liver function and improved histological features, and they may have been cured of infection (Yokosuka *et al.*, 1995). Since 1992, approximately 200,000 patients were treated by interferon in Japan, and 30% of those cleared HCV (VSR, virological sustained response).

IV. HISTOLOGICAL RESPONSE

Histology offers more important insight into disease prognosis and treatment response. Our previous study revealed that biochemical and virological responses to IFN therapy are associated with a significant improvement in liver histology during and shortly after IFN therapy (Omata *et al.*, 1981; Yokosuka *et al.*, 1995).

The short-term (less than 2 years after the end of IFN therapy) histological improvement indicates changes in inflammatory activity grade, especially in sustained responders (Manesis *et al.*, 1997; Reichard *et al.*, 1995). However, longterm effects on histological changes in fibrosis in particular remained unclear.

V. HISTOLOGICAL EVALUATION

Histological approaches to the assessment of IFN response have several limitations: factors influencing the interpretation of histological findings include sample variation, different or inconsistent definition of pathological findings, intra- and interobserver variations in evaluation of histology, and evaluation of an ordinal scale with nonconstant intervals in case of paired biopsies. Despite these difficulties, histological changes after IFN therapy have become important assessments of the treatment effect using several simple scoring systems.

The histologic response has been evaluated according to validated scoring systems, mostly using the semiquantitative score proposed by Knodell *et al.* (1981) or Scheuer (1991). Four categories are assessed according to Knodell: piecemeal necrosis and bridging necrosis, intralobular necrosis, portal inflammation, and fibrosis. Three categories were assessed in the scoring system of Scheuer: portal/periportal activity, lobular activity, and fibrosis. Instead of the complexed scoring systems using several categories, histological features of chronic hepatitis have been classified simply into two categories (necroinflammatory activity and fibrosis) based on the distinction of necroinflammatory activity (the grade of disease) from the fibrosis (the stage of disease) (Bedossa and Poynard, 1996; Desmet *et al.*, 1994; The French METAVIR Cooperative Study Group, 1994). The simple scoring systems proposed by the METAVIR group (Bedossa and Poynard, 1996) and by Desmet *et al.* (1994) offer more reproducibility than the use of a global numerical index in terms of intraobserver and interobserver variation. In Japan, we have modified our Inuyama classification to new Inuyama by incorporating the numerical scoring system (Omata, 1996).

VI. NATURAL COURSE OF FIBROSIS PROGRESSION

Using these simple semiquantitative scoring systems (activity grade from A0 to A3, fibrosis stage from F0 to F4), Poynard *et al.* (1997) evaluated the natural course of the fibrosis progression rate in patients with chronic hepatitis C as 0.133/year (Table I). They clarified that the fibrosis progression rate was much higher in patients who were alcohol drinkers (0.167 unit/year) and male.

TABLE I Fibrosis Progression Rate in Relation to Host Factor

Factor	Fibrosis progression rate (unit/year)
Poynard *et al.* (1996)	
All cases with HCV infection	0.133
Alcohol > 50 g/day	0.167
Alcohol < 50 g/day	0.143
Nonalcohol	0.125
Male	0.154
Female	0.111
Mathurin *et al.* (1998)	
Patients with normal ALT	0.050
Shiratori *et al.* (2000) using paired biopsy	
Nonalcohol	0.095

In contrast, Mathurin *et al.* (1998) showed that persistently normal ALT levels are correlated with the slow progression rate of fibrosis (0.05 unit/year). These values are based on a single, unpaired biopsy and a suspected (rather than proven) duration of infection from the patients' history, but we calculated the fibrosis progression rate using paired biopsy samples, which were obtained with a remote time interval (Shiratori *et al.*, 2000), indicating that the natural fibrosis progression rate calculated as the change of fibrosis staging per year is approximately 0.10 unit/year in untreated patients, except alcoholics (Table I).

VII. LONG-TERM EFFECT OF IFN ON HISTOLOGICAL IMPROVEMENT

The impact of treatment on the history of such a slowly progressive disease cannot be ascertained by evaluating results at the completion of treatment. Instead, the effect of treatment should be evaluated on a long-term basis, even though questions may arise as to the appropriateness of liver biopsy after treatment to evaluate the effects of therapy and to determine the progression/regression of liver disease. It is possible that repeated liver biopsies performed at remote time intervals from the completion of treatment may provide qualitative and quantitative information about the long-term response and may serve to define the extent of progression or regression in responders and nonresponders. Although Sobesky *et al.* (1999) showed amelioration of the fibrosis progression rate in 185 IFN-treated patients (calculated as 0.000 unit/year) in comparison with 102 untreated patients (0.13 unit/year), especially in the IFN responders, the interval of the paired biopsy was too short (20 months) to evaluate the long-term histological changes.

However, sustained regression of the necroinflammatory process was persistently observed in patients with a sustained virological response after 3–10 years, and Lau *et al.* (1998) showed histological improvement of activity and fibrosis in five sustained responders 6 to 13 years after IFN therapy, and Marcellin *et al.* (1997) showed loss of detectable intrahepatic HCV RNA and histological improvement in sustained virological responders at 1 to 7.6 years after IFN therapy. However, these reports were fragmentary. Therefore, we decided to analyze a larger number of patients with a longer duration of follow-up.

VIII. IMPROVEMENT OF HEPATIC FIBROSIS AND INHIBITION OF HEPATOCARCINOGENESIS BY INTERFERON THERAPY STUDY

In 1994, we initiated a nationwide control trial to see whether interferon therapy could reduce the incidence of HCC (Shiratori *et al.*, 2000; Yoshida *et al.*, 1999). We enrolled 2890 biopsy-proven patients (2400 treated and 490 untreated) (Fig. 2). The treated group showed half the incidence of the untreated group (Fig. 3). Of these 2890, we studied paired biopsy in 593 patients with an average interval of 5 years. In contrast to the plus (worsened) 0.10 fibrosis progression rate per year in 106 untreated patients, 183 virus-eradicated patients (VSR)

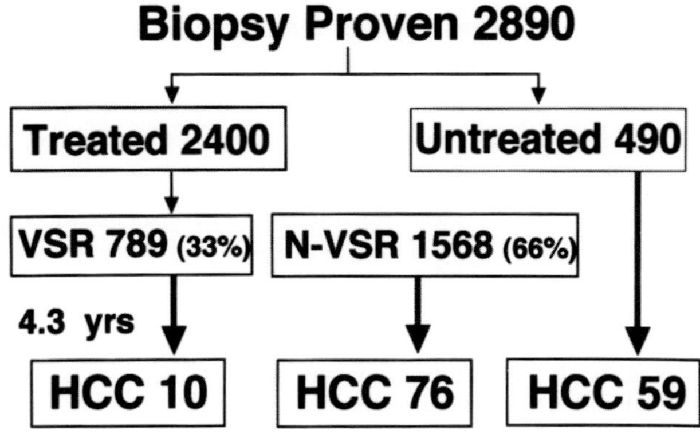

VSR : Virological Sustained Response

N-VSR : Non-VSR

HCC : Hepatocellular Carcinoma

FIGURE 2 Inhibition of hepatocarcinogenesis by the interferon therapy study, started in 1994. The virological sustained response is equivalent to the complete response.

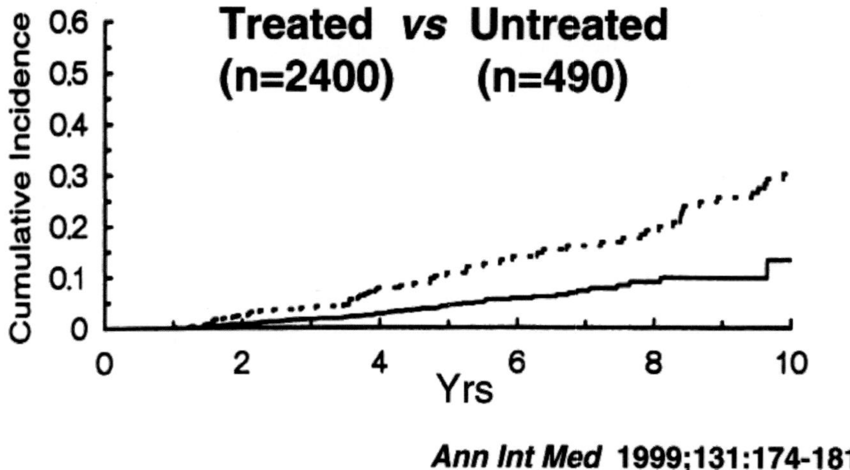

FIGURE 3 Incidence of HCC over time in control (n = 490) and interferon-treated (n = 2400) groups.

showed minus (improved) 0.28 units (Table II), clearly indicating that two to three times rapid resolution of hepatic fibrosis may occur if HCV were eradicated (Shiratori *et al.*, 2000). This may give us the overview of how HCV-related liver fibrosis progresses and could be regressed by treatment (Fig. 4).

Our biggest concern was whether the degree of resolution of hepatic fibrosis could occur similarly in mild and advanced, e.g., F4 cirrhotic, stages in patients who cleared HCV. Therefore, we analyzed the fibrosis progression (in this case, rather regression) rate per year. It turned out that fibrosis regression occurs similarly even in patients at the F4 stage if the virus was eradicated by interferon treatment (Table III). This indicates that even cirrhosis could be cured by antiviral treatment.

TABLE II Treated Cases and the Rate of the Natural Course of Hepatic Fibrosis Progression

Group	Fibrosis progression rate per year[a]
106 untreated	+0.10 ± 0.22
183 eradicated[b]	−0.28 ± 0.33
304 noneradicated[c]	+0.02 ± 0.36

[a]+,worsened; 2,improved.
[b]Viral elimination group (VSR).
[c]Viral nonelimination group (n-VSR).

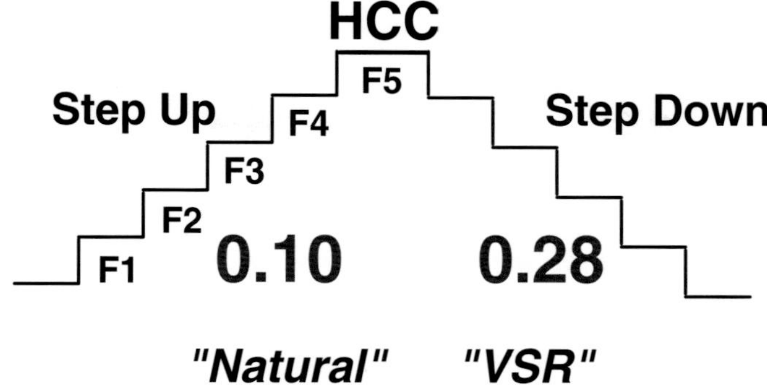

FIGURE 4 Histological examination of paired biopsy samples in our IHIT study (Shiratori *et al.*, 2000; Yoshida *et al.*, 1999) gives an idea of how hepatic fibrosis progresses when untreated and how it regresses when the virus is eradicated. Step up: natural course of type C chronic hepatitis as seen with a liver biopsy. Step down: speed of fibrosis amelioration following virus elimination with interferon treatment.

IX. CONCLUSION

Chronic hepatitis C is a common disease that is often asymptomatic and mild, but which may slowly progress to cirrhosis and eventually to hepatocellular carcinoma. We clarified the speed of hepatic fibrosis progression if left untreated, and also of resolution by virus eradication. We demonstrated an inhibition of HCC development with IFN therapy due to the attenuation of advanced hepatic fibrosis and even cirrhosis could be cured. A more effective therapy for patients with HCV infection might induce a more drastic reduction of HCC.

TABLE III Speed of Abatement of Fibrosis in Cases of Viral Elimination Classified by Extent of Fibrosis (F4–F1)

Improvement in VSR[a]	
From F4 (n = 24)	−0.283
From F3 (n = 45)	−0.374
From F2 (n = 69)	−0.284
From F1 (n = 42)	−0.152

[a]Virological sustained response.

REFERENCES

Bedossa, P., and Poynard, T. (1996). An algorithm for the grading of activity in chronic hepatitis C: The METAVIR Cooperative Study Group. *Hepatology* **24**, 289–293.

Chang, M. H., Chen, C. J., Lai, M. S., Hsu, H. M., Wu, T. C., Kong, M. S., Liang, D. C., Shau, D. C., and Chen, D. S. (1997). Universal hepatitis B vaccination in Taiwan and the incidence of hepatocellular carcinoma in children. *N. Engl. J. Med.* **336**, 1855–1859.

Desmet, V. J., Gerber, M., Hoofnagle, J. H., Manns, M., and Scheuer, P. J. (1994). Classification of chronic hepatitis: Diagnosis, grading and staging. *Hepatology* **19**, 1513–1520.

Di Bisceglie, A. M., Goodman, Z. D., Ishak, K. G., Hoofnagle, J. H., Melpolder, J. J., and Alter, H. J. (1991). Long-term clinical and histopathological follow-up of chronic posttransfusion hepatitis. *Hepatology* **14**, 969–974.

Fattovich, G., Giustina, G., Degos, F., Tremolada, F., Diodati, G., Almasio, P., Nevens, F., Solinas, A., Mura, D., Brouwer, J. T., Thomas, H., Njapoum, C., Casarin, C., Bonetti, P., Fuschi, P., Basho, J., Tocco, A., Bhalla, A., Galassini, R., Noventa, F., Schalm, S. W., and Realdi, G. (1997). Morbidity and mortality in compensated cirrhosis type C: A retrospective follow-up study of 384 patients. *Gastroenterology* **112**, 463–472.

Hoofnagle, J. H., and Di Bisceglie, A. M. (1997). The treatment of chronic viral hepatitis. *N. Engl. J. Med.* **336**, 347–356.

Knodell, R. G., Ishak, K. G., Black, W. C., Chen, T. S., Craig, R., Kaplowitz, N., Kiernan, T. W., and Wollman, J. (1981). Formulation and application of a numerical scoring system for assessing histological activity in asymptomatic chronic active hepatitis. *Hepatology* **1**, 431–435.

Lau, D. T., Kleiner, D. E., Ghany, M. G., Park, Y., Schmid, P., and Hoofnagle, J. H. (1998). 10-year follow-up after interferon-alpha therapy for chronic hepatitis C. *Hepatology* **28**, 1121–1127.

Lau, J. Y., Davis, G. L., Kniffen, J., Qian, K. P., Urdea, M. S., Chan, C. S., Mizokami, M., Neuwald, P. D., and Wilber, J. C. (1993). Significance of serum hepatitis C virus RNA levels in chronic hepatitis C. *Lancet* **341**, 1501–1504.

Lindsay, K. L. (1997). Therapy of hepatitis C: Overview. *Hepatology* **26**, 71s–77s.

Manesis, E. K., Papaioannou, C., Gioustozi, A., Kafiri, G., Koskinas, J., and Hadziyannis, S. J. (1997). Biochemical and virological outcome of patients with chronic hepatitis C treated with interferon alfa-2b for 6 or 12 months: A 4-year follow-up of 211 patients. *Hepatology* **26**, 734–739.

Marcellin, P., Boyer, N., Gervais, A., Martinot, M., Pouteau, M., Castelnau, C., Kilani, A., Areias, J., Auperin, A., Benhamou, J. P., Degott, C., and Erlinger, S. (1997). Long-term histologic improvement and loss of detectable intrahepatic HCV RNA in patients with chronic hepatitis C and sustained response to interferon-alpha therapy. *Ann. Intern. Med.* **127**, 875–881.

Mathurin, P., Moussalli, J., Cadranel, J. F., Thibault, V., Charlotte, F., Dumouchel, P., Cazier, A., Huraux, J. M., Devergie, B., Vidaud, M., Opolon, P., and Poynard, T. (1998). Slow progression rate of fibrosis in hepatitis C virus patients with persistently normal alanine transaminase activity. *Hepatology* **27**, 868–872.

Omata, M. (1996). "Live" Discussion on New Classification of Chronic Hepatitis, pp. 1–79. Churchill Livingstone, Japan.

Omata, M., Iwama, S., Sumida, S., Ito, Y., and Okuda, K. (1981). Clinico-pathological study of acute non-A, non-B post-transfusion hepatitis, histological features of liver biopsies in acute phase. *Liver* **1**, 201–208.

Poynard, T., Bedossa, P., and Opolon, P. (1997). Natural history of liver fibrosis progression in patients with chronic hepatitis C: The OBSVIRC, METAVIR, CLINIVIR, and DOSVIRC groups. *Lancet* **349**, 825–832.

Poynard, T., Leroy, V., Cohard, M., Thevenot, T., Mathurin, P., Opolon, P., and Zarski, J. P. (1996). Meta-analysis of interferon randomized trials in the treatment of viral hepatitis C: Effects of dose and duration. *Hepatology* **24**, 778–789.

Reichard, O., Glaumann, H., Fryden, A., Norkrans, G., Schvarcz, R., Sonnerborg, A., Yun, Z. B., and Weiland, O. (1995). Two-year biochemical, virological, and histological follow-up in patients with chronic hepatitis C responding in a sustained fashion to interferon alfa-2b treatment. *Hepatology* **21**, 918–922.

Scheuer, P. J. (1991). Classification of chronic viral hepatitis. a need reassessment. *J. Hepatol.* **13**, 372–374.

Seeff, L. B., Buskell-Bales, Z., Wright, E. C., Durako, S. J., Alter, H. J., Iber, F. L., Hollinger, F. B., G., G., Knodell, R. G., and Perrillo, R. P. (1992). Long-term mortality after transfusion-associated non-A, non-B hepatitis. *N. Engl. J. Med.* **327**, 1906–1911.

Shiratori, Y., Imazeki, F., Moriyama, M., Yano, M., Arakawa, Y., Yokosuka, O., Kuroki, T., Nishiguchi, S., Sata, M., Yamada, G., Fujiyama, S., Yoshida, H., and Omata, M. (2000). Histologic improvement of fibrosis in patients with hepatitis C who have sustained response to interferon therapy. *Ann. Intern. Med.* **132**, 517–524.

Shiratori, Y., Kato, N., Yokosuka, O., Imazeki, F., Hashimoto, E., Hayashi, N., Nakamura, A., Asada, M., Kuroda, H., Tanaka, N., Arakawa, Y., and Omata, M. (1997). Predictors of the efficacy of interferon therapy in chronic hepatitis C virus infection: Tokyo-Chiba Hepatitis Research Group. *Gastroenterology* **113**, 558–566.

Shiratori, Y., Shiina, S., Imamura, M., Kato, N., Kanai, F., Okudaira, T., Teratani, T., Tohgo, G., Toda, N., Ohashi, M., Ogura, K., Niwa, Y., Kawabe, T., and Omata, M. (1995). Characteristic difference of hepatocellular carcinoma between hepatitis-B and -C viral infection in Japan. *Hepatology* **22**, 1027–1033.

Sobesky, R., Mathurin, P., Charlotte, F., Moussalli, J., Olivi, M., Vidaud, M., Ratziu, V., Opolon, P., and Poynard, T. (1999). Modeling the impact of interferon alfa treatment on liver fibrosis progression in chronic hepatitis C: A dynamic view: The Multivirc Group. *Gastroenterology* **116**, 378–386.

Takano, S., Yokosuka, O., Imazeki, F., Tagawa, M., and Omata, M. (1995). Incidence of hepatocellular carcinoma in chronic hepatitis B and C : A prospective study of 251 patients. *Hepatology* **21**, 650–655.

The French METAVIR Cooperative Study Group (1994). Intraobserver and interobserver variations in liver biopsy interpretation in patients with chronic hepatitis C. *Hepatology* **20**, 15–20.

Tong, M. J., el-Farra, N. S., Reikes, A. R., and Co, R. L. (1995). Clinical outcomes after transfusion-associated hepatitis C. *N. Engl. J. Med.* **332**, 1463–1466.

Yokosuka, O., Kato, N., Hosoda, K., Ito, Y., Imazeki, F., Ohto, M., and Omata, M. (1995). Efficacy of longer interferon treatment in chronic liver disease evaluated by sensitive polymerase chain reaction assay for hepatitis C virus RNA. *Gut* **37**, 721–726.

Yoshida, H., Shiratori, Y., Moriyama, M., Arakawa, Y., Ide, T., Sata, M., Inoue, O., Yano, M., Tanaka, M., Fujiyama, S., Nishiguchi, S., Kuroki, T., Imazeki, F., Yokosuka, O., Kinoyama, S., Yamada, G., and Omata, M. (1999). Interferon therapy reduces the risk for hepatocellular carcinoma: National surveillance program of cirrhotic and noncirrhotic patients with chronic hepatitis C in Japan: IHIT Study Group. *Ann. Intern. Med.* **131**, 174–181.

CHAPTER 24

Possible Gene Therapy for Liver Cirrhosis

JIRO FUJIMOTO AND TAKAHIRO UEKI

First Department of Surgery, Hyogo College of Medicine, Nishinomiya 663-8501, Japan

Liver cirrhosis is the irreversible end result of chronic liver disease, characterized by diffuse disorganization of the normal hepatic structure by fibrous scaring and hepatocellular regeneration. It is a major cause of morbidity and mortality worldwide, with no effective therapy. The ideal strategy for the treatment of liver cirrhosis should include prevention of fibrogenesis, stimulation of hepatocyte mitosis, and reorganization of the liver architecture. We have developed a novel gene therapy approach for rat liver cirrhosis by the muscle-directed gene transfer of hepatocyte growth factor (HGF). In rats with lethal liver cirrhosis produced by dimethylnitrosamine (DMN), repeated transduction of the HGF gene into skeletal muscle induced a high plasma level of HGF and tyrosine phosphorylation of the c-Met/HGF receptor. HGF gene transduction inhibited fibrogenesis and hepatocyte apoptosis and also produced resolution of fibrosis in the cirrhotic liver. Thus, HGF gene therapy may be useful for the treatment of patients with liver cirrhosis.

I. INTRODUCTION

Various factors induce liver cirrhosis, including excessive alcohol intake, viral hepatitis, drug-induced hepatic injury, prolonged biliary obstruction, and the late

Copyright 2003, Elsevier Science (USA). All rights reserved.

stages of some parasitic diseases and some genetically transmitted metabolic disorders, such as Wilson's disease and hemochromatosis. Liver cirrhosis, the irreversible end result of fibrous scarring and hepatocellular regeneration, is a major cause of morbidity and mortality worldwide with no effective therapy. Hepatic dysfunction, esophageal varices, ascites, and liver cancer are the most serious complications and are often fatal (Cohn *et al.*, 1992). Cirrhosis can best be defined in terms of what is pathoanatomically certain. Cirrhosis is a chronic disease of the liver in which diffuse destruction and regeneration of hepatic parenchymal cell have occurred and in which a diffuse increase in connective tissue has resulted in disorganization of the lobular and vascular architecture. Therefore, ideal strategies for the treatment of liver cirrhosis should include resolution of fibrosis, stimulation of hepatic mitosis, and reorganization of the liver architecture.

II. LIVER FIBROSIS/LIVER CIRRHOSIS

A. Transforming Growth Factor (TGF)-β1

TGF-β1 is a cytokine whose widespread actions in promoting the production and deposition of extra cellular matrix are essential to normal tissue repair following injury (Sporn *et al.*, 1992). However, overproduction of TGF-β1, in response to injury or disease, is thought to be a major cause of tissue fibrosis in animals and in humans, especially in liver cirrhosis (Friedman, 1993; Blomhoff *et al.*, 1991).

B. Activation of TGF-β1 in Liver Fibrosis

Key events in liver fibrosis include the accumulation of extracellular matrix components, caused predominantly by the phenotypic transition from hepatic stellate cells to proliferating myofibroblast-like cells, which enhance the production of extracellular matrix components (Border *et al.*, 1994) and attenuate the degradation of extracellular matrix proteins (Nakamura *et al.*, 1986). Studies indicate that TGF-β1 plays a crucial role in the pathogenesis of liver cirrhosis/fibrosis (Friedman, 1993; Blomhoff *et al.*, 1991). TGF-β1 induces the phenotypic transition of hepatic stellate cells to proliferating myofibroblast-like cells and enhances their production of extracellular matrix components (Border *et al.*, 1994; Sanderson *et al.*, 1995). TGF-β1 is a potent growth inhibitor of epithelial and endothelial cells, including hepatocytes (Nakamura *et al.*, 1985), while it induces apoptotic cell death in hepatocytes (Oberhammer *et al.*, 1992) (Fig. 1).

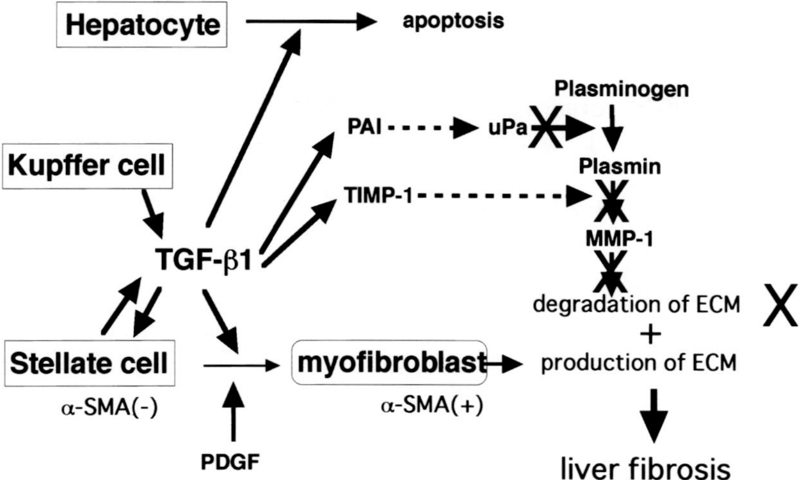

FIGURE 1 The mechanism of liver fibrosis. PAI, plasmin activator inhibitor; TMP, tissue inhibitor of metalloproteinase; MMP, matrix metalloproteinase; uPa, urokinase type plasmin activator; PDGF, platelet-derived growth factor.

III. HEPATOCYTE GROWTH FACTOR GENE THERAPY

A. Hepatocyte Growth Factor

During the past decade, hepatocyte growth factor, originally identified and cloned as a potent mitogen for hepatocytes (Nakamura *et al.*, 1984, 1989; Russell *et al.*, 1984), has been found to have mitogenic, motogenic, and morphogenic activities for a wide variety of cells (Boros *et al.*, 1995; Michalopoulus *et al.*, 1997; Matsumoto *et al.*, 1997). Several approaches, including transgenic animals, as well as *in vivo* infusion of HGF into animals, have revealed that HGF plays an essential role for both the development and the regeneration of liver (Boros *et al.*, 1995; Michalopoulus *et al.*, 1997; Matsumoto *et al.*, 1997; Schmidt *et al.*, 1995), and HGF has antiapoptotic and cytoprotective activities on hepatocytes (Baraelli *et al.*, 1996; Okano *et al.*, 1997). We hypothesized that overexpressed HGF could show such a beneficial effect on hepatocytes and counteract biological activities of TGF-β1 in cirrhotic liver (Fig. 2).

B. c-Met/HGF Receptor

The HGF receptor is a tyrosine kinase receptor encoded by the c-*met* protooncogene (Bottaro *et al.*, 1991). This protooncogene product (c-Met) is expressed by

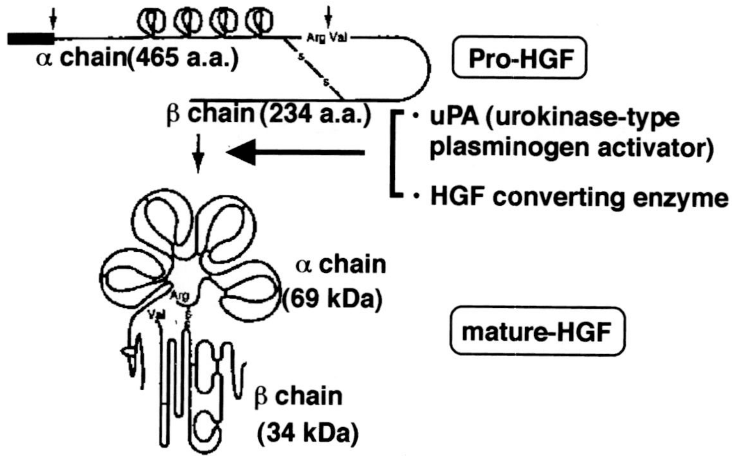

biological activities of HGF

mitogenic activity
motogenic activity
morphogenic activity
angiogenesis
anti apoptotic effect

FIGURE 2 Processing of HGF and biological activity.

both normal and malignant epithelial cells, as well as by several other cell types (Di Renzo *et al.*, 1991). The c-Met/HGF receptor is a heterodimer of two disulfide-linked chains, 50 kDa α and 145 kDa β subunits, that are generated by cleavage of a single 190-kDa precursor (Gonzatti-Haces *et al.*, 1989). Binding of HGF induces activation of the kinase and auto/transphosphorylation of specific tyrosine residues in the receptor β subunit (Bottaro *et al.*, 1991; Matsumoto and Nakamura, 1996).

C. HGF Gene Therapy for Rat Liver Cirrhosis

We have developed a novel gene therapy approach for rat liver cirrhosis by muscle-directed gene transfer of HGF (Ueki *et al.*, 1999).

1. Expression of HGF by *in Vivo* Gene Transfer to Muscle

Liver cirrhosis induced by dimethylnitrosamine mimics characteristics of liver cirrhosis in humans (Jezeqiul *et al.*, 1989; Pritehard *et al.*, 1989) and this model

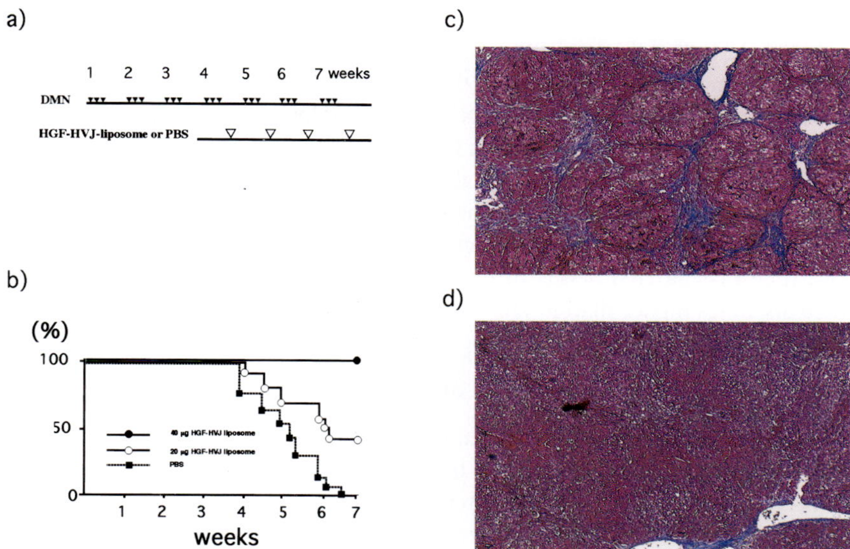

FIGURE 3 (a) Schedule of DMN administrations and HGF-HVJ liposome or PBS injection: DMN was administrated intraperitoneally 3 consecutive days for 7 weeks or until death on days indicated by arrowheads. HGF/HVJ-liposome or PBS was injected into skeletal muscles once a week after the 4 weeks of DMN treatment, as indicated by open arrowheads. (b) Survival of DMN-treated rats with or without HGF/HVJ-liposome injection. Survival of rats in a control group that were injected with PBS or rats in a control group that were injected with HGF-HVJ liposomes containing 20 or 40 mg of HGF DNA. Life table analyses presented as a Kaplan–Meier plot. (c and d) Change in the accumulation of fibrotic components in DMN-treated rat liver. Azan-Mallory staining liver section obtained from a rat treated with DMN for 6 weeks, not given (c, ×100) or given repetitive injection of HGF/HVJ-liposome containing 40 μg DNA (d, ×100). Azan-Mallory staining revealed that fibrous tissue components were hyperaccumulated in Glisson's sheath and reticulin fibers spread radically throughout the liver.

has been widely used. Following the initiation of DMN treatment, histological changes show progression of liver injury and liver cirrhosis/fibrosis (Fig. 3c). To achieve *in vivo* HGF gene expression in cirrhotic rats, we used the well-described method of combining DNA with a mixture of hemagglutinating virus of Japan (HVJ)-liposome (Kaneda *et al.*, 1989; Morishita *et al.*, 1997; Hirano *et al.*, 1998). Four weeks after DMN treatment, the HVJ-liposome containing the expression vector for human HGF (HGF/HVJ-liposome) was injected into skeletal muscle once a week (Fig. 3a).

Following a single injection of HGF/HVJ-liposome, human HGF protein was expressed in muscle tissue from 1 day after until 9 days after injection. Human HGF was specifically detected in the plasma of rats treated with HGF/HVJ-liposome 1 day after injection, and the level was maintained for 1 week, indicating secretion of human HGF from muscle to bloodstream. Moreover, repetitive injection of HGF/HVJ-liposome increased human HGF concentration in plasma. In phosphate-buffered saline (PBS)- injected control rats, plasma rat HGF reached the maximum value 1 week after the initial DMN treatment, while it then decreased even after continued DMN treatment. However, the plasma rat HGF value increased gradually in rats given HGF/HVJ-liposome, and the values reached a maximum 6 weeks after the initial DMN treatment. Results imply that the endogenous expression of rat HGF might be upregulated in an autocrine manner by the human HGF transgene product. Together with this increased endogenous HGF, total plasma HGF levels in rats given the HGF/HVJ-liposome were maintained six times higher than in rats without the transgene.

2. Expression and Phosphorylation of c-Met/HGF Receptor

Expression of c-Met was detected faintly by Western blotting in normal liver, while it increased along with DMN treatment, and relatively strong expression was seen in injured liver 4 weeks after DMN treatment. In control rats without transfection of the HGF gene, c-Met expression thereafter decreased, whereas c-Met expression increased markedly after the repetitive injection of HGF/HVJ-liposomes. Interestingly, it has been reported that an upregulation of c-Met mRNA was observed after cell stimulation by HGF (Hayashi *et al.*, 1996). Taken together, it is possible that c-Met expression might be enhanced after its ligand binding in HGF/HVJ-liposome-treated rat liver, and these results implicate an autoamplified cellular response after HGF binding, and subsequent c-Met activation may occur *in vivo*, as well as *in vitro*.

Because biologic activities of HGF depend on the tyrosine phosphorylation of c-Met, we analyzed the tyrosine phosphorylation of c-Met in the livers. Tyrosine phosphorylation of c-Met was not detected in normal livers and livers of rats subjected to DMN treatment for 2, 4, and 6 weeks without HGF/HVJ-liposome injection. In contrast, tyrosine phosphorylation of c-Met was seen in the liver of rats injected with HGF/HVJ-liposome. Our results indicate that the human HGF transgene product derived from skeletal muscle, as well as endogenously upregulated rat HGF, probably binds and activates the c-Met receptor in the cirrhotic livers, whereas activation of c-Met rarely occurs in rat livers without an HGF/HVJ-liposome injection.

3. Suppression of TGF-β1 Expression and Abrogation of Fibrosis in Liver

TGF-β1 is a major cytokine implicated in the pathogenesis of liver fibrosis and cirrhosis (Friedman, 1993; Blomhoff *et al.*, 1991). TGF-β1 stimulates hepatic stellate cells to transform into myofibroblast-like cells. It strongly enhances the production of extracellular matrix proteins, while it attenuates the degradation of extracellular matrix proteins (Friedman, 1993; Blomhoff *et al.*, 1991). In cirrhotic liver, desmin-positive hepatic stellate cells are increased in the fibrotic regions and

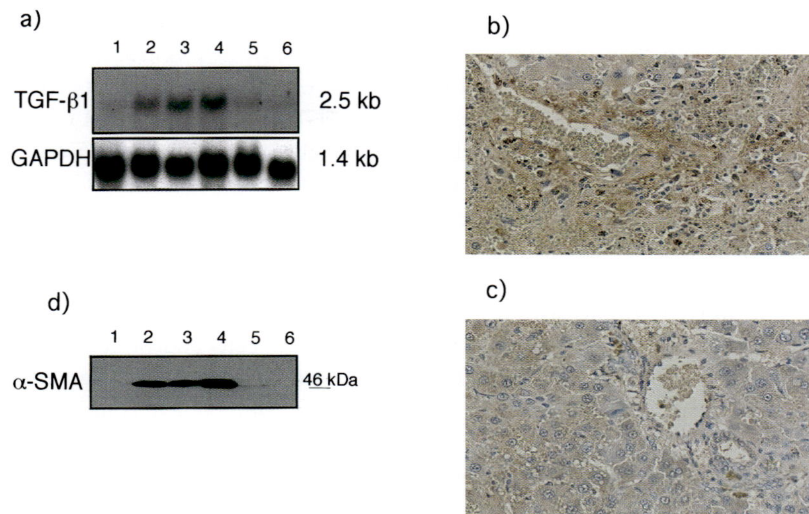

FIGURE 4 (a) Change in TGF-β1 mRNA expression in livers of normal and DMN-treated rats. RNAs were obtained from normal rat liver (lane 1) and livers of rats given DMN treatment 2 (lane 2), 4 (lane 3), and 6 (lane 4) weeks and from livers of rats given DMN treatment for 6 weeks and repetitive injection of HGF/HVJ-liposome containing 20 (lane 5) or 40 (lane 6) μg DNA. (b and c) Immunohistochemical staining of TGF-β1 in liver tissues. Liver sections were obtained from rats given DMN treatment for 6 weeks(b, ×200) and rats given DMN treatment for 6 weeks and repetitive injection of HGF/HVJ-liposome containing 40 μg DNA(c, ×200). (d) Change in α-SMA expression in livers of normal and DMN-treated rats given or not given the repetitive injection of HGF/HVJ-liposome. Liver extracts were obtained from a normal rat (lane 1) and rats given DMN treatment for 2 (lane 2), 4 (lane 3), and 6 (lane 4) weeks and from rats given DMN treatment for 6 weeks and repetitive injection of HGF/HVJ-liposome containing 20 or 40 (lanes 5 and 6) μg DNA. α-SMA expression was analysed by Western immunoblotting.

most of them coexpress α-smooth muscle actin (α-SMA), a marker of myofibroblast (Takahashi *et al.*, 1995), indicating phenotypic transition of hepatic stellate cells into myofibroblast-like cells. We analyzed expression of TGF-β1 and α-SMA in the cirrhotic rat liver. In normal rat liver, TGF-β1 mRNA was scarcely detectable, but it was strongly induced 2 weeks after DMN treatment, and enhanced transcription was noted during the progression of liver cirrhosis (Fig. 4a, lanes 1–4). Expression of TGF-β1 mRNA decreased markedly in the liver of rats that received HGF/HVJ-liposome injections (Fig. 4a, lanes 5 and 6).

Consistent with a change in TGF-β1 mRNA expression, immunohistochemical analysis identified few TGF-β1-positive cells in normal rat liver, whereas many TGF-β1-positive cells were distributed in the livers of DMN-treated control rats without transfection of the HGF gene (Fig. 4b). However, the number of TGF-β1-positive cells decreased markedly after the HGF/HVJ-liposome injections (Fig. 4c). Moreover, the expression of α-SMA, as detected by Western blotting and immunohistochemistry, correlated well with the expression of TGF-β1: α-SMA expression was not detected in normal rat livers, but was induced 2 weeks after DMN treatment and increased further as the progression of liver cirrhosis (Fig. 4d, lanes 1–4). In the liver of rats given HGF/HVJ-liposome, α-SMA expression was decreased markedly and faintly detectable (Fig. 4d, lanes 5 and 6). Consistently, only a few cells located in the periportal area were positive for α-SMA in normal livers, whereas the number of α-SMA-positive cells increased markedly in cirrhotic livers. Transfection of the HGF gene, however, decreased markedly the number of α-SMA-positive cells in DMN-treated rat livers. Thus, expression of the transgene for HGF suppressed the expression of TGF-β1, which might inhibit transition from hepatic stellate cells to myofibroblast-like cells and/or suppress the expression of α-SMA in the affected organ. However, HGF may upregulate matrix metalloproteinase (MMP) production (data not shown), which contributes to fibrolysis in the liver.

Consistent with the molecular and cellular events leading to liver cirrhosis, fibrous connective tissue components accumulated in Glisson's sheath and the formation of pseudolobules was obvious in the rat liver 6 weeks after DMN treatment (Fig. 3c). In contrast, repetitive HGF/HVJ-liposome treatment decreased the accumulated connective tissue components dramatically and the fibrous regions largely disappeared (Fig. 3d). The quantitative analysis for fibrous regions indicated that the fibrous components in HGF/HVJ-liposome-treated livers were reduced by more than 70% compared to control cirrhotic livers.

4. Prevention of Apoptosis and Promotion of Mitosis in Hepatocytes

Apoptosis in the liver was investigated by the TdT-mediated dUTP-biotin nick end labeling (TUNEL) method, which detects *in situ* DNA fragmentation. The liver section revealed no apoptotic liver cells in the normal liver, but a number of TUNEL-positive cells were present in DMN-treated cirrhotic liver and most of these cells were hepatocytes (Fig. 5a). Notably, few cells were seen undergoing apoptosis in the liver of rats given repetitive HGF gene therapy (Fig. 5b).

In normal rat livers, only a few mitotic hepatocytes were observed, as detected by immunohistochemical staining for proliferative cell nuclear antigen (PCNA), but several cells were stained with PCNA in DMN-treated rat liver. The number of PCNA-positive hepatocytes was much higher in the livers of rats given HGF gene therapy, and the labeling index showed that more than 45% of hepatocytes were in mitosis (Fig. 5c).

5. Abrogation of Lethality

Rats began to die 4 weeks after DMN treatment because of liver cirrhosis, and finally all rats died of hepatic dysfunction within 45 days (Fig. 3b). However, the

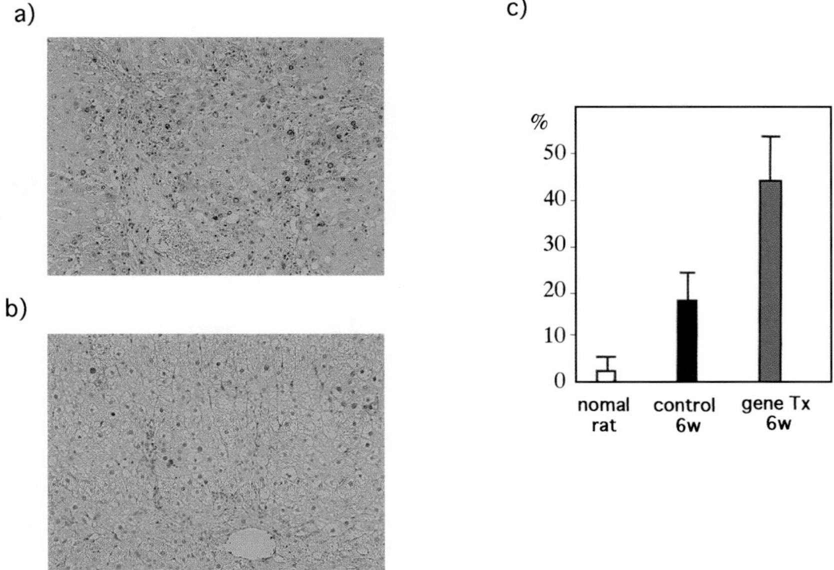

FIGURE 5 Distribution of apoptotic and mitotic cells in normal and DMN-treated rat livers. (a and b) Distribution of apoptotic cells in the liver (×100). Liver sections were obtained from rats given DMN treatment for 6 weeks without (a) or with repetitive injection of HGF/HVJ-liposome containing 40 μg DNA (b). (c) Change in labeling index of hepatocytes in normal and DMN-treated rats livers. The labeling index was determined by counting more than 2000 nuclei of hepatocytes using PCNA-stained tissue sections. Each value represents mean ± standard deviation.

survival rate was increased by HGF/HVJ-liposome treatment. Depending on the dosage of the HGF gene, the survival rate increased to as much as 45% in rats transfected with 20 μg DNA/rat. Of note, transfection with 40 μg DNA/rat completely abrogated mortality (Fig. 3d).

D. The Possibility of Gene Therapy for Liver Cirrhosis

The first successful trial of transgene expression in muscle involved the direct injection of DNA encoding various marker proteins and resulted in gene expression in the muscle fibers for more than 2 months (Woff *et al.*, 1990). Unfortunately, attempts in large animals revealed a low transfer efficiency in comparison with that obtained in mice. However, because the hemagglutinating virus of Japan (HVJ)-liposome-mediated gene transfer to muscle has shown a much higher efficiency than that obtained by naked DNA injection, the extension of this approach to large animals should be investigated further. Although the

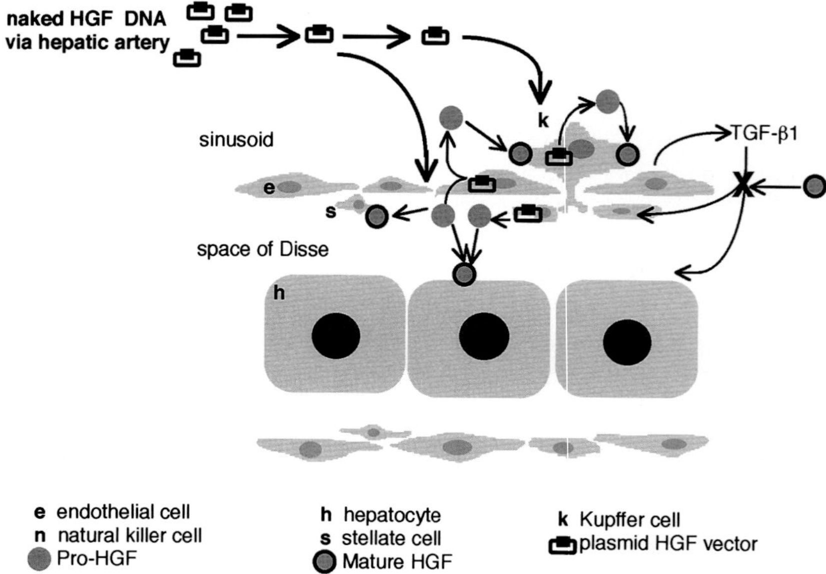

FIGURE 6 Proposed mechanisms of HGF gene therapy using naked plasmid DNA of the HGF gene. Naked HGF plasmids are delivered via the hepatic artery, and efficient HGF gene expression can be obtained in the liver using this approach. Nonparenchymal cells produce pro-HGF and secrete it into the sinusoid and the space of Disse. After conversion into the mature form, HGF acts on liver cells.

mechanism is not well understood, transduction of the HGF gene into skeletal muscle boosts endogenous HGF in rats. Gene therapy results in sustained plasma HGF levels about three times higher than normal, as the product of both transferred and endogenous HGF genes. HGF is a heparin-binding heterodimer related to plasminogen; it binds avidly to heparan sulfate, a glycosaminoglycan that forms proteoglycans along with the core protein. Proteoglycans are found on the surface of epithelial cells, including hepatocytes. Thus, HGF accumulates in the liver by binding to proteoglycans and is cleaved proteolytically from a single-chain precursor to generate the mature molecule. It has been reported that naked plasmid DNA in a large volume of physiological saline could achieve high levels of foreign gene expression *in vivo* (Zhang *et al.*, 1999). We have found that infusion of the naked HGF plasmid vector into the hepatic artery results in efficient HGF expression in a cirrhotic dog model (Fig. 6). This approach may also be clinically feasible because it is simple and no viral vector is needed.

Tumorigenicity has been reported in transgenic mice overexpressing HGF, in which the level of expression was more than 50-fold higher than in normal mice (Takayama *et al.*, 1997). We did not find any evidence of tumor formation in rats that received HGF gene therapy. Also, in transgenic mice that expressed HGF at levels similar to those used in our current experiment (two to three times higher than normal), the development of tumors was inhibited (Santoni-Rugiu *et al.*,

1996). Thus, these transgenic animal studies suggest that HGF shows an antitumor effect at levels below 10 times normal and with transient expression. Because of its profound effects on cell motility, HGF has been implicated in the invasion and metastasis of tumor cells. Whether or not HGF shows a motogenic effect on tumor cells *in vivo* should be addressed in the future.

We have developed a novel approach for liver cirrhosis using HVJ-liposome-mediated HGF gene therapy. It is simple and safe and can be performed without causing substantial inflammation or activation of cellular and humoral immunity. Another new approach, infusion of naked plasmid DNA via the hepatic artery, is also safe and can be performed without eliciting a strong host immune response. With these newly developed methods, HGF gene therapy may eventually be translated into a useful clinical regimen for the treatment of patients with liver cirrhosis, which is otherwise fatal and unresponsive to conventional therapy.

REFERENCES

Bardelli, A., Longati, P., Albero, D., Goruppi, S., Schneider, C., Ponzetto, C., and Comoglio, P. M. (1996). HGF receptor associates with anti-apoptotic protein BAG-1 and prevents cell death. *EMBO J.* **15**, 6205–6212.

Blomhoff, R., and Wake, K. (1991). Perisinusoidal stellate cells of the liver: Important roles in retinol metabolism and fibrosis. *FASEB J.* **5**, 271–277.

Border, W. A., and Nobel, N. A. (1994). Transforming growth factor β in tissue fibrosis. *N. Engl. J. Med.* **331**, 1286–1292.

Bors, P., and Miller, C. M. (1995). Hepatocyte growth factor: A multifunctional cytokine. *Lancet* **345**, 293–295.

Bottaro, D. P., Rubin, J. S., Faletto, D. L., Chan, A. M., Kmiecik, T. E., Vande Woude, G. F., and Aaronson, S. A. (1991). Identification of the hepatocyte growth factor as the c-met proto-oncogene product. *Science* **251**, 802–804.

Conn, H. O., and Atterbury, C. E. (1992). *In* "Diseases of the Liver" (L. Schiff and E. R. Schiff, eds.), pp.875–941. Lippincott, Philadelphia.

Di Renzo, M. F., Narsimhan, R. P., Olivero, M., Bretti, S., Giordano, S., Medico, E., Gaglia, P., Zara, P., and Comoglio, P. M. (1991). Expression of the Met/HGF receptor in normal and neoplastic human tissues. *Oncogene* **6**, 1997–2003.

Friedman, S. L. (1993). The cellular basis of hepatic fibrosis: Mechanisms and treatment strategies. *N. Engl. J. Med.* **328**, 1828–1835.

Gonzatti-Haces, M., Seth, A., Park, M., Copeland, T., Oroszlan, S., and Vande Woude, G. F. (1989). Characterization of the TPR-MET oncogene p65 and the MET protooncogene p140 protein thyrosine kinase. *Proc. Natl. Acad. Sci. USA* **85**, 21–25.

Hayashi, S., Morishita, R., Higaki, J., Aoki, M., Moriguchi, A., Kido, I., Sawa, Y., Matsumoto, K., Nakamura, T., Kaneda, Y., and Ogihara, T. (1996). Autocrine-paracrine effects of overexpression of hepatocyte growth factor gene on growth of endothelial cells. *Biochem. Biophys. Res. Commun.* **220**, 539–545.

Hirano, T., Fujimoto, J., Ueki, T., Yamamoto, H., Iwasaki, T., Morishita, R., Sawa, Y., Kaneda, Y., Takahashi, H., and Okamoto, E. (1998). Persistent gene expression in rat liver in vivo by repetitive transfections using HVJ-liposome. *Gene Ther.* **5**, 459–464.

Jezequel, A. M., Mancini, R., Rinaldesi, M. L., Ballardini, G., Fallani, M., Bianchi, F., and Orlandi, F. (1989). Dimethylnitrosamine-induced cirrhosis: Evidence for an immunological mechanism. *J. Heptol.* **8**, 42–52.

Kaneda, Y., Iwai, K., and Uchida, T. (1989). Increased expression of DNA co-introduced with nuclear in adult rat liver. *Science* **243**, 375–378.

Matsumoto, K., and Nakamura, T. (1996). Emerging multipotent aspects of hepatocyte growth factor. *J. Biochem.* **119**, 591–600.

Matsumoto, K., and Nakamura, T. (1997). Hepatocyte growth factor (HGF) as a tissue organizer for organogenesis and regeneration. *Biochem. Biophys. Res. Commun.* **239**, 639–644.

Michalopoulas, G. K., and DeFrances, M. C. (1997). Liver regeneration. *Science* **276**, 60–66.

Morishita, R., Sugimoto, T., Aoki, M., Kida, I., Tomita, N., Moriguchi, A., Maeda, K., Sawa, Y., Kanrda, Y., Higaki, J., and Ogihara, T. (1997). In vivo transfection of cis element "decoy" against nuclear factor-kB binding site prevents myocardial infarction. *Nature Med.* **3**, 894–899.

Nakamura, T., Nawa, K., and Ichihara, A. (1984). Partial purification and characterization of hepatocyte growth factor from serum of hepatectomized rats. *Biochem. Biophys. Res. Commun.* **122**, 1450–1459.

Nakamura, T., Hirosumi, J., Watanabe, M., Hattri, A., Nakamura, T., and Orimo, H. (1985). Inhibitory effect of transforming growth factor β on DNA synthesis of adult rat hepatocytes in primary culture. *Biochem. Biophys. Res. Commun.* **133**, 1042–1050.

Nakamura, T., Nishizawa, T., Hagiya, M., Seki, T., Shimonishi, M., Sugimura, A., Tashiro, K., and Shimizu, S. (1989). Molecular cloning and expression of human hepatocyte growth factor. *Nature* **342**, 440–443.

Oberhammer, F. A., Pavelka, M., Sharma, S., Tiefenbacher, R., Purchio, A. F., Bursch, W., and Schulte-Hermann, R. (1992). Induction of apoptosis in cultured hepatocyte and regressing liver by transforming growth factor b1. *Proc. Natl. Acad. Sci. USA* **89**, 5408–5412.

Okano, J., Shiota, G., and Kawasaki, H. (1997). Protective action of hepatocyte growth factor for acute liver injury caused by D-galactosamine in transgenic mice. *Hepatology* **26**, 1241–1249.

Pritchard, D. J., and Butler, W. H. (1989). Apoptosis: The mechanism of cell death in dimethylnitrosamine- induced hepatotoxicity. *J. Pathol.* **158**, 253–260.

Russell, W. E., McGowan, J. A., and Bucher, N. L. (1984). Partial characterization of a hepatocyte growth factor from platelets. *J. Cell Physiol.* **119**, 183–192.

Sanderson, N., Factor, V., Nagy, P., Kopp, J., Kondaiah, P., Wakefield, L., Roberts, A. B., Sporn, M. B., and Thorgeirsson, S. S. (1995). Hepatic expression of mature transforming growth factor b1 in transgenic mice results in multiple lesions. *Proc. Natl. Acad. Sci. USA* **92**, 2572–2576.

Santoni-Rugiu, E., Preisegger, K. H., Kiss, A., Audolfsson, T., Shiota, G., Schmidt, E., and Thorgeirsson, S. S. (1996). Inhibition of neoplastic development in the liver by hepatocyte growth factor in a transgenic mouse model. *Proc. Natl. Acad. Sci. USA* **93**, 9577–9582.

Schmidt, C., Bladt, F., Goedecke, S., Brinkmann, V., Zschiesche, W., Sharpe, M., Gherardi, E., and Birchmeier, C. (1995). Scatter factor/ hepatocyte growth factor is essential for liver development. *Nature* **373**, 699–702.

Sporn, M. B., and Roberts, A. B. (1992). Transforming growth factor-β: Recent progress and new challenges. *J. Cell Biol.* **119**, 1017–1021.

Takahashi, H., Fujimoto, J., Hanada, S., and Isselbacher, K. J. (1995). Acute hepatitis in rats expressing human hepatitis B virus transgenes. *Proc. Natl. Acad. Sci. USA* **92**, 1470–1474.

Takayama, H., LaRochelle, W. J., Sharp, R., Otsuka, T., Kriebel, P., Anver, M., Aaronson, S. A., and Merlino, G. (1997). Diverse tumorigenesis associated with aberrant development in mice over expressing hepatocyte growth factor/ scatter factor. *Proc. Natl. Acad. Sci. USA* **94**, 701–706.

Ueki, T., Kaneda, Y., Tsutsui, H., Nakanishi, K., Sawa, Y., Morishita, R., Matsumoto, K., Nakamura, T., Takahashi, H., Okamoto, E., and Fujimoto, J. (1999). Hepatocyte growth factor gene therapy of liver cirrhosis in rats. *Nature Med.* **5**, 226–230.

Wolff, J. A., Malone, R. W., Williams, P., Chong, W., Acsadi, G., Jani, A., and Felgner, P. L. (1990). Direct gene transfer into mouse muscle in vivo. *Science* **247**, 1465–1468.

Zhang, G., Budker, V., and Wolff, J.A. (1999). High level of foreign gene expression in hepatocyte after tail vein injections of naked plasmid DNA. *Hum. Gene Ther.* **10**, 1735–1737.

INDEX

D

E

I

M

V

W